AF412048

IET ENERGY ENGINEERING SERIES 89

Hydrogen Production, Separation and Purification for Energy

Other volumes in this series

Hydrogen Production, Separation and Purification for Energy

Edited by
Angelo Basile, Francesco Dalena,
Jianhua Tong and T. Nejat Veziroğlu

The Institution of Engineering and Technology

Published by The Institution of Engineering and Technology, London, United Kingdom

The Institution of Engineering and Technology is registered as a Charity in England & Wales (no. 211014) and Scotland (no. SC038698).

First published 2017

The Institution of Engineering and Technology
Michael Faraday House
Six Hills Way, Stevenage
Herts, SG1 2AY, United Kingdom

www.theiet.org

British Library Cataloguing in Publication Data
A catalogue record for this product is available from the British Library

ISBN 978-1-78561-100-1 (hardback)
ISBN 978-1-78561-101-8 (PDF)

Typeset in India by MPS Limited
Printed in the UK by CPI Group (UK) Ltd, Croydon

Contents

Preface

In the last few years, industrial research efforts have become more focused on the production of energy that could replace the use of hydrocarbon-based fuels. The statutes on the limits of the C1 block entry (CO and CO_2) in the air and the increasingly imminent depletion of oil stocks have addressed many research groups to find an adequate solution to this problem.

Among the various possible solutions, many research groups agree on the use of hydrogen as one of the most innovative alternative energy sources and consider it as the next-generation fuel. The high conversion efficiency (hydrogen has 2.75 times more energy than hydrocarbons), recyclability and nonpolluting nature make it the best fuel of the future. Hydrogen has the highest energy content per unit weight of any known fuel (142 kJ/g) and, above all, in comparison to the other known natural gases, it is environmentally safe; in fact, its combustion results only in water vapour and energy.

Nowadays global hydrogen production currently exceeds 1 billion m^3/day, of which 48% is produced from natural gas (by steam reforming of methane and other hydrocarbons), 30% from oil (non-catalytic partial oxidation of fossil fuels), 18% from coal and the remaining 4% by water electrolysis.

The aim of this volume is to provide an overview of worldwide research in the use of hydrogen in the energy development, its most innovative methods of production and the various steps necessary for the optimization of this product (from the optimization by the use of catalysts up to its purification processes). For this reason, the book is divided into three parts/sections corresponding to those that are the main steps of innovation for hydrogen production: the use of catalysts, the production reactions and reactors in which these take place and the purification processes.

In detail, Section 1 ("Catalyst preparation") opens with two chapters by Prof. Vincenzo Palma and his co-workers. Chapter 1 (V. Palma, C. Ruocco, M. Martino, E. Meloni and A. Ricca) deals with conductive structured catalyst which can both minimize heat transfer resistance towards the catalytic volume and enhance mass transfer mechanisms between solid and gas phases. Chapter 2 (V. Palma, M. Martino, E. Meloni, A. Ricca and C. Ruocco) highlights the importance of catalytic supports in the catalyst behaviour and, in particular, of the employment of bimetallic formulations. In fact, the synergic effect linked to the combination of two or more active species leads to an overall catalytic performances improvement in terms of activity and stability. Chapter 3 (C. Evangelisti, F. Bossola and V. Dal Santo) provides a recent state-of-the-art concerning catalyst based on low loaded noble and/or non-noble metals for hydrogen production processes (reforming and pyrolysis) starting from renewable raw materials, by-products or waste. In particular, this chapter highlights the most sustainable and critical raw materials-free systems. Chapter 4 (M. Huuhtanen, P.K. Seelam and R.L. Keiski) deals with Ni- and Cu-based catalysts in methanol and ethanol steam reforming. This contribution focuses on (a) the most commonly used methods, such as impregnation and precipitation techniques, for introducing the active Ni and Cu phases, and (b) the development of novel Ni and Cu catalysts (e.g. nanostructures) for H_2 production. Chapter 5 (A. Vita) analyses traditional methods and recent approaches to the design of noble metal (Rh, Pt, Ru) and Ni-based catalysts for the hydrogen production by steam reforming processes of different fuels (CH_4, ethanol, methanol) at relatively low temperatures (<600 °C). To be stressed that the development of robust catalysts for low temperature reforming processes is fundamental for the integration of reforming reactors with concomitant separation/purification step (membrane reactors (MRs)). Therefore, the preparation methods are described and compared, especially considering their contribution to develop catalysts with properties as high surface area, low particle size, resistance to sintering, resistance to carbon formation, opportune metal load. The last chapter of this first part, Chapter 6 (C. Cannilla, G. Bonura and F. Frusteri), focuses on supercritical water gasification (SCWG) processes to produce hydrogen. SCWG represents a suitable reaction medium for biomass gasification acting as a reactant, solvent catalyst and hydrogen donor. This chapter discusses the influence of the main reaction conditions as temperature, pressure and feed concentration on hydrogen yield.

Section 2 deals with the reactions and reactors for the production of hydrogen.

This part starts with Chapter 7 (S. Abramov, M. Shalygin, V. Teplyakov and A. Netrusov). It discusses about biofuels as starting materials for hydrogen production. This contribution discusses three different generations of starting materials and focuses on the prospects of commercially available membrane application by using different operating membrane module schemes for hydrogen recovery by energy saving processes. Chapter 8 (M. Alavi, A. Iulianelli, M.R. Rahimpour, R. Eslamloueyan, M. De Falco, G. Bagnato and A. Basile) highlights the importance of MRs and outlines the earlier state of the art about hydrogen production from reforming reactions in both conventional and MRs. Chapter 9 (J. Tong) focuses on the hydrogen production in three types of membrane micro-reactors: planar, hollow-fibre and monolithic configurations. After that, the concept of membrane micro-reactors as a fuel processor for portable polymer electrolyte membrane fuel cells is briefly overviewed. Finally, the modelling progress on the membrane micro-reactors is discussed. Chapter 10 (K. Ghasemzadeh, M.N. Nezhad and A. Basile) illustrates the developments occurred in the field of perovskite MRs during the last few years. Moreover, novel applications of the perovskite membranes in the MRs set-up are also discussed.

The last part of the book is inherent to the purification/separation process of hydrogen.

Chapter 11 (Y. Yampolskii and V. Ryzhikh) focuses on the polymeric membrane materials for hydrogen separation. Separation of hydrogen from its mixtures with N_2, CH_4 and other gases is one of the oldest tasks of membrane gas separation.

The prospect of improving membrane properties is considered in relation to the separation and purification of hydrogen. Chapter 12 (H.R. Rahimpour and M.R. Rahimpour) shows different types of membranes (such as metal, molecular sieving carbon, zeolites and ceramics) used for hydrogen separation from hydrogen-rich mixtures. This chapter also describes the performance of selected membranes in terms of hydrogen permeability and selectivity. Chapter 13 (P. Ribeirinha, M. Boaventura, J.M. Sousa and A. Mendes) provides a brief overview of the state-of-the-art of hybrid sorption-enhanced MRs within a historical perspective. A comprehensive description of the technology, available materials, applicability and comparative advantages is also presented. Chapter 14 (J.A. Lie, X. He, I. Kumakiri, H. Kita and M.-B. Hägg) concerns with the preparation of two types of carbon-based membranes (self-supported fibres and a thin microporous membranes on an inorganic support tube), status of development of such membranes and their potential applications. This section ends with Chapter 15 (G. Valenti). It explores the separation of hydrogen isotopes by cryogenic distillation. In fact, when combining in the diatomic molecular hydrogen, the isotopes yield molecules have slightly different physical and chemical characteristics. The diverse evaporation conditions can be exploited to separate the molecules of the same kind. Separation takes place at cryogenic conditions and, typically, in packed columns.

To conclude, the Editors would like to express special thanks to each one of the authors for their valuable contributions to this volume.

Francesco Dalena
Nejat T. Veziroglu
Jianhua Tong
Angelo Basile

About the editors

Angelo Basile, Full Professor in Chemical Engineering Principles and Senior Researcher at the Italian National Research Council, is responsible for research in ultra-pure hydrogen production using Pd-based membrane reactors. He is an editor of 32 books and 30 special issues, with over 140 scientific papers (*h*-index 38, over 3,660 citations and more than 90 book chapters). He is also an associate editor of the *International Journal of Hydrogen Energy* and editor-in-chief of the *International Journal of Membrane Science and Technology*, and member of the Editorial Board of 21 international journals.

Francesco Dalena is a Ph.D. doctor at the Chemistry Department, University of Calabria, Italy. His research topics include inorganic membrane reactors, chemical kinetics and the chemistry of artistic masterpieces.

Jianhua Tong is an Associate Professor in the Department of Materials Science and Engineering at Clemson University, South Carolina, USA. His research interests focus on advanced solid-state ionic materials and devices for energy conversion/storage/harvest, electrocatalytic reactions, and electrochemical sensors. He has published more than 60 peer-reviewed technical papers and 6 book chapters, and filed 14 patents. His research papers have been cited more than 3,200 times and his *h*-index is over 27.

T. Nejat Veziroğlu is President of the International Association for Hydrogen Energy, and Emeritus Professor at the University of Miami, Coral Gables, FL, USA. He has published some 350 scientific reports/papers and edited some 200 volumes on heat and mass transfer, renewable energies and hydrogen energy. He is a recipient of several international awards and honorary doctorates. In 2000, he was nominated for the Nobel Prize in Economics for both envisioning the hydrogen economy and striving towards its realization.

Chapter 1

Structured catalyst for process intensification in hydrogen production by reforming processes

Vincenzo Palma[1], Concetta Ruocco[1], Marco Martino[1], Eugenio Meloni[1] and Antonio Ricca[1]

Abstract

In the present chapter, a brief outlook on structured catalyst employment for process intensification in hydrogen production processes was provided.

The growing interest toward hydrogen, both as valuable chemical for other processes and, mainly, as energetic vector, increased the need to more sustainable routes for H_2 producing. Process intensification is "any chemical engineering development that leads to a substantially smaller, cleaner and more energy-efficient technology." In this direction, in the preset chapter, it was demonstrated that structured catalysts emerge as a viable solution in the process intensification of reforming processes for hydrogen production.

It is evidenced that structured catalysts (honeycomb, foams) improve heat and transfer mechanisms between gaseous and solid phases. However, to transfer a catalytic formulation on a structured carrier, several aspects should be taken in account: in principle, a deep affinity between catalyst formulation and substrate is mandatory, in order to assure good performances and acceptable catalyst lifetime. Moreover, structured carrier should be pretreated to assure chemical and physical compatibility with active species: the choice of the substrate material and the appropriate pretreatment technology, as well as the catalyst deposition method, will depend on the nature of carrier and active species, and on the process in which catalyst should be employed. Several experiences were reported in the field of hydrogen production reactions, evidencing the advantages in transferring a catalytic formulation on a structured carrier. In particular, by selecting a high thermal conductive structured substrate, an optimal thermal management in the catalytic volume was achieved. Such aspect on the one hand improves endothermic processes,

[1]Department of Industrial Engineering, University of Salerno, via Giovanni Paolo II 132, 84084 Fisciano (SA), Italy

by maximizing heat transfer from heating medium to the catalytic volume and by reducing the decreasing temperature peak due to the reaction endothermicity. On the other hand, temperature re-distribution in exothermic processes, apart to allow a better exploiting of the catalytic volume, helps to mitigate temperature rising of the process, minimizing risks of hot-spot phenomena. Such aspects, explored in detail in the present chapter, on the one hand improve catalytic performances and on the other increase catalyst lifetime, so resulting in an effective process intensification.

1.1 Introduction

The new global prospective in terms of energy production, finalized to minimize or solve problems due to the global warming, the lack of fossil resources, and the general increase of energy demand especially by the new emergent economies have become an important issue which has spurred the development and implementation of techniques to lessen the impact of the combustion of fossil fuels [1]. In this way, maintaining low emissions of polluting gases with the growing demand for energy worldwide is probably one of the biggest challenges that scientific research must address. Clean and alternative energies have become the areas of interest in which we have to lay the foundations for the development of sustainable energy [2]. Hydrogen is one of the most important raw materials in the chemical industry and the refining sector nowadays [3]. It is also becoming an attractive alternative to support energy consumption with a reduced environmental impact: a hydrogen-based system is considered to be feasible and advantageous to develop high quality energy services in a wide range of applications. Anyway, it should be emphasized that hydrogen must be produced first, as molecular hydrogen is an energy carrier but not an energy resource. Therefore, in the last years, much effort was employed in hydrogen production technology so as to obtain a stable source of hydrogen [4]. Currently, almost 96% of the world's hydrogen demand is supplied by traditional fossil fuels and about half of it comes from natural gas, the main composition of which is methane: the main technologies for hydrogen production from methane are steam reforming (SR), autothermal reforming (ATR), partial oxidation (POX), and dry reforming. An important factor that must be underlined is that the extended use of fossil fuels for most of the world's hydrogen production generates large quantities of CO_2 as by-product, and the amount of CO_2 produced per mole of H_2 is dependent on the technology used and the feedstock selection. From this standpoint, natural gas (mainly methane) is the preferred feedstock due to its high H/C ratio [3]. Beyond this, CO_2 formation is also due to the combustion of fuel by the burners used to provide the heat for the process.

SR technology is the oldest and the most feasible route to convert CH_4 into H_2 [5]. The endothermic reaction between H_2O and CH_4 typically proceeds over a nickel catalyst to produce syngas in a H_2/CO ratio of 3:1. Due to the high endothermicity of the reforming reaction, the reformer operates at high temperatures (800–1,100 °C) [6], and steam is usually fed in excess to reduce coke formation.

Steam-to-methane ratio of 2.5–3.0 is commonly used. The heat required to drive the process is generally supplied by burning some portion of the natural gas.

SR is a well-established technology, although continuous improvements have resulted in lower cost plants. Such improvements include catalysts less prone to coke and sulfur poisoning, materials for reformer tubes and better control of carbon limits. In addition, the development of process concepts based on a better understanding of the mechanisms of reaction, carbon formation, and sulfur poisoning have improved the cost efficiency of the process [7].

Carbon dioxide reforming (dry reforming) was also proposed for production of hydrogen. With the increasing interest in renewable energy sources and the reduction of greenhouse gases emission, CO_2 reforming of methane attracted attention in the last few years as an alternative to use biogas as fuel which is a mixture consisting mainly carbon dioxide and methane [8]. In this process, carbon dioxide reacts with methane to produce syngas with molar ratio $H_2/CO = 1$. Dry reforming reactions are usually followed by the water gas shift (WGS) reaction in order to maximize the hydrogen yield. However, this approach hardly finds its way for industrial use due to the high coke deposition and the low H_2/CO ratio.

Partial oxidation of methane uses sub-stoichiometric amounts of oxygen in a highly exothermic reaction to produce syngas in a H_2/CO ratio $= 0.5$. Catalysts are not required due to the very high temperature (up to 1,100 °C) of the process. However, the hydrogen yield per mole of methane input (and the system efficiency) can be significantly enhanced by use of catalysts and the reduced operating temperature. The main advantage of this approach is that it accepts all kind of hydrocarbon feeds. It is a particularly attractive process when dealing with feedstock of heavy hydrocarbons [8]. On the other hand, clear disadvantages are its lower efficiency as compared to the SR and the need of large quantities of pure oxygen, requiring a substantial investment in an adjoining O_2 plant. Some soot is normally formed in the process, which must be removed in a separate scrubber system downstream the reactor [3].

Autothermal reformers combine some of the best features of SR and POX systems: with the right mixture of input fuel, air and steam, the energy for the SR reaction is supplied by POX of methane, resulting in a thermally neutral process [9]. ATR is typically conducted at a lower pressure than POX. The main drawback of this process is that it requires an expensive and complex oxygen separation unit in order to feed pure oxygen to the reactor; otherwise, the product gas would be diluted with nitrogen [9]. In an attempt to decrease energy cost and expensive materials, as well as to ensure quick start up, low operation temperature ATR was proposed [10].

1.2   Steam reforming process

SR is the widest diffused process for the hydrogen production from natural gas, due to its higher hydrogen yield (with respect to other processes). The catalyst optimization for an SR system should take in account several aspects of the process, as thermodynamic constraints, and heat and mass transfer mechanisms toward and inside the reaction volume. Finally, the catalyst lifetime risks, linked to the active species poisoning or to the coke deposition, should be considered.

1.2.1   Thermodynamics

SR does not require oxygen, has a lower operating temperature than POX, dry and ATR and produces reformate with the higher H_2/CO ratio, which is beneficial for hydrogen production. However, SR is an endothermic process carried out in industrial furnace and sustained by external combustion; therefore, it has the highest flue gas emissions of the three processes [3], and high temperature operation results in cost disadvantages including the expensive tubular reformer, irreversible carbon formation in the reactor, and large energy consumption.

SR can be described by the strongly endothermic SR reaction (1.1):

$$C_nH_m + nH_2O \rightleftarrows nCO + \left(\frac{m}{2} + n\right)H_2 \tag{1.1}$$

In the case of methane, the system can be described by (1.2), which is always accompanied by the moderately exothermic WGS reaction (1.3):

$$CH_4 + H_2O \rightleftarrows CO + 3H_2 \quad \Delta H^{\circ}_{298K} = +206 \text{ kJ/mol} \tag{1.2}$$

$$CO + H_2O \rightleftarrows CO_2 + H_2 \quad \Delta H^{\circ}_{298K} = -41 \text{ kJ/mol} \tag{1.3}$$

As methane SR is a highly endothermic process, it is enhanced at high temperatures. The conversion of methane is complete at temperatures over 700 °C, S/C = 3 and atmospheric pressure. Higher pressures over 1 MPa (10 bar) shift the temperature for complete methane conversion to over 900 °C.

1.2.2   Reformers

The methane SR occurs in hundred (200–400) parallel, externally heated tubes (reformer tubes) filled with a nickel-based catalyst. After being preheated in heat exchangers or fired heaters to about 500 °C, the steam–hydrocarbon mixture is heated further over nickel catalyst in the tubes to the desired end temperature of normally 800–900 °C. Higher hydrocarbons (see (1.1)) are completely converted to carbon monoxide and hydrogen, methane only partially (depending on the reaction conditions).

The composition of the reformed gas at the reformer outlet reflects the equilibria of (1.1) and (1.2). For specified conditions (pressure, reformed gas temperature, steam–hydrocarbon ratio), the composition of the reformed gas can, therefore, be calculated from component balances and equilibrium conversion rates. The steam/hydrocarbon ratio is an essential parameter in the SR reaction and must not be lowered arbitrarily, because low ratios lead to carbon deposits on or in the catalyst.

Tubular reformers are mostly top and wall-fired box-type units, apart from older types or special designs, whereas in wall-fired reformers, rows of reformer tubes are heated mainly by the radiant sidewall; top- or bottom-fired reformers provide most of the heat through radiation of the burner flame and the hot flue gases. Top-fired reformers may, therefore, have several parallel rows of tubes, whereas wall-fired reformers can have only one row. Wall-fired reformers with a large number of tubes consist of several units adjacent to one another; waste heat from the flue gases is recovered in a common system. Modern reformers in large plants that produce ammonia and methanol synthesis gas from natural gas contain hundreds of tubes arranged in rows. Normally, the tubes have inside diameters of 75–125 mm, wall thicknesses of 10–20 mm, and fired lengths of 9–15 m, depending on reformer type (top or wall-fired). The fixed tube support is usually located outside the firebox at the top or bottom; when heated, the tube expands longitudinally from this support. Counterweights or spring hangers often relieve the tubes of about 50% of their dead weight. Tube wall thickness is calculated on the basis of internal pressure and of 100,000-h creep to rupture strength. More sophisticated calculations allowing for mechanical and thermal stresses, and in some cases even establishing a link between them, are being used increasingly [7].

Tubes are welded to the required overall length from centrifugally cast tube sections >3 m in length. The calculated "sound wall thickness" must be increased by an allowance for fabrication tolerances. The pressure drop inside a reformer tube, normally between 0.15 and 0.5 MPa, depends on the gas flow rate and geometry of the catalyst.

A special consideration is the lifetime of the expensive reformer tubes, made of highly alloyed chromium–nickel steel by centrifugal casting, because under the severe reaction conditions the material exhibits creep which finally leads to rupture [7]. The time to rupture for a specific material depends on the tube-wall temperature and on the internal pressure: this limits the reforming pressure. As the reforming reaction is endothermic and proceeds with volume increase, the negative effect of a pressure increase (lower conversion) has to be compensated by a higher reaction temperature and hence higher wall temperatures, but this is limited by the material. Another possibility to compensate is a higher steam surplus (steam/carbon ratio), but this is economically unfavorable. The standard material for a long time was HK 40 (20 Ni/25 Cr), but for replacements and new plants, HP modified (32–35 Ni/23–27 Cr stabilized with about 1.5% Nb) is being increasingly used on account of its superior high-temperature properties. With this latter tube material, a reforming pressure of 40 bar is possible at outer tube-wall temperatures of around 900 °C.

1.2.3 Catalysts

SR catalysts must meet stringent requirements such as high activity, reasonable life, good heat transfer, low pressure drop, high thermal stability, and excellent mechanical strength [2]. The metal catalysts active for SR of methane are the group VIII metals, usually nickel. Although other group VIII metals are active, they have some disadvantages; for example iron is oxidized rapidly, cobalt cannot withstand the partial pressures of steam, and the precious metals (rhodium, ruthenium, platinum, and palladium) are too expensive for commercial operation [3]. The supports for most industrial catalysts are based on ceramic oxides or oxides stabilized by hydraulic cement. The commonly used supports include α-alumina, magnesia, calcium aluminate, or magnesium aluminate.

These catalysts are usually in the form of thick-walled Raschig rings, with 16-mm diameter and height and a 6–8-mm hole in the middle. However, especially in modern steam reformers with high heat flux, catalysts with high geometric surface type are preferred. The access to the inner surface of catalysts continuously decreases with the increase in temperature, due to diffusion limitation. This is compensated by introducing catalysts with a higher surface to volume ratio. Typical shapes of this type of catalysts are spoked wheels, gear wheels, or rings with several holes. These catalysts are additionally advantageous with regard to their low pressure drop. A general distinction is made between impregnated and precipitated catalysts. The latter are the more traditional types which are produced by coprecipitation of all constituents. Precipitated catalysts have a Ni-surface which is by one order of magnitude higher than that of impregnated catalysts, and for this reason, these have an excellent activity already at moderate temperatures. But they are more prone to sintering at higher temperatures and do not have the mechanical strength of supported catalysts. Impregnated catalysts are manufactured by separately impregnating a support with nickel as the active constituent. The main advantage of this type of catalyst is that the support can be calcined at high temperatures before impregnation, resulting in predetermined excellent mechanical properties which are important especially at high temperature application. As the conditions of SR have generally become more severe, impregnated catalysts predominate by far at present. Frequently used support materials are alumina (the α-modification is preferred today), magnesium oxide, or spinel type substances (e.g., magnesium aluminum oxide).

In some cases, catalysts are stabilized by addition of a hydraulic binder (calcium aluminum oxide). Crushing strength is decreased by the hydration of magnesium oxide at low temperature ($<500\,^{\circ}$C) and by changes from the γ to the α-modifications of alumina above 500 $^{\circ}$C. A silicon dioxide content of <0.3 wt% in the catalyst is essential, because silicon dioxide becomes volatile in the presence of steam and may foul the downstream heat exchangers through which the reformed gas flows.

Nickel is at first present in the form of nickel oxide; after the system has been heated, it is reduced in the reformer tube by hydrogen formed when natural gas is added to steam. Some catalysts require a special method of reduction for which a definite steam/hydrogen ratio must be maintained (Figure 1.1) [11].

1.2.4   Catalyst poisons and desulfurization

Catalytic activity is affected seriously even by very low concentrations of catalyst poisons in inlet gases to reformer tubes. Such catalyst poisons include sulfur, arsenic, copper, vanadium, lead, and chlorine or halogens in general. Sulfur, found in practically all hydrocarbon feedstock, in particular decreases catalyst activity [8].

Earlier, desulfurization systems used impregnated activated carbon as adsorbent and operated at ambient temperature; however, as the efficiency of these systems differs according to the particular sulfur compounds in the gas and because the residual sulfur content in the exhaust gases is always lowering, zinc oxide desulfurization systems operating at 350–400 $^{\circ}$C are generally preferred today.

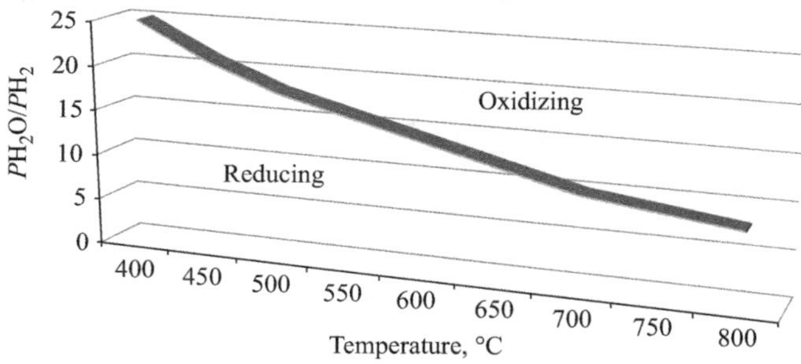

Figure 1.1   Reduction conditions for nickel oxide

Iron-based adsorbents are used in only a few cases where special conditions (i.e., low operating temperature, possibility of regeneration, sulfur composition) prevail. Zinc oxide reactors are very reliable in absorbing hydrogen sulfide and, with limitations, also in other sulfur compounds, such as carbonyl sulfide and mercaptans; however, optimum desulfurization conditions are achieved only in presence of certain space velocities and effective flowrates, and for a sulfur content up to 30%, so that almost complete conversion of zinc oxide to zinc sulfide is possible.

Particular organic sulfur compounds, such as mercaptans and thiophenes, require hydrogenation over cobalt–molybdenum, or nickel–molybdenum catalysts, typically arranged in a separate vessel. Hydrogen or hydrogen-containing gas are added to the process feedstock at temperatures of about 350–380 °C, so that the organic sulfur compounds are converted to hydrogen sulfide and the corresponding saturated hydrocarbons. Hydrogen sulfide produced in the hydrogenation stage is then absorbed by zinc oxide. The hydrogenation reactor can be used simultaneously to hydrogenate unsaturated hydrocarbons in the raw gas. This reaction is strongly exothermic. As the temperature range for the hydrogenation stage is limited to 250–400 °C, unsaturated hydrocarbon content is limited as well. Ammonia, carbon monoxide, or carbon dioxide impurities in hydrogen affect desulfurization and can lead to undesirable side reactions such as methanation [11]. An economic solution particularly for natural gases with low sulfur content is the combination of hydrogenation and zinc oxide absorption. This means that both catalysts can be combined in one bed in the same reactor. Residual sulfur content is normally <0.2 mg/m^3.

1.2.5   Carbon deposition

Carbon deposition is a frequently experienced phenomenon with SR. The reactions that may cause solid carbon formation include Boudouard reaction (1.4), hydrocarbon decomposition (1.5), and heterogeneous water gas reaction (1.6).

$$2CO \rightleftarrows CO_2 + C \tag{1.4}$$

$$C_nH_m \rightleftarrows nC + \frac{m}{2}H_2 \tag{1.5}$$

$$CO + H_2 \rightleftarrows H_2O + C \tag{1.6}$$

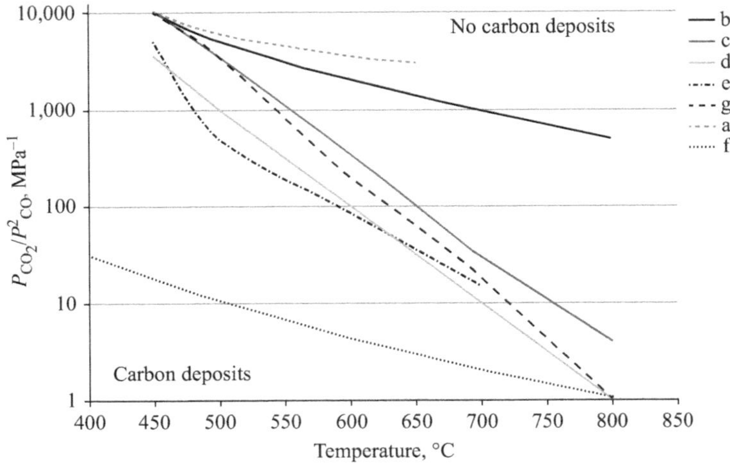

Figure 1.2   Equilibrium constant of the Boudouard reaction (1.4). (a) Condition of rich gas from naphtha at inlet to tubular reformer; (b) equilibrium reached for reaction (1.3) only; (c) equilibria reached for reactions (1.2) and (1.3); (d) according to Reference 14; (e) according to Reference 11; (f) for nickel–uranium catalyst according to Reference 15; (g) for sulfided catalyst according to Reference 11

Hydrocarbon decomposition is practically the exclusive cause for carbon formation in the course of reactions in the SR process. Therefore, carbon deposition is observed mainly in the first half of the reformer tubes where hydrocarbon concentrations are still high and the approach to equilibrium low.

Investigations showed [3] that up to 600 °C the methane-steam system is driven toward carbon formation only at very low S/C ratios (<0.6) and high pressure (30 atm/~3 MPa).

Figures 1.2 and 1.3 (both in logarithmic scale) show experimental "equilibrium constants" for the Boudouard reaction and methane decomposition measured by various authors, as well as calculated figures for thermodynamic equilibrium according to Moser and Bogdan [12]. Figure 1.3 also shows the "working line" for methane SR according to Balandin [13], by applying two catalysts with different activities. Points on the curves indicate the location of the measuring point as a proportion of tube length (viewed from the gas inlet). In SR of natural gas, the temperature at the catalyst inlet is such that carbon might theoretically be deposited. This does not occur, however, below about 650 °C, but only at higher temperature when an inactive catalyst fails to decompose sufficient methane. Generally, the formation of carbon deposits is not inevitable even at operating temperatures where it might be expected.

Figures 1.2 and 1.3 also illustrate the more favorable initial situation and possible further reforming of a naphtha-based rich gas with added carbon dioxide (to obtain

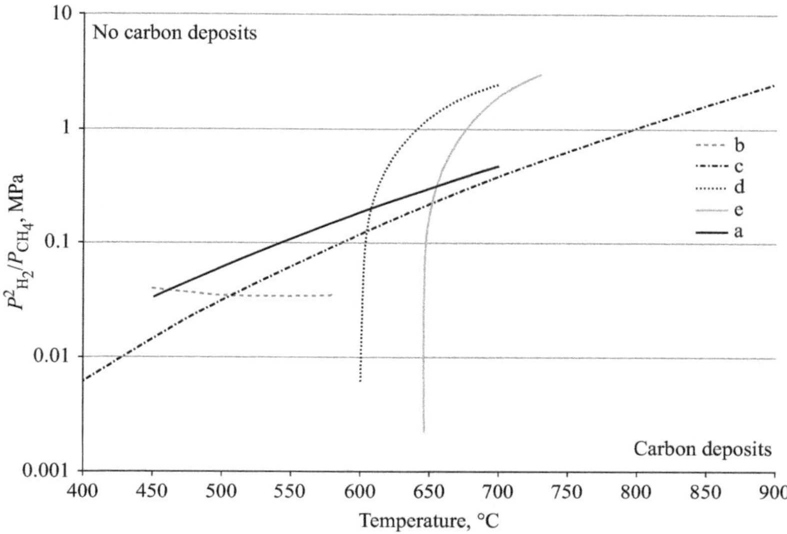

Figure 1.3   Methane decomposition (CH₄ → C + 2H₂). (a) Equilibria reached for reactions (1.2) and (1.3); (b) equilibrium reached for reaction (1.3) only; (c) equilibrium constant for carbon in graphite form according to Reference 12; (d) and (e) working lines for methane reforming in a reactor tube; (d) high-activity catalyst; (e) low-activity catalyst with red hot tube section from about 650 °C to equilibrium; figures at the data points are proportions of tube length as viewed from inlet

high-CO gases) if catalyst activity remains unimpaired (line (c) in Figure 1.2 or line (a) in Figure 1.3) or if no further methane conversion occurs (line (b) in Figures 1.2 and 1.3).

1.3   Supported catalysts

During a catalytic process, some or all of the reactants adsorb onto active sites of the catalyst where bonds are rapidly made or broken. For a heterogeneous solid catalyst processing a liquid and/or gas, the adsorption of reactants is called chemisorption, characterized by kinetics and reaction energies equivalent to a chemical reaction. Frequently chemisorbed species decompose to an intermediate that is rapidly converted to other intermediates or the final product. After the reaction is complete, the catalyst returns to its original state, and there is no net change of the catalyst. The process of chemisorption of reactants requires adsorption on the surface of the catalyst: so to maximize the reaction rate, the catalytic surface area should also be maximized. This is achieved by dispersing the catalytic species onto a high surface area inorganic carrier: supported catalysts play a significant role in many industrial processes. The support also stabilizes the dispersion of the active component (e.g., metals supported on oxides): active phase–support interactions, which are dictated by the surface chemistry of the support for a given active phase, are responsible for the dispersion and the chemical state of the latter. Although supports are often considered to be inert, this is not generally the case, as in many reactions, supports may actively interfere with the catalytic process. Typical examples for the active interplay between support and active phase are bifunctional catalysts such as highly dispersed noble metals supported on the surface of an acidic carrier.

To achieve the high surface areas and stabilize the highly disperse active phase, supports are typically porous materials having high thermo-stability. For application in industrial processes, they must also be stable toward the feed, and they must have a sufficient mechanical strength.

1.3.1 Supports

Carriers such as Al_2O_3, SiO_2, TiO_2, CeO_2, ZrO_2, C, and combinations of these materials are commonly used. All the mentioned supports have different surface properties and are used in applications dependent on the requirement for acidity, inertness to solubility, interactions with reactants, affinity for catalytically active components, and resistance to components in the gas phase: so the choice of the support must take into account all these requirements. For example, high surface area Al_2O_3 is not well suited for combustion reactions in which SO_2/SO_3 are present due to the formation of $Al_2(SO_4)_3$: in such cases, high area TiO_2 and/or ZrO_2 are used because of their inertness.

Carbons in various forms (charcoal, activated carbon) can be applied as supports unless oxygen is required in the feed at high temperatures, for example for supporting precious metals in hydrogenation reactions. In this case, precious metal recovery is achieved simply by burning the carbon.

Silicon carbide, SiC, can also be used as a catalyst support with high thermal stability and mechanical strength [16].

Generally, the most common carrier is gamma alumina (γ-Al_2O_3), characterized by an internal area of about 200–300 m^2/g. As its surface is highly hydroxylated, the H^+ sites provide acidity required for many reactions and exchange sites for catalytic metal cations.

Zeolites are combinations of Al_2O_3 and SiO_2, and their peculiarity is the crystalline structures with precisely defined pore structures in the molecular size range of 0.4–1.5 nm or 4–15 Å. A related group of materials known as mesoporous silica-alumina has extended the range of pore sizes attainable in ordered SiO_2–Al_2O_3 supports to 4 nm (40 Å). They are commonly used in the chemical and petroleum industry due to their surface acidity and ability to exclude molecules larger than the pore diameter. For this reason, they are often referred to as molecular sieves.

Monolithic supports with unidirectional macrochannels are used in automotive emission control catalysts where the pressure drop has to be minimized [16]. The channel walls are nonporous or may contain macropores and mesopores. For the above application, the monoliths must have high mechanical strength and low thermal expansion coefficients (TECs) to give sufficient thermal shock resistance. Nowadays, the preferred materials of monolith structures are ceramics (cordierite or SiC) or high quality corrosion-resistant steel. Cordierite is a natural aluminosilicate ($2MgO \cdot 2Al_2O_3 \cdot 5SiO_2$). The accessible surface area of these materials corresponds closely to the geometric surface area of the channels. High surface area is created by depositing a layer of a mixture of up to 20 different inorganic oxides, which include transitional alumina as a common constituent.

This so-called *washcoat* develops internal surface areas of 50–300 m^2/g.

1.3.2   Rate-limiting steps for a supported catalyst

Supporting a catalytic component introduces a physical size constraint dictated by the pore size of the carrier: so a key parameter is the accessibility of the reactants to the active catalytic sites within the high surface area carrier. For example, considering a hydrogenation reaction in which Ni is located in extremely small pores (i.e., 1 nm or 10 Å), the H_2 molecule has easy access but a large molecule, having a size comparable to the diameter of the pore, would experience great resistance moving toward the active sites. So if large amounts of Ni are present in pores and are not accessible to the molecules to be hydrogenated, the reaction rate will not be enhanced to its fullest potential. Is now evident that the carrier with its geometric sizes and its pore size distribution must be carefully designed to permit the reagents and products to move in the pores with minimum resistance. In heterogeneous catalysis, seven fundamental steps in the conversion of a reagent molecule(s) to product(s) are underlined and must be taken into account in the definition and development of a structured catalyst. These steps are the following:

1.   Bulk diffusion of all reactants to the outer surface of the catalyzed carrier from the external reaction media.
2.   Diffusion through the porous network to the active sites.
3.   Chemisorption onto the catalytic sites (or adjacent sites) by one or more of the reactants.
4.   Conversion and formation of the chemisorbed product.
5.   Desorption of the product from the active site.
6.   Diffusion of the products through the porous network to the outer surface of the catalyzed carrier.
7.   Bulk diffusion of the products to the external fluid.

Steps 1, 2, 6, and 7 depend on the physical properties of the catalyzed carrier and are not activated processes, in the sense that no intermediate chemical complex is formed. For this reason, we use the term apparent activation energy which is a term useful for comparing temperature dependence. Steps 3 through 5 are chemically activated, as there is the intermediate complexes formation during conversion to products, and depend on the chemical nature of the interaction of the reactants and products with the active sites.

Step 1 is referred to as *bulk mass transfer* and describes the transfer of reactants from the bulk fluid to the surface of the catalyzed carrier. When this step is rate limiting, reactant molecules arriving at the external surface of the catalyst are converted instantaneously resulting in zero concentration of reactants at the surface; in this case, the internal surface of the catalyst is not used. Such as mass transfer–controlled process is non-activated, and we assign an apparent activation energy of less than 2 kcal/mol. Rates vary only slightly with temperature ($T^{3/2}$), which allows it to be distinguished from other rate-limiting steps.

Step 7 is similar to Step 1 except that the products diffuse from the external surface of the catalyst particle into the bulk fluid. The temperature dependence of this phenomena is relatively weak and has an apparent activation energy similar to that observed in Step 1 when it is rate limiting.

When only the external surface of the catalyst particle is participating in the catalysis, it is said to have a low effectiveness factor. The effectiveness factor is defined as the actual rate divided by the maximum rate achievable when all catalytically active sites participate in the reaction. In the case of bulk mass transfer, the effectiveness factor approaches zero.

Steps 2 and 6 are both pore diffusion processes with apparent activation energies between 2 and 10 kcal/mol. This apparent activation energy is stated to be about 1/2 that of the chemical rate activation energy. The concentration of reactants decreases from the outer perimeter toward the center of the catalyst particle for Step 2. In this case, some of the interior of the catalyst is used, but not fully. Therefore, the effectiveness factor is greater than zero but considerably less than one. These reactions are moderately influenced by temperature but to a greater extent than bulk mass transfer.

Steps 3, chemisorption of the reactant(s), 4, chemical reaction forming the adsorbed product, and 5, desorption of the product(s) from the active site(s), are dependent on the chemical nature of the molecule(s) and the nature of their interaction with the active site(s). Activation energies are typically greater than 10 kcal/mol for kinetically or chemically controlled reactions.

Chemical kinetic phenomena are controlling when all transport processes are fast relative to the reactions occurring at the surface of the active species so the effectiveness factor is one. All available sites are being utilized, and the concentration of reactants and products is uniform throughout the particle. These reaction processes are affected by temperature more than either transport mechanisms.

1.3.3   Catalyst deactivation

The first indication of catalyst deactivation is a significant change in the activity/selectivity of the process. Catalyst deactivation occurs in all processes, but it often can be controlled if its causes are well known. This subject is very extensive, but the most common deactivation modes for heterogeneous catalysts are pictorially shown in Figure 1.4, and can be distinguished in:

- Sintering of the active components
- Carrier sintering
- Poisoning

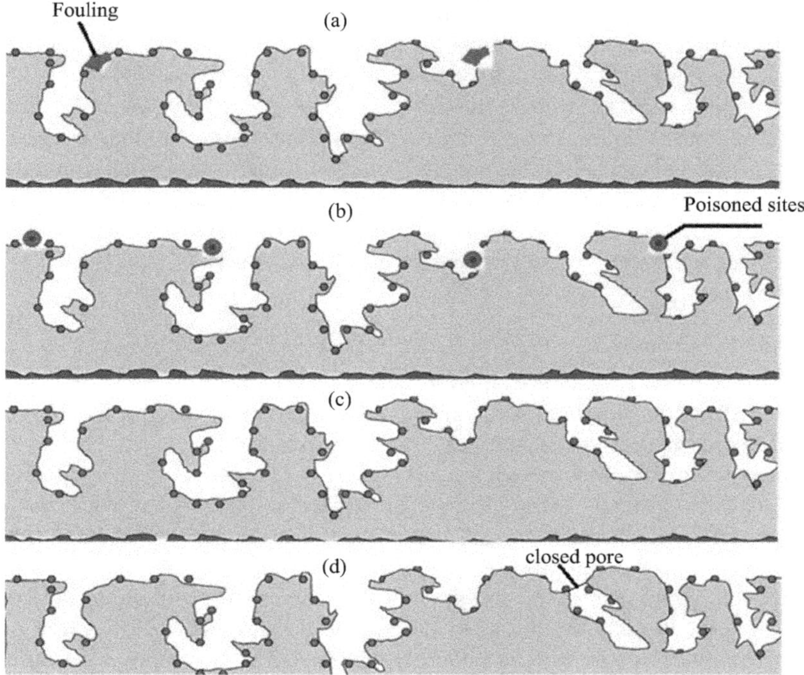

Figure 1.4　Idealized poisoning mechanism: (a) catalyst fouling, (b) catalyst poisoning, (c) catalyst sintering, and (d) carrier sintering

1.4　Structured catalysts: an overview

As previously described, structured catalysts are one-piece unitary structures characterized by wide pores or channel structures able to reduce pressure drop in high flow conditions. They are mainly defined by two parts: the structure, or carrier, responsible to the physical characteristics (geometry, strength, heat and mass transfer, etc.) and the active phase, usually located on the external surface of the structure, responsible of catalytic behavior. The structured carrier classifies structured catalysts in three groups: metal-gauze catalysts, foams monoliths, and honeycomb monoliths.

Metal-gauze catalysts are realized by metals structures; the low surface/volume ratio results in extremely short residence time; therefore, they are typically employed fast kinetics processes, or if there is a need to stop a reaction mechanism before reaching thermodynamic equilibrium. For this reason, similar catalysts were widely employed in nitrogen oxidation process from ammonia [17]; anyway, experiences were reported for wire-mesh catalysts utilization in CO preferential oxidation [18].

Open-cells foams, also named sponges, are monolithic blocks with a cellular structure and more or less isotropic mechanical properties in which space is filled by filaments (struts) forming a continuous network which encloses cavities (cells), interconnected by open-windows (pores) [19]. Cells are the void part enclosed in the struts, forming usually pentagonal or hexagonal polyhedrons: cells approach a spherical shape (constituting foam pores), and cells size is usually expressed in pores per linear inches. Struts, whose dimension is typically in the order of hundreds of microns, may be solid or hollow, depending on the forming method and may have different cross section geometry: circular (most common), triangular, etc. Struts are interconnected by means of nodes. Foams properties may be summarized in three parameters: void fraction ε, expressed in (1.7), external surface area S_V (exposed surface/foam volume) and pores mean diameter [20].

$$\varepsilon = 1 - \frac{\rho_{\text{foam}}}{\rho_{\text{material}}} \tag{1.7}$$

The 3D chaotic structure assures high turbulences of flux inside the foams, without dramatic effects on pressure drop; such characteristic enhances gas–catalyst interactions, improving both mass and heat transfer mechanisms.

Honeycomb monoliths are continuous structures constituted by many parallel or zig-zag narrow channels. They are the preferable solution for a wide range of environmental applications in the gas phase, including the huge automotive market. The reasons for their popularity are the low pressure drop at high flow rates, the dust tolerance, the high mechanical strength and the easiness of positioning (horizontal, vertical, tilted). In addition, the mass production has resulted in affordable prices [21]. Monolithic catalysts were intensely employed in automotive exhaust gas treatment, evidencing an appreciable thermal shock resistance and thermal stability.

Ceramic honeycombs are obtained by extrusion technique, channels have a triangular, square or hexagonal shapes. Metal honeycombs are realized by rolling or stacking alternate corrugated and flat strips: parallel channels have a triangular-like shape and are generated by corrugated strips. Corrugation is obtained by crimping a metal foil on a pair of rollers having sinusoidal or triangular teeth. The shape and the dimensions of rollers are responsible of cells pitch and width that in turn may affect fluid dynamic and heat transfer rate [22], and in turn catalyst performances [23]. Honeycomb monoliths were characterized by two main parameters: the cell density (expressed in cell per square inch, CPSI) and the wall thickness (or web thickness). Ceramic monoliths are characterized by cell density between 50 and 400 CPSI and a wall thickness between 0.1 and 1.0 mm; metallic monoliths are characterized by cell density up to 1,600 CPSI and a wall thickness between 0.02 and 0.1 mm.

A different category should be devoted to microreactors (or reactors of microchannels), referred to chemical reactors in which channels have at least one dimension smaller than 1 mm. Microreactors are constituted by units, each realized by several interconnected microchannels. Units are usually grouped in stacks, that arranged in a special housing and connected each other form a device. Several advantages are obtained by microreactors, as linear dimensions decreasing, surface/volume ratio increasing, volume reduction, and improved flexibility. However, it is not negligible that scale-up of such technologies involves additional costs to conventional processes, mainly due to micromachining. For these reasons, microreactors find a promising applicability in small-scale processing, such as the substitution of batch processes by continuous processes, process intensification, work safety, changes in product quality, and production delocalization.

1.4.1 Main advantages

One of the most critical parameters for packed bed reactors is particle size. On the one hand, small particles promote catalyst effectiveness, which impacts activity, selectivity, and stability alike, and the on the other, highly thick packed beds, typical of endothermic processes, produces excessive pressure drop along the catalytic volume, that in turn may counter, in catalytic process performances, advantages achieved by catalyst exploiting. Egg-shell catalysts partially overcome such difficulties, in which high dimension particles were covered by catalyst in a thin layer: such solution is able to reduce pressure drops without affecting catalyst performances. The main disadvantages of egg-shell catalysts concert the relevant idle fraction of the overall volume reaction; moreover, this configuration has no improvements toward mass transfer mechanisms in gas phase.

Structured catalysts are often considered as the structuring the extra-particle space, so the characteristic length of diffusion and reaction is the wall thickness and the characteristic length for the momentum balance and the extra-particle mass balance is the channel or pores diameter. Such dimensions are not correlated, so structured catalyst design could independently optimize both parameters. The possibility to choose the extra-particle length-scale independently introduces the extra degree of freedom that allows maximum catalyst effectiveness at minimal pressure drop. In addition, the appropriate selection of geometry and the shape of the interstitial voids results in the optimization of the external mass transfer can be optimized as well [24].

Most reactions are characterized by a relevant increasing or decreasing enthalpy significantly different from zero, potentially involving large heat flows. In order to keep temperature to desired value (estimated as optimal for catalytic performances), considerable amount of heat should be removed or supplied to the system. Geometrically, a structured packing is much better suited for conductive heat transport than a randomly packed bed, as in the latter particles only have point contacts, which diminishes heat transfer [21]. Several studies demonstrated that high conductive foams allowed flatter thermal profiles than traditional systems, both in the radial [25] and axial directions [10,26]. The "redistribution" of temperature in the catalytic volume is considered one of the main advantages of structured catalysts: generically, low carrier void fraction promotes flatter thermal profiles [27]. Structured catalysts, therefore, assure a better availability of the catalyst in the reaction, whereas the synergic effect of enhanced effective bed thermal conductivity and the improved solid–gas convective interaction significantly reduce heat transfer resistance, allowing lower furnace wall temperature required for a given heat flux [28]. Improved heat transfer toward catalytic volume assures a faster catalytic system, while the reduced thermal gradient in the catalytic volume leads to more stable performances [29]. The phenomenon is more evident in highly endothermic reactions, in which catalytic stability improvement is remarked in stressing operating condition [30] both in terms of catalytic performances and chemical–physical catalyst properties [31].

If on the one hand honeycomb monolithic carriers enhance conductive heat transfer in the solid phase [32], the random porous network of foam support results in a more efficient heat and mass transfer between solid and gaseous phase [33]. As an interesting compromise, monolithic carriers may be characterized by multi-channels capillary walls that link adjacent channels [34]: by alternately plugging channels at the inner and outer section of the monolith, gas is forced to cross the porous walls, achieving a deep contact between gas and solid that improve mass and heat transfer between the two phases [5].

Several experiences are available in the literature evidencing catalytic improvements led to catalyst deposition on structured carriers. Sangsong *et al.* [35], in their study on dry reforming of methane, demonstrated that by transferring of a common formulation (Ni/MgO–Al$_2$O$_3$) from pellet shape to stainless steel plate substrate, ten times higher activity was observed, due to the improvement of heat and mass transfer performance, that moreover effectively prevent carbon formation.

1.5   Structured catalysts preparation

Several routes were developer for structured catalysts preparation: the choice of a method rather than another depends by a wide number of factors, such as (but not limited to) involved materials (substrate, chemical support, active phase), catalyzed process, operating conditions, expected catalyst deterioration. Avila *et al.* [23] proposed an excellent classification (schematized in Figure 1.5) of monolithic catalysts based on preparation procedure, reporting four categories: coated catalysts, impregnated catalysts, extruded catalysts, and incorporated catalysts. Coated and impregnated catalysts will be deeply discussed in following sections; some words should be spent on extruded and incorporated catalysts.

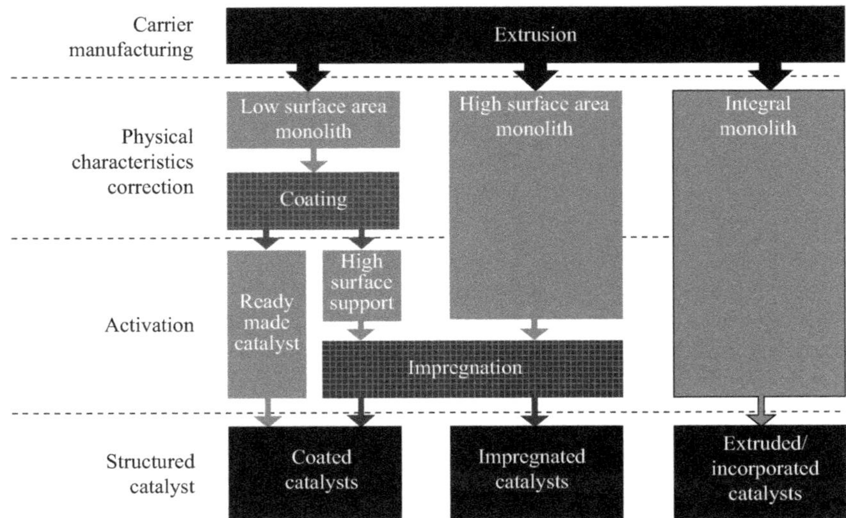

Figure 1.5   Preparation methods for ceramic monolithic catalysts

Integral monoliths were obtained by including active phase (as compound or precursor) in the composition of the dough of an extruded monolith: in this way, active species become a part of the monolith, and a very homogeneous distribution is achieved in the whole bulk volume. Such solution appears quite expensive in terms of active phase; anyway, it is advantageous in case of application in which catalyst erosion or abrasion may occur (the processed gas may contain solid or dust in suspension), being responsible of deactivation. In integral monolith preparation, great attention should be paid to porous structure, in order to avoid, or at least reduce, diffusion limitations phenomena. An appreciable technology is the activated carbon templating [36], in which support precursors and active carbon previously impregnated with the active phase precursors were kneaded together (with water). Once shaped, the monolith was treated at very high temperature in air to burn out the active carbon so obtaining the final catalyst. Such technology assures an excellent catalyst distribution on structured catalysts with respect coating and impregnation methods [37].

1.5.1   Substrate materials

Approaching the preparation of structured catalysts, as a first step, the adequate carrier should be selected. The choice of the substrate materials depends on three main requested properties:

- Properties related to the fabrication process, such as deformability, in order to allow the inserting of catalytic system in the reactor;
- Properties related to the kind of coating and the requested coating method: therefore, chemical compatibility and surface roughness should be taken in account;
- Properties related to the catalyst employment (thermal and chemical resistance under the operating conditions, etc.).

On the basis of the operation field in which catalyst should be employed, carriers should be characterized by a very low TEC, a high melting point and thermal stability, and an appropriate surface roughness or structure porosity.

Cordierite ($2MgO \cdot 2Al_2O_3 \cdot 5SiO_2$) is the most common substrate material for the ceramic carriers due to the very low TEC; its structures are characterized by macropores that facilitate the anchoring of catalytic layer. Anyway, it is possible to find ceramic monoliths made by a wide range of materials (based on α- and γ-alumina, mullite, titania, zirconia, silicon nitride, silicon carbide) depending on the final applications.

In a wide range of application fields, carrier thermal conductivity plays an important role in catalytic performances. For this reason, metal substrates are arousing growing interest in catalyst development, both for structured catalyst and microchannels reactors. Experiences of metallic structured catalysts dates back to the 1950s, when prototypes of stainless steel, chromel, nikrothal and nichrome wires, grids, mats, and crimped ribbons were coated with noble metal catalysts and employed in environmental catalytic processes, exploiting their good resistance to the high electrical currents used to heat up the catalytic systems by the Joule effect [19]. Particular attention was devoted to aluminum and steel as carrier materials [38,39]: the first one for the relevant heat conductibility, the second one for the better resistance to high temperature and pressure stress. Metallic materials have an intrinsic mechanical strength; therefore, it is possible to obtain struts or wall thickness significantly lower than ceramic monoliths. Such features greatly increase the monoliths void fraction, resulting in reduced pressure drops and in a considerable reduction of thermal inertia, and then transitory times in operating conditions variation. On the other hand, the very poor adhesion between monolithic wall and catalytic coating, and the thermal and chemical stability of substrate in operating conditions result relevant bottlenecks in metallic monolith catalysts. Therefore, the chemical affinity between substrate and catalyst deposition become a fundamental parameter in the possibility (or convenience) to employ metal carriers. From this point of view, the direct adhesion of metal active phase on substrate layer appears more favorable than the monolithic coating with alumina, ceria or zirconia washcoat. It is also relevant to remark that catalyst-substrate affinity may depend on operating temperature, as carrier stability at high temperature should be taken in account [40]. Furthermore, steel monoliths suffer of metal dusting, in particular when employed in reforming processes. It was observed that several steel alloys (e.g., AISI 304, AISI 316) in the presence of CO, CO_2, and H_2 in a temperature range between 400 and 850 °C showed the formation of a layer constituted by graphite and small metal particles (carbides and oxides) [41,42]. Such difficulties clearly overlap reforming processes conditions, causing both reactor wall modification and affecting gas-phase composition. It is also evidenced that the increasing in nickel content [43] and the presence of tungsten [32], in the steel alloy improve metal dust resistance. Domínguez *et al.* [44] studied the effect of the enamel protection [45] on a stainless steel monolith: an enamel barrier between the substrate and the catalyst is placed, in order to reduce interaction between process gas mixture and metallic support, so improving corrosion resistance and reducing metal dust. FeCrAl alloy carriers are often selected as catalytic substrate: after treatments at very high temperature, aluminum migrates from the bulk to the surface, resulting in the formation of a very open structure of long randomly oriented whiskers. On the other hand, aluminum alloys are often employed in low temperature field, by anodizing carrier surface. By choosing the electrolyte and the electrochemical parameters (temperature, time, electrolyte concentration, voltage and current density), it is possible to achieve an excellent control of surface area and pore size of the alumina coating [46].

1.5.2   *Substrate activation*

The structured carried becomes a structured catalyst by deposing a catalytic formulation on the carrier surface. The optimization of a structured catalyst therefore concerns the proper choice of substrate, the identification of a good catalytic formulation, but also the good adherence of the catalyst on carrier: so, the interface constituted by the carrier-coating discontinuity arouses a great importance in structured catalyst preparation. A structured catalyst should be characterized by a good homogeneity of the coating on the entire surface of the structural material, a proper thickness (or loading) of deposited catalyst, and an appreciable adhesion or peel strength during handling and use of structured system. Therefore, from the structural material point of view, carrier should have a chemical and physical compatibility with the catalytic phase. Catalytic layer should be compact and resistant to the operating conditions and preserve the catalytic properties evaluated in powder formulation. Such aspects are candidate to determine a sufficient coating thickness and homogeneity; carrier thermal treatments are able to increase surface roughness, promoting mechanical anchorage; finally, the addition of an intermediate layer (primer) on the carrier surface could improve chemical compatibility between support and catalytic phase.

The structured carrier should be characterized by a high specific surface area (SSA), in order to promote active species dispersion and offer a high specific surface to chemical reaction. If structured carrier surface area in not large enough before deposing active species, a preliminary stage is required, aimed to increase SSA. It consists in covering carrier surface with a slurry characterized by an intrinsic higher surface area: such procedure consists in carrier coating.

The most diffused technique for catalyst depositing on structured carrier is the dip coating (or washcoating); it consists in the introduction of substrate in a liquid suspension or solution (sol–gel) containing the catalytic formulation, so depositing a layer of a high-surface-area (>10 m^2/g) oxide(s). Washcoat may be defined as a slurry of particles with a comparable size to support macropores dimensions. Such method can be applied to almost all substrate and allowed to depose a wide range of catalysts, since liquid phase could be suspension, colloid or sol containing solids or precursors. The main advantages of washcoating is that washcoat may be constituted by ready-made catalysts since active elements may be incorporated into the coating layer, so support activation can be achieved in one step; alternatively, washcoating aim is to improve chemical and physical features of the carrier, achieving a surface layer more adequate for further adhesion of active species.

Washcoating consists in several steps: substrate immersion in liquid, permanence in liquid, extraction, removing of excess liquid, drying, and calcination. All these steps, as well as the slurry properties, are responsible of coating quality. Usually, a washcoat layer thickness in the range of 10–80 μm is desired [47]. In most cases, slurry is a suspension of solid particles, containing the catalyst to depose on carrier surface; therefore, slurry optimization concerns both solid and liquid phases.

A slurry is considered stable when particle settling is negligible: such condition is reached if particles velocity is very low, and the drag force of liquid on particles is able to delete gravity force applied to the particles. According to Newtonian fluids prediction on creeping flow regime, the particle terminal velocity linearly depends on particle square diameter and the difference between particles and fluid density, and inversely depends on fluid viscosity. Such assumption underlines that to increase washcoat stability, it is possible to reduce particle diameter: dedicated studies concluded that particle size distribution below 10 μm assures a good stability [48]. It is anyway worth to underline that an excessive reduction of diameter promotes particle aggregation (gelling) [49] and then flocculation phenomena. Another relevant parameter for slurry stability is the fluid viscosity. In an ideal system, viscosity of a suspension only depends on the amount of particle in the fluid: the higher the particle content, the higher is the viscosity; in the real systems, particle interaction cannot be neglected. From this point of view, the addition of thickeners, such as inorganic colloids (alumina, silica, etc.) or organic compounds (polyvinyl alcohol, polyvinylpyrrolidone, ammonium methacrylate, etc.), may increase slurry viscosity. On the other hand, pH modification of the suspension may lead to the polarization of oxide particles, resulting in a mutual repelling of particles (having the same charge) and reducing the aggregation phenomena. Moreover, pH also affects the aggregate size, resulting in the viscosity determination. The addition of long-chain surfactants containing hydrophilic and hydrophobic groups may be responsible for slurry stabilization, due to their adsorption on particle surface. Of course, to select the additives aimed to stabilize slurry, the interaction between different additives and their effect on catalytic performance should be taken in account. In sol–gel technology, one important factor is the ageing time allowing the gelation (peptization) of the sol. It can vary from a few minutes to several weeks, depending on the concentrations in the sol and the characteristic size of the object to coat. The conditions during sol formation have to be chosen in order to obtain oligomers with desired degree of branching [47].

The washcoat properties, and its modification by surfactants, strictly depends on structured carrier to activate, and on the other hand, washcoating methodology depends on slurry properties. The amount of deposed catalyst on inner walls (and into their porosity) of channels in a monolith structure clearly depends by the washcoat solid content: the solid particles fixed to the wall as a thin filter cake. On the other hand, a fundamental step in filling monolithic channels is the absorption of the water constituting the slurry on the monolithic walls, on which the catalytic cake grows. The filling mechanism of slurry along channels of a monolith disposed vertical way occurs by a capillary mechanism: if the immersion velocity is too slow, the solvent will rise by capillarity through the wall faster than the suspension front and no solid cake will be formed on the wall. If monolith wetting is too fast, suspension cannot enter and rise in the channel. Due to the capillary mechanism of slurry transfer in channels, great attention should be paid to the surface tension, the fluid/solid wetting angle and the channel diameter. The viscosity and channels size control the washcoat rising rate, and in turn the optimal immersion rate: usually, a monolith velocity of 2–6 cm/min is recommended, with a residence time of about 5 min. The amount and quality of monoliths coating strictly depends on channels size and shape. Since an accumulation of slurry is expected in the channels inner corning, the accumulation is higher for the more acute angle corners. Moreover, it was demonstrated [49] that reducing size of channels promotes slurry adhesion on monolith walls. It is anyway clear that viscosity is the main parameter affecting washcoating results. Obviously, low viscosity promotes easy coating and homogeneity; however, low viscosity means low particle in the slurry, and so low loading on the structure walls: as a result, repeated coating are required to reach desired loading. There is so a diffuse trend to increase viscosity, aimed to minimize the number of coatings, but too high viscosity may cause problems both in channels wetting and removing washcoat in excess, causing slurry accumulations that may lead to channels plugging. On the other hand, the contact angle between carrier surface and the slurry, depending on the energy surface of the two phases, determine appropriate coating. Very low contact angles assure a well coating of the substrate, while substrate hydrophobicity was obtained if contact angle overcomes 90°. To avoid such conditions, an appropriate surface roughness is required; moreover, surfactants addition in slurry may modify the liquid surface tension.

Once concluded the monolith washcoating, the slurry in excess should be removed, in order to obtain a thin uniform washcoat loading. Elimination could be mainly obtained by air blowing through carrier channels or pores, or by centrifugation. The first one, sometimes conducted by only gravitational removing of excess, is of course the easiest, and it is commonly applied for big monoliths; it is worth to underline that such procedure requires attention in order to avoid inhomogeneity in catalyst thickness; on the other hand, centrifuging technique assures higher washcoat removing.

Final steps in coating procedure are the structured catalyst drying and calcination. The drying mainly aims to remove water from slurry layer, such step may cause cracks and catalyst detachments from the substrate. To prevent this unwanted may be reduced by adding surfactants to the washcoat, aimed to reduce surface tension. Calcination is the last step, in which any organic compound was removed from washcoat. After calcination in possible to evaluate the real amount of catalyst deposed on substrate surface in a washcoating step: if the amount is lower than required, washcoating cycles may be repeated up to obtain the desired loading.

An alternative method for washcoat deposition is the slurry percolation, in which substrates were placed inside a tube where an excess of slurry was poured. The slurry was drained at a controlled velocity while percolating through carrier. The linear velocity of percolating slurry is similar to the dipping velocity described above. The highly reproducible percolation and the possibility of a semi-continuous procedure make such technique interesting for industrial applications [50].

Although less common, other methods may be employed for substrate coating. In chemical vapor deposition, substrate is exposed to a vapor phase in which precursors with high vapor pressure are dispersed, that decompose and depose on carrier surface; such method may be applied to the most complex carrier geometries. In physical vapor deposition, adhesion is achieved by a mechanical (cathodic sputtering) and thermal methods (evaporation and electron-beam evaporation) [47]. Other technologies based on electrochemical deposition may be employed, such as electrophoretic deposition, in which a suspension of charged particles are attracted by an electric field to the surface to be coated, and the electrochemical deposit of an ionic solution [51], in which ions are attracted by substrate surface charged oppositely (of course, a conductive substrate is required). In the case of flat carriers, spray technologies (traditional thermal spray or supersonic cold gas dynamic spray technologies) may be employed, in which catalytic powder was forced against substrate wall by a carrier gas [52,53]. These techniques, applicable in the preparation of a limited number of structured catalysts, assure a very uniform catalyst deposition.

1.6   Structured catalysts in reforming processes

The two most diffused technologies for H_2 production are hydrocarbon fuels processing and H_2O electrolysis. The gap in hydrogen costs obtained by these two different processes is very large; therefore, hydrocarbons fuel processing still remains the best solution for a period of transfer toward a hydrogen-based economy. The purpose of a fuel processor is to convert a hydrocarbon fuel (natural gas, gasoline, diesel) into a H_2-rich stream to fed a fuel cell system.

1.6.1   Reforming processes: an overview

A typical fuel processor consists of three main steps: a reforming unit, in which syngas is produced from hydrocarbons, a WGS unit, and a preferential oxidation unit, aimed to remove CO from syngas. There are three primary techniques used to produce hydrogen from hydrocarbon fuels: SR in which hydrocarbon reacts with steam at high temperatures; POX in which hydrocarbon reacts with lean oxygen; and ATR that results from a combination of the previous technologies in which hydrocarbon reacts with both steam and oxygen [54].

1.6.2   Steam reforming

SR is the widely diffused technology to produce hydrogen mainly due to its highest yield. SR is a catalytic endothermic process in which a hydrocarbon (e.g., methane) reacts with steam to produce mainly hydrogen and carbon monoxide (see (1.2)).

In reforming processes, the choice and optimal setup of the catalytic system may greatly affect both the conversion degree as well as the selectivity of the reaction. Obviously, the catalyst selection should be made according to the defined operating conditions and based on the selected fuels. Several studies have demonstrated that nickel [55] as well as noble metals (Pt, Rh, Ru) [56] supported on Al_2O_3 or rare earth oxides (CeO_2, ZrO_2, La_2O_3, Y_2O_3) show good activity toward reforming reactions [57–59]; improvements in stability and selectivity are achieved from bimetallic catalytic systems [60].

The process endothermicity implies that very high reaction temperatures and heat fluxes toward the reaction system are required to achieve high hydrocarbon conversion. From this point of view, thermal management in the catalytic volume becomes a crucial aspect. A typical thermal profile along a reformer tube is reported in Figure 1.6: the SR reaction is more extreme at the beginning of the reformer, leading to a dramatic temperature reduction; temperature progressively increased by external heating once reforming reaction rate decreases for the reduction of hydrocarbon content. In the simulation proposed by Vakhshouri and Hashemi [61], the temperature gap between inner and outer sections may reach 500 °C. Moreover, since catalyst heating was carried out through tube wall, a radial thermal profile is also noticeable (a temperature difference up to 190 °C was estimated) in the whole catalytic volume.

The process intensification in SR processes covers both thermal management and reactions rate. Moreover, it is widely accepted that reactions take place on

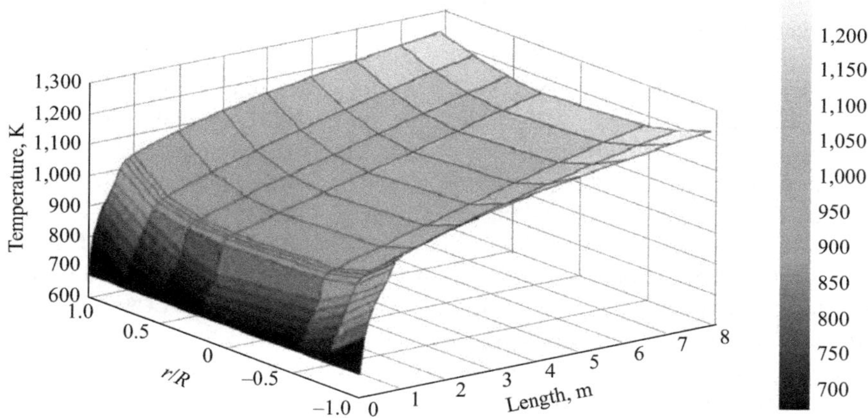

Figure 1.6   2-D thermal profile in a typical steam reformer tube [61]

catalytic wall [62]; therefore, the monolithic temperature distribution and the heat transfer rate between solid phase and gaseous stream are key parameters in the system performances behavior. Several studies [63] have demonstrated that high thermal conductivity supports may allow for a flatter axial thermal profile along the catalytic bed, thus resulting in a higher average temperature at the outlet section of the reactor, and consequently in larger hydrocarbon conversion [64]. In addition, the highly conductive supports ensure a more uniform radial temperature profile, thus resulting in a better heat transfer and reduction of hot-spot phenomena [65].

Experiences on natural gas SR were carried out by Roh *et al.* [28] that compared performances of traditional Ru/Al_2O_3 pellets catalysts to a metallic monolithic catalyst coated by the same catalytic formulation. Obtained results evidenced that metallic substrate allows a proper thermal management, resulting in a relevant reduction in heat transfer resistance from heating medium to the reaction volume: the advantages provided by structured catalyst is more evident at very high gas rate, thus remarking the increasing of Nusselt number in a monolithic structure. The increased heat flux on the one hand allow lower temperature for heating medium, on the other hand suggest faster reactions rate, and then a reduced required catalyst volume.

Hiramitsu *et al.* [66] tested a honeycomb monolith, realized by a 30-μm nickel foil, in the SR of methane. The use of an integral monolith was possible as nickel (in metallic form) was highlighted as one of the most active species, and in particular the most used in commercial reforming catalyst, due to its low cost. Despite the absence of any promoters or chemical supports led to a relevant coke deposition on catalytic wall and the POX of superficial nickel, and thus to a not negligible deactivation of the system, the high conductivity of the metallic structure and the high cell density allows an effective redistribution of the temperature along the catalytic volume, assuring a flat thermal profile. The employment of nickel foams is often considered as structure carrier for reforming processes; it was however demonstrated that the coverage of sponge surface with promoters (Al_2O_3, ZrO_2, CeO_2, etc.) leads to beneficial effects in terms of activity and stability of the system, enhancing the resistance to coke formation [67].

Fukuhara *et al.* [68] proposed a study on a structured catalyst activated by nickel deposition achieved by electroless plating method. A metallic honeycomb was covered by washcoating method with a thin layer of alumina, then nickel was deposed by electroless method, achieving 70–150-nm nickel clusters. The catalytic performances of prepared structured catalyst were compared to the ones achieved by a commercial catalyst (METH134, Clariant Catalysts K.K) for the methane dry reforming reaction. Experimental achievements evidenced that structured catalyst assured higher conversion, in particular at more severe conditions (high temperature), evidencing the improved mass and transfer mechanisms due to the monolithic structure. In particular, the wide surface/volume ratio determined by a structured carrier allows a better dispersion of active phase on catalytic surface, thus allowing to achieve better performances (both in terms of conversion and coke resistance) also by employing lower active phase loadings.

Metal monolithic catalysts may be employed in smart configurations, by managing reforming catalytic volume and heating gas, resulting in a catalytic reactor itself. Mei *et al.* [69] proposed a study on a metal monolith catalytic reactor for methane SR–combustion coupling.

In such configuration, heat duty of the SR, realized in the external zone of the reactor, was supplied by the combustion reaction taking place in the inner volume. Model studied by authors evidenced that the effectiveness of the proposed system. The proper channel arrangement or catalyst distribution, such as more SSA of the reforming part or non-uniform catalyst distribution of the combustion side, can optimize temperature distribution in both sides, thus increasing hydrocarbon reforming rate.

Chang and Lee [70] proposed an interesting comparison between powder and honeycomb $Ni–CeO_2$ based catalysts for the SR and biogas reforming reactions. By transferring catalytic formulation on structured carrier, a relevant increasing in catalytic performances was observed not only in terms of methane conversion but also in terms of selectivity to hydrogen. The enhanced performances were more evident in biogas reforming, for which a noticeable increasing in hydrogen selectivity was found (from 0% to 40%). Such behavior may be strictly linked to the thermal management along the catalytic bed: the more uniform thermal profile (both in axial and radial direction) helps catalytic system to address reaction pathways toward desired products.

Smorygo *et al.* [71] compared performances of Fe–Cr and Ni–Al foams activated with several formulations for the internal SR of SOFC fuel cell. It was observed that Fe–Cr foams showed a fast deactivation (with respect to the Ni–Al foams), evidencing an interaction with active components, due to Fe and Cr cation transfer to the active layer surface, which creates favorable conditions for the methane molecule cracking.

An interesting evaluation of foam catalysts utilization in industrial SR processes was carried out by Faure *et al.* [72] that spent efforts to evaluate the stability of catalytic alumina foams under process conditions. Mechanical resistance of foams catalyst appeared to be increased after catalyst deposition, but water may cause limited corrosion of the surface at grain boundaries during aging treatments. Special foams were produced, characterized by an increasing density (decreasing pore size) from center to external diameter and/or along foam axis, thus controlling the active phase concentration in the catalytic bed.

The transferring of catalytic formulation on foams catalyst does not involve any loss in catalytic properties, while the enhancements in heat and mass transfer mechanisms with respect to pellets catalysts may result in a dramatic increasing in reaction rate (per unit of active species mass), as demonstrated by Richardson *et al.* [73] in their study on catalytic systems for dry reforming reactions. In their work, it was also evidenced that specific reaction rate may be affected by foams porosity, since higher pore density implied higher reforming rate.

Silicon carbide was often considered as structured carriers in heterogeneous catalysis, due to chemical and physical characteristics (high thermal conductivity, high mechanical strength, low specific weight, chemical inertness). The high thermal conductivity facilitates flat temperature profiles, that in turn could be reflected in a more uniform catalyst exploiting in the catalytic process, resulting in very stable performances [74].

The improvements linked to structured catalyst was evidenced also in SR of ethanol. Palma *et al.* [26] evidenced that by transferring a catalytic formulation on SiC foams helps to increase kinetics of the involved reactions. As a consequence, total ethanol conversion and very high hydrogen yield can be achieved at very short contact times, without effects on catalyst stability. Moreover, it was also demonstrated that the employing of a high conductive foam carrier helps to reduce of 33% the heating medium temperature. SiC catalysts were also employed in pressurized ATR of methane by Li *et al.* [25]. It was observed that, beside the better performances with respect pellets catalyst, a not negligible deactivation occurred after 500 h in test conditions. Apart of a weak carbon deposition on catalytic surface, a not negligible shrinkage of active coating, and subsequently a stripping off from SiC substrate was observed; moreover, a weak degradation of SiC foam occurred. Similar results were also observed by Sang *et al.* [75] in their study on AISI 316 stainless steel foam catalysts for the reforming of simulated biogas. Excellent performances were observed in a wide range of operating conditions, in terms of operating temperature, contact time, and CO_2/CH_4 ratio. On the other hand, characterization of fresh and spent catalysts evidenced that Fe in AISI 316 foam was oxidized or recombined with other elements during methane reforming of CO_2 at high temperature, however very stable performances were observed in 100 h test.

An interesting catalytic configuration was proposed by Palma *et al.* [2], which carried out SR of methane on SiC wall-flow monolith activated by nickel. Wall flow monoliths were honeycomb structures characterized by porous walls: once channels are plugged alternately at the inner and outer section, gas is forced to cross the porous wall, assuring an intimate contact between gas and solid phase, maximizing heat and mass transfer rates between reactants and catalyst. Such arrangement, coupled to the intrinsic thermal conductivity of SiC, assured a relevant intensification of reforming process, requiring a reduced nickel amount and resulting able to reach high performances at lower heating medium temperature.

1.6.3 Partial oxidation

Hydrocarbons POX is a process in which an oxidation takes places in lack of oxygen. It is an exothermic reaction (1.8) characterized by a very fast kinetic, allowing very short contact times: such feature results in reduced plant size with respect to SR process; on the other hand, a lower hydrogen yield is expected.

$$CH_4 + \frac{1}{2}O_2 \rightleftarrows 2H_2 + CO \quad \Delta H^\circ_{298K} = -35.6 \text{ kJ/mol} \tag{1.8}$$

A proposed reaction mechanism resumes POX reaction as the subsequence of total oxidation of a part of fuel, producing steam and carbon dioxide [76]. The generated combustion heat is partially exploited to sustain endothermic reactions of steam and dry reforming.

In principle, POX reaction could be carried out in homogeneous phase, without any catalyst, since in temperature between 1,300 and 1,500 °C hydrocarbons total conversion could be achieved. Due to the absence of catalysts, no fuel pretreatments (such as desulfurization) are required, resulting in a cheaper process. On the other hand, extreme temperatures are not devoted to process management, leading to unwanted phenomena. Therefore, growing interest arouses the catalytic POX that assures similar performances at lower temperature (800–1,000 °C). Catalyst's role is to increase process kinetics and to deny unwanted reactions, such as hydrocarbons cracking.

The POX is characterized by several criticisms, mainly linked to the reactants homogeneity and temperature profiles. The local lack of one reactant with respect to another may generate coke formation or hot spots, resulting in catalyst deactivation. Foams structures may improve continuous mixing of process stream, improving composition homogeneity along reaction volume sections. Moreover, the employing of conductive supports may help to distribute temperature in the system, resulting in a more controlled process [77]. In a general concept, hydrocarbons POX requires catalytic formulations very similar to SR ones; it was however evidenced that since POX was carried out at very short contact time, rhodium was candidate as most promising active species, due to its very high specific activity [78]; it is however fundamental the extremely high dispersion of rhodium of catalyst to achieve adequate performances.

Aartun *et al.* [79] proposed a comparative study between a stainless steel microchannel reactor and an alumina foam catalyst, both activated by Rh deposition, for the POX and ATR of propane. The Rh-impregnated foams evidenced a pronounced temperature peak just in the inner section of the catalytic volume. The microchannel monolith displays both smaller axial temperature gradients and lower differences between maximum catalyst temperature and furnace temperature than the foams. Microchannel reactor was characterized by the intrinsic higher heat conductivity of the Fecralloy systems; moreover, the channel geometry reduced back-mixing phenomena along the catalytic volume. On the other hand, the 3D chaotic structure of foam catalyst improved catalytic performances of the system, evidencing higher hydrocarbon conversion and selectivity to hydrogen with respect of the microchannel structure. The better thermal management recorded in micro-channel reactor suggests the absence of hot-spot phenomena, as a consequence a better stability in catalyst behavior was observed, especially in start-up and shut-down cycles.

Since the POX may present a relevant temperature peak in the catalytic volume, corresponding to the prevalence of total oxidation reactions, the choice of adequate materials for structured catalyst appears crucial in catalytic system design. Too high temperature may cause sintering phenomena of active phase and the carrier damage or melting. By covering metallic supports with proper coatings prior depositing active phase on the one hand may improve both active species dispersion ant their resistance to sintering phenomena, on the other may reduce risks of thermal damage for the carrier [80].

1.6.4   Auto-thermal reforming

The ATR is a combination of SR and POX, in which hydrocarbon reacts both with steam and oxygen (see (1.9)).

$$CH_4 + xH_2O + yO_2 \rightleftarrows aCO + bCO_2 + cCH_4 + dH_2O + eH_2 + fC_{(S)} \quad (1.9)$$

In principle, ATR may be considered as the sequence of the two previous reactions: POX and SR. It is a self-sustained catalytic process, in which the exothermicity of hydrocarbon oxidation reaction supplies to the system the heat needed for the SR reaction. Exothermicity of the reaction as well as hydrogen yield depend strictly on feed ratio values x and y. The temperature profile along the catalytic volume has a sharp rise to the peak in the POX zone and then a decrease due to the endothermic reactions to a relatively low and flat level in the SR zone. The non-uniform axial temperature distribution could cause hot-spots phenomena, which induce the potential risk of local catalyst deactivation due to thermal-induced mechanisms such as sintering. It was often remarked that temperature gradient greater than 150 °C in the axial direction may induce in a fast catalytic deactivation [81].

It is a diffused concept that to minimize hot-spots high conductive catalyst support structures, and proper flow configurations may be a solution, by promoting effective heat transfer along the catalytic volume and achieving a more uniform axial temperature distribution. In this direction, metallic monoliths, foams, wire-gauzes, or microchannel reactors assure better behavior with respect to traditional pellet catalysts. In principle, flow with high turbulence can improve hot-spots since the turbulent flow enhances the heat transfer coefficient between the flow and solid walls. On the other hand, monolithic channels dimensions may strongly affect the peak temperature inside the catalytic volume [82].

The structured catalyst preparation by washcoating of a ceramic monolith for ATR of methane was investigated by Vita *et al.* [8]: in their work, a 400 CPSI cordierite monolith was covered by a CeO_2 layer by solution-combustion synthesis method [83], while active species (Rh, Pt, or Ni) were deposed by wet impregnation method. The prepared structured catalysts presented a very uniform catalytic layer, with thickness between 20 and 25 µm, characterized by a high mechanical strength. The use of monoliths helped to improve the catalytic performance also in very stressing operating conditions: the large open frontal area, the high surface area/volume ratio, the good heat and mass transfer properties, and the low pressure drop characterizing monoliths contributed to overcome the main problems of packed bed reactors. In particular, the Rh- and Pt-based monoliths showed high catalytic activity and stability also at low contact time (up to 400 $NL/g_{cat}/h$) without evidencing deactivation, whereas Ni-based sample showed good performances up to 200 $NL/g_{cat}/h$, while at lower contact time the formation of encapsulated carbon and carbon whiskers were observed due to fast sintering.

A very interesting study on structured catalysts for ATR of methane process was carried out by Ciambelli *et al.* [84]. In the proposed study, two structured carriers, cordierite monolith and alumina foam, were activated by the same formulation and procedure (commercial CeO_2–Al_2O_3 slurry, 10 wt% Nickel loading). Activity tests, carried out at different feed ratios, evidenced globally better performances of the foam samples, that resulted in higher methane conversion (beside an overall lower temperature) and a better approach to thermodynamic equilibrium also at very low contact time, and a very flat thermal profile in a wide range of process stream rate. Authors reported that generally, when heat and mass transport limitations are controlling, such as in the catalytic POX reaction, the random porous network of a foam catalyst carrier may determine an improvement in the gas temperature profile and species diffusion, helping to realize a proper catalyst bed design. It was also remarked that foams had a higher surface-to-volume ratio with respect to honeycomb monolith catalyst, resulting in a thinner washcoat layer, that may be responsible of lower eventual diffusional effects.

In a further investigation, Palma *et al.* [85] remarked the influence of flux geometry in the adiabatic ATR of methane. It was demonstrated that foams catalysts assured better performances than monolithic catalysts, due to a better management of heat and process stream along the catalyst: foams catalysts allowed to operate ad very low contact time (up to 10 ms), assuring the total hydrocarbon conversion. It was also demonstrated that by cutting monolithic catalyst in five bricks, spaced few millimeters each other, the process stream mixing after each brick was assured, leading to a massive improvement of catalytic performances. Of course, the discretization of the catalyst negatively affects the axial thermal profile of the catalytic volume.

1.6.5   Purification stages

During downstream reforming processes, WGS is typically carried out, in which carbon monoxide reacts with steam to produce further hydrogen (see (1.3)). It is a weakly exothermic reaction, therefore thermodynamically promoted at low temperature, at which the process is characterized by too low kinetics.

Industrially, the process is divided in two stages, the first one carried out at higher temperature (350–450 °C) on iron and chrome catalysts, in which up to 95%

of CO is converted; the second one carried out at low temperature (180–220 °C) on copper and zinc catalysts to achieve up to 99.5% of CO conversion. Such scheme requires a cooling unit between the two WGS stages, dramatically affecting the overall plant size.

The conventional configuration of a WGS section within a fuel processor makes use of adiabatic reactors, that is without active cooling to a second medium. Because of the temperature rise associated with the exothermal WGS reaction, typically two adiabatic reactors with intermediate heat exchange are required to reach sufficiently low CO levels. The redistribution of temperature from the exit to inner zone of the catalytic volume appears a promising solution for the process. The concept of isothermal adiabatic operation of the WGS reactor for a fuel processor focused on back mixing of the reaction heat throughout a continuous catalyst structure, the catalyst bed is nearly isothermal while the adiabatic character is maintained. This single stage isothermal adiabatic reactor replaces the conventional double staged adiabatic operation with intermediate heat exchange [86].

Process flux geometry could improve heat transfer in the catalytic volume. Palma *et al.* [63] demonstrated that radial flux configuration in an adiabatic WGS reactor led to a relevant increasing of CO conversion in particular at highest flow rates. The radial configuration assures a lower gas velocity, so enhancing conductive heat transfer (from the exit to the inner section) with respect to convective heat transfer. Moreover, the flux direction also affected system performances, since centrifugal configuration resulted in higher reaction rate than centripetal one.

Depending on the final utilization of produced syngas, further purification stages are required. If H_2 is addressed to PEM fuel cell CO content must be reduced to 10–50 ppm, and consequently a further step of preferential oxidation of CO (CO-PROX) is required. In this stage, residual carbon monoxide present in the syngas is selectively oxidized by low amount of oxygen (see (1.10)).

$$CO + \frac{1}{2}O_2 \rightleftarrows CO_2 \quad \Delta H^\circ_{298K} = -283.0 \text{ kJ/mol} \tag{1.10}$$

On the one hand, the reaction is highly exothermic, and on the other, process selectivity is strongly affected to process temperature. Therefore, a part of the catalytic formulation (Cu, Pt and Au on CeO_2 or Al_2O_3 are widely employed [87–89]), the thermal management along the reaction volume plays a fundamental role.

Barbato *et al.* [90] evidenced the role of cell density and bulk material in Cu-based catalysts for CO-PROX at low temperature. By reducing cell density, surface layer was increased, allowing a higher catalyst loading, resulting in better catalytic performances. By increasing the wall thermal conductivity, so employing the temperature profiles are flattened with an increase of the selectivity to CO_2. Similar conclusions were obtained by Milt *et al.* [91], evidencing that by transferring their catalytic formulation (Au/TiO_2) on stainless steel monolith, a reduction in catalyst activation temperature threshold was observed. Moreover, the good heat transfer ability of FeCrAl support can depress the reverse WGS effectively, which is beneficial for purification of CO in hydrogen-rich gases via preferential oxidation [92].

Nomenclature

Symbols

$\Delta H^{\circ}_{298\ K}$	standard enthalpy of reaction
T	temperature
p	pressure
p_A	partial pressure of A
ε	void fraction
S_V	external surface area
ρ	density
r	radial position
R	radius
x	steam/methane ratio
y	oxygen/methane ratio
a	yield of carbon monoxide
b	yield of carbon dioxide
c	yield of methane
d	yield of steam
e	yield of hydrogen
f	yield of coke

Acronyms

ACT	activated carbon templating
AISI	American Iron and Steel Institute
ATR	auto-thermal reforming
CO-PROX	carbon monoxide preferential oxidation
CPSI	cells per square inch
CVD	chemical vapor deposition
HGS	high geometric surface
P	density
PEM	proton exchange membrane
POX	partial oxidation
PPI	pores per inch
PVD	physical vapor deposition
PVOH	polyvinyl alcohol
SCS	solution-combustion synthesis
SOFC	solid oxides fuel cell
SSA	specific surface area

SR	steam reforming
T	temperature
TEC	thermal expansion coefficient
WGS	water–gas shift

References

[1] LeValley TL, Richard AR, Fan M. The progress in water gas shift and steam reforming hydrogen production technologies – a review. International Journal of Hydrogen Energy. 2014;39:16983–7000.

[2] Palma V, Ricca A, Meloni E, Martino M, Miccio M, Ciambelli P. Experimental and numerical investigations on structured catalysts for methane steam reforming intensification. Journal of Cleaner Production. 2016;111, Part A:217–30.

[3] Angeli SD, Monteleone G, Giaconia A, Lemonidou AA. State-of-the-art catalysts for CH_4 steam reforming at low temperature. International Journal of Hydrogen Energy. 2014;39:1979–97.

[4] Wang F, Qi B, Wang G, Li L. Methane steam reforming: kinetics and modeling over coating catalyst in micro-channel reactor. International Journal of Hydrogen Energy. 2013;38:5693–704.

[5] Palma V, Miccio M, Ricca A, Meloni E, Ciambelli P. Monolithic catalysts for methane steam reforming intensification: experimental and numerical investigations. Fuel. 2014;138:80–90.

[6] Matsumura Y, Nakamori T. Steam reforming of methane over nickel catalysts at low reaction temperature. Applied Catalysis A: General. 2004;258:107–14.

[7] Rostrup-Nielsen J, Dybkjaer I, Christiansen LJ. Steam reforming opportunities and limits of the technology. In: de Lasa HI, Doğu G, Ravella A, editors. Chemical Reactor Technology for Environmentally Safe Reactors and Products. Dordrecht: Springer Netherlands; 1992. p. 249–81.

[8] Vita A, Cristiano G, Italiano C, Pino L, Specchia S. Syngas production by methane oxy-steam reforming on Me/CeO_2 (Me = Rh, Pt, Ni) catalyst lined on cordierite monoliths. Applied Catalysis B: Environmental. 2015;162:551–63.

[9] Vita A, Cristiano G, Italiano C, Specchia S, Cipiti F, Specchia V. Methane oxy-steam reforming reaction: performances of Ru/gamma-Al_2O_3 catalysts loaded on structured cordierite monoliths. International Journal of Hydrogen Energy. 2014;39:18592–603.

[10] Palma V, Ricca A, Ciambelli P. Structured catalysts for methane auto-thermal reforming in a compact thermal integrated reaction system. Applied Thermal Engineering. 2013;61:128–33.

[11] Armor JN. New catalytic technology commercialized in the USA during the 1980s. Applied Catalysis. 1991;78:141–73.

[12] Moser MD, Bogdan PL. Catalytic reforming. In Handbook of Heterogeneous Catalysis. Wiley-VCH Verlag GmbH & Co. KGaA; Weinheim, Germany; 2008.

[13] Balandin AA. Modern state of the multiplet theory of heterogeneous catalysis. In: D.D. Eley HP, Paul BW, editors. Advances in Catalysis. Academic Press; New York, USA; 1969. p. 1–210.

[14] Chauvel A, Delmon B, Hölderich WF. New catalytic processes developed in Europe during the 1980s. Applied Catalysis A: General. 1994;115:173–217.

[15] Misono M, Nojiri N. Recent progress in catalytic technology in Japan. Applied Catalysis. 1990;64:1–30.

[16] Palma V, Ciambelli P, Meloni E, Sin A. Catalytic DPF microwave assisted active regeneration. Fuel. 2015;140:50–61.

[17] Pura J, Kwaśniak P, Jakubowska D, *et al.* Investigation of degradation mechanism of palladium–nickel wires during oxidation of ammonia. Catalysis Today. 2013;208:48–55.

[18] Marban G, Lopez I, Valdessolis T, Fuertes A. Highly active structured catalyst made up of mesoporous Co_3O_4 nanowires supported on a metal wire mesh for the preferential oxidation of CO. International Journal of Hydrogen Energy. 2008;33:6687–95.

[19] Montebelli A, Visconti CG, Groppi G, *et al.* Methods for the catalytic activation of metallic structured substrates. Catalysis Science and Technology. 2014;4:2846–70.

[20] Ochońska-Kryca J, Iwaniszyn M, Piatek M, *et al.* Mass transport and kinetics in structured steel foam reactor with Cu-ZSM-5 catalyst for SCR of NO_x with ammonia. Catalysis Today. 2013;216:135–41.

[21] Gascon J, Van Ommen JR, Moulijn JA, Kapteijn F. Structuring catalyst and reactor – an inviting avenue to process intensification. Catalysis Science and Technology. 2015;5:807–17.

[22] Dai C, Lei Z, Li Q, Chen B. Pressure drop and mass transfer study in structured catalytic packings. Separation and Purification Technology. 2012;98:78–87.

[23] Avila P, Montes M, Miró EE. Monolithic reactors for environmental applications: a review on preparation technologies. Chemical Engineering Journal. 2005;109:11–36.

[24] Kreutzer MT, Kapteijn F, Moulijn JA. Shouldn't catalysts shape up? Structured reactors in general and gas–liquid monolith reactors in particular. Catalysis Today. 2006;111:111–18.

[25] Li C, Xu H, Hou S, *et al.* SiC foam monolith catalyst for pressurized adiabatic methane reforming. Applied Energy. 2013;107:297–303.

[26] Palma V, Ruocco C, Castaldo F, Ricca A, Boettge D. Ethanol steam reforming over bimetallic coated ceramic foams: effect of reactor configuration and catalytic support. International Journal of Hydrogen Energy. 2015;40:12650–62.

[27] Fuqiang W, Jianyu T, Huijian J, Yu L. Thermochemical performance analysis of solar driven CO_2 methane reforming. Energy. 2015;91:645–54.

[28] Roh H-S, Lee DK, Koo KY, Jung UH, Yoon WL. Natural gas steam reforming for hydrogen production over metal monolith catalyst with efficient heat-transfer. International Journal of Hydrogen Energy. 2010;35:1613–19.

[29] Roh H-S, Koo KY, Jung UH, Yoon WL. Hydrogen production from natural gas steam reforming over Ni catalysts supported on metal substrates. Current Applied Physics. 2010;10:S37–9.

[30] Roy PS, Park CS, Raju ASK, Kim K. Steam-biogas reforming over a metal-foam-coated (Pd–Rh)/(CeZrO$_2$–Al$_2$O$_3$) catalyst compared with pellet type alumina-supported Ru and Ni catalysts. Journal of CO$_2$ Utilization. 2015;12:12–20.

[31] Roy PS, Park N-K, Kim K. Metal foam-supported Pd–Rh catalyst for steam methane reforming and its application to SOFC fuel processing. International Journal of Hydrogen Energy. 2014;39:4299–310.

[32] Horita T, Xiong Y, Kishimoto H, Yamaji K, Sakai N, Yokokawa H. Application of Fe–Cr alloys to solid oxide fuel cells for cost-reduction: oxidation behavior of alloys in methane fuel. Journal of Power Sources. 2004;131: 293–8.

[33] Italiano C, Balzarotti R, Vita A, *et al.* Preparation of structured catalysts with Ni and Ni–Rh/CeO$_2$ catalytic layers for syngas production by biogas reforming processes. Catalysis Today. 2016; 273:3–11.

[34] Lee M, Wu Z, Wang B, Li K. Micro-structured alumina multi-channel capillary tubes and monoliths. Journal of Membrane Science. 2015;489:64–72.

[35] Sangsong S, Phongaksorn M, Tungkamani S, Sornchamni T, Chuvaree R. Dry methane reforming performance of Ni-based catalyst coated onto stainless steel substrate. Energy Procedia. 2015;79:137–42.

[36] Blanco J, Petre AL, Yates M, Martin MP, Suarez S, Martin JA. Novel one-step synthesis of porous-supported catalysts by activated-carbon templating. Advanced Materials. 2006;18:1162–5.

[37] Portela R, García-Sánchez VE, Villarroel M, Rasmussen SB, Ávila P. Influence of the pore generation method on the metal dispersion and oxidation activity of supported Pt in monolithic catalysts. Applied Catalysis A: General. 2016;510:49–56.

[38] Palma V, Pisano D, Martino M, Ricca A, Ciambelli P. High thermal conductivity structured carriers for catalytic processes intensification. Chemical Engineering Transactions. 2015;43:2015.

[39] Park D, Moon DJ, Kim T. Preparation and evaluation of a metallic foam catalyst for steam-CO$_2$ reforming of methane in GTL-FPSO process. Fuel Processing Technology. 2014;124:97–103.

[40] Qi J, Sun Y, Xie Z, Collins M, Du H, Xiong T. Development of Cu foam-based Ni catalyst for solar thermal reforming of methane with carbon dioxide. Journal of Energy Chemistry. 2015;24:786–93.

[41] Holland ML, De Bruyn HJ. Metal dusting failures in methane reforming plant. International Journal of Pressure Vessels and Piping. 1996;66:125–33.

[42] Zhang J, Boddington K, Young DJ. Oxidation, carburisation and metal dusting of 304 stainless steel in CO/CO$_2$ and CO/H$_2$/H$_2$O gas mixtures. Corrosion Science. 2008;50:3107–15.

[43] Zeng Z, Natesan K. Corrosion of metallic interconnects for SOFC in fuel gases. Solid State Ionics. 2004;167:9–16.

[44] Domínguez MI, Pérez A, Centeno MA, Odriozola JA. Metallic structured catalysts: influence of the substrate on the catalytic activity. Applied Catalysis A: General. 2014;478:45–57.

[45] Serres T, Dreibine L, Schuurman Y. Synthesis of enamel-protected catalysts for microchannel reactors: application to methane oxidative coupling. Chemical Engineering Journal. 2012;213:31–40.

[46] Sanz O, Martínez TLM, Echave FJ, *et al.* Aluminium anodisation for Au–CeO_2/Al_2O_3–Al monoliths preparation. Chemical Engineering Journal. 2009;151:324–32.

[47] Meille V. Review on methods to deposit catalysts on structured surfaces. Applied Catalysis A: General. 2006;315:1–17.

[48] Meille V, Pallier S, Rodriguez P. Reproducibility in the preparation of alumina slurries for washcoat application – role of temperature and particle size distribution. Colloids and Surfaces A: Physicochemical and Engineering Aspects. 2009;336:104–9.

[49] Almeida LC, Echave FJ, Sanz O, Centeno MA, Odriozola JA, Montes M. Washcoating of metallic monoliths and microchannel reactors. In: E.M. Gaigneaux MDSHPAJJAM, Ruiz P, editors. Studies in Surface Science and Catalysis: Elsevier; Amsterdam, The Netherlands; 2010. p. 25–33.

[50] Cristiani C, Finocchio E, Latorrata S, *et al.* Activation of metallic open-cell foams via washcoat deposition of Ni/$MgAl_2O_4$ catalysts for steam reforming reaction. Catalysis Today. 2012;197:256–64.

[51] Rosen BA, Gileadi E, Eliaz N. Electrodeposited Re-promoted Ni foams as a catalyst for the dry reforming of methane. Catalysis Communications. 2016; 76:23–8.

[52] Bozorgtabar M, Rahimipour M, Salehi M. Effect of thermal spray processes on anatase–rutile phase transformation in nano-structured TiO_2 photocatalyst coatings. Surface Engineering. 2010;26:422–7.

[53] Wang F, Qi B, Wang G, Cui W. Catalyst coating deposition behavior by cold spray for fuel reforming. International Journal of Hydrogen Energy. 2014;39:13852–8.

[54] Palma V, Ricca A, Ciambelli P. Performances analysis of a compact kW-scale ATR reactor for distributed H_2 production. Clean Technologies and Environmental Policy. 2012;15:63–71.

[55] Ayabe S, Omoto H, Utaka T, *et al.* Catalytic autothermal reforming of methane and propane over supported metal catalysts. Applied Catalysis A: General. 2003;241:261–9.

[56] Li D, Nakagawa Y, Tomishige K. Methane reforming to synthesis gas over Ni catalysts modified with noble metals. Applied Catalysis A: General. 2011;408:1–24.

[57] Lisboa JS, Terra LE, Silva PRJ, Saitovitch H, Passos FB. Investigation of Ni/Ce–ZrO_2 catalysts in the autothermal reforming of methane. Fuel Processing Technology. 2011;92:2075–82.

[58] Yang N, Wang R. Sustainable technologies for the reclamation of greenhouse gas CO_2. Journal of Cleaner Production. 2015;103:784–92.

[59] Palma V, Castaldo F, Ciambelli P, Iaquaniello G. Steam reforming of ethanol to H_2 over bimetallic catalysts: crucial roles of CeO_2, steam-to-carbon ratio and space velocity. Chemical Engineering Transactions. 2013; 35:1369–74.

[60] Cimino S, Lisi L, Russo G, Torbati R. Effect of partial substitution of Rh catalysts with Pt or Pd during the partial oxidation of methane in the presence of sulphur. Catalysis Today. 2010;154:283–92.

[61] Vakhshouri K, Hashemi MMYM. Simulation study of radial heat and mass transfer inside a fixed bed catalytic reactor. International Journal of Chemical, Molecular, Nuclear, Materials and Metallurgical Engineering. 2007; 10:97–104.

[62] Irani M, Alizadehdakhel A, Pour AN, Hoseini N, Adinehnia M. CFD modeling of hydrogen production using steam reforming of methane in monolith reactors: surface or volume-base reaction model? International Journal of Hydrogen Energy. 2011;36:15602–10.

[63] Palma V, Palo E, Ciambelli P. Structured catalytic substrates with radial configurations for the intensification of the WGS stage in H_2 production. Catalysis Today. 2009;147,Supplement:S107–12.

[64] Halabi MH, de Croon MHJM, van der Schaaf J, Cobden PD, Schouten JC. Reactor modeling of sorption-enhanced autothermal reforming of methane. Part I: Performance study of hydrotalcite and lithium zirconate-based processes. Chemical Engineering Journal. 2011;168:872–82.

[65] Palma V, Ricca A, Ciambelli P. Monolith and foam catalysts performances in ATR of liquid and gaseous fuels. Chemical Engineering Journal. 2012; 207–208:577–86.

[66] Hiramitsu Y, Demura M, Xu Y, Yoshida M, Hirano T. Catalytic properties of pure Ni honeycomb catalysts for methane steam reforming. Applied Catalysis A: General. 2015;507:162–8.

[67] Li Y, Wang Y, Zhang Z, Hong X, Liu Y. Oxidative reformings of methane to syngas with steam and CO_2 catalyzed by metallic Ni based monolithic catalysts. Catalysis Communications. 2008;9:1040–4.

[68] Fukuhara C, Hyodo R, Yamamoto K, Masuda K, Watanabe R. A novel nickel-based catalyst for methane dry reforming: a metal honeycomb-type catalyst prepared by sol–gel method and electroless plating. Applied Catalysis A: General. 2013;468:18–25.

[69] Mei H, Li C, Ji S, Liu H. Modeling of a metal monolith catalytic reactor for methane steam reforming–combustion coupling. Chemical Engineering Science. 2007;62:4294–303.

[70] Chang ACC, Lee K-Y. Biogas reforming by the honeycomb reactor for hydrogen production. International Journal of Hydrogen Energy. 2016; 41:4358–65.

[71] Smorygo O, Mikutski V, Marukovich A, *et al.* Structured catalyst supports and catalysts for the methane indirect internal steam reforming in the intermediate temperature SOFC. International Journal of Hydrogen Energy. 2009;34:9505–14.

[72] Faure R, Rossignol F, Chartier T, *et al.* Alumina foam catalyst supports for industrial steam reforming processes. Journal of the European Ceramic Society. 2011;31:303–12.

[73] Richardson JT, Garrait M, Hung JK. Carbon dioxide reforming with Rh and Pt–Re catalysts dispersed on ceramic foam supports. Applied Catalysis A: General. 2003;255:69–82.

[74] Liu H, Li S, Zhang S, *et al.* Catalytic performance of novel Ni catalysts supported on SiC monolithic foam in carbon dioxide reforming of methane to synthesis gas. Catalysis Communications. 2008;9:51–4.

[75] Sang L, Sun B, Tan H, Du C, Wu Y, Ma C. Catalytic reforming of methane with CO_2 over metal foam based monolithic catalysts. International Journal of Hydrogen Energy. 2012;37:13037–43.

[76] Chang Y-F, Heinemann H. Partial oxidation of methane to syngas over Co/MgO catalysts. Is it low temperature? Catalysis Letters. 21:215–24.

[77] Al-Hamamre Z, Voß S, Trimis D. Hydrogen production by thermal partial oxidation of hydrocarbon fuels in porous media based reformer. International Journal of Hydrogen Energy. 2009;34:827–32.

[78] Christian Enger B, Lødeng R, Holmen A. A review of catalytic partial oxidation of methane to synthesis gas with emphasis on reaction mechanisms over transition metal catalysts. Applied Catalysis A: General. 2008;346:1–27.

[79] Aartun I, Silberova B, Venvik H, *et al.* Hydrogen production from propane in Rh-impregnated metallic microchannel reactors and alumina foams. Catalysis Today. 2005;105:469–78.

[80] Verlato E, Barison S, Cimino S, *et al.* Catalytic partial oxidation of methane over nanosized Rh supported on Fecralloy foams. International Journal of Hydrogen Energy. 2014;39:11473–85.

[81] Xu X, Li P, Shen Y. Small-scale reforming of diesel and jet fuels to make hydrogen and syngas for fuel cells: a review. Applied Energy. 2013;108:202–17.

[82] Beretta A, Donazzi A, Livio D, *et al.* Optimal design of a CH_4 CPO-reformer with honeycomb catalyst: combined effect of catalyst load and channel size on the surface temperature profile. Catalysis Today. 2011;171:79–83.

[83] Aliotta C, Liotta LF, La Parola.V, *et al.* Ceria-based electrolytes prepared by solution combustion synthesis: the role of fuel on the materials properties. Applied Catalysis B: Environmental. 2016;197:14–22.

[84] Ciambelli P, Palma V, Palo E. Comparison of ceramic honeycomb monolith and foam as Ni catalyst carrier for methane autothermal reforming. Catalysis Today. 2010;155:92–100.

[85] Palma V, Ricca A, Ciambelli P. Methane auto-thermal reforming on honeycomb and foam structured catalysts: the role of the support on system performances. Catalysis Today. 2013;216:30–7.

[86] van Dijk HAJ, Boon J, Nyqvist RN, van den Brink RW. Development of a single stage heat integrated water–gas shift reactor for fuel processing. Chemical Engineering Journal. 2010;159:182–9.

[87] Gu C, Lu S, Miao J, Liu Y, Wang Y. Meso–macroporous monolithic CuO–CeO$_2$/γ/α-Al$_2$O$_3$ catalysts for CO preferential oxidation in hydrogen-rich gas: effect of loading methods. International Journal of Hydrogen Energy. 2010;35:6113–22.

[88] Lang S, Türk M, Kraushaar-Czarnetzki B. Novel PtCuO/CeO$_2$/α-Al$_2$O$_3$ sponge catalysts for the preferential oxidation of CO (PROX) prepared by means of supercritical fluid reactive deposition (SFRD). Journal of Catalysis. 2012;286:78–87.

[89] Martinez-Arias A, Fernández-García M, Gálvez O, *et al.* Comparative study on redox properties and catalytic behavior for CO oxidation of CuO/CeO$_2$ and CuO/ZrCeO$_4$ catalysts. Journal of Catalysis. 2000;195:207–16.

[90] Barbato PS, Di Benedetto A, Landi G, Lisi L. CuO/CeO$_2$ based monoliths for CO preferential oxidation in H$_2$-rich streams. Chemical Engineering Journal. 2015;279:983–93.

[91] Milt VG, Ivanova S, Sanz O, *et al.* Au/TiO$_2$ supported on ferritic stainless steel monoliths as CO oxidation catalysts. Applied Surface Science. 2013;270: 169–77.

[92] Zeng SH, Liu Y. Nd- or Zr-modified CuO–CeO$_2$/Al$_2$O$_3$/FeCrAl monolithic catalysts for preferential oxidation of carbon monoxide in hydrogen-rich gases. Applied Surface Science. 2008;254:4879–85.

Chapter 2

Bimetallic supported catalysts for hydrocarbons and alcohols reforming reactions

Vincenzo Palma[1], Concetta Ruocco[1], Marco Martino[1], Eugenio Meloni[1] and Antonio Ricca[1]

Abstract

This chapter gives a brief resume of the state of the art on bimetallic catalysts for the reforming of hydrocarbons and alcohols. As part of hydrogen production technologies, reforming is the most widely used process, thanks to its low cost and mature technology. The bimetallic catalysts show an improvement in activity, selectivity, and stability with respect to the monometallic systems; their use allows to design systems with unique characteristics, totally different from those of monometallic catalysts. The most diffused reforming technology concerns methane conversion and is mainly used in industrial processes for the production of ammonia, methanol, and the C5–C12 hydrocarbon fractions. However, the reforming of non-fossil sources is the fastest growing research field, due to the renewability and low environmental impact. This chapter is divided into three main sections, reforming of methane, hydrocarbons, and alcohols, which are focused on the description of the main bimetallic catalyst systems, reaction mechanisms proposed and, where possible, comparisons on the performance.

2.1 Introduction

The environmental concerns on world warming phenomenon drive the research activities toward technologies for energy production different from fossil fuels burning. Hydrogen, clean energy carrier, is a promising green energetic carrier, which can be produced from various feed-stocks. Particularly, in the context of a "hydrogen economy," reforming of hydrocarbons and alcohols is of significant industrial relevance.

Bimetallic catalysts have been extensively studied for reforming processes, because they have shown improved activity, selectivity, and stability with respect

[1]Department of Industrial Engineering, University of Salerno, via Giovanni Paolo II 132, 84084 Fisciano (SA), Italy

to monometallic catalysts [1,2]. The synergistic effects, typical of bimetallic catalytic systems, induce properties remarkably different from the corresponding monometallic catalysts due to a unique interaction between the oxide support and the metal particles [3–6] the superior performance of the bimetallic catalyst can be also attributed to the better metal dispersion and/or formation of stable solid solutions or metals alloys [7].

Commonly, non-noble-based catalysts (Ni-containing systems in most cases) are preferred to noble metals for the reforming of hydrocarbons and alcohols due to their lower costs and higher availability [8–10]. However, Ni-based catalyst, despite their good catalytic activity, suffer from deactivation due to sintering, coke formation, or sulfur poisoning [11,12]. Conversely, noble metals (Pt, Ru, Rh, and Pd) are highly resistant to coke formation [13–17]. In this scenario, several authors [18–21] found that the addition of small amounts of noble metals to Ni-based catalysts can improve NiO reducibility, enhance Ni dispersion, and assure a superior coke resistance. In order to improve the carbon resistance of Ni-based catalysts, some researchers also proposed the addition of a second non-noble metals, such as Co [10,22]. On the other hand, adding Cu to Ni-containing catalysts was shown to improve the water gas shift activity during reforming processes, resulting in higher H_2 yields [23,24].

The support choice also plays a key role in the activity and long-term operation of the final catalysts. Al_2O_3 is commonly used as a support for reforming catalysts [25,26]. However, such oxide, especially when combined with Ni, causes a faster catalyst deactivation. Conversely, cerium oxide as well as CeO_2-containing materials have been intensively studied for their oxygen storage and release capacity [27,28] which is shown to considerably reduce coke selectivity. Other authors proposed to neutralize the acidity of the support by adding alkaline metals or to choice less acidic materials (La_2O_3, MgO, and ZrO_2) [10,29]. It may also be useful to employ high surface oxides, such as SiO_2, able to improve metals dispersion [30].

In the following sections, a review of the state of the art on steam reforming (SR) processes over bimetallic catalysts is provided.

2.2   Reforming of methane

Reforming of methane, especially SR process, is one of the most feasible routes for industrial hydrogen production [31].

Morales-Cano *et al.* [20] investigated the role of Rh, Ir, and Ru addition (0.5–1 wt%) on the catalytic behavior of a 3 wt% Ni/α-Al_2O_3 catalyst for methane SR (MSR) ($T = 500$ °C, H_2O/CH_4 ratio $= 4$, gas hourly space velocity (GHSV) $= 11,195$ h^{-1}). The Ni–Rh and Ni–Ir catalysts, due to the formation of a noble/non-noble metal alloy, maintain larger metal surface area and higher catalytic activity during operation with respect to the pure monometallic catalysts. In fact, nickel diffuses into the cubic face centered structures of Rh and Ir, thus preventing sintering phenomena. Conversely, the formation of a Ni–Ru alloy is less favored and worse performances than the pure Ru/Al_2O_3 catalyst were observed.

Table 2.1   Products selectivity and methane conversion (X) at different temperatures over Ni and Ni–Cu catalysts supported over Al_2O_3; CH_4/H_2O ratio = 1/3

T (°C)	H_2 (%) Ni	H_2 (%) Ni–Cu	CO_2 (%) Ni	CO_2 (%) Ni–Cu	CO (%) Ni	CO (%) Ni–Cu	X (%) Ni	X (%) Ni–Cu
500	91.2	99.1	5.8	0.7	2.9	0.3	77.0	86.3
550	90.2	98.1	6.8	1.1	3.1	0.8	83.6	82.7
600	95.3	98.8	2.1	0.8	2.6	0.4	78.2	86.2
650	95.4	99.0	1.4	0.6	3.2	0.3	83.9	89.2
700	95.7	98.8	1.3	0.8	3.0	0.4	88.9	88.7

The activity of a commercial Ni/Al_2O_3 and a $Ni–Cu/Al_2O_3$ catalyst for MSR was compared in a work of Khzouz *et al.* [32]. The increase of temperature, as suggested by thermodynamic equilibrium, positively affected methane conversion in both cases. However, mean conversions (Table 2.1) were higher over the bimetallic catalyst. Furthermore, the Ni–Cu catalyst displayed lower CO and CO_2 selectivities: the formation of an alloy between the two metals could promote the blocking and decrease of the sites involved in the carbon growth, as explained by Boudouard reaction.

Therefore, Cu addition minimized carbon selectivity and increased the stability of the final catalyst.

MSR was also studied over γ-alumina supported Ni catalysts modified with Au or Ag (1 wt%) [33]. Under a CH_4:H_2O:Ar = 1:4:0.7 stream and at 550 °C, the Ni–Au/Al sample displayed the highest methane conversion (84%). On the other hand, Ag addition drastically reduced catalyst activity, and methane conversion was lower than 5%. Nonetheless, both the bimetallic catalysts showed no carbon deposition, whereas for the Ni/Al_2O_3 catalyst the deposited carbon was 3.2%. At the same temperature and a CH_4/H_2O ratio = 1 [34], the employment of a $MgAl_2O_4$ support for a Ni–Au catalyst, despite caused a reduction of almost 20% of the initial catalyst activity, resulted in an improvement of stability with respect to the monometallic Ni/Al_2O_4 catalyst.

Foletto *et al.* [35] investigated the effect of different Pt loadings (0.05, 0.1, 0.2, and 0.3 wt%) on methane conversion of $Ni/MgAl_2O_4$ catalysts between 400 and 600 °C and at a H_2O/CH_4 ratio of 4. The sample without platinum showed a lower conversion compared to the Pt containing-catalysts. In particular, the highest CH_4 conversion was recorded over the 0.1 wt% Pt sample. However, increasing Pt loadings resulted in a worse distribution of the active species and, as a consequence, a lessened CH_4 conversion. In fact, the 0.1 wt% Pt sample showed a surface area of almost 80 m^2/g, whereas for the $Ni/MgAl_2O_4$ catalyst, a value of 50 m^2/g was measured: the addition of low Pt amounts caused an increase of specific area, which also resulted in an improved activity.

Bimetallic catalysts are extensively employed also for oxidative SR (OSR) of methane, selected in order to internally supply the heat required for

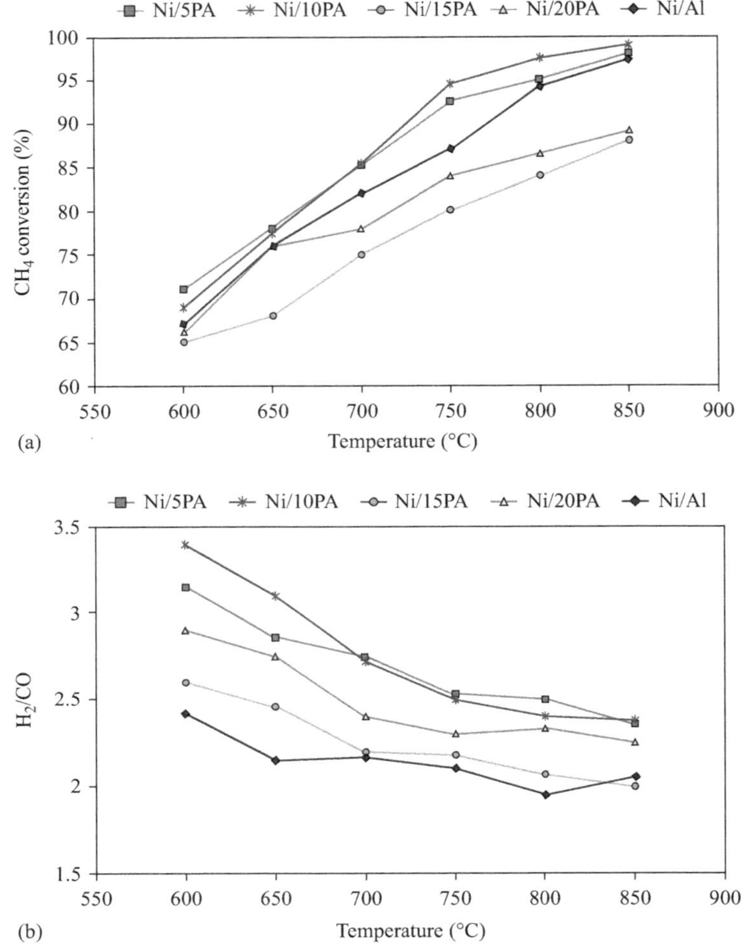

Figure 2.1   *CH$_4$ conversion (a) and H$_2$/CO ratio (b) over Ni/xPr-Al$_2$O$_3$ catalyst*
(x = 0 Ni/Al, x = 5 Ni/5PA and so on) as a function of reaction
temperature; CH$_4$:H$_2$O:O$_2$=2:1:1

endothermic MSR and increase, as a consequence, the energy efficiency of the process [19,36].

Wang *et al.* [37] investigated the application of Ni/xPr–Al$_2$O$_3$ (x = 5, 10, 15, and 20 wt%) catalysts for auto-thermal reforming of methane. As shown in Figure 2.1, the CH$_4$ conversion increased with Pr content from 5% to 10%, whereas a further increase in Pr loading led to the decline of CH$_4$ conversion. The H$_2$/CO molar ratio obtained over Pr-containing catalysts is higher than that over Ni/Al$_2$O$_3$ supported catalyst under the same operating conditions. This effect can be explained considering that the oxygen storage capacity of praseodymium oxide promoted the water gas shift reaction, thus increasing the H_2/CO ratio. Moreover, the monometallic catalyst deactivated after 10 h, whereas the activity Pr-based catalysts was not affected for 48 h of time-on-stream.

The effect of Pt addition to Ni/Al_2O_3 catalysts on the inhibition of hot spot formation was studied in a work of Li *et al.* [38]. Under a $CH_4/H_2O/O_2 = 4/3/2$ stream, a marked exothermic peak was observed at the catalyst bed inlet, which can be mitigated by Pt addition. However, the performances of the bimetallic catalysts were strongly dependent from the preparation method, which affected the interaction between Pt and Ni: Pt addition by a sequential impregnation method had a more positive effect than the preparation by a co-impregnation method. For the co-impregnated catalysts, the maximum of temperature was closer to the catalyst bed inlet with the decrease of W/F, which is the ratio between the catalytic mass and the molar flow rate. Conversely, no exothermic peak was observed at 0.24 (g h)/mol when the impregnated samples were employed and, at 0.07 (g h)/mol, the maximum temperature was almost 80 °C lower than that observed for the co-impregnate catalyst.

The influence of additives ($Ce_{0.5}Zr_{0.5}O_2$ and La_2O_3) and preparation method (impregnation and co-precipitation) on the performances of bimetallic NiPd catalysts supported on Al_2O_3 was also investigated under a feed composition of CH_4: $H_2O:O_2:He = 1:1:0.75:2.5$ [39]. La_2O_3 contents increasing from 5 to 20 wt% assured higher H_2 yields. However, the best results were recorded over the sample prepared by sequential impregnation (Ni earlier than Pd) and containing 10 wt% of $Ce_{0.5}Zr_{0.5}O_2$:H_2 yield, defined as $F_{H_2,out}/(2F_{CH_4,in} + F_{H_2O,in})$, reached 75%, and methane conversion was 95% in a wide temperature range (750–950 °C). Such results can be linked to the catalyst physiochemical properties: the co-impregnation method caused the presence of large NiO particles, wider NiO particle size distribution, nonuniform Zr distribution across the support, ZrO_2 sintering. Conversely, a more-uniform structure was observed for the impregnated catalyst.

One of most important options for enhancing activity and stability in methane dry reforming reactions is to use bimetallic catalysts [40,41]. The combination of two metals, in fact, may minimize Boudouard and methane decomposition reactions, which are the main causes of catalyst deactivation.

Ay *et al.* [42] studied DRM at 700 °C ($CH_4/CO_2/Ar = 1/1/3$) over impregnated Ni, Co, and Ni–Co catalysts (metal loads of 8% and 8% for Ni and 4% for Co in the bimetallic sample) supported over CeO_2. The activities of ceria-based Ni-containing catalysts decreased with increasing calcination temperature (from 700 to 900 °C) accompanied by a decrease in coke deposition. Although Ni/CeO_2 and $Ni–Co/CeO_2$ catalysts exhibited comparable high activities, Co/CeO_2 catalysts showed very low activity: this result was attributed to strong metal–support interaction.

Such effect was confirmed with Transmission Electron Microscopy (TEM) images (Figure 2.2) showing, especially for the monometallic Co sample, a layer of support coating the metal particles which was shown to have a detrimental effect on catalyst activity. However, coke deposition rate was almost the same for the Ni–Co and Ni catalysts.

The results of stability tests, carried out by Luisetto *et al.* [5] over co-precipitated Ni and Co catalyst with slightly different metal loadings (7.5% for

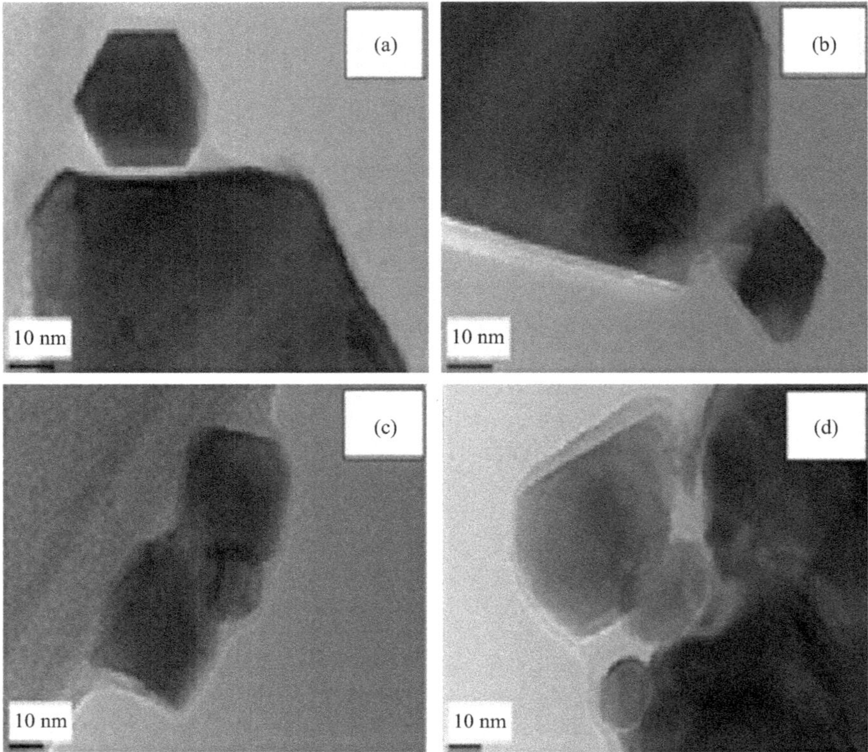

Figure 2.2   HRTEM of reduced (a and b) Co/CeO$_2$ and (c and d) Ni-Co/CeO$_2$ catalysts, both calcined at 700°C

monometallic and 3.75% for both Ni and Co bimetallic sample) are reported in Figure 2.3. The Ni/CeO$_2$ catalyst showed the highest deactivation with a CH$_4$ conversion decrease of 8%. Conversely, cobalt-containing catalysts showed a stable behavior during 20 h on stream, with a very low reduction in conversion (0.5% in the case of Co/CeO$_2$ and 2% for the bimetallic sample).

Moreover, for the latter two samples a weight loss of almost 6% was recorded during thermos-gravimetric analysis while the Co–Ni sample exhibited a weight variation of 25%. Such results suggest that Co is efficient in preventing catalyst deactivation also in the presence of a strong interaction with Ni (Co–Ni alloy).

Ni–Co catalysts supported over CeO$_2$–ZrO$_2$ (80:20 wt%) were also employed for methane dry reforming [43]. The catalysts were synthetized using ethylene glycol or hydrothermal method; the total nickel and Co loadings ranged from 3 to 18 wt% while the Ni/Co weight ratio was fixed to 1.5. The catalysts prepared by hydrothermal method displayed the highest H$_2$/CO ratios and very low values of carbon accumulation (T from 500 to 800 °C and CH$_4$/CO $=$ 1). Moreover,

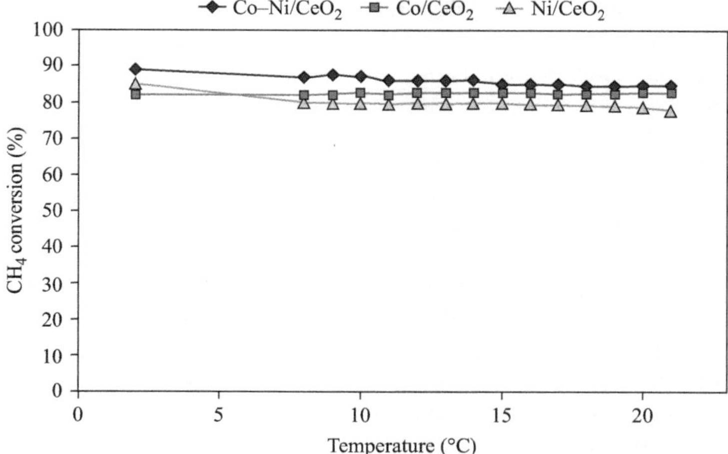

Figure 2.3   CH₄ conversion as a function of time-on-stream at 750°C; CH₄/CO₂/Ar = 1/1/3

the increase of Ni and Co loading caused a progressive decrease of metals dispersion: metal loading higher than 6% should be avoided, as they cause metal phase segregation which resulted in lower interaction with the CeZr support.

Other authors [44,45] compared the performances for DRM of Ni–Co catalysts supported over ceria–zirconia and prepared using two different approaches. Aw *et al.* [44] prepared the catalysts by freeze–drying and NO calcination (FD) or oven-drying and air calcination. The sublimation of the frozen liquid, carried out in liquid N_2, reduced the morphological distortion which occurs when a wet sample dries by means of normal evaporation. As a result, high surface area and small pore size were measured: both the phenomena assured higher activity for dry reforming of methane. During stability tests carried out at 750 °C under a 50% CH_4/50% CO_2 stream, the FD samples displayed the highest CH_4 conversion (90%) and H_2/CO ratio (0.83%–0.85%).

Conversely, two solvothermal approaches (HT, hydrothermal aging in NH_4OH aqueous solution and glycothermal aging in ethylene glycol) were followed by Crnivec *et al.* [45]. The materials aged in alkaline water exhibit large crystallite size and structural segregation of the CeO_2 and ZrO_2 crystalline phases. On the contrary, the samples synthesized by glycothermal reduction show the properties of a homogeneous solid solution with highly developed nanocrystalline structure. Scanning electron microscopy (SEM) analysis displayed spherical particles ranging from 20 to 30 nm in the HT CeZr sample, whereas for EG CeZr samples crystallite dimensions between 10 and 15 nm were identified.

The support crystallite sizes also affected active species dispersion: the catalysts containing small active metal particles showed superior resistance to carbon accumulation on the surface and display certain selectivity for the

Reverse Water Gas Shift (RWGS) reaction; catalysts with large active metal particle sizes promoted methane dehydrogenation reaction and consequentially accumulate considerable amounts of carbon on their surface.

The catalytic performances of NiCo deposited on a dual support $CeZrO_2/\beta$-SiC or $CeZrO_2/\gamma$-Al_2O_3 were also investigated [46]. The supports were prepared by dry impregnation, wet impregnation (WI) and a deposition precipitation (DP) method. The samples supported over $CeZrO_2/\gamma$-Al_2O_3 DP showed the highest stability (at a CH_4/CO_2 ratio of 1.5): during tests performed at 750 °C for 550 h, the coke selectivity was only 0.56 wt%. On the other hand, 2.59 and 7.72 wt% were recorded over $CeZrO_2$/SiC prepared, respectively, by WI and DP. The promising performances of all the samples containing alumina are related to the synergistic impact of the bimetallic system and the effect of dual support to effectively preserve the dispersed $CeZrO_2$ and NiCo phases. Moreover, the formation of surface spinels, such as $NiAl_2O_4$ and $CoAl_2O_4$, could induce a positive effect on the suppression of carbon deposition, thus contributing to remarkably low coke content for the DP catalyst supported on $CeZrO_2/\gamma$-Al_2O_3.

The addition of Co to $Ni/MgAlO_x$ catalysts [47] was also shown to stabilize the final system, resulting in small sizes of metal nanoparticles that are uniformly dispersed on catalyst surfaces. Sajjadi *et al.* [48] prepared Ni–Co/Al_2O_3–MgO–ZrO_2 nanocatalysts with various amounts of Mg (5, 10, and 25 wt%) via sol–gel method. The increase of MgO loading reduced the sample crystallinity enhancing, at the same time, the dispersion of active species. All the samples, tested at 850 °C for 1,440 min at $CH_4/CO_2 = 1$, did not deactivate during time on stream (TOS). In fact, the promoting effect of zirconia in gasification of adsorbed intermediates as well as the synergy of Ni–Co by improving metal support interaction and separating of metal ensembles in smaller metal particles declined the amount of carbon deposition without deactivation. Moreover, MgO addition decreased carbon formation by reducing CO disproportionation (Boudouard reaction) and promoting the activation of CO_2 on the surface. As a consequence, 25 wt% sample assured the highest CO and H_2 selectivities.

Methane dry reforming was also investigated over bimetallic catalyst containing Ni and a noble or non-noble metal instead of cobalt.

Huang *et al.* [49] studied the effect of the addition of small Mo quantities (0.3, 0.5, and 0.7 wt%) to 12 wt% Ni supported on SBA-15 molecular sieves. The highest stability (250 h) was recorded over the catalyst containing 0.5 wt% Mo, whereas CH_4 conversion dropped after just 60 h over the monometallic Ni catalysts ($T = 800$ °C, $CH_4/CO_2 = 1$). Moreover, the deactivation rate followed the trend of carbon deposition: $Ni \gg 0.3Mo > 0.7Mo > 0.5Mo$. The results of TEM analysis (not shown) revealed that two types of carbon were deposited on Ni and Ni–Mo (0.5 wt%) catalysts: in the first case, catalyst is mostly in the form of shell-like carbon, whereas whisker carbon species, with a hollow internal channel, is mainly formed on the surface of bimetallic catalyst. The main factor contributing to the prevention of carbon deposition is the metal–support interaction. Figure 2.4(a) shows that metal particle away from support which is easily encapsulated by carbon layers. In the latter configuration, in fact, carbon species derived from the

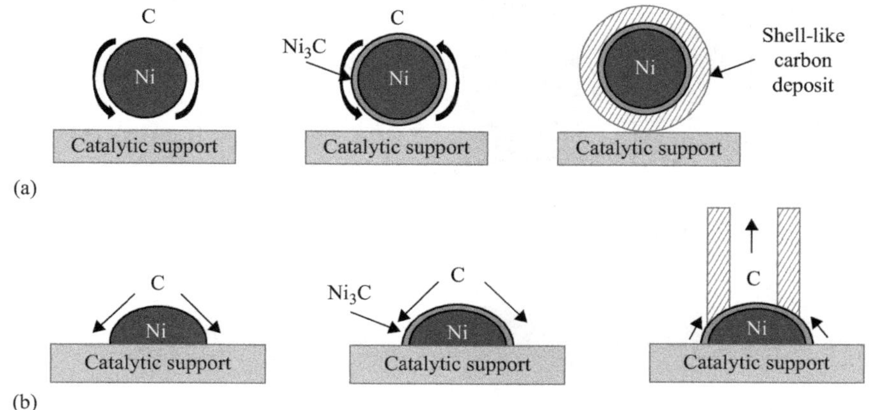

Figure 2.4   Influence of the metal–support interaction on the mode of growth of carbon deposit. Metal is away from support (a) and Metal is strongly interacting with the support (b)

dissociative adsorption of CH_4 are uniformly distributed on the surface of Ni particles. After the formation of Ni_3C on the surface of metal particle, the Ni particle is slowly encapsulated by carbon layers.

Conversely, the strong interaction between metal phase and support (Figure 2.4(b)) prevented metal particle being carried away from the supported surface during the growth process of carbon nanotubes. For the bimetallic systems, Mo addition improved metal–support interaction thus resulting in lower carbon selectivity for the bimetallic samples.

Silica (Mobil Composition of Matter No. 41 (MCM-41) mesoporous molecular sieves) was also employed as support for Ni–Ti, Ni–Mn, and Ni–Zr catalysts [50]. The catalytic activity of Ni–Ti and Ni–Mn at $CH_4/CO_2/He = 1$:1:2 and 750 °C was significantly lower than that of the monometallic sample. On the contrary, the Zr-containing catalysts displayed enhanced initial activity. Zirconia addition, in fact, improved the catalyst activity due to the ability of Zr^{4+} species to activate CO_2 as well as to increase thermal stability; moreover, the formation of well-dispersed Ni particles was assured by the anchoring effect of Zr^{4+} species.

Menegazzo *et al.* [51] investigated the performances of Pt–Ni and Pd–Ni supported on zirconia at 800 °C and $CH_4/CO_2 = 1$. The addition of Pd by co-impregnation improves the catalytic activity of the monometallic Ni catalyst, although less than expected. On the contrary, Pt–Ni catalyst behaves like the monometallic sample. Moreover, the addition of noble metals hindered the accumulation of coke deposits on catalyst surface.

Methane dry was also investigated over Ni/Al_2O_3 catalysts containing Au or Pt (0.2 wt%) as additives [52]. Noble metals positively affected NiO reducibility and resulted in lower metals particles. The bimetallic catalysts maintained slightly higher activity than the monometallic samples in the overall range of temperature investigated. The observed improvement of activity may be ascribed to the formation

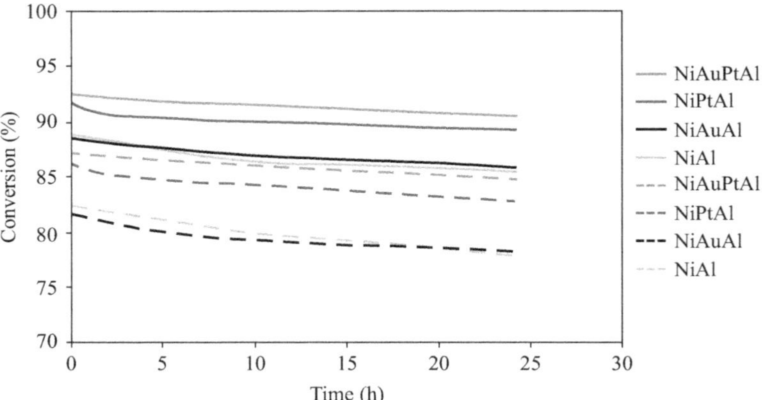

Figure 2.5   CH₄ and CO₂ conversion for Al₂O₃ supported catalysts during 24 hours of TOS

of well-dispersed bulk alloyed active phases and/or to synergistic interaction between nanosized metals. However, as shown Figure 2.5, the highest CH_4 and CO_2 conversion at 750 °C and $CH_4/CO_2 = 1$, was recorded over the trimetallic NiAuPt catalyst. Moreover, the addition of CeO_2 and, to a less extent, of MgO (10 wt%) improved catalytic conversion, stability, and coke resistance.

Wu *et al.* [53] investigated the activity of Rh–Ni catalysts supported over boron nitride (BN) and γ-Al₂O₃. The conversions of CH_4 with CO_2 on Rh–Ni/BN catalysts were higher than those on Rh–Ni/γ-Al₂O₃ catalysts. The lack of acidic sites reduced carbon formation over BN support and, at 800 °C and $CH_4/CO_2 = 1$, the maximum CH_4 and CO_2 conversions (72% and 81%, respectively) was recorded for an Rh/Ni ratio of 0.01.

The impact of Pt addition to MgO supported nickel-based catalysts was also studied [54]. The formation of a solid solution NiO–MgO and the basicity of the catalyst assured high catalytic stability for both monometallic and bimetallic catalysts at 700 °C and $CH_4/CO_2 = 1$. In terms of catalyst activity, the sample without Pt showed a lower conversion with respect to the bimetallic catalysts. Conversely, the highest CH_4 conversion was recorded over 0.1 wt% Pt–Ni/MgO, which also displayed a wider metallic surface area. The worst performances recorded for increasing Pt contents suggested that the distribution of the noble metal as well as the interaction between the active species and the support were less favored.

2.3   Reforming of other hydrocarbons

Bimetallic catalysts were also reported as suitable systems for reforming of higher hydrocarbons or hydrocarbon mixtures.

Natural gas contains significant amounts of higher hydrocarbons, mainly ethane. Such species, especially during high temperature operations, can easily decompose leading to coke formation. In this context, the degree of carbon deposition can be mitigated through the selection of proper catalysts. In particular, ethane SR is usually carried out over monometallic systems [55,56]. On the other hand, the interest toward the reforming of higher hydrocarbons, performed over both monometallic and bimetallic systems, is related to their potential use as fuels for portable H_2 generation. In fact, they can be transported in a liquid form and they have a high gravimetric and volumetric hydrogen density [57].

Propane is a potential candidate for hydrogen production due to its common storage and its well-established infrastructure in modern cities [58]. Up to date, steam, oxidative and dry reforming as well as partial oxidation are the main routes for H_2 or syngas production from C_3H_8. Bimetallic catalysts, prepared by WI in which a second metal (1 wt% of Co, Pt, Ag, and Ru) was added to a 10 wt% Ni/CeO_2 sample, were employed for SR of propane between 350 and 450 °C [59]; the feed mixture flowrate was $C_3H_8:N_2:H_2O = 6.8:61.6:55.6$ mL/min. The C_3H_8 conversion over the bimetallic samples increased in the order $Co < Ag < Pt < Ru$. However, H_2 yield did not follow the same trend, and the highest value was recorded over the Ag-containing catalyst. Therefore, in order to verify the effect of Ag loading, samples with different Ag contents were prepared. The conversion rate of C_3H_8 slightly dropped with increasing Ag addition, whereas the H_2 yields increased with the Ag addition up 1 wt%, however decreased at 2 wt% Ag.

Steam and oxidative reforming of propane was studied over $Ni–Mo/Al_2O_3$ catalysts [60]. The tests were performed at 450 °C; steam-to-carbon molar ratio (S/C) and oxygen-to-carbon molar ratio (O_2/C) were equal, respectively, to 3 and 0.3. Under SR conditions (Figure 2.6(a)), the 0.1Mo sample reached the highest conversion (76%), whereas a much lower conversion (50%) was recorded over 0.5Mo catalyst. The monometallic sample attained approximately 70% conversion in the first hour, but the conversion decreased gradually with time. O_2 addition (Figure 2.6(b)) improved the performances of both 15Ni and 0.1Mo samples. However, despite oxygen cofeeding, the monometallic catalyst lost activity due to coke formation, whereas no carbon was detected after test over the 0.1Mo catalyst.

Siahvashi *et al.* [61] studied dry reforming of propane over $Mo–Ni/Al_2O_3$ catalysts from 550 to 650 °C. The influence of reactants ratio on products distribution was investigated, finding that that H_2 and CO formation rates have initially increased and reached the peak at stoichiometric reactant ratio of 3, followed by a gradual decrease. Moreover, the initial increase in CO formation rate was followed by a weak decline, suggesting that at lower feed ratio the carbon deposited on the surface is oxidized to CO. A further improvement of the performances of the above catalysts was reached by K addition [62], which reduced the acid:base ratio: as a result, a reduction in carbon formation was achieved.

The long-term stability of $Co–Ni/Al_2O_3$ catalysts under dry reforming conditions was investigated at 650 °C [63]. The product H_2:CO ratio was generally invariant with time over the 72 h period for different $CO_2:C_3H_8$ feed ratio values (4–7) but remained within a band between 0.4 and 0.6. Moreover, the solid carbon content (total organic compound (TOC)) was reduced from 50% to approximately 0% by increasing the feed ratio from 4 to 7.

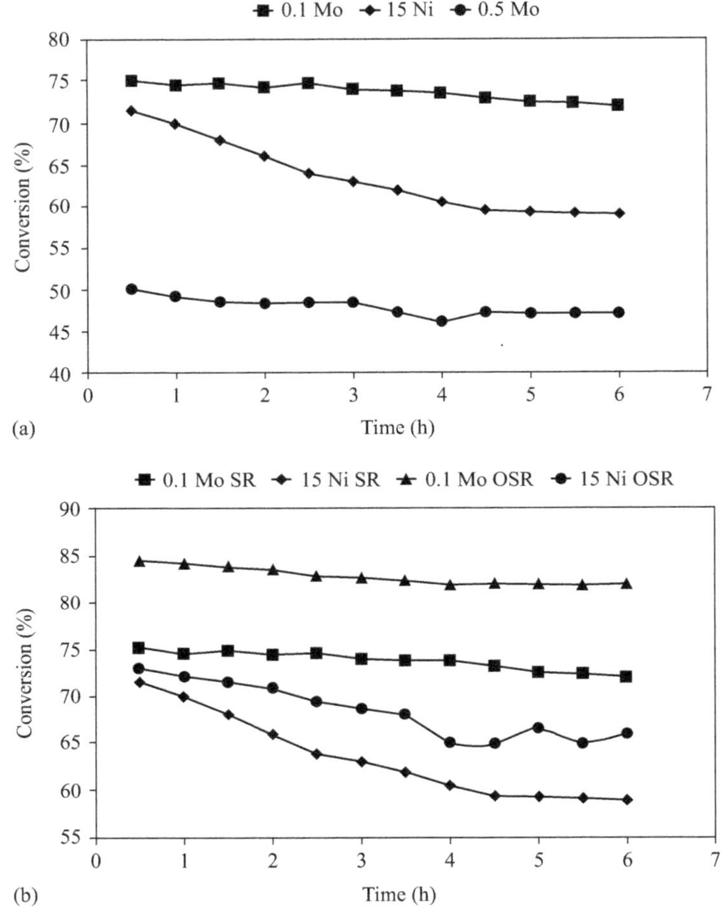

Figure 2.6   *(a) Propane steam reforming over 15 wt% Ni/Al$_2$O$_3$ (15Ni) and over bimetallic 0.5 wt% Mo (0.5Mo) and 0.1 wt% Mo (0.1Mo)-15 wt% Ni/Al$_2$O$_3$; (b) comparison between steam and oxidative steam reforming conditions; S/C = 3, O$_2$/C = 0.3, T = 450°C*

Monometallic and bimetallic Ni and Rh catalysts supported on La–Al$_2$O$_3$ were employed for OSR of *n*-butane at 700 °C at S/C = 2 and O$_2$/C = 0.5 [64]. For the Ni/La–Al$_2$O$_3$ catalyst, after 100 h of TOS; the H$_2$ production was 2.6 times lower than that recorded for the other two samples, which approached thermodynamic equilibrium. However, the amount of unconverted *n*-butane was smaller for the Ni–Rh sample compared to Rh alone at the same rhodium concentration. Moreover, for the bimetallic catalyst, there was less CO and more CH$_4$ in the reformate. On the basis of these results, the bimetallic formulation was also deposited on less acidic supports (Ce$_{0.42}$Zr$_{0.58}$O$_2$ and CeMgO$_x$) and tested under SR conditions (*T* = 900 °C, S/C = 3). The lowest amounts of propane, propene, and unconverted butane were measured for the ceria–zirconia supported sample. Moreover, the catalysts containing CeO_2 tend to form less Co and more CO_2 than those containing alumina. Finally, SEM analysis (not shown) revealed a lack of carbon whiskers on $Ce_{0.42}Zr_{0.58}O_2$, which proves that this material is able to gasify carbon better than lanthanum–alumina and ceria–magnesia supports.

Al-Musa *et al.* [65] studied SR of iso-octane over mono and bimetallic catalysts $Cu_{20-x}–Co_x$ catalysts supported on CeO_2 (S/C = 3). At 700 °C, the increase of Co loading (from 0 to 20 wt%) led to a growth in H_2 and CO yields: the monometallic Co/CeO_2 sample achieved a 75% of H_2 yield (calculated as the product between ethanol conversion and H_2 selectivity). At the same time, CH_4 and higher hydrocarbon formation was favored over the Cu-rich samples.

Ni and Pt/Ni catalysts supported on different oxides (CeO_2, $CeO_2–ZrO_2$, and $Gd_2O_3–CeO_2$) were employed for autothermal reforming of *n*-hexadecane [66]. The increase of Pt loading (from Pt/Ni = 3/97 to 10/90), at T ranging from 700 and 1,000 °C, suppressed hydrocarbon formation on all three supports. The bed temperature as well as hydrogen yield decreased with increasing carbon-to-oxygen molar ratio (from 0.8 to 1.2). For each support, hydrogen yield generally increased with Pt passing from 3% to 10%. In particular, for the ceria-zirconia supported sample, the hydrogen yield was pronouncedly increased with the addition of low platinum amount. However, a further increase of Pt/Ni ratio did not affect significantly H_2 yield. For the $Gd_2O_3–CeO_2$-based sample, no significant improvement in hydrogen yield was noted until the Pt/Ni ratio reached 10/90. Finally, when CeO_2 was chosen as support, the hydrogen yield was generally less than that of the other two supports for all Pt/Ni molar ratios.

$Mo–Ni/Al_2O_3$ catalysts (15 wt% Ni and Mo ranging from 0 to 0.5 wt%) were also employed for the OSR of a simulated liquefied petroleum gas, with a propane/butane ratio of 1 [67]. The addition of low Mo loadings (0.05 and 0.1 wt%) increased hydrocarbon formation and H_2 yields with respect to monometallic sample during stability tests at 450 °C, S/C = 3, and O_2/C = 0.3. Despite higher Mo contents increased H_2 productivity, a lower conversion of the fuel feeds was measured. After 18 h of stability tests, carbon deposition was visually detected over the monometallic sample, whereas a very high resistance to coking was observed over the 0.05 wt% Mo catalyst.

Xie *et al.* [68] investigated the sulfur tolerance of mono and bimetallic Ni and Rh catalysts supported over $CeO_2–Al_2O_3$ for SR of a hydrocarbons mixture comprising only normal paraffins with an average carbon number of 13. The reaction was performed at two different temperatures (550 and 800 °C), and it was found that, over both the monometallic Rh sample and the bimetallic catalysts, sulfur accumulation is unfavorable at 800 °C. However, at 550 °C, significantly higher sulfur content was measured in the $Rh–Ni/CeO_2–Al_2O_3$ sample. Such result can be explained considering that Rh addition strongly enhanced Ni dispersion, generating more surface nickel sites for sulfur adsorption.

Ru–Ni catalysts, prepared at different metals loadings supported over ceria–alumina, were also employed for SR of kerosene [69]. Experimental results of the tests performed at 800 °C indicated complete conversion of the feeding fuel over all the samples and a maximum in hydrogen yield was measured at an S/C ratio of 3.89.

However, over the 1 wt% Ru–5 wt% Ni sample, the CH_4 selectivity was almost seven times lower than that measured over the sample containing 15 wt% Ni and the latter sample displayed negligible coke formation during 10 h of TOS. Moreover, at the same operative conditions, the 1 wt% Ru–15 wt% Ni showed superior activity with respect to the monometallic 1 wt% Ru and 15 wt% Ni due to the sintering prevention and the enhanced reducibility of NiO particles, which is promoted by the noble metal.

Koike *et al.* [70] studied SR of biomass tar compounds (benzene and toluene) in the presence of Ni/Mg/Al and Ni–Fe/Mg/Al catalysts. Under H_2O/C_7H_8 and H_2O/C_6H_6 ratios of 11.9 and 9.9, the bimetallic catalyst exhibited higher conversion than Ni/Mg/Al (about twice) at 600 °C: the formation of a Ni–Fe alloy promoted the reactants adsorption on catalyst surface and their subsequent conversion. Moreover, Fe addition was effective in the minimization of carbon deposition due, on the one hand to the suppression of benzene and toluene decomposition and, on the other hand, of CO disproportionation reactions.

2.4   Reforming of alcohols

In order to reduce the dependence of world energetic system from fossil fuels, a great effort is rising to find alternative and clean energy sources. From this point of view, the employment of biomass-derived alcohols for H_2 production via reforming is an attractive resolution. For example, bio-oil [71], produced by degradation of biomass, contains hundreds kinds of oxygenates, and catalytic reforming is the most popular route to recover hydrogen from this clean source.

Methanol, produced either from renewables or from fossil fuels, has the advantage of containing only one C atom: the absence of C–C bonds reduces the number of by-products during reforming, thus allowing very low temperature operations (200–350 °C).

Bobadilla *et al.* [72] employed monometallic and bimetallic Ni–Sn/CeO$_2$–MgO–Al$_2$O$_3$, prepared via impregnation or polyol method, for methanol SR at 350 °C and S/C = 2. In the second case, nanoparticle-based catalysts were synthetized, which showed CH_3OH conversion as well as hydrogen yields slightly higher than the impregnated samples. However, for the bimetallic catalysts, no matter the preparation method, Sn addition reduced catalytic activity: Sn poisoned Ni defects and edge sites, thus increasing CO uptake and reducing both Water Gas Shift (WGS) activity and H_2 yield. However, the presence of Sn, especially for the impregnated samples, minimized coke deposition. In fact, NiSn alloy formation prevents the nickel carbide formation and, as a consequence, the carbon formation.

Methanol SR was also investigated over $Ni_{0.2}$–$Cu_{0.8}$ or $Ni_{0.8}$–$Cu_{0.2}/ZrO_2$ catalysts [73], annealed at two different temperatures (350 and 400 °C). For the catalyst treated at 350 °C, a noncrystallized mass was detected by SEM analysis, whereas slices of crystals are clearly visible after the annealing at higher temperature.

The reaction was performed between 200 and 350 °C at a stoichiometric feed ratio (S/C = 1). The catalysts calcined at 350 °C displayed a mean H_2 yield three times higher than the samples treated at 400 °C: probably, water is adsorbed better on the amorphous ZrO_2 owing to the higher content of hydroxyl groups on its surface. Moreover, starting from 250 °C, the hydrogen-to-carbon dioxide ratio is very close to stoichiometric one on the Cu-rich sample, demonstrating almost complete absence of side reactions, while significant amounts of CO were detected over the $Ni_{0.8}$–$Cu_{0.2}$/ZrO_2 catalyst.

Bimetallic catalysts containing Pd–Zn and Cu–Zn catalysts supported on Al_2O_3 were employed for methanol SR at 250 °C, and the effect of catalyst regeneration on activity of Cu and Pd-based catalysts was compared [74]. For the Cu–Zn catalyst, a drop of 40% in methanol conversion was observed after 60 h of TOS. Conversely, the noble metal-containing sample only suffered of an initial drop of activity of 17%. Both the catalysts, after stability tests, were oxidized in air at 420 °C, re-reduced and then exposed to the reforming mixture. The Pd–Zn catalyst regains its original activity within the first hour of TOS, whereas copper-based sample was not regenerated and continued to loss activity with TOS. However, after oxidation at higher temperature, worse performances were recorded also over the Pd–Zn catalyst, as the Pd–Zn alloy is stable until 420 °C.

The activity of Pd–Zn/ZnO catalysts, prepared by calcination of ZnO precursor in different atmospheres (O_2, air, N_2, and H_2), was investigated at 180 °C and S/C = 1.5 [75]. The performances of the catalyst increased by switching from oxidative to reducing atmosphere. Moreover, the selectivity to CO dropped significantly after calcination in N_2 or H_2 with respect to the others atmospheres. Overall, the highest activity and selectivity was measured over the sample treated in hydrogen stream, which determined the formation of different active sites, mainly oxygen defects on the surface, thus improving the adsorption of H_2O and CH_3OH molecules.

Men *et al.* [76] studied methanol SR over Pd–In/Al_2O_3 catalysts, prepared at different Pd:In ratios and metals loadings. At T ranging from 325 to 425 °C and under a water/methanol ratio of 1.5, CH_3OH conversion increased with increasing Pd:In ratio from 5:10 to 10:10. However, a drastic increase in CO content from 0.7% to 5% was observed. Methanol conversion also increased with Pd loading, and the highest conversion as well as the least amounts of CO were recorded over the 15Pd–30In sample.

OSR of methanol was also investigated over Cu/ZrO_2, Ni/ZrO_2 and bimetallic Cu–Ni/ZrO_2 catalysts at low temperatures (250–400 °C) [77]. The steam-to-carbon ratio was fixed to 0.17, whereas the oxygen-to-carbon ratio to 0.34. The highest methanol conversion in the whole temperature interval was recorded over the Cu–Ni/ZrO_2 sample. The increase of temperature up to 300 °C rose H_2 yield which, however, after this temperature remained unchanged over the three samples. At $T > 300$ °C, despite methanol conversion increased, no variation in hydrogen productivity was observed over Ni-based catalysts, probably due to O_2 co-feeding, which caused H_2 oxidation. Cu addition reduced methane selectivity, whereas, as a drawback, the presence of bimetallic Ni-rich particles was responsible for higher CO formation.

Further, the effect of bimetallic phase content on the performances of the bimetallic catalysts supported on ZrO_2 was investigated [78]. The catalysts were prepared at 80% Cu and 20% Ni to obtain 3, 15, and 30 wt% of total metallic phase. The sample having 15 wt% of bimetallic phase showed 90% of CH_3OH conversion at 400 °C and the highest H_2 yield. TEM images (not shown) revealed the formation of a core–shell structure (a core of Cu articles with a Ni layer) with the increase of Cu/Ni phase content, beneficial for hydrogen production. However, the 30 wt% sample, despite characterized by a further increase in the frequency of Ni/Cu core–shell structures, was shown to perform worse than the catalyst with lower loading (15 wt%) in terms of H_2 selectivity.

Bioethanol is mainly obtained from sugar and starch-based materials, or from lignocellulosic waste materials [79], it is considered a much more attractive source of hydrogen with respect to methanol, thanks to a cheap production technology, to the easy storage and handling, to a higher hydrogen content and a reduced toxicity. However, as previously mentioned, the presence of C–C bonds is the source of several by-products during the reforming process so, many studies have been reported with the intent of studying the mechanisms of reaction and improve the yields. The Cu-based catalysts, used in methanol SR, are not useful for ethanol SR due to the ineffectiveness to break the C–C bond; on the contrary Ni, Co and some noble metal-based catalysts are very active but are not selective or too expensive; therefore, many of the published works are focused on the use of bimetallic and alloy catalysts, to exploit a desirable synergic effect. Among the first examples, in 1992 Luengo *et al.* [80] reported on the activity of a Ni/Cu/Cr catalyst supported on porous α-Al_2O_3 (4% Ni, 0.75% Cu, and 25% Cr) for ethanol gasification in the temperature range of 573–823 K, with a water/ethanol mole ratio between 0.4 and 2.0, a space velocity between 2.5 and 15 h^{-1}. The data revealed a good activity of these catalysts, both in terms of conversion and selectivity to H_2 and CO especially at lower temperatures. The Ni/Cu catalytic system was extensively investigated in subsequent years by Mariño *et al.* [81]; the XRD analysis performed on the Cu/Ni[K]γ-Al_2O_3 catalysts, after the impregnation step, showed three different phases for copper, $Cu(NO_3)_2$ and CuAl and/or CuNiAl hydrotalcite-type compound, and a nickel phase of NiAl hydrotalcite-type. The ratio between the two phases of copper depended on the nickel content, whereas the hydrotalcite-type was favored by adding nickel. The calcination step (400–800 °C) produced a CuO segregated phases and/or copper phases called "surface spinel," the segregated phase was observed only in samples containing a copper loading higher than 4.7% calcined at low temperature; moreover, in all samples the $NiAl_2O_4$ phase was present, whereas the NiO phase has never been detected, the adding of KOH does not affect the catalytic structure [82]. The activity tests, performed on ethanol SR, with a variable copper loading 0–6.36 wt% (nickel content 4 wt%, potassium 0.15%) showed that the best performances were obtained at low copper content. The data obtained enabled to identify the copper as the active agent, the nickel as the cleavage promoter of the C–C bond, and the increaser of hydrogen selectivity, whereas the potassium as the neutralizer of the acid sites of the alumina. A further comparative study demonstrated [83] that the nickel favors the ethanol gasification and reduces the acetaldehyde and acetic acid production, so a higher content of nickel seems to be beneficial. Moreover, the presence of nickel favors the segregation of copper ions on the catalytic surface and induces the decomposition of the by-products (see (2.1) and (2.2)).

$$CH_3COH \rightarrow CO + CH_4 \tag{2.1}$$

$$CH_3COOH \rightarrow CO_2 + CH_4 \tag{2.2}$$

Klouz *et al.* [84] tried to optimize an ethanol reforming process, using Ni–Cu/ SiO_2 catalysts containing the 16.7% of Cu and 1.7% of Ni, in order to directly feed a solid polymer fuel cell. The authors studied the influence of the temperature, of the molar ratio H_2O/EtOH, of the contact time and of the feeding of oxygen; the best condition provided high temperatures (600 °C), a H_2O/EtOH ratio of 1.6 with the addition of oxygen in an O_2/EtOH ratio of 0.5, and a contact time close to 1 min kg/mol. Fierro *et al.* [85] highlighted the importance of the alloy formation between copper and nickel showing the results of a comparative study on the performance of Ni/SiO_2, Ni–Cu/SiO_2 and other Pt, Pd, Ru, and Rh-based catalysts, in the OSR. The conditions used were similar to those chosen from Klouz (700 °C, H_2O/EtOH ratio of 1.6, O_2/EtOH ratio of 0.68); Ni–Cu/SiO_2 showed high activity and selectivity to hydrogen production, in contrast the Ni-based catalyst deactivated rapidly due to coke formation. In a recent study, it has been reported the effect of different H_2 reduction temperatures on the morphology of the 5% Ni–Cu/ SiO_2 catalysts obtained by treatment of the impregnated precursor with a solution of 5% NaBH$_4$ [86]. The H_2 treatment at 350 °C was not sufficient to reduce the nickel phase; on the contrary, the reduction at 650 °C generated mixed phases of Cu-rich and Ni-rich alloy particles.

Bergamaschi *et al.* [87] reported the good performance of a nickel–copper catalyst supported on zirconia microspheres, for ethanol SR. The comparative study on the activities of the monometallic catalysts 6% Ni/ZrO_2 and 3% Cu/ZrO_2 and the bimetallic 6% Ni–3% Cu/ZrO_2 showed a huge advantage in using the Ni–Cu alloy; at 550 °C and with a molar ratio 3:1 of H_2O/EtOH, the conversion of the ethanol was complete for the three catalysts; however, the hydrogen selectivity was 38% and 33%, respectively, for copper and nickel derivate, whereas it was 60% for the bimetallic catalyst.

Furtado *et al.* [88] investigated the role of the support for Ni–Cu-based catalysts; they compared the performance of the catalysts containing the 10% of nickel and 1% of copper supported on α-Al$_2$O$_3$, Nb$_2$O$_5$, ZnO, and CeZrO$_4$, for ethanol SR at atmospheric pressure, at 400 °C with a water/ethanol molar ratio of 10:1. The Ni–Cu/ZnO catalyst showed an excellent initial ethanol conversion (90%), but it deactivated rapidly, stabilizing after 6 h at 15% of conversion; the best performances were obtained with the Ni–Cu/CeZrO$_4$ catalyst, with an average conversion of 43% during 8 h of reaction, the high amount of hydrogen produced was justified with the occurrence of parallel reactions such as the ethanol decomposition.

The distribution of the reaction products indicated the responsibility of the acid sites of the supports in the ethanol dehydration reaction, making evident the relation between the acidity of the support, the efficiency of the catalyst, and the responsibility of nickel in the cleavage of the C–C bonds, increasing the production of C1 compounds. High activity and stability has been reported, for low temperature reforming of ethanol (250–300 °C), by using a Ni–Cu (28% copper loading) catalyst with the nickel-Raney structure [89]. The kinetic data were compatible with a two-step model, the dehydrogenation of ethanol to acetaldehyde in a first-order reaction (activation energy of 149 kJ/mol) followed by a first-order decarbonylation.

The effect of the Cu/Ni ratio in the CuNiZnAl mixed oxide catalysts has also been studied for the OSR of bioethanol [90]. The authors reported that both the CuZnAl and NiZnAl catalysts exhibited an ethanol conversion close to 100% at 300 °C; however, the first one favored the dehydrogenation of ethanol to acetaldehyde, whereas the Ni derivate produced a mixture of hydrogen, CO, CO_2, and methane, whereas the addition of nickel to the copper/zinc system favored the C–C bond rupture and improved the gasification of ethanol. The reaction pathway over CuNiZnAl proceeded through the acetaldehyde intermediate, and the hydrogen yield was between 2.5 and 3.5 mole per mole of converted of ethanol, depending on the reaction conditions. The effects of the prereduction operation on the activity of the NiZnAl catalyst have been studied by Barroso *et al.* [91], for the ethanol SR. The prereduction step was of primary importance at reaction temperatures below 400 °C, on the contrary the activity was independent from the prereduction treatment over 450 °C, with a quantitative ethanol conversion. This effect was attributed to the role of the Ni^0 species and the capacity of the catalyst to self-activate under reforming conditions. The selectivity to hydrogen was, in all cases, better with the prereduced catalyst; however, the difference was not enough to justify the pre-reduction treatment.

In addition to the Ni–Cu catalytic systems, many studies have been reported on the use of bimetallic and polymetallic Ni catalysts in combination with lanthanum, cobalt, zirconium, yttrium, gallium, and especially with noble metals, in ethanol reforming processes, with the main objective of reducing the coke formation. Interesting results were obtained with the Ni–La catalytic system; the activity tests of a comparative study [92] on various Ni–M (M = La, Co, Cu, Zr, and Y) catalysts, supported on alumina/silica in the ethanol SR reaction, indicated that the Ni–La system provides the highest selectivity to hydrogen and lowest selectivity to carbon monoxide, a good long-term stability and resistance to coke formation at low temperature. Among the different metals tested, the lanthanum was the best inhibitor of the crystal growth of nickel and beneficial in the reduction of nickel oxide. The activity of the $Ni–Fe/La_2O_2CO_3$ was compared with the monometallic counterpart catalysts in ethanol SR [93]. The results showed a better performance of the bimetallic derivate; moreover, the use of lanthanum oxycarbonate is a really interesting strategy to reduce the coke formation, in fact the $La_2O_2CO_3$ species can react with carbon cleaning the deposits on the nickel surface (see (2.3)).

$$La_2O_2CO_3 + C \rightarrow La_2O_3 + 2CO \tag{2.3}$$

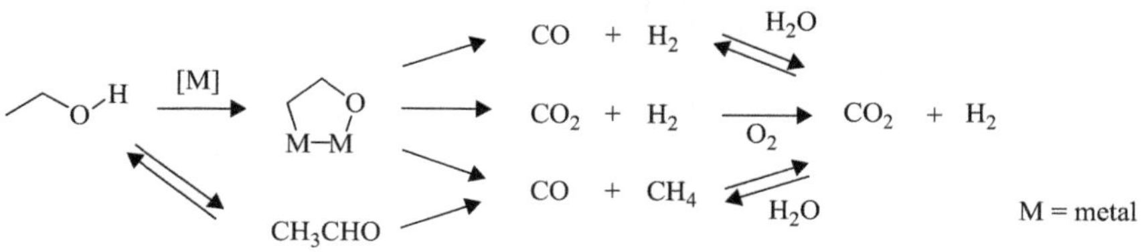

$CH_3CH_2OH \xrightarrow{[M]}$ oxametallacycle (M–M) $\rightarrow$

- $CO + H_2 \underset{H_2O}{\rightleftharpoons}$
- $CO_2 + H_2 \xrightarrow{O_2} CO_2 + H_2$
- $CO + CH_4 \underset{H_2O}{\rightleftharpoons}$

CH_3CHO

M = metal

Figure 2.7   Reaction pathway for the ethanol oxidative steam reforming over Ni-Rh catalysts

Beneficial effects have been obtained by k-doping of $Ni/LaFeO_3$ catalysts [94]. The doping prevents the sintering of the nickel particles, thanks to the electron donation of potassium to nickel, increasing the ability to break the C–C bond. The studies on the Ni/Ga/Mg/zeolite Y catalytic system, for ethanol SR, showed that the simultaneous addition of nickel and gallium may depress the sintering between the Ni and the support, retarding the catalytic deactivation [95]. The synergic effect increased the ethanol conversion and hydrogen production, making the system stable up to 59 h.

For what concern the association of nickel with noble metals, interesting results have been reported for $Ni–Rh/CeO_2$ catalysts [96] for low-temperature reforming of ethanol (below 450 °C). With this catalysts, the OSR was found to be more efficient than SR, thanks to a 100% of ethanol conversion at 375 °C with the highest hydrogen and carbon dioxide selectivity and a lowest CO selectivity; on the contrary, to achieve a 100% of ethanol conversion with the SR, were required higher temperatures, at least 450 °C, with the formation of large amount of carbon monoxide. The best formulation provided an amount of 5% of nickel and 1% of rhodium, increasing the Ni content caused a decreasing of the H_2 selectivity and an increasing of CH_4 and acetaldehyde selectivity. The role of rhodium was the cleavage of the C–C and C–H bonds, whereas the role of nickel was to help the conversion of CO by the water gas shift mechanism. The authors proposed also a reaction pathway (Figure 2.7), for the oxidative reforming, which provided the formation of an oxametallacycle intermediate subsequently reformed to hydrogen and dioxide and decomposed in a mixture of methane and carbon monoxide.

The effect of the properties of the ceria support on the catalytic activity has also been studied [4]. The crystallite size of the ceria support had a strong influence on the dispersion of the Ni–Rh system, the smaller the crystallite size the higher the Rh dispersion, with a consequent higher catalytic activity. Interesting results on the performance of a Rh promoted Ni/CeO_2ZrO_2 catalyst has been reported [97]. The nickel percentage was of 30 wt%, whereas the rhodium of 1 wt%; in ethanol SR was achieved a conversion of 86% and a hydrogen selectivity of 73%, in OSR hydrogen yield and selectivity reduced because of the partial oxidation of oxygenated compounds present in the feed.

The influence of the addition of nickel to the $Rh/Y_2O_3–Al_2O_3$ catalytic system, in presence of methyl-2-propan-1-ol as impurity, was studied for ethanol SR [98].

The incorporation of nickel did not modify the basic properties of the support but induced a rearrangement of the acid sites; the appearance of the nickel aluminate induced an increasing of the Lewis acid sites of weak strength, accordingly with a decreasing of the coke formation and an increasing of the catalyst stability; moreover the presence of nickel facilitated the accessibility and stabilized the particle size of rhodium.

A more complex catalytic system was $Ni/La_2O_3–Al_2O_3$ modified with noble metals (Pt and Pd), investigated in the SR of ethanol [99]. The presence of lanthanum oxide in the support prevented the formation of the inactive nickel aluminate, whereas the noble metals induced the decreasing of the reduction temperature of the nickel oxide species, simplifying the reduction of the promoted catalysts; moreover, the bimetallic catalyst showed a higher ethanol conversion and hydrogen production than the $Ni/La_2O_3–Al_2O_3$ catalyst. Very similar results were reported by the same group for the $Ni/CeO_2–Al_2O_3$ doped with noble metals (Pt, Ir, Pd, and Ru) [18]; the best performance, in ethanol SR at 600 °C, was attributed to the Ni–Pd derivate (Ni 6.39 wt% and Pd 0.48 wt%).

Among the most interesting catalytic systems for ethanol SR, there are those that make use of platinum–nickel alloys. Studies conducted on $Pt–Ni/\delta\text{-}Al_2O_3$ (Ni 10–15 wt% and Pt 0.2–0.3 wt%) indicated that the best performances were obtained with a nickel loading of 15% and a platinum content of 0.3%, both in terms of ethanol conversion and selectivity to hydrogen, up to 773 K [100]. The kinetic studies, with an integral reactor, at 723 K, under conditions extending up to 70% of ethanol conversion, showed a reaction order of 1.01 and −0.09, respectively, for ethanol and steam partial pressures; the apparent activation energy, in the temperature range 673–823 K, was 59.3 ± 2.3 kJ/mol, for the $0.3Pt–15Ni/\delta\text{-}Al_2O_3$. Diffuse Reflectance Infrared Fourier Transform-Mass Spectrometry (DRIFT–MS) analyses have been used to study mechanistic aspect of the ethanol SR reaction, performed on 2.5 wt% Pt 13 wt% $Ni/\gamma\text{-}Al_2O_3$ [101]. These studies indicated two possible mechanisms, the main reaction pathway consisted in a dehydrogenation of the ethanol followed by acetaldehyde decomposition; the active sites of the decomposition step were rapidly deactivated in the first minutes on-stream by the coke formation, due to the dehydrogenation of the C_XH_Y intermediates; the second reaction pathway, which became prevailing once the acetaldehyde decomposition was deactivated, consisted in the decomposition of an acetate intermediate, formed over the surface of alumina. The comparison between the performance of the bimetallic catalyst and the two monometallic counterparts ($Pt/\gamma\text{-}Al_2O_3$ and $Ni/\gamma\text{-}Al_2O_3$) showed that the first possessed a higher activity and stability, probably due to the higher ability to gasify the methyl group formed in the decomposition of the acetate species. The catalytic activity and stability of PtNi alloy supported on modified $\gamma\text{-}Al_2O_3$ with CeO_2 and La_2O_3 in ethanol/glycerol mixture SR has been investigated [102]. The best performances were obtained in presence of ceria, whereas lanthanum and cerium oxides seemed reduce the acidity of alumina, limiting the extension of the dehydration reactions; moreover, the presence of ceria significantly improved the removal of coke by gasification. The role of platinum in the activity and stability of $PtNi/CeO_2–Al_2O_3$ catalysts, in ethanol SR, was investigated with XPS studies [103].

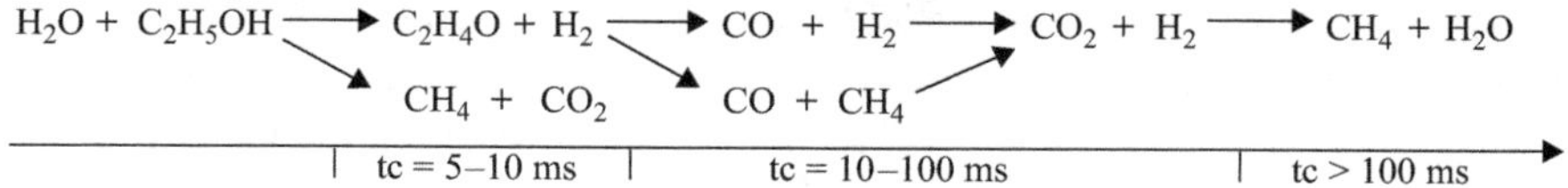

Figure 2.8   Reaction pathway for the ethanol oxidative steam reforming over Ni-Rh catalysts

The interaction between platinum and the existing species at the surface of the catalyst (CeO_2, NiO, and $NiAl_XO_Y$) was not at electronic level, and no alloy formation was observed; the platinum enhanced the reducibility of the nickel and cerium species, showing a higher overall exposition of the metal surface. A platinum loading, greater than 1 wt%, stabilized the $Ni/CeO_2–Al_2O_3$ system in terms of lower carbon deposition and less production of methane and carbon monoxide; this effect was attributed to the ability of platinum to hydrogenate the coke precursors, formed on nickel particles. The interaction Pt–Ce seemed to play a crucial role in the dispersion and stabilization of the platinum particles and thus in the ability of platinum to modify the catalytic behavior of the Ni species. The evolution of the products distribution, as function of contact time (0–600 ms), was reported and a possible reaction pathway was proposed (Figure 2.8), for the ethanol SR over 3 wt% Pt–10 wt % Ni/CeO_2 catalyst at 370 °C [7]. The set of reactions involved included the following steps: ethanol adsorption followed by dehydrogenation to acetaldehyde, decomposition and reforming to hydrogen, methane, carbon oxides, and finally CO-WGS and CO_2 methanation reactions.

A strategy to reduce the deactivation of the $Pt–Ni/CeO_2$ catalyst, in the low-temperature SR of ethanol, has been proposed [104]. The increase of the water-to-ethanol molar ratio from 3 to 6, considerably increased the durability of the catalyst; the increase of the water fed should not be considered an added cost, since in real bioethanol stream the amount of ethanol is in the range of 10–18 wt%, depending on the biomass.

An important class of catalysts, for ethanol reforming, is the system based on Ni–Co in which the high activity of nickel is coupled to the high selectivity of the cobalt, in the attempt to induce a synergic effect. An evidence of this effect has been reported, by comparing the activity of cobalt and nickel alone and in together, supported on alumina, for ethanol SR [105]. At low temperature, the Ni/Al_2O_3 catalyst exhibited a better activity, whereas the Co/Al_2O_3 catalyst showed high selectivity to hydrogen and low selectivity to methane, under the same conditions the bimetallic catalyst exhibited a higher activity and selectivity. Evidence for alloy formation as well as mixed oxides in the bimetallic system obtained by addition of nickel to Na-promoted ZnO-supported cobalt catalyst, were found by TEM analysis coupled to Electron Energy Loss Spectroscopy (HRTEm-EELS) analysis, after running the ethanol SR [106]. The hydrogen productivity of the bimetallic catalyst was high than those obtained with the corresponding monometallic counterpart, moreover no significant improvement was observed with a similar bimetallic Co–Cu catalyst. A series of hydrotalcite like $Co–Ni/MgAlCO_3$ catalysts, prepared via coprecipitation method, were tested for ethanol SR [107]. The analytical results demonstrated that the particle size and reducibility were influenced by the degree of formation of the hydrotalcite-like structure and increased with the Co content. The presence of nickel stabilized the layer structure better than the cobalt, due to the different ionic radii. The initial activity declines with the increasing of nickel content; however, the best activity and stability was reached with 30 wt% Co–10 wt% Ni, whereas the hydrogen yield was close to the thermodynamic equilibrium for all the catalysts. The studies on the possibility to carry out the OSR of ethanol, in absence of prereduction step, suggested that the Co–Ni catalysts are potentially active and selective to hydrogen in this conditions [108].

Good resistance to sintering and to coke formation have been reported for the Ni–Co bimetallic catalysts supported on perovskite-type oxide on $LaFeO_3$, tested for ethanol SR [109]. The characterization results indicated the formation of a solid solution alloy for the Ni–Co system; the comparative study between the mono-metallic and bimetallic catalysts showed a similar anti-coke deposition ability, but a superior anti-sintering ability of the bimetallic derivate.

Recently, Ni–Co alloy/$MgAlO_X$ nanosheets vertically supported on macro-pores' walls of monolithic γ-Al_2O_3 catalysts, obtained by in situ growth through hydrothermal process, have been reported [110]. These catalysts were tested for the ethanol SR at weight hourly space velocity of 240,000 mL/g_{cat}/h, with a water/ethanol ratio of 3, at 650 °C and compared to the corresponding monometallic counterpart; in all cases, the conversion and selectivity to hydrogen were good and near to 100% and 60%, respectively; however, the bimetallic catalysts also showed a good stability, maintaining these performance over a period of 30 h. Also in this case, the good performances were attributed to the excellent resistance to the sintering and to the coke formation.

A published work that seems to contradict the results so far exposed, studied the effect of the Co, Fe, and Rh addition on coke deposition over Ni/$CeZrO_4$ for SR of ethanol [111]. The best resistance to coke formation was attributed to the Ni–Rh catalyst; moreover, the authors claimed a poor resistance of the Ni–Co derivate, the activity of which maintained only 6 h.

Cobalt was used also in conjunction with other metals, such as Fe, Cu, Cr, Na [112], and noble metals [113] for SR of ethanol. The best results were obtained with iron or chromium doping, probably thanks to the promotion of the redox exchange with the cobalt; the iron loading promoted SR without promoting acet-aldehyde decomposition. The studies on the Pt–Co/ZnO catalytic systems showed that the addition of platinum increased the ability in C–C bond breaking and reduced the temperature reaction for a complete conversion [114].

The differences of performance of ceria-supported Pt–Ni and Pt–Co catalysts, for low-temperature ethanol SR, in terms of activity, stability, and durability, by evaluating the effect of preparation method, GHSV, water-to-ethanol molar ratio and dilution ratio, have been recently reported [115]. The best performances were obtained through impregnation method; moreover, it seemed to be more convenient to add the noble metal in the second impregnation step, probably thanks to the higher availability of the platinum at the gas–solid interface, promoting the ethanol adsorption and the hydrogenation of the CH_X coke precursors. The Pt/Ni catalyst showed a good activity and selectivity, whereas its stability was negatively affected by coke formation; however, by increasing the water content in the feed mixture it was possible to favor the gasification of coke. The Pt–Co catalyst appeared more promising due to the higher selectivity to hydrogen and a better durability due to the low selectivity to coke, also at stoichiometric water-to-ethanol molar ratio. This behavior has been explained in terms of reaction mechanism, the cobalt species promoted the oxidative reactions but not the methanation [116].

Among the catalytic systems that use neither nickel nor cobalt, really interesting is the bimetallic Ru–Pt supported nanoparticles, derived from organometallic cluster precursors [117]. The high catalytic efficiency, in ethanol SR, was attributed to the size of the metallic nanoparticles; moreover, the activity and selectivity were better than those a commercial catalyst. The TPD experiments suggested that the Pt lowered the hydrogen desorption temperature and increased the efficiency in C–C bond dissociation, probably following a reaction pathway similar to that proposed for Rh–Pt catalysts, previously mentioned, which involves the formation of an intermediate five-member oxometallocycle [118].

Similar results have been reported for the Rh–Pd/CeO$_2$ catalyst in ethanol SR [119]. The rhodium was responsible for the breaking of the carbon–carbon bond, whereas the palladium favored the water gas shift reaction and the H$_2$–H$_2$ recombination reaction. SR of *n*-propanol was studied at 450 and 500 °C and at an S/C of 4 over Ru–Ni catalysts supported on CeO$_2$–Al$_2$O$_3$ [120]. The impact of CeO$_2$ and metals loading as well as the effect of preparation procedure where investigated in depth. The impregnation method was followed for all the samples; however, after Rh deposition, some catalysts were calcined at 500 °C for 4 h, whereas other samples were only dried at 120 °C for 8 h. Calcination of ruthenium precursor leads to formation of large RuO$_2$ crystallites that reduces into poorly dispersed ruthenium metal particles: as a result, a detrimental effect was observed on the catalyst activity. Concerning the support, low ceria loadings promoted dehydration of *n*-propanol, which caused a quick catalyst deactivation. On the other hand, the increase in Ni content favored SR reaction while Ru addition slightly increased methane selectivity. The catalyst with 3 wt% Ru, 10 wt% Ni, and CeO$_2$ loading of 3 or 10 wt% was shown to be the most active and selective at 450 °C. At 500 °C, H$_2$ production was further enhanced, due to the contribution of MSR reaction.

In order to revalorize a by-product of biodiesel production, glycerol reforming for H$_2$ production was extensively studied. Together with the classical SR process, a growing interest is focused on the aqueous phase reforming (APR) [121], which allows for relatively low process temperatures (200–260 °C) that will not result in decomposition reactions and operates under mild conditions (15–60 bar).

Over Co–Ni/Al$_2$O$_3$ catalysts, tested between 500 and 550 °C under stoichiometrically SR conditions [122], the presence of acidic sites on the support was responsible for carbon deposition occurrence (TOC ranging between 18% and 30%). However, catalyst regeneration through temperature-programed reduction–temperature programed oxidation cycles restored the physiochemical properties at the same level of the fresh catalyst.

Co–Ni/Al$_2$O$_3$ catalysts, having different Co loadings (0, 4, and 12 wt%), were also tested at 300 °C, 500 °C, and 700 °C under a water/glycerol molar ratio of 6 [123]. Co addition promoted H$_2$ production and unfavored CO$_2$ generation by decreasing the reaction temperature, whereas CH$_4$ formation was favored at high temperatures. Increasing the Co loading from 4 to 12 wt%, the H$_2$ production increases slightly at low temperature and displays more stability during the reaction. However, the H$_2$ production at 500 and 700 °C is lesser, showing a larger proportion of the remaining compounds.

CeZr–CoRh catalysts, prepared at different CeO$_2$/ZrO$_2$ ratios (P sample: 0.25, I sample: 1.86, and R sample: 4), were also tested for SR of glycerol (650 °C for 24 h and water/glycerol molar ratio of 9) [124]. Within the first 8.5 h of TOS (Figure 2.9), glycerol was completely converted over all the three samples. However, the ability to convert glycerol is progressively lost. After 8.5 h, X (global conversion) decreases to 67%, 64%, and 77% for P, I, and R sample, respectively. The increase of CeO$_2$ amount improved both X and X$_G$ (conversion to non-condensable products), whereas X$_L$ (conversion to condensable products) decrease followed an opposite order. In all the cases, X$_G \gg$ X$_L$ but this difference decreases with TOS. Moreover, for all the samples, the formation of condensable products is not observed in the first 5 h of reaction, whereas the highest H$_2$ yields were recorded. In terms of noncondensable species, the P sample showed a CH$_4$ selectivity slightly higher than the other samples, probably due to its lower oxygen storage capacity that promotes carbon hydrogenation instead of its oxidation.

Bobadilla *et al.* [125] studied the effect of MgO addition (loading ranging from 0 to 30 wt%) to NiSn catalysts supported on Al$_2$O$_3$ and employed for glycerol SR. The catalyst containing 30 wt% of MgO presented the highest and stable glycerol conversion at 650 °C and water-to-glycerol molar ratio of 12. Moreover, H$_2$ production was notably increased when Al$_2$O$_3$ is loaded with 5 wt% of MgO, whereas it is only slightly improved for higher MgO contents. The beneficial effects of MgO addition are related to the modification of Ni–Al$_2$O$_3$ interactions: the formation of MgAl$_2$O$_4$ species inhibits the incorporation of nickel in the Al$_2$O$_3$ phase, thus improving Ni dispersion. Moreover, MgO addition reduced the concentration of acidic centers, thus avoiding secondary reactions which lead to liquid intermediates formation. The best catalytic behavior as well as the lowest carbon formation rate was measured over the sample containing 10 wt% of MgO. The performances of the latter catalyst were also investigated after the addition of 15 wt% of CeO$_2$ [126]. Ceria addition improved H$_2$ and CO$_2$ yields, reducing, at the same time, CH$_4$ and C$_2$H$_4$ selectivity. The simultaneous introduction of Mg and Ce, in fact, decreased notably the CH$_4$/H$_2$ molar ratio inhibiting the methanation reaction and diminished the CO/CO$_2$ molar ratio favoring the water gas shift reaction.

For 5 wt% Ni–15 wt% Co catalysts supported on γ-Al$_2$O$_3$ and tested under a 5 wt% glycerol in water stream (APR conditions), at 220 °C and 25 bar, the addition of low CeO$_2$ amounts (2 wt%) was favorable for the suppression of methane selectivity [127]. For higher loadings, lower activity was recorded, probably due to the presence of CeO$_2$ over the support which occupied the active sites. Moreover,

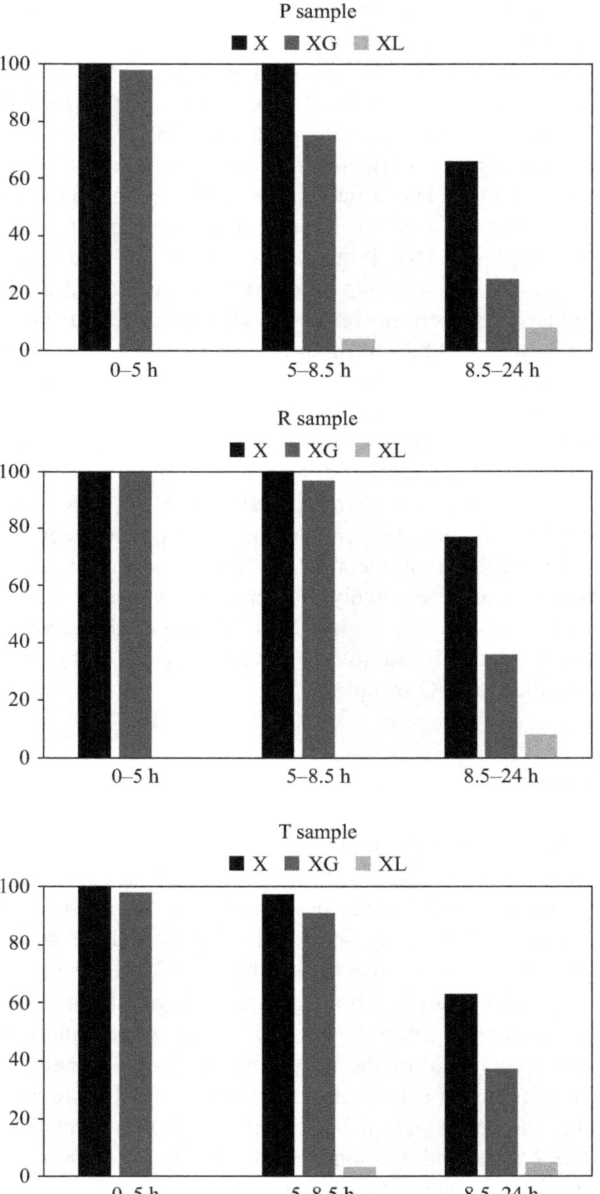

Figure 2.9   Mean conversions (X, XG, XL) over the R, I and P samples during 24 hours of glycerol steam reforming at 650°C, S/C = 9

CeO_2 addition avoided sintering as well as the formation of filamentous carbon with the Ni particles on the filament top.

PtFe, PtCo, and PtNi catalysts supported on γ-Al_2O_3 and prepared by sol–gel method and conventional impregnation were employed for the APR of glycerol (250 °C, 50 bar and 10 wt% $C_3H_8O_3$ in water) [128]. The catalyst prepared by sol–gel reached the highest performances, showing increasing activity in the order PtFe < PtCo ≪ PtNi. The latter sample, in fact, reached the highest glycerol conversion (74%), whereas a value almost nine times lower was recorded over the other two samples. PtNi catalyst also provided the highest H_2 yield $\left(0.51\ F_{H_2,\text{out}}/F_{\text{glycerol,in}}\right)$ and gaseous products flow, indicating a lower selectivity to liquid by-product formation; however, the alkanes yield was not negligible.

The reforming of model bio-oil mixtures, which simulate bioresources-derived solutions, was also investigated. The SR of a mixture of butanol, acetone, ethanol (6:3:1 mass ratio), and water (60 wt%) was studied over monometallic and bimetallic Co and Ir-containing catalysts [129]. At 600 °C, the Co/ZnO catalyst showed stable catalytic performances during 90 h of TOS and a conversion of the mixture of 97%. Conversely, total conversion for almost 40 h was recorded over the bimetallic Co–Ir/ZnO sample, whereas, for higher times, the conversion achieved was similar to that of the monometallic Co-based sample. However, the H_2 and CH_4 concentrations were, respectively, higher and lower than the Co/ZnO catalyst (70% vs 65% for H_2 and 6% vs 13% for CH_4). Moreover, the carbon formation rate after the test on the bimetallic sample (0.38 $mg_{\text{coke}}/g_{\text{cat}}$/h) was almost 1 half than that measured for the Co/ZnO sample.

2.5   Conclusions

The reforming processes are the most widespread technologies for hydrogen and syngas production; among the raw materials used, natural gas is the primary fossil fuel, whereas, in the renewable field, bio-ethanol has the most benefits. Although noble metals-based catalysts provide excellent performance in MSR, the most common catalytic systems make use of the cheaper nickel. The Ni-based catalysts display a good activity but suffer from sintering and coke formation, thus claiming the design of more efficient catalysts. Similarly, and perhaps more urgent, it is the need for alternative catalysts for the reforming of ethanol, where the necessity to couple high conversions and selectivity requires the use of more complex systems. To that end, huge potential lies in bimetallic systems, having properties greatly different from the corresponding monometallic catalysts. In this chapter, we have presented the results of several articles focused on the study and optimization of sophisticated bimetallic catalysts for hydrocarbons and alcohols reforming; the experimental results suggest that the presence of metallic alloys improves the performance in the process. Many metals combinations have been investigated and proposed; however, the need to couple high efficiency and low costs, inevitably makes the mixed noble–non noble alloy systems more attractive. A good example is the 5% Ni–1% Rh/CeO_2 catalyst that provides a good performance in oxidative low-temperature reforming of ethanol, reaching total conversion at 375 °C with high hydrogen selectivity and low carbon monoxide formation. These results were explained in terms of synergistic effect due to the combination of rhodium and nickel: it was suggested that Rh facilitated the C–C and C–H bonds cleavage, whereas Ni helped the CO conversion by water gas shift mechanism. Similarly, for MSR, it was reported that iridium–nickel and rhodium–nickel systems, thanks to the alloy formation, are able to prevent the sintering phenomena while gold–nickel catalysts depress coke formation.

Nomenclature

APR	aqueous phase reforming
DRIFT-MS	Diffuse Reflectance Infrared Fourier Transform-Mass Spectrometry
F	molar flowrate (mol/min)
GHSV	gas hourly space velocity
HRTEM	high resolution transmission electron microscopy
LPG	liquefied petroleum gas
MDR	dry reforming of methane
MSR	methane steam reforming
O_2/C	oxygen-to-carbon molar ratio
OMSR	oxidative steam reforming of methane
OSR	oxidative steam reforming
RWGS	Reverse Water Gas Shift
S/C	steam-to-carbon molar ratio
SEM	scanning electron microscopy
SR	steam reforming
TEM	Transmission Electron Microscopy
TOC	total organic compound
TOS	time on stream
TPO	temperature programmed oxidation
TPR	temperature programmed reduction
WGS	Water Gas Shift
WHSV	weight hourly space velocity

References

[1]   V.L. Dagle, R. Dagle, L. Kovarik, *et al.*, *Appl. Catal., B: Environ.* 185 (2016) 142–152.

[2]   V. Palma, C. Ruocco, F. Castaldo, A. Ricca, D. Boettge, *Int. J. Hydrogen Energy* 40 (2015) 12650–12662.

[3]   T. Wang, M.D. Porosoff, J.G. Chen, *Catal. Today* 233 (2014) 61–69.

[4]   J. Kugai, V. Subramani, C. Song, M.H. Engelhard, Y.-H. Chin, *J. Catal.* 238 (2006) 430–440.

[5]   I. Luisetto, S. Tuti, E. Di Bartolomeo, *Int. J. Hydrogen Energy* 37 (2012) 15992–15999.

[6]   J. Estephane, S. Aouad, S. Hany, *et al.*, *Int. J. Hydrogen Energy* 40 (2015) 9201–9208.

[7]   V. Palma, F. Castaldo, P. Ciambelli, G. Iaquaniello, *Appl. Catal., B: Environ.* 145 (2014) 73–84.

[8]   T.-C. Feng, W.-T. Zheng, K.-Q. Sun, B.-Q. Xu, *Catal. Commun.* 73 (2016) 54–57.

[9]   L. De Rogatis, T. Montini, A. Cognigni, L. Olivi, P. Fornasiero, *Catal. Today* 145 (2009) 176–185.

[10] A. Gutierrez, R. Karinen, S. Airaksinen, R. Kaila, A.O.I. Krause, *Int. J. Hydrogen Energy* 36 (2011) 8967–8977.

[11] S. Jens, *Catal. Today* 111 (2006) 103–110.

[12] H. Wan, X. Li, S. Ji, B. Huang, K. Wang, C. Li, *J. Nat. Gas Chem.* 16 (2007) 139–147.

[13] J. Chen, C. Yao, Y. Zhao, P. Jia, *Int. J. Hydrogen Energy* 35 (2010) 1630–1642.

[14] P. Ferreira-Aparicio, A. Guerrero-Ruiz, I. Rodriguez-Ramos, *Appl. Catal., A: Gen.* 170 (1998) 177–187.

[15] K. Jabbour, N. El Hassan, S. Casale, J. Estephane, H. El Zakhem, *Int. J. Hydrogen Energy* 39 (2014) 7780–7787.

[16] M. Nagai, K. Nakahira, Y. Ozawa, Y. Namiki, Y. Suzuki, *Chem. Eng. Sci.* 62 (2007) 4998–5000.

[17] P. Ciambelli, V. Palma, A. Ruggiero, *Appl. Catal., B: Environ.* 96 (2010) 18–27.

[18] L.P.R. Profeti, E.A. Ticianelli, E.M. Assaf, *Int. J. Hydrogen Energy* 34 (2009) 5049–5060.

[19] Y. Mukainakano, K. Yoshida, S. Kado, K. Okumura, K. Kunimori, K. Tomishige, *Chem. Eng. Sci.* 63 (2008) 4891–4901.

[20] F. Morales-Cano, L.F. Lundegaard, R.R. Tiruvalam, H. Falsig, M. Skov Skjøth-Rasmussen, *Appl. Catal., A: Gen.* 498 (2015) 117–125.

[21] T. Miyata, D. Li, M. Shiraga, *et al.*, *Appl. Catal., A: Gen.* 310 (2006) 97–104.

[22] L. Huang, C. Rongrong, C. Deryn, A.T. Hsu, *Int. J. Hydrogen Energy* 35 (2010) 1138–1146.

[23] T.-J. Huang, T.-C. Yu, S.-Y. Jhao, *Ind. Eng. Chem. Res.* 45 (2005) 150–156.

[24] P.-H. Liao, H.-M. Yang, *Catal. Lett.* 121 (2008) 274–282.

[25] A.G. Gil, Z. Wu, D. Chadwick, K. Li, *Appl. Catal., A: Gen.* 506 (2015) 188–196.

[26] P. Wu, X. Li, S. Ji, B. Lang, F. Habimana, C. Li, *Catal. Today* 146 (2009) 82–86.

[27] C. Liang, Z. Ma, H. Lin, *et al.*, *J. Mater. Chem.* 19 (2009) 1417–1424.

[28] A.B. Kehoe, D.O. Scanlon, G.W. Watson, *Chem. Mater.* 23 (2011) 4464–4468.

[29] M. Ni, D.Y.C. Leung, M.K.H. Leung., *Int. J. Hydrogen Energy* 32 (2007) 3238–3247.

[30] D. Zhao, J. Sun, Q. Li, G.D. Stucky, *Chem. Mater.* 12 (2000) 275–279.

[31] C. Agrafiotis, H. von Storch, M. Roeb, C. Sattler, *Renewable Sustainable Energy Rev.* 29 (2014) 656–682.

[32] M. Khzouz, J. Wood, B. Pollet, W. Bujalski, *Int. J. Hydrogen Energy* 38 (2013) 1664–1675.

[33] M. Dan, M. Mihet, A.R. Biris, *et al.*, *React. Kinet. Mech. Catal.* 105 (2012) 173–193.

[34] Y.-H. Chin, D.L. King, H.-S. Roh, Y. Wang, S.M. Heald, *J. Catal.* 244 (2006) 153–162.

[35] E.L. Foletto, R.W. Alves, S.L. Jahn, *J. Power Sources* 161 (2006) 531–534.

[36] C. Ni, L. Pan, Z. Yuan, L. Cao, S. Wang, *J. Rare Earths* 32 (2014) 184–188.

[37] Y. Wang, J. Peng, C. Zhou, *et al.*, *Int. J. Hydrogen Energy* 39 (2014) 778–787.

[38] B. Li, S. Kado, Y. Mukainakano, *et al.*, *Appl. Catal., A: Gen.* 304 (2006) 62–71.

[39] I.Z. Ismagilov, E.V. Matus, V.V. Kuznetsov, *et al.*, *Int. J. Hydrogen Energy* 39 (2014) 20992–21006.

[40] G. Zhang, L. Hao, Y. Jia, Y. Du, Y. Zhang, *Int. J. Hydrogen Energy* 40 (2015) 12868–12879.

[41] J. Cheng, W. Huang, *Fuel Process. Technol.* 91 (2010) 185–193.

[42] H. Ay, D. Üner, *Appl. Catal., B: Environ.* 179 (2015) 128–138.

[43] P. Djinovic, I.G. Osojnik Crnivec, B. Erjavec, A. Pintar, *Appl. Catal., B: Environ.* 125 (2012) 259–270.

[44] M.S. Aw, I.G. Osojnik Crnivec, P. Djinovic, A. Pintar, *Int. J. Hydrogen Energy* 39 (2014) 12636–12647.

[45] I.G. Osojnik Crnivec, P. Djinovic, B. Erjavec, A. Pintar, *Chem. Eng. J.* 207–208 (2012) 299–307.

[46] M.S. Aw, M. Zorko, P. Djinovic, A. Pintar, *Appl. Catal., B: Environ.* 164 (2015) 100–112.

[47] H. Wang, J.T. Miller, M. Shakouri, *et al.*, *Catal. Today* 207 (2013) 3–12.

[48] S.M. Sajjadi, M. Haghighi, F. Rahmani, *J. Nat. Gas Sci. Eng.* 22 (2015) 9–21.

[49] T. Huang, W. Huang, J. Huang, P. Ji, *Fuel Process. Technol.* 92 (2011) 1868–1875.

[50] D. Liu, X.Y. Quek, W.N.E. Cheo, R. Lau, A. Borgn, Y. Yang, *J. Catal.* 266 (2009) 380–390.

[51] F. Menegazzo, M. Signoretto, F. Pinna, P. Canton, N. Pernicone, *Appl. Catal., A: Gen.* 439–440 (2012) 80–87.

[52] H. Wu, G. Pantaleo, V. La Parola, *et al.*, *Appl. Catal., B: Environ.* 156–157 (2014) 350–361.

[53] J.C.S. Wu, H.-C. Chou, *Chem. Eng. J.* 148 (2009) 539–545.

[54] F. Meshkani, M. Rezaei, *Int. J. Hydrogen Energy* 35 (2010) 10295–10301.

[55] X. Huang, R. Reimert, *Fuel* 106 (2013) 380–387.

[56] C. Veranitisagul, N. Koonsaeng, N. Laosiripojana, A. Laobuthee, *J. Ind. Eng. Chem.* 18 (2012) 898–903.

[57] I. Kang, J. Bae, *J. Power Sources* 159 (2006) 1283–1290.

[58] T. Hou, B. Yu, S. Zhang, *et al.*, *Appl. Catal., B: Environ.* 168–169 (2015) 524–530.

[59] M. Matsuka, K. Shigedomi, T. Ishihara, *Int. J. Hydrogen Energy* 39 (2014) 14792–14799.

[60] Z.O. Malaibari, E. Croiset, A. Amin, W. Epling, *Appl. Catal., A: Gen.* 490 (2015) 80–92.

[61] A. Siahvashi, D. Chesterfield, A.A. Adesina, *Chem. Eng. Sci.* 93 (2013) 313–325.

[62] A. Siahvashi, A.A. Adesina, *Catal. Today* 214 (2013) 30–41.

[63] F.M. Althenayan, S. Yei Foo, E.M. Kennedy, B.Z. Dlugogorski, A.A. Adesina, *Chem. Eng. Sci.* 65 (2010) 66–73.

[64] M. Ferrandon, A. Jeremy Kropf, T. Krause, *Appl. Catal., A: Gen.* 379 (2010) 121–128.

[65] A.A. Al-Musa, Z.S. Ioakeimidis, M.S. Al-Saleh, A. Al-Zahrany, G.E. Marnellos, M. Konsolakis, *Int. J. Hydrogen Energy* 39 (2014) 19541–19554.

[66] J. Xie, X. Sun, L. Barrett, *et al.*, *Int. J. Hydrogen Energy* 40 (2015) 8510–8521.

[67] Z.O. Malaibari, A. Amin, E. Croiset, W. Epling, *Int. J. Hydrogen Energy* 39 (2014) 10061–10073.

[68] C. Xie, Y. Chen, Y. Li, X. Wang, C. Song, *Appl. Catal., A: Gen.* 390 (2010) 210–218.

[69] A. Derya Deniz Kaynar, D. Dogu, N. Yasyerli, *Fuel Process. Technol.* 140 (2015) 96–103.

[70] M. Koike, D. Li, H. Watanabe, Y. Nakagawa, K. Tomishige, *Appl. Catal., A: Gen.* 506 (2015) 151–162.

[71] F. Xu, Y. Xu, H. Yin, X. Zhu, Q. Guo, *Energy Fuels* 23 (2009) 1775–1777.

[72] L.F. Bobadilla, S. Palma, S. Ivanova, *et al.*, *Int. J. Hydrogen Energy* 38 (2013) 6646–6656.

[73] A.A. Lytkina, N.A. Zhilyaeva, M.M. Ermilova, N.V. Orekhova, A.B. Yaroslavtsev, *Int. J. Hydrogen Energy* 40 (2015) 9677–9684.

[74] T. Conant, A.M. Karima, V. Lebarbier, *et al.*, *J. Catal.* 257 (2008) 64–70.

[75] K.M. Eblagon, P.H. Concepción, H. Silva, A. Mendes, *Appl. Catal., B: Environ.* 154–155 (2014) 316–328.

[76] Y. Men, G. Kolb, R. Zapf, M. O'Connell, A. Ziogas, *Appl. Catal., A: Gen.* 380 (2010) 15–20.

[77] R. Pérez-Hernández, G. Mondragon Galicia, D. Mendoza Anaya, J. Palaciosa, C. Angeles-Chavez, J. Arenas-Alatorre, *Int. J. Hydrogen Energy* 33 (2008) 4569–4576.

[78] R. Pérez-Hernández, A. Gutiérrez-Martínez, M.E. Espinosa-Pesqueira, M.L. Estanislao, J. Palacios, *Catal. Today* 250 (2015) 166–172.

[79] F. Talebnia, D. Karakashev, I. Angelidaki, *Bioresour. Technol.* 101 (2010) 4744–4753.

[80] C. Luengo, G. Ciampi, M.O. Cencing, C. Steckelberg, M.A. Laborde, *Int. J. Hydrogen Energy* 17 (1992) 677–681.

[81] F. Mariño, G. Baronetti, M. Jobbargy, M. Laborde, *Appl. Catal., A: Gen.* 238 (2003) 41–54.

[82] F.J. Mariño, E.G. Cerrella, S. Duhalde, M. Jobbargy, M.A. Laborde, *Int. J. Hydrogen Energy* 23 (1998) 1095–1101.

[83] F. Mariño, M. Boveri, G. Baronetti, M. Laborde, *Int. J. Hydrogen Energy* 26 (2001) 665–668.

[84] V. Klouz, V. Fierro, P. Denton, *et al.*, *J. Power Sources* 105 (2002) 26–34.

[85] V. Fierro, O. Akdim, C. Mirodatos, *Green Chem.* 5 (2003) 20–24.

[86] L.-C. Chen, S.D. Lin, *Appl. Catal., B: Environ.* 148–149 (2014) 509–519.

[87] V.S. Bergamaschi, F.M.S. Carvalho, C. Rodrigues, D.B. Fernandes, *Chem. Eng. J.* 112 (2005) 153–158.

[88] A.C. Furtado, C. Gonçalves Alonso, M. Pereira Cantão, N. Regina Camargo, F. Machado, *Int. J. Hydrogen Energy* 34 (2009) 7189–7196.

[89] D.A. Morgenstern, J.P. Fornango, *Energy Fuels* 19 (2005) 1708–1716.

[90] S. Velu, N. Satoh, S. Chinnakonda Gopinath, K. Suzuki, *Catal. Lett.* 82 (2002) 145–152.

[91] M.N. Barroso, M.F. Gómez, L.A. Arrúa, M.C. Abello, *React. Kinet. Catal. Lett.* 97 (2009) 27–33.

[92] L. Zhang, W. Li, J. Liu, C. Guo, Y. Wang, J. Zhang, *Fuel* 88 (2009) 511–518.

[93] Q. Shi, Z. Peng, W. Chen, N. Zhang, *J. Rare Earths* 29 (2011) 861–865.

[94] L. Zhao, Y. Wei, Y. Huang, Y. Liu, *Catal. Today* 259 (2016) 430–437.

[95] B.S. Kwak, J.S. Lee, J.S. Lee, B.-H. Choi, M.J. Ji, M. Kang, *Appl. Energy* 88 (2011) 4366–4375.

[96] J. Kugai, V. Subramani, C. Song, *Catal. Lett.* 101 (2005) 255–264.

[97] T. Mondal, K.K. Pant, A.K. Dalai, *Appl. Catal., A: Gen.* 499 (2015) 19–31.

[98] A. Le Valant, N. Bion, D. Duprez, F. Epron, *Appl. Catal., B: Environ.* 97 (2010) 72–81.

[99] L.P.R. Profeti, J.A.C. Dias, J.M. Assaf, E.M. Assaf, *J. Power Sources* 190 (2009) 525–533.

[100] F. Soyal-Baltacioğlu, A. Erhan Aksoylu, Z. Ilsen Önsan, *Catal. Today* 138 (2008) 183–186.

[101] M.C. Sanchez-Sanchez, R.M. Navarro Yerga, D.I. Kondarides, X.E. Verykios, J.L.G. Fierro, *J. Phys. Chem. A* 114 (2010) 3873–3882.

[102] M. El Doukkali, A. Iriondo, P.L. Arias, J.F. Cambra, I. Gandarias, V.L. Barrio, *Int. J. Hydrogen Energy* 37 (2012) 8298–8309.

[103] M.C. Sanchez-Sanchez, R.M. Navarro, I. Espartero, A.A. Ismail, S.A. Al-Sayari, J.L.G. Fierro, *Top. Catal.* 56 (2013) 1672–1685.

[104] V. Palma, F. Castaldo, C. Ruocco, P. Ciambelli, G. Iaquaniello, *J. Power Technol.* 95 (2015) 54–66.

[105] L.-P. Mao, X. Hu, G.X. Lu, *Lanzhou Ligong Daxue Xuebao* 35 (2009) 60–64.

[106] N. Homs, J. Llorca, P. Ramírez de la Piscina, *Catal. Today* 116 (2006) 361–366.

[107] L. He, H. Berntsen, E. Ochoa-Fernández, J.C. Walmsley, E.A. Blekkan, D. Chen, *Top. Catal.* 52 (2009) 206–217.

[108] M. Muñoz, S. Moreno, R. Molina, *Int. J. Hydrogen Energy* 39 (2014) 10074–10089.

[109] Z. Wang, C. Wang, S. Chen, Y. Liu, *Int. J. Hydrogen Energy* 39 (2014) 5644–5652.

[110] Y. Yue, F. Liu, L. Zhao, L. Zhang, Y. Liu, *Int. J. Hydrogen Energy* 40 (2015) 7052–7063.

[111] J.Y.Z. Chiou, C. Liang Lai, S.-W. Yu, H.-H. Huang, C.-L. Chuang, C.-B. Wang, *Int. J. Hydrogen Energy* 39 (2014) 20689–20699.

[112] A. Casanovas, M. Roig, C. de Leitenburg, A. Trovarelli, J. Llorca, *Int. J. Hydrogen Energy* 35 (2010) 7690–7698

[113] Y. Sekine, A. Kazama, Y. Izutsu, M. Matsukata, E. Kikuchi, *Catal. Lett.* 132 (2009) 329–334.

[114] J.Y.Z. Chiou, W.Y. Wang, S.Y. Yang, C.L. Lai, H.H. Huang, C.B. Wang, *Catal. Lett.* 143 (2013) 501–507.

[115] V. Palma, F. Castaldo, P. Ciambelli, G. Iaquaniello, *Clean Technol. Environ. Policy* 14 (2012) 973–987.

[116] V. Palma, F. Castaldo, P. Ciambelli, G. Iaquaniello, G. Capitani, *Int. J. Hydrogen Energy* 38 (2013) 6633–6645.

[117] A.C.W. Koh, L. Chen, W.K. Leong, *et al.*, *Int. J. Hydrogen Energy* 34 (2009) 5691–5703.

[118] P.-Y. Sheng, A. Yee, G.A. Bowmaker, H. Idriss, *J. Catal.* 208 (2002) 393–403.

[119] M. Scott, M. Goeffroy, W. Chiu, M.A. Blackford, H. Idriss, *Top. Catal.* 51 (2008) 13–21.

[120] M. Wang, C.-T. Au, S.-Y. Lai, *Int. J. Hydrogen Energy* 40 (2015) 13926–13935.

[121] P.J. Dietrich, F.G. Sollberger, M. Cem Akatayb, *et al.*, *Appl. Catal., B: Environ.* 156–157 (2014) 236–248.

[122] C. Kui Cheng, S.Y. Foo, A.A. Adesina, *Catal. Today* 164 (2011) 268–274.

[123] E.A. Sanchez, R.A. Comelli, *Int. J. Hydrogen Energy* 39 (2014) 8650–8655.

[124] L.M. Martínez, T.M. Araque, J.C. Vargas, A.C. Roger, *Appl. Catal., B: Environ.* 132–133 (2013) 499–510.

[125] L.F. Bobadilla, A. Penkova, F. Romero-Sarria, M.A. Centeno, J.A. Odriozola, *Int. J. Hydrogen Energy* 39 (2014) 5704–5712.

[126] L.F. Bobadilla, A. Penkova, A. Álvarez, *et al.*, *Appl. Catal., A: Gen.* 492 (2015) 38–47.

[127] N. Luo, K. Ouyang, F. Cao, T. Xiao, *Biomass Bioenergy* 34 (2010) 489–485.

[128] M. El Doukkali, A. Iriondo, J.F. Cambra, *et al.*, *J. Mol. Catal. A: Chem.* 368–369 (2013) 125–136.

[129] W. Cai, P.R. de la Piscina, N. Homs, *Bioresour. Technol.* 107 (2012) 482–486.

Chapter 3

Catalysts for hydrogen production from renewable raw materials, by-products and waste

Claudio Evangelisti[1], Filippo Bossola[1,2] and Vladimiro Dal Santo[1]

Abstract

Catalytic materials used in hydrogen production processes starting from different raw materials (e.g. fossil (oil, gas and coal), renewables (primary and secondary bio-based raw materials) and waste materials (municipal solid waste (MSW), refuse-derived fuel (RDF), agro-food residues, manure)) are reviewed highlighting the most relevant advances of the last 5 years. The best results obtained mainly in reforming reactions and supercritical water gasification processes, in terms of improved performances such as higher hydrogen yield, lower by-products, coke, tar formation, as well as milder reactions conditions, are discussed and compared taking into account the balance between costs and performances of the used catalytic materials. Moreover, still open issues for the application of these processes (e.g. catalysts stability, low resistance to N, S poisoning) have been pointed out.

3.1 Introduction

Hydrogen can be considered as the most versatile and flexible energy carrier available today since it can be produced starting from a number of energy sources and raw materials, can be stored even in large scale, and, finally, can be converted again into other energy vectors upon request. Hydrogen applications range from mobility to stationary cogeneration, from grid balancing to power, and to gas processes. Nevertheless, traditional usage of hydrogen as a commodity in oil refining, food industry, ammonia synthesis should be mentioned. Hydrogen can be obtained by many different raw materials: fossil (oil, gas and coal), renewables (lignocellulosic biomass, secondary bio-based raw materials, such as alcohols,

[1]CNR—Istituto di Scienze e Tecnologie Molecolari, Via Golgi 19, Milano 20133, Italy
[2]Dipartimento di Scienza e Alta Tecnologia, Università dell'Insubria, Como 20133, Italy

ketones, and acids), waste materials (MSW, RDF, agro-food residues, manure, etc.), simple water and by using different kind of energy sources and vectors: fossils, solar, tidal, eolic, hydro energy sources, and electrical power and heat as vectors. Related processes involve pyrolysis, gasification, reforming reactions, partial oxidations, electrolysis, photocatalytic and photoelectrocatalytic water splitting and fermentations. Almost all these processes could be intensified by some catalytic steps, in which mostly heterogeneous catalysts based on supported metal nanoparticles are involved. The development of active, selective and stable catalysts plays a key role in improving economics, materials, energy efficiency and sustainability of the overall hydrogen production processes. In the following paragraphs, the principal hydrogen production processes (mainly reforming and pyrolysis) starting from renewable raw materials, by-products and waste will be reviewed with special focus on catalytic materials, highlighting the most relevant advances of the last 5 years.

3.2   Primary raw materials

Among different primary raw materials, the use of inedible biomass, such as the lignocellulosic materials and microalgae, is highly promising for the production of bio-renewable hydrogen owing to their minimal impact on the food security and low life cycle greenhouse CO_2 gas emissions. Supercritical catalytic water gasification (SCWG), also known as hydrothermal gasification, is a promising technology for the production of hydrogen from biomass materials. Supercritical water ($T_c = 374\ °C$ and $P_c = 22.1$ MPa) is used both as reaction medium and reactant with a strong ability to break down hydrocarbons and carbohydrates, resulting in the production of pressurized gases mainly rich in H_2, CO, CO_2 and CH_4. Actually, the generation of ions (H^+ and OH^-) and free radicals at high density in supercritical conditions promotes hydrolysis and pyrolysis reactions of biomass components. In the presence of heterogeneous catalysts, and depending on the operating conditions, the complex mixture can be further reformed to simple gases such as H_2, CO, CO_2 and CH_4 produced by water gas shift (WGS), methanation, hydrogenation and other reactions. The role of catalyst in SCWG is to reduce the operating costs (decreasing temperature and pressure of the process), improving hydrogen selectivity as well as decrease char and tar formation.

3.2.1   *Lignocellulose*

Lignocellulosic biomass is the most abundant type of biomass on the earth. It is the non-edible part of the plants consisting of lignin, cellulose, hemicellulose, extractives and inorganic materials [1,2]. The first three components are the main constituents, comprising as high as 98% of the total material by weight. Lignin (15%–25%) is a complex cross-linked amorphous copolymer derived from random polymerization of phenolic phenyl propane monomers. On the other hand, cellulose (30%–60%) and hemicellulose (20%–40%) are both polymeric carbohydrates [3]. In particular, cellulose is a linear homopolymer of β-D-glucose linked by β-1-4

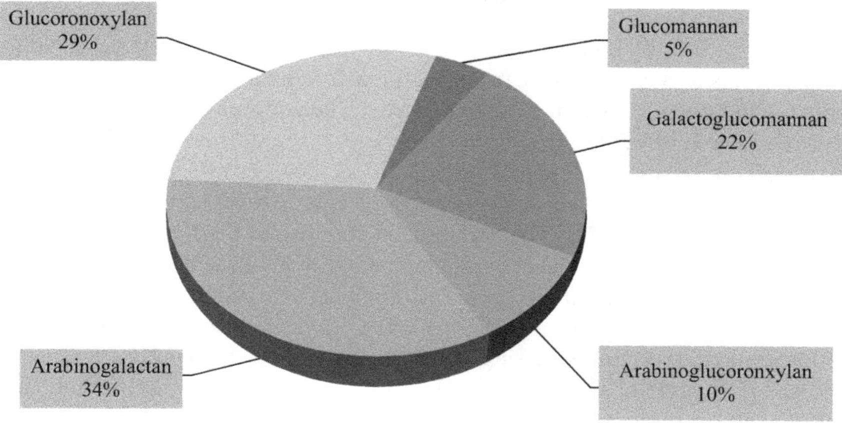

Figure 3.1   Average composition of hemicellulose (wt.%) [4]

glycosidic linkages with high polymerization degree (>10,000). Hemicellulose consists of heteropolysaccharides of glucose, galactose, mannose, xylose, glucuronic acid and arabinose with shorter polymerization degree than cellulose (150–250). Figure 3.1 shows the average composition of the major components of hemicellulose in plants: Mannans, mainly present in softwoods, and xylans, typical of hardwoods [4].

In supercritical water conditions, dissolved cellulose and hemicellulose are breakdown to simple C_5 and C_6 sugars, whereas lignin component is dissociated to phenolic monomers and oligomers compounds (i.e. guaiacols and syringols). Transition metal-based heterogeneous catalysts (i.e. Ni, Pt, Pd, Ru and Rh) deposited on different supports inorganic supports (i.e. AC, metal oxides) have been widely used in SCWG of lignocellulosic components (e.g. lignin, cellulose, hemicellulose) because of their high activity. The ideal catalyst for SCWG of lignocellulosic feedstocks should exhibit high reactivity for the cleavage of C–C bonds together with low activity towards C–O bond cleavage, so as to limit the methane formation. Moreover, a high efficiency in WGS reaction (decreasing CO selectivity) and a reasonable hydro-thermal stability are required. Among the noble metals, Ru-based supported catalysts were found to be very effective for SCWG [5–8]. Elliott *et al.* [9] studied the gasification of p-cresol, as a model compound of lignin, at 400 °C and reported that Ru, Rh and Ni showed higher catalytic performances than those for Pt and Pd metals based on the weight of catalyst. Sato *et al.* [10] reported the activity of the catalyst in the order Ru > Rh > Pt > Pd for the gasification of alkylphenols as model compounds of lignin. Recently, Yamaguchi *et al.* reported lignin gasification behaviour over various metal salts supported on charcoal in supercritical water at 400 °C. The order of activity for the gasification among the metal catalysts was following: Ru > Rh > Pt > Pd ≈ Ni, confirming the high activity of Ru-based catalysts for the SCWG of lignin [7]. Osada *et al.* [11]

showed the efficiency of Ru/TiO_2 catalyst for decomposition of lignocellulosic components for SCWG. Organosolv lignin and cellulose were gasified at 400 °C with gas yields of 30% and 70%, respectively, affording in both cases CH_4 as main product without the formation of solid residues. The addition of sulphur to the feed has a detrimental effect to the gas yield because of the poisoning of Ru/TiO_2 catalyst, thereby reducing the H_2 yield [12,13]. In the presence of NaOH or a Ni/Al_2O_3 catalyst, the lignin and cellulose gasification efficiency resulted lower with respect to that obtained by Ru/TiO_2. The activity and stability of the Ru/TiO_2 catalyst for lignin SCWG was also evaluated and compared with Ru-based catalysts supported on γ-Al_2O_3 and activated carbon [14]. The initial activity of the catalysts was in the order of $Ru/TiO_2 > Ru/\gamma$-$Al_2O_3 > Ru/AC$. The Ru/TiO_2 catalyst maintained high gasification activities for three subsequent uses, whereas Ru/γ-Al_2O_3 and Ru/AC decreased their activity after repetitive use. Onwudili *et al.* reported and enhancement of H_2 selectivity over CH_4 when CaO was added to Ru supported on α-Al_2O_3 spheres for cellulose SCWG at 550 °C and 36 MPa [15]. Yamaguchi *et al.* [16] reported the lignin gasification over unreduced Ru trivalent-salts (ruthenium(III) chloride or ruthenium(III) nitrosyl nitrate) supported on titanium oxide and charcoal at 400 °C. The order of gasification activity was $Ru/C \approx Ru(NO)(NO_3)_3/C \approx Ru(NO)(NO_3)_3/TiO_2 > RuCl_3/C \approx RuCl_3/TiO_2$. The trend in the catalytic behaviour was justified by the formation of small Ru particles by reduction of the nitrosyl nitrate salt during the lignin gasification when compared with those observed for the chloride salt. Ni-based catalysts (also in the form of Ni Raney nickel) have been widely studied in SCWG due to their comparable activity to that of noble metal catalysts (e.g. Ru and Rh), and their relatively low-cost [7]. Yoshida *et al.* [17] studied the gasification of cellulose/lignin sulfonate mixtures with a commercially available nickel catalyst (Ni-5132P, Engelhardt) in supercritical water at 400 °C and 25 MPa. A strong deactivation of the catalyst was observed; however, the magnitude of the negative effect decreased at higher catalyst/feedstock ratios. The authors proposed a possible role of tarry products from the reaction between cellulose and lignin for the observed deactivation. Azadi *et al.* [18] studied the catalytic activity and hydrogen selectivity of different Ni-based catalysts (i.e. Ni/α-Al_2O_3, Ni/hydrotalcite, Raney nickel) in SCWG of different lignocellulosic biomass at 380 °C. It was found that Ni/α-Al_2O_3 and Ni/hydrotalcite did not improve significantly the H_2 selectivity with respect to Raney nickel catalyst. The effect of the support on the gasification performances of Ni-based catalysts was also reported by Minowa and Ogi [19]. Reduced Ni catalysts supported on Al_2O_3, SiO_2–Al_2O_3, aluminium silicate, SiO_2, kieselguhr and MgO were studied for cellulose SCWG (200–350 °C, 8–22 MPa). The catalysts with different supports had different activity, and the activity depended not only on the kind of support materials, but also on the overall catalyst size. MgO-supported catalyst showed the highest gas yield, but no effect on the H_2 selectivity was registered. MgO-supported Ni catalysts were also studied by Sato *et al.* [20] for the gasification of lignin in SCW (250–400 °C). MgO promoted the decomposition of lignin to reactive intermediates, whereas Ni-catalysed reaction between intermediates and water to form gases. However, the stability of magnesium-supported nickel catalyst needs to be improved. Furusawa *et al.* [21] evidencing the presence of Ni particles and NiO–MgO phase after metal reduction by hydrogen and further investigated the structural features Ni/MgO catalyst prepared by impregnation method. The catalytic data for SCWG of lignin showed an optimal Ni metal particle size for this reaction. Recently, Ruppert *et al.* [22] explored the potential of Ni/ZrO_2 catalysts for SCWG of cellulose. Several factors, such as the crystalline phase type of the support, the NiO particle size, as well as the metal–support interaction and the surface area stability, were found to exert a significant impact on the catalytic activity. The highest hydrogen yield was obtained with catalysts containing tetragonal zirconia and small NiO crystallites. Although supported reduced transition metals are usually used as catalysts for lignocellulosic biomass gasification, few examples of metal oxides have been employed as catalysts for the SCWG of lignin and/or cellulose. Park and Tomiyasu [23] reported for the first time the use of RuO_2 catalyst for the SCWG of cellulose at 450 °C and around 44 MPa; the main components of the product gas were H_2, CH_4 and CO_2. Recently, Yamamura *et al.* showed the high activity of RuO_2 catalyst for the nearly quantitative SCWG of cellulose (400 °C at 30 MPa and 500 °C at 50 MPa), which was higher of those achieved with other metal oxides (MoO_3, NiO and ZrO_2). Watanabe *et al.* [24] reported cellulose gasification with ZrO_2 catalyst (400–440 °C, 30–35 MPa) affording a hydrogen yield almost twice as much as observed without catalyst. Hao *et al.* [25] reported the SCWG of cellulose promoted by CeO_2, nano-CeO_2 and nano-$(CeZr)_xO_2$ (500 °C, 27 MPa). The results showed that the catalytic activities were nano-$(CeZr)_xO_2$ > nano-CeO_2 > CeO_2 particle. However, the metal oxides led to lower gas yield than activated carbon-supported noble metal catalysts (i.e. Ru/C and Pd/C). Yanik *et al.* [26] investigated the catalytic gasification (CG) of natural biomasses (lignocellulosic and proteinous materials) in supercritical water (500 °C). Besides K_2CO_3 and Raney-Ni, which are commonly used catalysts in SCWG, Trona (a natural mineral) and red mud (a by-product containing: Fe_2O_3, Al_2O_3, SiO_2, TiO_2, Na_2O and CaO) were also used as catalysts. The results showed that the catalysts enhanced the WGS reaction and formation of CH_4, likely due to the H_2 production rather than methanation, pointing out iron-based catalysts as active materials for SCWG of biomass. Recently, a method to obtain CO and H_2 by WGS reaction involving reacting biomass (lignin, lignocellulose, cellulose, hemicellulose or combination thereof) with a polyoxometalate catalyst (i.e. $H_5PV_2Mo_{10}O_{40}$) has been reported [27]. The process involves the catalytic formation of formaldehyde, formic acid and related hemiacetals/acetals in mild reaction conditions (<150 °C) that, in presence of an acid, result in the formation of CO. Hydrogen is further electrochemically generated by using a Pt electrode. Significantly, the polyoxometalate solution can be reused without need for any catalyst recovery procedures. Table 3.1 summarizes the metal-based catalysts used for H_2 production form lignocellulosic biomass, highlighting feedstock and operating parameters (e.g. temperature, feedstock ratio to metal, water density and reaction time), as well as the best performances in terms of H_2 production obtained.

Table 3.1 H_2 production by SCWG of lignocellulosic biomass with supported metal catalysts

Feedstock	Operating conditions[*]	Catalyst	Maximum H_2 yields	Reference
Lignin	400 °C; 37.1 MPa; 7.5 wt.%; 0.5 g cm^{-3}; 1 h	Ru/C, Rh/C, Pt/C, Pd/C, Ni/C	Gas yield: 18.7%; H_2 sel.: 30.6% (RuCl$_3$/C)	[10]
Lignin, cellulose	400 °C ; 6 wt.% (Ru); 57 wt.% (Ni); 0.33 g cm^{-3}; 15 min	Ni/Al$_2$O$_3$ Ru/TiO$_2$	Lignin: gas yield: 31.1%; H_2 sel.: 14% (Ru/TiO$_2$) Cell.: gas yield: 74.4%; H_2 sel.: 9% (Ru/TiO$_2$)	[11]
Lignin	400 °C; 37.1 MPa; 1.3 wt.% (Ru/TiO$_2$) 3.3 wt.% (Ru/C, Ru/γ-Al$_2$O$_3$) 0.5 g cm^{-3}; 180 min	Ru/TiO$_2$ Ru/γ-Al$_2$O$_3$ Ru/C	Gas yield: 100%; H_2 sel.: ca. 5% (Ru/TiO$_2$)	[14]
Cellulose	550 °C; 37 MPa; 5 wt.%; 10 min	Ru/Al$_2$O$_3$ (CaO)	Gas yield: 17%; H_2 sel.: 46%	[15]
Lignin	400 °C; 22.1 MPa; 7.5 wt.%; 0.5 g cm^{-3}; 1 h	Ru salts/TiO$_2$ Ru salts/C	Gas yield: 20.2%; H_2 sel.: ca. 19.7% (RuCl$_3$/TiO$_2$) Gas yield: 61.8%; H_2 sel.: ca. 6.2% (Ru(NO)(NO$_3$)$_3$/TiO$_2$)	[16]
Cellulose, lignin and cellulose–lignin mixture	400 °C; 25 MPa; 0.08 g$_{cat}$; 0.166 g cm^{-3}; 20 min	Ni-5132P (Engelhardt)	Gas yield: 58%; H_2 yield: 54% (Cell/lignin 3:1)	[17]
Lignin and cellulose	380 °C; 23 MPa; 60 wt.% (Ni) 3 wt.% (Ru); 15 min	Ni Raney Ni/α-Al$_2$O$_3$ Ni/HT Ru/AC Ru/γ-Al$_2$O$_3$	Ratio of the produced H_2 gas to the amount of hydrogen available in the feed: Lignin: 1.43 (Ni/HT); 0.22 (Ru/C) Cellulose: 1.63 (Ni/HT); 1.04 (Ru/γ-Al$_2$O$_3$)	[18]
Cellulose	350 °C; 22 MPa; 20 wt.%; 0.234 g cm^{-3}; 1 h	Ni supported on different supports	Gas yield: 90%; H_2 sel.: ca. 23% (Ni/MgO) [19]	[19]
Lignin	450 °C; 44 MPa; 5–20 wt.%, 0.3 g cm^{-3}; 2 h	Ni/MgO	H_2 yield: 11.7% (Ni/MgO, 10 wt.%)	[20]
Lignin	400 °C; 37 MPa; 10 wt.%; 0.3 g cm^{-3}; 2 h	Ni/MgO	Gas yield: ca. 29%; H_2 sel.: ca. 28%	[21]
Cellulose	700 °C; 0.8 wt.%; 4 h	Ni/ZrO$_2$	H_2 yield: 13 mmol g^{-1} (cellulose)	[22]

[*]Operating conditions refer to temperature (°C) and pressure (MPa), amount of feedstock/metal wt.%, water density (g cm^{-3}) and reaction time.

3.2.2   Algae

Microalgae are a promising renewable feedstock to convert atmospheric CO_2 into potential biofuels and chemicals because of their high photosynthetic efficiency, as well as fast growth rates (10–30 g dry cell m^{-2} d^{-1}) [28]. SCWG of microalgal biomass provides an interesting way to convert the wet biomass to a fuel-rich gas containing H_2 and/or CH_4. Among the conventional thermochemical methods (such as incineration, pyrolysis, gasification), this process does not require high energy to dry up the biomass, and it is well suited for processing aquatic biomass. Supported Ni and Ru-based have been investigated as effective catalysts for SCWG of algae. Minowa *et al.*, for the first time, reported SCWG hydrothermal gasification of a freshwater microalga (*Chlorella vulgaris*) using a commercially available silica-alumina Ni catalyst under mild conditions (350 °C, 18 MPa) [29]. Carbon conversion ranging from 35% to 70% was obtained in three consecutive experimental runs, with a maximum H_2 selectivity of 35% in the first run which rapidly decreased in the further runs. Moreover, a significant amount of nickel was leached from the support, which could have compromised the suitability of the recycled streams, thus inhibiting microalgae growth [30]. Stucki *et al.* showed that the complete gasification of cyanobacterium microalgae (*Spirulina platensis*) to methane-rich gas products is now possible in supercritical water using supported Ru/C and Ru/ZrO_2 catalysts. At 400 °C and 30 MPa feed carbon gasification greater than 50% can only be achieved with high catalyst loadings. In these conditions, methane-rich gas products were obtained (>40%) with a maximum H_2 concentration of ca. 18%. Guan *et al.* reported the SCWG of a marine microalga (*Nannochloropsis* sp.), with a Ru/C catalyst at 410 °C. Longer reaction times, higher catalyst loadings and water densities, as well as lower algae loadings, provided higher gas yields. The catalyst loading had the most significant impact on both the yield and composition of the gaseous products. Significantly, the yield of H_2 was very sensitive to the algae loading; a reduction of about a factor of 4 as the algae loading increased from 1.8 (H_2 yield: ca. 11.5 mmol g^{-1}) to 13.5 wt.% was observed. Moreover, a significant decrease of the activity was observed after the first catalytic run. Onwudili *et al.* [31] reported SCWG of different types of algal biomass (*C. vulgaris, S. platensis* and *Saccharina latissima*) at 500 °C, 36 MPa with NaOH and/or Ni/Al_2O_3. The maximum H_2 yield was 15.1 mol kg^{-1} (*Saccharina*), in the presence of NaOH alone, whereas yield decreased slightly (14.2 mol kg^{-1}) when both NaOH and Ni/Al_2O_3 were used. Elliott *et al.* [32] reported continuous-flow process of several wet algae feedstock in a bench-scale reactor operating nominally at 350 °C and 21 MPa with a Ru catalyst supported on partially graphitized carbon extrudate. High conversions were obtained even with high slurry concentrations of *Spirulina* strains, but less positive results were achieved with the other strains. The product gas had high CH_4 content, whereas H_2 was present in very low amount (<2%). On the contrary, the experimental data clearly showed evidence of catalyst deactivation. Duman *et al.* [33] studied steam gasification of different algal biomass in a dual-bed microreactor in a two-stage process.

Gasification of tar derived from algae pyrolysis was performed with Fe_2O_3–CeO_2 and red mud catalysts obtaining an efficiency ranging from 53% to 70%. It was observed that the characteristic of algae gasification was dependent on its components and the catalysts used. The maximum hydrogen yields obtained were 1,036 mL g_{algae}^{-1} for *Fucus serratus*. The widespread presence of sulphur-based compounds in algae elements, although in less amount in the terrestrial biomass, often lead to the poisoning of metal-based catalysts during SCWG. In this respect, the development of effective catalysts with long life represents a new challenge in engineering and catalyst design for SCWG of algal biomass.

3.3   Secondary raw materials

Biomass is the ideal candidate to substitute fossil raw materials, because it is renewable and CO_2-neutral. However, one of the major problems associated with the direct use of raw biomass for hydrogen production is its very variable and complicated composition, along with very low-energy density, which makes its transportation costly [34]. Thus, the conversion of biomass into more simply and easier-to-process compounds is fundamental. Bio-ethanol, bio-methanol and bio-oil with all its components are generally considered to be the most promising secondary raw materials, since they can be produced from many different types of biomass, ranging from agricultural residues to municipal waste, by very well-known methods, such as fast pyrolysis, fermentation and gasification, because they possess an higher energy density, if compared to raw biomass [35–37]. The final production of hydrogen is generally achieved by catalytic reforming processes, such as, among others, steam reforming (SR) and aqueous phase reforming (APR) [38–40]. The main difference between these two processes basically relies on whether the water is fed in the vapour phase or liquid phase. Of course, in the last case, test units able to withstand pressures as high as 30/60 bar must be used. SR (see (3.1)) is the most studied reaction for the conversion of secondary raw materials, mainly due to the large industrial background and good hydrogen production rates achievable. One of the major drawbacks of SR is the need to vaporize both the water and the hydrocarbons, which could trigger undesired polymerization and/or decomposition reactions of heavy compounds, such as sugars or lignin-derived molecules, eventually even leading to a complete clogging of the test unit. On the other hand, APR offers the possibility to also reform heavy hydrocarbons, such as the sugars present in the bio-oil, without the need to vaporize the reactant mixture, thus preventing any undesired side-reactions [41]. Moreover, APR occurs at temperature in which the WGS reaction (see (3.2)) is favourable. However, lower hydrogen production rates with respect to SR are generally achieved. High CO_2 selectivity and long-term stability are the most important challenges to be overcome for a large-scale hydrogen production from secondary raw materials, especially from high molecular weight hydrocarbons. More efficient processes and, in particular, more active and stable catalysts are thus fundamental if we want to switch to more sustainable hydrogen production processes.

$$C_nH_mO_k + (n - k)H_2O \rightarrow nCO + (n + m/2 - k)H_2 \quad \text{SR} \tag{3.1}$$

$$CO + H_2O \rightarrow CO_2 + H_2 \quad\quad\quad \text{WGS} \tag{3.2}$$

3.3.1   Bio-oil

One of the most convenient and simple way of processing biomass is the fast pyrolysis, which basically consists in the degradation of biomass at around 500 °C in absence of oxygen with high heating rate and low residence time in the reactor (1–2 s), to yield a liquid fuel (hereafter bio-oil), as well as solid (bio-char) and non-condensable gases [42]. Bio-oil is a dark to brown viscous liquid, with a composition highly dependent on the biomass source (Figure 3.2) [43]. Nonetheless, the main components are water, carboxylic acids, phenols, acetone, ethylene glycol (EG), sugars and unreacted lignin [44]. Promising results have been obtained in the past years in the study of the reforming of single model molecules, such as acetic acid and EG [45]. On the other hand, the direct use of crude bio-oil has been only recently explored, mostly because, although stable at room temperature, it polymerizes upon heating even at moderate temperature (80 °C) and decomposes at higher temperature [46].

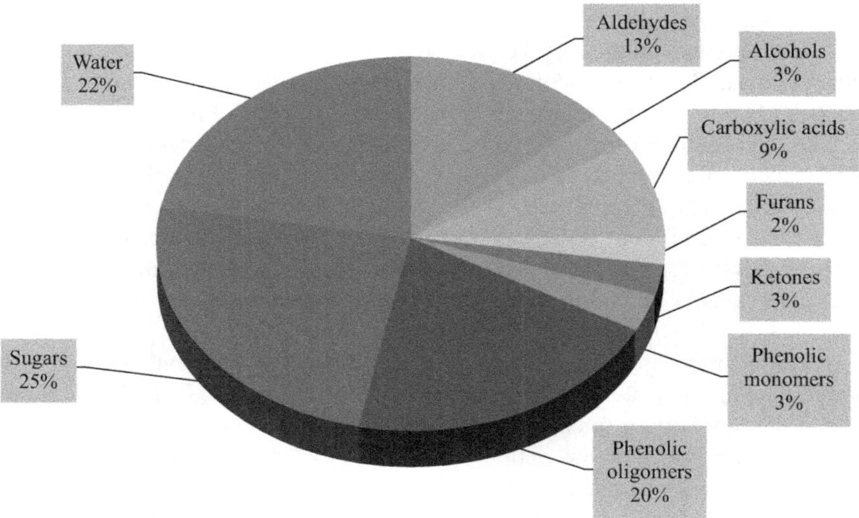

Figure 3.2   Average composition of crude bio-oil (wt.%) [43]

Table 3.2   Catalysts for SR of crude bio-oil

Type of oil	Catalyst	$T_R(°C)$	GHSV (h^{-1})	S/C	H_2 yield (%)	Stability (h)	Reference
Saw dust	Ni/CNT	350–550	12,000	2–6.1	92.5	6	[53]
Beech wood	Ru/MgO/ Al₂O₃	550–800	3,000–17,000	7.2	60	>45	[56]
Rice hull	Z417/CaO	600	–	–	85	2	[57]

3.3.1.1   Crude bio-oil

As previously mentioned, there are few works about the direct reforming of crude bio-oil or its fractions, because of the general poor stability of the catalysts at long time-on-stream. Besides working on different reactor schemes and adding pre-reformers [37,47], as well as hydrogenation reactors [48], most of the research effort has been focused on the development of stable catalytic materials, able to inhibit coke formation and/or sintering of the active metallic phase. The metals used as catalysts in SR of whole bio-oil so far investigated are Pt, Rh and Ni, with a major focus on the last one [49–55]. For example, Hou *et al.* proposed a carbon nanotube-supported Ni catalyst prepared by homogeneous deposition–precipitation (DP) method, which showed good performances at low temperature (550 °C), reaching 92.5% of hydrogen yield. Such a good activity was mainly ascribed to the narrow and uniform metal nanoparticles distribution [53]. Among the catalysts prepared with noble metals, is worth a mention the Ru/MgO/Al₂O₃ catalyst prepared by wet impregnation by Basagiannis and Verykios [56], which showed a remarkable stability (up to 45 h t.o.s.) at high gas hourly space velocity (GHSV) $(3,000–17,000\ h^{-1})$ in the reforming of the aqueous fraction of the bio-oil. Sorbents for CO_2, like dolomite or CaO, have been explored for enhancing the catalytic performances of a commercial Z417 catalyst by Yan *et al.* A remarkable hydrogen yield of 85% was obtained with CaO, although deactivation occurred very fast with a total loss of sorption power (ca. 2 h) [57]. Table 3.2 summarizes the discussed catalysts for SR, whereas for a more complete list of all the catalytic materials available in the open literature see Chattanathan *et al.* [37] and Trane *et al.* [44].

APR of crude bio-oil or its fractions is generally carried out with Pt-based catalyst supported on alumina. For instance, Pan *et al.* [58] used a Pt/Al₂O₃ for the APR of the low-boiling fraction of bio-oil with good results in terms of hydrogen production (65 vol.% of the outlet gas). Chen *et al.* had investigated the same catalyst and found that such good performances were due to very small particles (around 1.6 nm). The size sensitivity was ascribed to a high number of atoms at the edges of the nanoparticles, believed to be the main responsible of the C–C cleavage activity [59].

3.3.1.2   Bio-oil fractions

Due to the complexity and very variable composition of bio-oil, most of the works in the open literature focus their attention on the selection of model molecules, as representative as possible of the whole matrix [39]. Besides water, the major moieties of the molecules in the bio-oil are carboxylic acids, ketones and hydroxyls groups, thus making acetic acid, acetone and EG the most studied model molecules. Only recently and in very few papers, sugars, such as xylitol and sorbitol, have been used as feedstock in APR reactions [41,60,61]. SR is the most investigated process when low boiling point molecules are used, such as acetic acid and acetone, whereas for sugars and polyols, which may undergo degradation or polymerization at high temperature, APR is preferred. In the following paragraphs, recent advances in the development of active catalyst for hydrogen production from acetic acid, acetone and EG are presented and discussed.

Acetic acid

Acetic acid is the most abundant carboxylic acid in the aqueous fraction of the bio-oil (up to 12%) [43] and can be considered as a representative model compound for all the carboxylic acid groups. In (3.3), is represented the overall equation for steam and APR of acetic acid. The positive reaction enthalpy value indicates that this reaction is endothermic and favoured at high temperature [62,63].

$$CH_3COOH + 2H_2O \rightarrow 2CO_2 + 4H_2 \quad \Delta H = +32.21 \text{ kJ mol}^{-1} \qquad (3.3)$$

Several studies have been reported for acetic acid SR, and the main conclusion point is that the high tendency to decomposition forming carbonaceous deposits, which lead to high coke deposition rates, is the limiting factor preventing a large-scale application, especially in fixed bed reactors [64]. Noble metals (Pt, Rh and Ru) have demonstrated to be the most suitable ones, even if, due to their high cost, other metals and/or promoters are being studied (e.g. Ni, Fe, Co, La_2O_3 and CeO_3), always with the aim of minimizing the coke deposition rates without compromising the overall activity [62,63,65–73]. Despite being very active and selective towards H_2, even noble metal-based catalysts suffer rather quick loss of activity during SR of acetic acid. For example, the Pt/ZrO_2 catalyst proposed by Takanabe *et al.* [72] was stable for only 3 h. Besides good hydrogen production rate (turnover frequency (TOF) of 10 s^{-1}), the Rh-based catalyst presented by Lemonidou *et al.*, prepared by wet impregnation method on a La_2O_3-modified CeO_2–ZrO_2 support, showed remarkably enhanced stability up to 15 h t.o.s., with only 8% loss of activity. Such an improvement was mainly ascribed to the low metal nanoparticles sintering, thanks to La_2O_3, and to increased surface oxygen availability, due to fast CH_x intermediates swift gasification [62,65]. One of the most active metals in reforming acetic acid is Ru [73]. Recently, Bossola *et al.* have prepared a Ru-based catalyst for SR supported on MgAl(O) mixed oxide via a novel colloidal procedure, obtaining well formed, crystalline metal nanoparticle with a very sharp metal nanoparticles distribution (1.6 nm). The minor presence of defective sites on the metal nanoparticles, along with strong metal–support interaction, resulted in a low coke deposition rate, which allowed the catalyst to steadily work at 100% conversion for up to 20 h t.o.s., without any appreciable loss of activity [66]. Significantly, small Ru nanoparticles supported on TiO_2 were also found to be active in APR performed in a batch reactor, likely due to the more efficient activation of water and acetic acid [73]. Noteworthy, de Vlieger *et al.* [69] proposed a Ru catalyst supported on carbon nanotube, which showed promising performances in the APR reaction, achieving near complete acetic acid reforming under commercially relevant conditions (400 °C, 250 bar, 7 h t.o.s.). The addition of second metals has demonstrated to be useful for enhancing the stability, as shown by Zhang *et al.*, [74] in which the addition of Co to a Ni/MgO catalyst improved the stability up to 20 t.o.s. with 100% acetic acid conversion. Alkali metals have shown significant properties both reducing the coke deposition rate (due to the inhibition of the acid catalysed condensations of the acetic acid intermediates) and improving the reduction of the metallic phase, typically Ni. Iwasa *et al.* [75] studies revealed that 10 wt.% of K improved both conversion and hydrogen yield in Ni incorporated in smectite-type material catalyst (K-SM(Ni)). Similar results were obtained by Wang *et al.*, [68] in which a Ni-based catalyst supported on coal ash (ZhunDong Ash, ZDA), which is rich in Fe and alkali metals, achieved 100% and 89.6% of conversion and hydrogen yield, respectively. An interesting synthetic approach for the preparation of a Ni-based catalyst was proposed by Resende *et al.*, [76] in which a perovskite-like precursor ($LaNiO_3$), synthesized by precipitation method, was thermally reduced, giving rise to Ni^0 nanoparticles. The so-prepared catalyst showed good activity up to 24 h t.o.s. in the oxidative SR (OSR), due to the

Table 3.3   Catalysts for acetic acid reforming

Catalyst	Reaction	T_R (°C)	S/C	Space velocity	Performances	Stability (h)	Reference
Pt/ZrO_2	SR	600	5	GHSV: 40,000 h^{-1}	Conversion: 100% H$_2$ yield: 75%	3	[72]
$Rh/La_2O_3/$ CeO_2–ZrO_2	SR	650	3	GHSV: 28,000 h^{-1}	Conversion: 95% H$_2$ sel.: 100%	15	[65]
Ru/MgAl(O)	SR	700	3	WHSV: 6 h^{-1}	Conversion: 100% H$_2$ yield: 73%	20	[66]
Ru/CNT	APR	400	10	WHSV: 448 h^{-1}	Conversion: 95% H$_2$ sel.: 19%	7	[69]
Ni–Co/MgO	SR	600	4	GHSV: 8,600 h^{-1}	Conversion: 100% H$_2$ sel.: 80%	20	[74]
K– SM(Ni)	SR	450	3.3	120 cm^3 min^{-1}*	Conversion: 90% H$_2$ sel.: 80%	5	[75]
Ni–Fe/ZDA	SR	700	9.2	WHSV: 4 h^{-1}	Conversion: 100% H$_2$ yield: 89.6%	11	[68]
$LaNiO_3$	OSR	600	6	400 mL min^{-1}*	Conversion: 60% H$_2$: 50%**	24	[76]

GHSV = gas hourly space velocity (h^{-1}); WHSV = weight hourly space velocity (h^{-1}).
*Feed flow rate.
**Percentage of hydrogen in the gaseous products.

participation of the support in the activation of the acetic acid. Table 3.3 summarizes the catalytic performances of the catalysts presented in this paragraph.

Acetone
Although recognized to be one of the most important reaction intermediates in the SR of acetic acid [77], as well as it is representative of bonds that are characteristic of most of the biomass-derived oxygenates, specifically the CH_3 and the C–C bonds, there are very few works in the open literature about the reforming of acetone (see (3.4)). Like acetic acid, acetone reforming is an endothermic reaction, therefore, is favoured at high temperatures (>450 °C), and coke formation is the main issue to be addressed before reaching long-term stability [78].

$$CH_3COCH_3 + 5H_2O \rightarrow 3CO_2 + 8H_2 \quad \Delta H = +58.62 \text{ kJ mol}^{-1} \qquad (3.4)$$

Besides noble metals, Ni-based catalysts are extensively studied because of the high C–C bond cleavage activity and relatively low cost. Navarro *et al.* prepared Ni catalyst supported on La-modified Al_2O_3 with the addition of a second metal (Cu and Pt), in the attempt of promoting metal nanoparticle stability and hindering coke deposition [79]. They found that the addition of small amounts of Pt resulted in lower coke deposition rates and improved thermal stability, probably due to the better dispersion of the metallic phase together with a better H-atom mobility on the PtNi alloyed nanoparticles. Conversion and H_2 selectivity were 100% and 55%, respectively. On the other hand, the CuNi catalyst showed slightly better selectivity towards hydrogen (60%) if compared to the Ni catalyst, but a much lower conversion to gaseous products (56.5%) and high coke deposition rates. This may be ascribed to the presence of very small metallic Cu nanoparticles, which have been demonstrated to have an acidic behaviour [80,81]. In order to further improve catalyst stability and H_2 selectivity, Sun *et al.* focused their efforts in the minimization of coke formation and methane selectivity, respectively. To do so, they developed a Co catalyst supported on graphitized activated carbon (g-AC) and studied through density functional theory (DFT) calculations the role of the metal in the reaction. The Co/a-AC catalyst, prepared by incipient wetness impregnation, exhibited exceptional stability, up to 70 h t.o.s., with a conversion close to 100% and H_2 selectivity always above 80%. Such performances were mostly ascribed to enhanced oxygen mobility on Co nanoparticles and lack of acid/base sites on the inert support. Moreover, the theoretical calculation suggested that the more facile scission of the C–H bonds compared to C–C bonds on the Co nanoparticles suppressed the methane formation [78]. Table 3.4 summarizes the catalytic performances of the catalyst discussed in the paragraph.

Ethylene glycol
EG is currently attracting attention because it can be considered the model molecule for many of the polyols present in the bio-oil or in other biomass-derived feedstock [43]. The most harnessed way to produce hydrogen from EG is via APR, followed by conventional SR (the overall reaction is reported in (3.5)) [82]. Supercritical water reforming (SWR) has been only recently proposed

Table 3.4   Catalysts for SR of acetone

Catalyst	T_R (°C)	S/C	Space velocity	Performances	Stability (h)	Reference
PtNi/Al$_2$O$_3$	600	6	GHSV: 10,180 h^{-1}	Conversion: 100% H$_2$ sel.: 55%	17	[79]
CuNi/Al$_2$O$_3$	600	6	GHSV: 10,180 h^{-1}	Conversion: 56.5% H$_2$ sel.: 60%	17	[79]
Co/g-AC	450	5	Contact time: 0.094 g s mL^{-1}	Conversion: 100% H$_2$ sel.: >80%	70	[78]

GHSV = gas hourly space velocity (h^{-1}); LHSV = liquid hourly space velocity (h^{-1}); contact time = weight of the catalyst/flow rate of the gas.

with significant results [83]. As per many other reforming reactions, the most studied metals are Pt, Pd, Ni, Co and Fe. Only very recently Rh has been tested [84–89].

$$C_2H_6O_2 + 5H_2O \rightarrow 2CO_2 + 5H_2 \quad \Delta H = +91 \text{ kJ mol}^{-1} \tag{3.5}$$

In the APR reaction, maximum hydrogen selectivity (100%) was achieved by Wang *et al.* [88] with a Pt–Co catalyst supported on single-walled carbon nanotubes, prepared by incipient wetness impregnation method on HNO$_3$-activated nanotubes. Remarkable hydrogen production was obtained by Chu *et al.* with a Co/ZnO catalyst by co-precipitation method, which exhibited a H$_2$ TOF of 101.4 min^{-1}, among the highest values found in the open literature. Carbon monoxide was not detected. The catalytic tests were carried out at 225 °C, weight hourly space velocity of 0.59 h^{-1} and 8 h t.o.s. [89]. Another catalyst which showed excellent activity in the APR reaction was Pd/Fe$_2$O$_3$, with H$_2$ TOF of 101.1 min^{-1} and a conversion of 99.6%. Such good performances were ascribed to the ability of Fe$_2$O$_3$ to promote the WGS reaction, which is considered to be the rate-limiting step [86]. Mn was used as a promoter in a Pt-based catalyst supported on an ordered mesoporous carbon (CMK-3), achieving almost the double of the conversion compared to the unpromoted catalyst (39.7%), and higher hydrogen selectivity (40.2%) [89]. As far as concerns the SR of EG, an effective study has been done by Tupy *et al.* [85], in which they found that a PtNi bimetallic catalyst supported on carbon resulted to be more active and selective towards hydrogen if compared with the same catalysts prepared on Al$_2$O$_3$ and TiO$_2$. A similar bimetallic catalyst, namely PtNi/Al$_2$O$_3$, prepared by wet co-impregnation method, was tested in SWR (450 °C and 250 bar). Considering the high space velocity, both H$_2$ selectivity and conversion resulted to be good. This was probably due to the presence of stable PtNi bimetallic nanoparticles, which suppressed the methane formation [83]. Table 3.5 reports the discussed catalyst, along with their catalytic performances.

Table 3.5   Catalysts for ethylene glycol reforming

Catalyst	Reaction	T_R (°C)	S/C	Space velocity	Performances	Stability (h)	Reference
PtCo/ SWCNT	APR	225	15	60 µL min^{-1}*	H$_2$ sel.: 100% H$_2$TOF: 2.35 min^{-1}	–	[88]
Co/ZnO	APR	225	30	WHSV: 0.59 h^{-1}	Conversion: 5.2% H$_2$ TOF: 101.4 min^{-1}	8	[87]
Pd/Fe$_2$O$_3$	APR	225	30	WHSV: 3.6 h^{-1}	Conversion: 99.6% H$_2$ TOF: 101.1 min^{-1}	6	[86]
Pt – Mn/ CMK-3	APR	250	15	WHSV: 2 h^{-1}	Conversion: 39.7% H$_2$ sel.: 40.2%	–	[89]
PtNi/C	SR	230	4	200 cm^3 min^{-1}*	Conversion: 15% H$_2$ sel.: 70%	>20	[85]
PtNi/Al$_2$O$_3$	SWR	450	30	WHSV: 17.8 h^{-1}	Conversion: 42% H$_2$ sel.: 80%	–	[83]

GHSV = gas hourly space velocity (h^{-1}); SWCNT = single walled carbon nanotubes; WHSV = weight hourly space velocity (h^{-1}).
*Feed flow rate.

3.3.2   Bio-methanol

There are many reasons why bio-methanol is considered to be among the most important secondary raw materials for the future. Besides being an H$_2$ carrier, and as such produced by CO$_2$ reduction [90], bio-methanol can be easily produced from many types of biomass by very well-known technologies [91]. Moreover, it is liquid at room temperature and possesses a high H/C ratio; hence, it has a low propensity in soot formation [92]. Many technologies have been investigated for the production of H$_2$, namely methanol decomposition [93], partial oxidation [94], SR [95–97] and OSR [98,99]. For the sake of brevity, and because of the higher number of works in the open literature published in the last years, in this paragraph only catalysts for SR and OSR will be discussed. For a more exhaustive list of all the catalysts so far prepared and tested, see Yong *et al.* [92] and Davidson *et al.* [95]. In (3.6) and (3.7) are reported the overall reactions for the SR and OSR of methanol, respectively [92].

$$CH_3OH + H_2O \rightarrow CO_2 + 3H_2 \quad \Delta H = +49.4 \text{ kJ mol}^{-1} \tag{3.6}$$

$$CH_3OH + (1-n)H_2O + 0.5nO_2 \rightarrow CO_2 + (3-n)H_2$$

$$\Delta H = +49.4 - 242n \text{ kJ mol}^{-1} \tag{3.7}$$

Cu/ZnO and Cu/ZnO/Al$_2$O$_3$ are among the first and most studied catalyst for SR of methanol [97,100]. Recently, Sanches *et al.* have investigated the impact of the preparation procedure on the catalytic performances of Cu/ZnO/Al$_2$O$_3$ catalysts, as well as the effect of the addition of small amounts of promoters, such as Zr and Y. The sample prepared by co-precipitation method with Zr displayed the best catalytic performances, both in terms of conversion and H$_2$ yield. The addition of Zr promoted the formation of CuZn alloy, which is considered to be responsible for the high catalytic activity [101]. High conversion was very recently achieved by Wang *et al.* [102] with a Cu/ZnO/Al$_2$O$_3$ catalyst coating generated by an innovative method named cold gas dynamic spray. Generally, Pd–Zn-based catalysts have attracted attention because of their high activity, low CO selectivity and high stability. A catalyst with good activity deserving a note is the one prepared by impregnation with a Pd acetate solution [103]. Extensive studies have been carried out in order to fully understand the actual reforming mechanism, especially about the role of the Pd–Zn alloy. It has been proposed that a synergistic effect between intermetallic PdZn and ZnO is fundamental for high activity of Pd–Zn-based catalysts [104]. OSR has been proposed in the last years as an effective way for H$_2$ production from methanol, because in this way lower coke deposition rates, as well as lower CO selectivity, are generally achieved. For example, Mierczynski *et al.* have proposed an Au–Ni catalyst supported on multi-walled carbon nanotubes, which even at relatively high temperature (300 °C) showed no CO production. The spillover effect between metallic gold and Ni oxide sites was ascribed as the main reason beyond such low CO selectivity [99]. Table 3.6 reports the discussed catalysts, along with their catalytic performances.

Table 3.6   Catalysts for bio-methanol reforming

Catalyst	Reaction	T_R (°C)	S/C	Space velocity	Performances	Stability (h)	Reference
Cu/ZrO$_2$/ZnO/ Al$_2$O$_3$	SR	250	3	90 mL min^{-1}	Conversion: 72.6% H$_2$ yield:84%	–	[101]
Cu/ZnO/Al$_2$O$_3$	SR	270	1.3	WHSV: 1.09 h^{-1}	Conversion: 90.45% H$_2$ yield:–	20	[102]
Pd/ZnO	SR	250	1.1	WHSV: 2.1 h^{-1}	Conversion: 95% H$_2$ yield:–	–	[103]
Au–Ni/MWCNT	OSR	300	1/1/0.4*	GHSV: 26,700 h^{-1}	Conversion: 99.8% CO sel.: 0%	–	[99]

GHSV = gas hourly space velocity (h^{-1}); WHSV = weight hourly space velocity (h^{-1}).
*Molar ratio between H$_2$O/Methanol/O$_2$.

3.3.3 Bio-ethanol

Compared to other secondary raw materials, ethanol has several advantages as feedstock for H_2 production, thanks to its low toxicity and safe storage. Moreover, new development on agricultural waste conversion technologies is making bio-ethanol one of the most important commodities for the future production of bio-fuels and bio-hydrogen [105]. However, due to the presence of a C–C bond, ethanol reforming is more difficult if compared, for example, to methanol as higher temperature is required (typically $> 450\ °C$) [95]. Generally, catalytic reforming of ethanol is carried out by SR (see (3.8)), because of the higher efficiency and higher H_2 production rates achievable if compared, for example, to auto-thermal reforming. So far, APR has been mainly carried out in batch reactors, which are not ideal for a real industrial H_2 production [106,107].

$$CH_3CH_2OH + 3H_2O \rightarrow 2CO_2 + 6H_2 \qquad \Delta H = +173.4\ kJ\ mol^{-1} \qquad (3.8)$$

Based on the composition, the catalysts for ethanol SR and APR are divided into two groups: noble metal catalysts (Pt, Pd, Rh, Ru and Ir), which are highly selective towards hydrogen, and non-noble metal catalysts (Ni and Co), which are attractive for their lower cost and low CH_4 selectivity. Supports with basic character are generally more suitable for long-term stability (such as MgO, ZnO, CeO_2, La_2O_3, hydrotalcites, and promoted Al_2O_3 and ZrO_2), because of their higher resistance to coke deposition [108,109]. Due to the enormous number of papers on the catalytic hydrogen production from ethanol, in this paragraph is provided only a very short review on some of the most relevant catalysts so far proposed, with a focus on the most recent ones, whereas for a more complete overview, see Davidson *et al.* [95], Contreras *et al.* [106] and Ni *et al.* [108]. Rh is a metal often chosen for the preparation of catalysts for reforming reaction, due to its high C–C bond cleavage activity [109]. Chen *et al.* [110] obtained with an iron-promoted Rh catalyst (Rh–Fe/Ca–Al_2O_3) a carbon monoxide-free hydrogen production at low temperature via SR, thanks to the presence of Fe_xO_y, which promoted the WGS reaction and improved the catalyst stability. Chiou *et al.* investigated the reaction pathways of ethanol SR using Pt, Ir and Co catalysts supported on CeO_2. Pt resulted to be the most active catalyst thanks to its superior C–C bond cleavage activity. However, such a good activity leads to high CO and CH_4 selectivities, which are undesired products [111]. As mentioned before, non-noble metal-based catalysts have lower selectivity towards CH_4. In a recent work, Han *et al.* [112] proposed a Ni–Al_2O_3–ZrO_2 catalyst synthesized by epoxide-driven sol–gel method, obtaining good results in the SR of ethanol for a Ni-based catalyst, in particular concerning coke resistance, which was mainly ascribed to the very high Ni surface available. A good stability, along with stable total conversion (350 h), was achieved by Shi *et al.* with a Fe-promoted Ni catalyst supported on $La_2O_2CO_3$, prepared by calcining $La_2(CO_3)_3$ [113]. Noteworthy, Banach *et al.* used a Co/ZnO–Al_2O_3 catalyst prepared by co-precipitation method for the SR of ethanol simulating a fermentation broth.

Table 3.7   Catalysts for SR of acetone

Catalyst	T_R (°C)	S/C	Space velocity	Performances	Stability (h)	Reference
Rh–Fe/ Ca–Al$_2$O$_3$	350	1.5	WHSV: 0.54 h^{-1}	Conversion: 100% H$_2$ yield: 68.3%	>250	[110]
Pt/CeO$_2$	400	12.3	WHSV: 1.34 h^{-1}	Conversion: 100% H$_2$ yield: 30.3%	–	[111]
Ni–Al$_2$O$_3$–ZrO$_2$	500	1/6/24.5*	23.14 mL h^{-1} g$_{cat}$$^{-1}$	Conversion: 100% H$_2$ sel.: 89%	15	[112]
Co/ZnO–Al$_2$O$_3$	420	42	100 cm^3 min^{-1}	Conversion: 100% H$_2$ sel.:95%	2	[114]
Ni–Fe/ La$_2$O$_2$CO$_3$	500	25	0.05 mL min^{-1}	Conversion: 100% H$_2$ sel.: 95%	350	[113]

WHSV = weight hourly space velocity (h^{-1}).
*Feed molar composition (EtOH/H$_2$O/N$_2$).

Although the catalyst poor stability, this catalyst revealed to be among the most promising Al$_2$O$_3$-supported catalysts, due to the total conversion and high H$_2$ selectivity achieved at relatively low temperature (420 °C) [114]. Table 3.7 reports the discussed catalysts for the SR of ethanol, along with their catalytic performances.

3.4   Waste

MSW, defined as the materials discarded in urban areas, includes predominantly kitchen garbage, paper, wood, textile, leather, plastics, glass, metals and garden waste. Consequently, MSW contains a high fraction of materials, which can be converted in fuels (solid, liquid and gaseous) as an alternative process to more common energy recovery by simple incineration (combustion). Very often MSW undergoes to mechanical sorting and processing aimed at separating its combustible fraction, giving as final product the so-called RDF that can be combusted or treated as MSW. With respect to traditional disposal technologies, such as composting, landfill and incineration, catalytic gasification or pyrolysis (CG or CP) is attractive since it can avoid some drawbacks of standard technologies and possessing some inherent advantages. Composting and landfill deplete land resources and result in the disposal of useful raw materials; incineration, besides allowing heat recovery, produces large volumes of flue gas and hazardous fly ash waste streams. Conversely, CG permits higher efficiency in energy production. In general, steam gasification is a combination of two steps: the first one is a thermochemical decomposition (i.e. non-catalytic pyrolysis) of MSW with production of tar, char and volatiles (in (3.9)). This step is usually performed at temperatures ranging from 300 to 700 °C, or even higher. The second step includes reactions of CO, CO_2, H_2 and H_2O with the hydrocarbon gases and carbon in MSW, thereby producing gaseous products. The catalytic steam gasification mechanism of MSW might be described by the following reactions shown in (3.10)–(3.15):

$$C_xH_yO_z \rightarrow CO_2 + H_2O + CH_4 + CO + H_2 + C_nH_m$$
$$+ \text{ tar} + \text{char} \qquad \text{MSW pyrolysis} \qquad (3.9)$$

$$C_xH_yO_z + H_2O \rightarrow CO_2 + H_2 \qquad \text{MSW steam reforming} \qquad (3.10)$$

$$H_2O + CO \rightarrow H_2 + CO_2 \qquad \text{WGS} \qquad (3.11)$$

$$C_nH_m + H_2O \rightarrow CO_2 + H_2 \qquad \text{Hydrocarbons SR} \qquad (3.12)$$

$$C + H_2O \rightarrow CO + H_2 \qquad \text{C gasification} \qquad (3.13)$$

$$C + CO_2 \rightarrow 2CO \qquad \text{Reverse Boudouard reaction} \qquad (3.14)$$

$$\text{Tar} + H_2O \rightarrow CO_2 + H_2 \qquad \text{Tar steam gasification} \qquad (3.15)$$

There are some open issues in CG that still need proper solutions, such as efficient tar removal from fuel gas, development of catalyst with improved performance, quality of the obtained syngas, suitable for desired final applications (energy generation in turbines or gas engines, hydrogen production, chemical feedstock synthesis, such FT or others) [115,116]. From a process point of view, two main set-ups are used: single-step catalytic pyrolysis/gasification, with catalyst mixed with fed both in fixed and fluidized bed reactors, and two stage plants comprising a traditional pyrolysis reactor followed by a downstream reformer reactor acting on pyrolysis gases (usually a fixed bed reactor). As catalysts regards, due to economical and scale constrains, they are usually based on cheap readily available or commercial materials. Typical examples are natural minerals containing active elements, such as Ca, Mg basic oxides, dolomite and olivine. Supported transition metals or metal oxides (Ni, Fe, etc.), even if more complex systems, are claimed more active and/or stable by some authors. Calcium oxide (CaO) has been widely used in coal, biomass and waste gasification processes, owing to its low cost and convenience, in order to further improve hydrogen yields from syngas since it acts as CO_2 sorbent improving WGS (see (3.11)), thus giving higher H_2 yields [116–121]. Moreover, when blended with feedstock it favours devolatilization during gasification or pyrolysis [117]. Other natural occurring materials, employed as catalysts, include dolomite and olivine, which are well known for their tar removal performance [118–121]. Arena *et al.* tested their use in pilot-scale bubbling fluidized bed gasifier (BFBG) treating five waste plastics. Olivine revealed as the best bed catalyst for the cracking reactions of tar, allowing one to obtain a drastic reduction in tar content, together with an increase in the content of H_2 and CO in the syngas [116]. However, the same authors [122] showed how olivine suffered a progressive reduction of the catalytic action correlated to the loss of metals, responsible of polymer dehydrogenation, thus preventing the possibility to recover its catalytic capacity by thermal or mechanical. Nickel-based catalysts have been extensively studied to reduce tar formation and are preferred over PGMs-based catalysts, such as Rh, Ru or Pt, because of their lower cost, availability and efficiency of tar removal. The use of a commercial nickel-based catalyst (C11-NK), developed for reforming moderately heavy petroleum fractions employed in an integrated two-stage process of pyrolysis/catalytic SR lab-scale plant, was reported for the first time by Czernik *et al.* [123]. Samples of several types of plastics, such as polyethylene, polypropylene, polystyrene, poly(ethylene terephthalate), nylon, polyurethane and poly(vinyl chloride) (PVC), were treated and yielded up to 80% of the stoichiometric potential of hydrogen production when polyethylene was the fed. NiO supported on γ-Al_2O_3 was employed in the catalytic steam gasification of MSW for syngas production in a lab scale two-stage fixed-bed reactor by Luo *et al.* Compared with MSW catalytic pyrolysis, the introduction of steam leads to more tar and char participating in steam gasification, which resulted in a rapid increase of syngas yield and carbon conversion efficiency. The NiO/γ-Al_2O_3 catalyst revealed better catalytic performance for the cracking of tar than calcined dolomite. The highest H_2 content (54.22%) and gas yield (1.75 N m^3 kg^{-1}) were achieved at 900 °C, $S/C = 2.41$ [124]. A 5.5 wt.% Ni-based catalyst supported on γ-alumina, prepared by wet impregnation and calcination was also employed in pilot-scale FBG processing wood pellets (100 wt.% of pine wood), biomass/plastic pellets (20 wt.% of polyethylene chips and 80 wt.% of pine wood sawdust) and olive husk pellets (100 wt.% of olive husk) [125]. The effect of the catalyst presence on the hydrogen yield and decreased tar production was higher than the presence of steam [126]. Similar results were reported by Blanco *et al.* using Ni/SiO_2 catalysts prepared by a sol–gel method in the catalytic pyrolysis/gasification of RDF. The effect of Ni:citric acid (CA) ratio during preparation was deeply investigated and using optimized systems with Ni:CA ratios of 1:3 low tar concentration of 0.2 mg_{tar} g_{RDF}^{-1} was attained together with a high hydrogen concentration (58 vol.%), and low CH_4 (2.2 vol.%) and C_2–C_4 concentrations (0.8 vol.%) [127]. Conversely, Corella *et al.* [128] reported Ni/Olivine catalysts with low activity in tar elimination and quick deactivation when employed in biomass gasification in a circulating fluidized-bed and on a BFBG. A bimetallic Ni−Mn−Al catalyst, prepared by a co-precipitation method, was employed in a two stage, fixed steam pyrolysis-reforming reaction system [129] with improved hydrogen yield: 94.4, 91.8, 81.8 mmol $g_{plastic}^{-1}$ for waste high-density polyethylene (HDPE), HDPE/PVC, motor oil containers, respectively [130]. On the other hand, Li *et al.* [115] reported catalytic steam gasification of MSW to hydrogen-rich fuel gas in a combined fixed bed reactor using a more complex trimetallic catalysts. The dried MSW (original one had a moisture content of 9.08%) was a mixture of kitchen garbage (45.98 wt.%), wood and leaves (25.89 wt.%), paper (9.85 wt.%), textile (1.88 wt.%) and plastic (16.40 wt.%). The trimetallic catalyst, nano-NiLaFe/γ-Al_2O_3, was prepared by DP method and consisted in 28–35 nm trimetallic nanoparticles supported on alumina, with a 21 wt.% loading of tri-metallic oxide in catalysts and the mass fractions of NiO, Fe_2O_3 and La_2O_3 of 8.6%, 7.4% and 5.9%, respectively. The catalyst bed was

Table 3.8   Catalysts for waste catalytic gasification

Catalyst	Feed	Gas yield	H$_2$ yield (%)	Reaction conditions	Reference
Ni C11-NK	Waste polypropylene	100% (C mass balance)	80	$T_{\text{pyrolysis}}$: 650 °C; $T_{\text{reforming}}$: 850 °C, S/C: 4.6, G_{C1}VHSV: 1,600 h^{-1}	[123]
NiO/γ-Al$_2$O$_3$	MSW	1.75*	54.22	T: 900 °C, S/C: 2.41	[124]
Ni/SiO$_2$	RDF	71.2 (% yield)	58*	$T_{\text{pyrolysis}}$: 600 °C; $T_{\text{reforming}}$: 800, steam/RDF: 1.75	[127]
Ni/Olivine	Biomass	–	15.3***	T: 827 °C, WHSV: 0.33 (kg$_{\text{biomass}}$/h) kg$_{\text{S+D}}^{-1}$	[128]
Ni−Mn−Al	Waste HDPE	170.1** (wt.%)	94.4 mmol g^{-1}	$T_{\text{pyrolysis}}$: 500 °C; $T_{\text{reforming}}$: 800, RDF/catalyst: 0.5	[130]
nano-NiLaFe/ γ-Al$_2$O$_3$	MSW	2.18 (% yield)	53.9	T: 800 °C; S/M ratio: 1.33	[115]

*Nm^3 kg^{-1}.
**wt.%.
***H$_2$ concentration (vol.%).

placed downstream the MSW gasifier, and it improved both the gas and hydrogen yield, whereas substantially decreasing tar yield if compared to non-catalytic pyrolysis, as shown in Table 3.8.

These good performances are mainly due to enhancement of the cracking of tar and hydrocarbons (CH$_4$ to C$_x$H$_y$) in vapour leading to valuable gases. Particularly, the content of H$_2$ in gas components was enhanced significantly, whereas that of CH$_4$ was decreased markedly. The authors also performed a detailed study on the influence of reaction conditions, and the optimal values of *S/M* and *C/M* were respectively found to be 1.33 and 0.5, whereas higher temperature improved gas quality and yield. Another kind of waste materials is represented by food and agro wastes, which have an enormous environmental and economic impacts on the society, since about 1.3 billion tons of food waste are generated in the world annually. Apart from standard high-temperature (catalytic) gasification, SCWG gained attention due to the low operating temperatures and lowest tar formation. In this process, the most diffused catalysts are simple alkali, like Na, K and Ca hydroxides or (hydrogeno)carbonates, and, generally, potassium alkali showed the best performances [131,132]. On the other hand, Matsumura *et al.* reported on the activity of a suspended activated carbon catalyst in SCWG of different biomass feedstock [133].

3.5   Conclusions and perspectives

Analysing the survey of data presented in this review, it readily appears how the use of catalytic materials in hydrogen production processes results in improved performances, such as higher hydrogen yield, lower by-products, coke, tar formation, as well as milder reactions conditions if compared to non-catalytic processes. Ni-based catalysts are still the most diffused due to the good balance between costs and performances, even if Ru-based systems found applications in lignocellulose SCWR, and noble metals are widely used in bio-oils SR. However, some issues are still open in some applications, such as poor catalysts stability, low resistance to N, S poisoning, non-detailed characterization of catalysts structure (in particular for complex formulations).

Abbreviations and acronyms

APR	aqueous phase reforming
BFBG	bubbling fluidized bed gasifier
CFB	circulating fluidized bed
FBG	fluidized bed gasifier
HDPE	high density polyethylene
MSW	municipal solid waste
OSR	oxidative steam reforming
PGM	platinum group metal
PVC	poly(vinyl chloride)
SCWG	supercritical catalytic water gasification
SR	steam reforming
SWR	supercritical water reforming
RDF	refused-derived fuels
GWS	water gas shift

References

[1]   Azadi P., Inderwildi O.R., Farnood R., King D.A. 'Liquid fuels, hydrogen and chemicals from lignin: a critical review'. *Renew. Sustainable Energy Rev.*, 21 (2013) 506–523.

[2]   Nanda S., Azargohar R., Kozinski J.A., Dalai A.K. 'Characteristic studies on the pyrolysis products from hydrolyzed Canadian lignocellulosic feed-stocks'. *Bioenergy Res.*, 7 (2014) 174–191.

[3]   Nanda S., Mohanty P., Pant K.K., Naik S., Kozinski J.A., Dalai A.K. 'Characterization of North American lignocellulosic biomass and biochars in terms of their candidacy for alternate renewable fuels'. *Bioenergy Res.*, 6 (2013) 663–677.

[4]　Sjostrom E. *Wood Chemistry*, 2nd ed. New York, NY: Academic Press; 1999.

[5]　Resende F.L.P., Savage P.E. 'Effect of metals on supercritical water gasification of cellulose and lignin'. *Ind. Eng. Chem. Res.*, 49 (2010) 2694–2700.

[6]　Elliott D.C. 'Catalytic hydrothermal gasification of biomass'. *Biofuels, Bioprod. Biorefin.*, 2 (2008) 254–265.

[7]　Yamaguchi A., Huyoshi N., Sato O. 'Gasification of organosolv-lignin over charcoal supported noble metal salt catalysts in supercritical water'. *Top. Catal.*, 11 (2012) 889–896.

[8]　Osada M., Sato O., Watanabe M. 'Water density effect on lignin gasification over supported metal catalysts in supercritical water'. *Energy Fuels*, 20 (2006) 930–935.

[9]　Elliott D.C., Sealock L.J., Backer E.G. 'Chemical processing in high-pressure aqueous environments. 2. Development of catalysts for gasification'. *Ind. Eng. Chem. Res.*, 32 (1993) 1542–1548.

[10]　Sato T., Osada M., Watanabe M., Shirai M., Arai K. 'Gasification of alkylphenols with supported noble metal catalysts in supercritical water'. *Ind. Eng. Chem. Res.*, 42 (2003) 4277–4282.

[11]　Osada M., Sato T., Watanabe M., Adschiri T., Arai K. 'Low-temperature catalytic gasification of lignin and cellulose with a ruthenium catalyst in supercritical water'. *Energy Fuels*, 18 (2004) 327–333.

[12]　Osada M., Hiyoshi N., Sato O., Arai K., Shirai M. 'Effect of sulfur on catalytic gasification of lignin in supercritical water'. *Energy Fuels*, 21 (2007) 1400–1405.

[13]　Osada M., Hiyoshi N., Sato O., Arai K., Shirai M. 'Reaction pathway for catalytic gasification of lignin in presence of sulfur in supercritical water'. *Energy Fuels*, 21 (2007) 1854–1858.

[14]　Osada M., Sato O., Arai K. 'Stability of supported ruthenium catalysts for lignin gasification in supercritical water'. *Energy Fuels*, 20 (2006) 2337–2343.

[15]　Onwudili J.A., Williams P.T. 'Hydrogen and methane selectivity during alkaline supercritical water gasification of biomass with ruthenium–alumina catalyst'. *Appl. Catal. A: Gen.*, 132–133 (2013) 70–79.

[16]　Yamaguchi A., Hiyoshi N., Sato O. 'Lignin gasification over supported ruthenium trivalent salts in supercritical water'. *Energy Fuels*, 22 (2008) 1485–1492.

[17]　Yoshida T., Oshima Y., Matsumura Y. 'Gasification of biomass model compounds and real biomass in supercritical water'. *Biomass Bioenergy*, 26 (2004) 71–78.

[18]　Azadi P., Khan S., Strobel F., Azadi F., Farnood R. 'Hydrogen production from cellulose, lignin, bark and model carbohydrates in supercritical water using nickel and ruthenium catalysts'. *Appl. Catal. B: Environ.*, 117–118 (2012) 330–338.

[19]　Minowa T., Ogi T. 'Hydrogen production from cellulose using a reduced nickel catalyst'. *Catal. Today*, 45 (1998) 411–416.

[20] Sato T., Furusawa T., Ishiyama Y., *et al.* 'Effect of water density on the gasification of lignin with magnesium oxide supported nickel catalysts in supercritical water'. *Ind. Eng. Chem. Res.*, 45 (2006) 615–622.

[21] Furusawa T., Sato T., Sugito H., *et al.* 'Hydrogen production from the gasification of lignin with nickel catalysts in super-critical water'. *Int. J. Hydrogen Energy*, 32 (2007) 699–704.

[22] Ruppert A.M., Niewiadomski M., Grams J., Kwapinski W. 'Optimization of Ni/ZrO_2 catalytic performance in thermochemical cellulose conversion for enhanced hydrogen production'. *Appl. Catal. B: Environ.*, 145 (2014) 85–90.

[23] Park K.C., Tomiyasu H. 'Gasification reaction of organic compounds catalyzed by RuO_2 in supercritical water'. *Chem. Commun.*, 6 (2003) 694–695.

[24] Watanabe M., Inomata H., Arai K. 'Catalytic hydrogen production from biomass (glucose and cellulose) with ZrO_2 in supercritical water'. *Biomass Bioenergy*, 22 (2002) 405–410.

[25] Hao H.X., Guo L.J., Zhang X.M., Guan Y. 'Hydrogen production from catalytic gasification of cellulose in supercritical water'. *Chem. Eng. J.*, 110 (2005) 57–65.

[26] Yanik J., Ebale S., Kruse A., Saglam M., Yuksel M. 'Biomass gasification in supercritical water: II. Effect of the catalyst'. *Int. J. Hydrogen Energy*, 33 (2008) 4520–4526.

[27] Neumann R., Bidyut-Bikash S., *Catalytic Formation of Carbon Monoxide (CO) and Hydrogen (H_2) from Biomass*. WO 2015/063763 Al. 2015.

[28] Inderwildi O., King D. 'Quo vadis biofuels'. *Energy Environ. Sci.*, 2 (2009) 343–346.

[29] Minowa T., Sawayama S. 'A novel microalgal system for energy production with nitrogen cycling'. *Fuel*, 78 (1999) 1213–1215.

[30] Tsukahara K., Kimura T., Minowa T., *et al.* 'Microalgal cultivation in a solution recovered from the low-temperature catalytic gasification of the microalga'. *J. Biosci. Bioeng.*, 91 (2001) 311–313.

[31] Onwudili J.A., Lea-Langton A.R., Ross A.B., Williams P.T. 'Catalytic hydrothermal gasification of algae for hydrogen production: composition of reaction products and potential for nutrient recycling'. *Bioresour. Technol.*, 127 (2013) 72–80.

[32] Elliott D.C., Hart T.R., Neuenschwander G.C., Rotness L.J., Olarte M.V., Zacher A.H. 'Chemical processing in high-pressure aqueous environments. 9. Process development for catalytic gasification of algae feedstocks'. *Ind. Eng. Chem. Res.*, 51 (2012) 10768–10777.

[33] Duman G., Uddin M.A., Yanik J. 'Hydrogen production from algal biomass via steam gasification'. *Bioresour. Technol.*, 166 (2014) 24–30.

[34] Trane-Restrup R., Jensen A.D. 'Steam reforming of cyclic model compounds of bio-oil over Ni-based catalysts: product distribution and carbon formation'. *Appl. Catal. B: Environ.*, 165 (2015) 117–127.

[35] Shamsul N.S., Kamarundin S.K., Rahman N.A., Kofli N.T. 'An overview on the production of bio-methanol as potential renewable energy'. *Renew. Sustainable Energy Rev.*, 33 (2014) 578–588.

[36] Gupta A., Verma J.P. 'Sustainable bio-ethanol production from agro-residues: a review'. *Renew. Sustainable Energy Rev.*, 41 (2015) 550–567.

[37] Chattanathan S.A., Adhikari S., Abdoulmoumine N. 'A review on current status of hydrogen production from bio-oil'. *Renew. Sustainable Energy Rev.*, 16 (2012) 2366–2372.

[38] Vagia E.C., Lemonidou A.A. 'Thermodynamic analysis of hydrogen production via autothermal steam reforming of selected components of aqueous bio-oil fraction'. *Int. J. Hydrogen Energy*, 33 (2008) 2489–2500.

[39] Vagia E.C., Lemonidou A.A. 'Thermodynamic analysis of hydrogen production via steam reforming of selected components of aqueous bio-oil fraction'. *Int. J. Hydrogen Energy*, 37 (2007) 212–223.

[40] Davda R.R., Shabaker J.W., Huber G.W., Cortright R.D., Dumesic J.A. 'A review of catalytic issues and process conditions for renewable hydrogen and alkanes by aqueous-phase reforming of oxygenated hydrocarbons over supported metal catalysts'. *Appl. Catal. B: Environ.*, 56 (2005) 171–186.

[41] Kirilin A.V., Tokarev A.V., Manyar H., *et al.* 'Aqueous phase reforming of xylitol over Pt–Re bimetallic catalyst: effect of the Re addition'. *Catal. Today*, 223 (2014) 97–107.

[42] Bridgwater A.V. 'Review of fast pyrolysis of biomass and product upgrading'. *Biomass Bioenergy*, 38 (2012) 68–94.

[43] Staš M., Kubička D., Chudoba J., Pospíšil M. 'Overview of analytical methods used for chemical characterization of pyrolysis bio-oil'. *Energy Fuels*, 28 (2014) 385–402.

[44] Trane R., Dahl S, Skjøth-Rasmussen M.S., Jensen A.D. 'Catalytic steam reforming of bio-oil'. *Int. J. Hydrogen Energy*, 37 (2012) 6447–6472.

[45] Wang H., Male J., Wang Y. 'Recent advances in hydrotreating of pyrolysis bio-oil and its oxygen-containing model compounds'. *ACS Catal.*, 3 (2013) 1047–1070.

[46] Oasmaa A., Kuoppala E. 'Fast pyrolysis of forestry residue. 3. Storage stability of liquid fuel'. *Energy Fuels*, 17 (2003) 1075–1084.

[47] Preau A. *Process for Upgrading a Pyrlysis Oil, in Particular in a Refinery*. WO2011020966A1. 2011.

[48] Huber G.W. *Production of Hydrogen, Liquid Fuels, and Chemicals from Catalytic Processing of Bio-oils*. WO2010033789A2. 2010.

[49] Czernik S., French R., Feik R., Chornet E. 'Hydrogen by catalytic steam reforming of liquid byproducts from biomass thermoconversion processes'. *Ind. Eng. Chem. Res.*, 41 (2002) 4209–4215.

[50] Czernik S., Evans R., French R. 'Hydrogen from biomass-production by steam reforming of biomass pyrolysis oil'. *Catal. Today*, 129 (2007) 265–268.

[51] Li H., Xu Q., Xue H., Yan Y. 'Catalytic reforming of the aqueous phase derived from fast-pyrolysis of biomass'. *Renew. Energy*, 34 (2009) 2872–2877.

[52] Wu C., Huanq Q., Sui M., Yan Y., Wang F. 'Hydrogen production via catalytic steam reforming of fast pyrolysis bio-oil in a two-stage fixed bed reactor system'. *Fuel Process. Technol.*, 89 (2008) 1306–1316.

[53]   Hou T., Yuan L., Ye T., Gong L., Tu J., Yamamoto M., Torimoto Y., Li Q. 'Hydrogen production by low-temperature reforming of organic compounds in bio-oil over a CNT-promoting Ni catalyst'. *Int. J. Hydrogen Energy*, 34 (2009) 9095–9107.

[54]   Domine M.E., Iojoiu E.E., Davidian T., Guilhaume N., Mirodatos C. 'Hydrogen production from biomass-derived oil over monolithic Pt- and Rh-based catalysts using steam reforming and sequential cracking processes'. *Catal. Today*, 133–135 (2008) 265–573.

[55]   Rioche R., Kulkarni S., Meunier F.C., Breen J.P., Burch R. 'Steam reforming of model compounds and fast pyrolysis bio-oil on supported noble metal catalysts'. *Appl. Catal. B: Environ.*, 61 (2005) 130–139.

[56]   Basagiannis A.C., Verykios X.E. 'Steam reforming of the aqueous fraction of bio-oil over structured $Ru/MgO/Al_2O_3$ catalysts'. *Catal. Today*, 127 (2007) 256–264.

[57]   Yan C., Hu E., Cai C. 'Hydrogen production from bio-oil aqueous fraction with in situ carbon dioxide capture'. *Int. J. Hydrogen Energy*, 35 (2010) 2612–2616.

[58]   Pan C., Chen A., Liu Z., Chen P., Lou H., Zheng X. 'Aqueous-phase reforming of the low-boiling fraction of rice husk pyrolyzed bio-oil in the presence of platinum catalyst for hydrogen production'. *Bioresour. Technol.*, 125 (2012) 335–339.

[59]   Chen A., Chen P., Cao L., Lou H. 'Aqueous-phase reforming of the low-boiling fraction of bio-oil for hydrogen production: the size effect of Pt/Al_2O_3'. *Int. J. Hydrogen Energy*, 40 (2015) 14798–14805.

[60]   Isahak W.N.R.W., Hisham M.W.M., Yarmo M.A., Hin T.Y. 'A review on bio-oil production from biomass by using pyrolysis method'. *Renew. Sustainable Energy Rev.*, 16 (2012) 5910–5923.

[61]   Kirilin A.V., Tokarev A.V., Murzina E.V., Kustov L.M., Mikkola J., Murzin D.Y. 'Reaction products and transformations of intermediates in the aqueous-phase reforming of sorbitol'. *ChemSusChem*, 3 (2010) 708–718.

[62]   Vagia E.C., Lemonidou A.A. 'Investigations on the properties of ceria–zirconia-supported Ni and Rh catalysts and their performance in acetic acid steam reforming'. *J. Catal.*, 269 (2010) 388–396.

[63]   Basagiannis A.C., Verykios X.E. 'Catalytic steam reforming of acetic acid for hydrogen production'. *Int. J. Hydrogen Energy*, 32 (2007) 3343–3355.

[64]   Kechagiopoulos P.N., Voutetakis S.S., Lemonidou A.A., Vasalos I.A. 'Hydrogen production via steam reforming of the aqueous phase of bio-oil in a fixed bed reactor'. *Energy Fuels*, 20 (2006) 2155–2163.

[65]   Lemonidou A.A, Vagia E.C., Lercher J.A. 'Acetic acid reforming over Rh supported on La_2O_3/CeO_2-ZrO_2: catalytic performance and reaction pathway analysis'. *ACS Catal.*, 3 (2013)1919–1928.

[66]   Bossola F., Evangelisti C., Allieta M., Psaro R., Recchia S., Dal Santo V. 'Well-formed, size-controlled ruthenium nanoparticles active and stable for acetic acid steam reforming'. *Appl. Catal. B: Environ.*, 181 (2016) 599–611.

[67] Pant K.K., Mohanty P., Agarwal S., Dalai A.K. 'Steam reforming of acetic acid for hydrogen production over bifunctional Ni–Co catalysts'. *Catal. Today*, 207 (2013) 36–43.

[68] Wang S., Zhang F., Cai Q., Zhu L., Luo Z. 'Steam reforming of acetic acid over coal ash supported Fe and Ni catalysts'. *Int. J. Hydrogen Energy*, 40 (2015) 11406–11413.

[69] de Vlieger D.J.M., Lefferts L., Seshan K. 'Ru decorated carbon nanotubes – a promising catalyst for reforming bio-based acetic acid in the aqueous phase'. *Green Chem.*, 16 (2014) 864–874.

[70] Hoang T.M.C., Geerdink B., Sturm J.M., Lefferts L., Seshan K. 'Steam reforming of acetic acid – a major component in the volatiles formed during gasification of humin'. *Appl. Catal. B: Environ.*, 163 (2015) 74–82.

[71] Hu X., Lu G. 'Comparative study of alumina-supported transition metal catalysts for hydrogen generation by steam reforming of acetic acid'. *Appl. Catal. B: Environ.*, 99 (2010) 289–297.

[72] Takanabe K., Aika K., Seshan K., Lefferts L. 'Catalyst deactivation during steam reforming of acetic acid over Pt/ZrO$_2$'. *Chem. Eng. J.*, 120 (2006) 133–137.

[73] Nozawa T., Mizukoshi Y., Yoshida A., Naito S. 'Aqueous phase reforming of ethanol and acetic acid over TiO$_2$ supported Ru catalysts'. *Appl. Catal. B: Environ.*, 146 (2014) 221–226.

[74] Zhang F., Wang N., Yang L., Huang L. 'Ni–Co bimetallic MgO-based catalysts for hydrogen production via steam reforming of acetic acid from bio-oil'. *Int. J. Hydrogen Energy*, 39 (2014) 18688–18694.

[75] Iwasa N., Yamane T., Arai M. 'Influence of alkali metal modification and reaction conditions on the catalytic activity and stability of Ni containing smectite-type material for steam reforming of acetic acid'. *Int. J. Hydrogen Energy*, 36 (2011) 5904–5911.

[76] Resende K.A., Ávila-Neto C.N., Rabelo-Neto R.C., Noronha F.B., Hori C.E. 'Hydrogen production by reforming of acetic acid using La–Ni type perovskites partially substituted with Sm and Pr'. *Catal. Today*, 242 (2015) 71–79.

[77] Hu X., Lu G. 'Investigation of the steam reforming of a series of model compounds derived from bio-oil for hydrogen production'. *Appl. Catal. B: Environ.*, 88 (2009) 376–385.

[78] Sun J., Mei D., Karim A.M., Datye A.K., Wang Y. 'Minimizing the formation of coke and methane on Co nanoparticles in steam reforming of biomass-derived oxygenates'. *ChemCatChem*, 15 (2013) 1299–1303.

[79] Navarro R.M., Guil-Lopez R., Gonzalez-Carballo J.M., *et al.* 'Bimetallic MNi/Al$_2$O$_3$–La catalysts (M = Pt, Cu) for acetone steam reforming: role of M on catalyst structure and activity'. *Appl. Catal. A: Gen.*, 474 (2014) 168–177.

[80] Zaccheria F., Scotti N., Marelli M., Psaro R., Ravasio N. 'Unravelling the properties of supported copper oxide: can the particle size induce acidic behaviour?'. *Dalton Trans.*, 42 (2012) 1319–1328.

[81] Scotti N., Dangate M., Gervasini A., Evangelisti C., Ravasio N., Zaccheria F. 'Unraveling the role of Low coordination sites in a Cu metal nanoparticle: a step toward the selective synthesis of second generation biofuels'. *ACS Catal.*, 4 (2014) 2818–2826.

[82] Izquierdo U., Wichert M., Kolb G., *et al.* 'Micro reactor hydrogen production from ethylene glycol reforming using Rh catalysts supported on CeO_2 and La_2O_3 promoted α-Al_2O_3'. *Int. J. Hydrogen Energy*, 39 (2014) 5248–5256.

[83] de Vlieger D.J.M., Chakinala A.G., Lefferts L., Kersten S.R.A., Seshan K., Brilman D.W.F. 'Hydrogen from ethylene glycol by supercritical water reforming using noble and base metal catalysts'. *Appl. Catal. B: Environ.*, 111–112 (2012) 536–544.

[84] Huber G.W., Shabaker J.W., Evans S.T., Dumesic J.A. 'Aqueous-phase reforming of ethylene glycol over supported Pt and Pd bimetallic catalysts'. *Appl. Catal. B: Environ.*, 62 (2006) 226–235.

[85] Tupy S.A., Chen J.G., Vlachos D.G. 'Comparison of ethylene glycol steam reforming over Pt and NiPt catalysts on various supports'. *Top. Catal.*, 56 (2013) 1644–1650.

[86] Liu J., Sun B., Hu J., Li H., Qiao M. 'Aqueous-phase reforming of ethylene glycol to hydrogen on Pd/Fe_3O_4 catalyst prepared by co-precipitation: metal–support interaction and excellent intrinsic activity'. *J. Catal.*, 274 (2010) 287–295.

[87] Chu X., Liu J., Sun B., *et al.* 'Aqueous-phase reforming of ethylene glycol on Co/ZnO catalysts prepared by the coprecipitation method'. *J. Mol. Catal. A: Chem.*, 335 (2011) 129–135.

[88] Wang X., Li N., Pfefferle L.D., Haller G.L. 'Pt–Co bimetallic catalyst supported on single-walled carbon nanotubes: effect of alloy formation and oxygen containing groups'. *J. Phys. Chem. C*, 114 (2010) 16996–17002.

[89] Kim H., Park H.J., Kim T., *et al.* 'Hydrogen production through the aqueous phase reforming of ethylene glycol over supported Pt-based bimetallic catalysts'. *Int. J. Hydrogen Energy*, 37 (2012) 8310–8317.

[90] Li C., Yan X., Fujimoto K. 'Development of highly stable catalyst for methanol synthesis from carbon dioxide'. *Appl. Catal. A: Gen.*, 469 (2014) 306–311.

[91] Shamsul N.S., Kamarudin S.K., Rahman N.A., Kofti N.T. 'An overview on the production of bio-methanol as potential renewable energy'. *Renew. Sustainable Energy Rev.*, 33 (2014) 578–588.

[92] Yong S.T., Ooi C.W., Chai S.P., Wu X.S. 'Review of methanol reforming-Cu-based catalysts, surface reaction mechanisms, and reaction schemes'. *Int. J. Hydrogen Energy*, 38 (2013) 9541–9552.

[93] Spassova I., Tsontcheva T., Velichkova N., Khristova M., Nihtianova D. 'Catalytic reduction of NO with decomposed methanol on alumina-supported Mn–Ce catalysts'. *J. Colloid Interf. Sci.*, 374 (2012) 267–277.

[94] Li C.L., Lin Y.C. 'Catalytic partial oxidation of methanol over copper–zinc based catalysts: a comparative study of alumina, zirconia, and magnesia as promoters'. *Catal. Lett.*, 140 (2010) 69–76.

[95] Davidson S.D., Zhang H., Sun J., Wang Y. 'Supported metal catalysts for alcohol/sugar alcohol steam reforming'. *Dalton Trans.*, 43 (2014) 11782–11802.

[96] Yaakob Z., Kamarudin S.K., Daud W.R.W., Yosfiah M.R., Lim K.L., Kazemian H. 'Hydrogen production by methanol-steam reforming using Ni-Mo-Cu/gamma-alumina trimetallic catalysts'. *Asia-Pacific J. Chem. Eng.*, 5 (2010) 862–868.

[97] Sá S., Silva H., Brandão L., Sousa J.M., Mendes A. 'Catalysts for methanol steam reforming – a review'. *Appl. Catal. B: Environ.*, 99 (2010) 43–57.

[98] Chen W.H., Syu Y.J. 'Thermal behavior and hydrogen production of methanol steam reforming and autothermal reforming with spiral preheating'. *Int. J. Hydrogen Energy*, 36 (2011) 3397–3408.

[99] Mierczynski P., Vasilev K., Mierczynska A., Maniukiewicz W., Szynkowska M.I., Maniecki T.P. 'Bimetallic Au–Cu, Au–Ni catalysts supported on MWCNTs for oxy-steam reforming of methanol'. *Appl. Catal. B: Environ.*, 185 (2016) 281–294.

[100] Peppley B.A., Amphlett J.C., Kearns L.M., Mann R.F. 'Methanol–steam reforming on $Cu/ZnO/Al_2O_3$ catalysts. Part 2. A comprehensive kinetic model'. *Appl. Catal. A: Gen.*, 179 (1999) 31–49.

[101] Sanches S.G., Flores J.H., de Avillez R.R., da Silva M.I.P. 'Influence of preparation methods and Zr and Y promoters on Cu/ZnO catalysts used for methanol steam reforming'. *Int. J. Hydrogen Energy*, 37 (2012) 6572–6579.

[102] Wang G., Wang F., Li L., Zhao M. 'Methanol steam reforming on catalyst coating by cold gas dynamic spray'. *Int. J. Hydrogen Energy*, 41 (2016) 2391–2398.

[103] Karim A.M., Conant T., Datye A.K. 'Controlling ZnO morphology for improved methanol steam reforming reactivity'. *Phys. Chem. Chem. Phys.*, 10 (2008) 5584–5590.

[104] Friedrich M., Penner S., Heggen M., Armbrüster M. 'High CO_2 selectivity in methanol steam reforming through ZnPd/ZnO teamwork'. *Angew. Chem. Int. Ed.*, 52 (2013) 4389–4392.

[105] Gupta L., Verma J.V. 'Sustainable bio-ethanol production from agro-residues: a review'. *Renew. Sustainable Energy Rev.*, 41 (2015) 550–567.

[106] Contreras J.L., Salmones J., Colín-Luna J.A., *et al.* 'Catalysts for H_2 production using the ethanol steam reforming: a review'. *Int. J. Hydrogen Energy*, 39 (2014) 18835–18853.

[107] Nozawa T., Yoshida A., Hikichi S., Naito S. 'Effects of Re addition upon aqueous phase reforming of ethanol over TiO_2 supported Rh and Ir catalysts'. *Int. J. Hydrogen Energy*, 40 (2015) 4129–4140.

[108] Ni M., Leung D.Y.C., Leung M.K.H. 'A review on reforming bio-ethanol for hydrogen production'. *Int. J. Hydrogen Energy*, 32 (2007) 3238–3247.

[109] Mattos L.V., Jacobs G., Davis B.H., Noronha F.B. 'Production of hydrogen from ethanol: review of reaction mechanism and catalyst deactivation'. *Chem. Rev.*, 112 (2012) 4094–4123.

[110] Chen L., Choong C.K.S., Zhong Z., *et al.* 'Carbon monoxide-free hydrogen production via low-temperature steam reforming of ethanol over iron-promoted Rh catalyst'. *J. Catal.*, 276 (2010) 197–200.

[111] Chiou J.Y.Z., Siang J., Yang S., *et al.* 'Pathways of ethanol steam reforming over ceria-supported catalysts'. *Int. J. Hydrogen Energy*, 37 (2012) 13667–13673.

[112] Han S.J., Bang Y., Seo J.G., Yoo J., Song I.K. 'Hydrogen production by steam reforming of ethanol over mesoporous $Ni-Al_2O_3-ZrO_2$ xerogel catalysts: effect of Zr/Al molar ratio'. *Int. J. Hydrogen Energy*, 38 (2013) 1376–1383.

[113] Shi Q,. Peng Z., Chen W., Zhang N. '$La_2O_2CO_3$ supported Ni–Fe catalysts for hydrogen production from steam reforming of ethanol'. *J. Rare Hearths*, 29 (2011) 861–865.

[114] Banach B., Machocki A., Rybak P., Denis A., Grzegorczyk G., Gac W. 'Selective production of hydrogen by steam reforming of bio-ethanol'. *Catal. Today*, 176 (2011) 28–35.

[115] Li J., Liao S., Dan W., Jia K., Zhou X. 'Experimental study on catalytic steam gasification of municipal solid waste for bioenergy production in a combined fixed bed reactor'. *Biomass Bioenergy*, 46 (2012) 174–180.

[116] Arena U., Zaccariello L., Mastellone M.L. 'Fluidized bed gasification of waste-derived fuels'. *Waste Manage.*, 30 (2010) 1212–1219.

[117] Zhou C., Stuermer T., Gunarathne R., Yang W., Blasiak W. 'Effect of calcium oxide on high-temperature steam gasification of municipal solid waste'. *Fuel*, 122 (2014) 36–46.

[118] Guan Y., Luo S., Liu S., Xiao B., Cai L. 'Steam catalytic gasification of municipal solid waste for producing tar-free fuel gas'. *Int. J. Hydrogen Energy*, 34 (2009) 9341–9346.

[119] Arena U., Zaccariello L., Mastellone M.L. 'Gasification of a plastic waste in a fluidized bed of olivine'. In Werther J., Nowak W., Wirth K.E., Hartge E.U. (eds.). Proceedings of CFB9 – Ninth International Conference on Circulating Fluidized Beds. 2008, May 13–16, Hamburg, Germany. pp. 691–696.

[120] Arena U., Zaccariello L., Mastellone M.L. 'Tar removal during the fluidized bed gasification of a plastic waste'. *Waste Manage.*, 29 (2009) 783–791.

[121] Arena U., Di Gregorio F. 'Energy generation by air gasification of two industrial plastic wastes in a pilot scale fluidized bed reactor'. *Energy*, 68 (2014) 735–743.

[122] Mastellone M.L., Zaccariello L. 'Metals flow analysis applied to the hydrogen production by catalytic gasification of plastics'. *Int. J. Hydrogen Energy*, 38 (2013) 3621–3629.

[123] Czernik S., French R.J. 'Production of hydrogen from plastics by pyrolysis and catalytic steam reforming'. *Energy Fuels*, 20 (2006) 754–758.

[124] Luo S., Zhou Y., Yi C. 'Syngas production by catalytic steam gasification of municipal solid waste in fixed-bed reactor'. *Energy*, 44 (2012) 391–395.

[125] Miccio F., Piriou B., Ruoppolo G., Chirone R. 'Biomass gasification in a catalytic fluidized reactor with beds of different materials'. *Chem. Eng. J.*, 154 (2009) 369–374.

[126] Ruoppolo G., Ammendola P., Chirone R., Miccio F. 'H_2-rich syngas production by fluidized bed gasification of biomass and plastic fuel'. *Waste Manage.*, 32 (2012) 724–732.

[127] Blanco P.H., Wu C., Onwudili J.A., Dupont V., Williams P.T. 'Catalytic pyrolysis/gasification of refuse derived fuel for hydrogen production and tar reduction: influence of nickel to citric acid ratio using Ni/SiO_2 catalysts'. *Waste Biomass Valorization*, 5 (2014) 625–636.

[128] Corella J., Toledo J.M., Padilla R. 'Olivine or dolomite as in-bed additive in biomass gasification with air in a fluidized bed: which is better?'. *Energy Fuels*, 18 (2004) 713–720.

[129] Wu C., Williams P.T. 'Hydrogen production by steam gasification of polypropylene with various nickel catalysts'. *Appl. Catal. B: Environ.*, 87 (2009) 152–161.

[130] Wu C., Nahil M.A., Miskolczi N., Huang J., Williams P.T. 'Processing real-world waste plastics by pyrolysis-reforming for hydrogen and high-value carbon nanotubes'. *Environ. Sci. Technol.*, 48 (2014) 819–826.

[131] Amuzu-Sefordzi B., Huang J. 'Effects of increasing alkali catalysts concentration on hydrogen gas yield during the supercritical water gasification of food waste'. In Psaras P.A., Dale H. (eds.). *Advanced Materials Research*. Trans Tech Publications, Switzerland; 2015. pp. 905–910.

[132] Nanda S., Isen J., Dalai A.K., Kozinski J.A. 'Gasification of fruit wastes and agro-food residues in supercritical water'. *Energy Convers. Manage.*, 110 (2016) 296–306.

[133] Matsumura Y., Hara S., Kaminaka K., *et al.* 'Gasification rate of various biomass feedstocks in supercritical water'. *J. Jpn. Pet. Inst.*, 56 (2013) 1–10.

Chapter 4

Ni- and Cu-based catalysts for methanol and ethanol reforming

Mika Huuhtanen[1], Prem Kumar Seelam[1]
and Riitta L. Keiski[1]

Abstract

Steam reforming of light alcohols such as methanol and ethanol can be one solution in the transfer towards hydrogen economy. The increasing need of hydrogen pushes the scientists in academia and industry to develop new and efficient catalysts for production of hydrogen. An extensive number of articles have been published on methanol and ethanol steam reforming catalysts based on the transition and precious metals (e.g. Cu, Ni, Pd and Pt) supported on various metal and mixed oxides (e.g. Al_2O_3, ZnO, TiO_2, ZrO_2, CeO_2, CeO_2–ZrO_2) as well as on carbon supports (e.g. active carbon (AC), carbon nanotubes (CNTs)). Catalysts' activity, selectivity and tolerance towards deactivation are the main questions in which the answers are needed to be found. In this chapter, the recently developed nickel- and copper-based catalysts are presented for steam reforming of light alcohols.

4.1 Introduction

Novel catalysts for hydrogen production by steam reforming of light alcohols (e.g. methanol and ethanol) can play an important role in developing and boosting the change towards hydrogen economy in the future. The rapidly increasing demand of hydrogen to be used as a renewable fuel in fuel cells and as a reactant in chemical reactions and processes enforces the researchers in academia and in industry to develop new catalytic materials and components for hydrogen production [1–4]. Nowadays, the energy used is mainly produced from fossil fuels, and the resources of these are diminishing rapidly. It is estimated that the currently known crude oil and natural gas resources will be consumed depending on the consumption increase and development by around 2070 and 2090–2100, respectively [5]. In addition, the combustion of oil- and natural gas-based fuels

[1]University of Oulu, Faculty of Technology, Environmental and Chemical Engineering Research Unit, Oulu, Finland

leads to the formation of environmentally harmful compounds such as NO_x, SO_x and CO_2 to the atmosphere [6]. Light alcohols (such as methanol and ethanol) possess advantages and also a few drawbacks when used as reactants in H_2 production. One of the main advantages is the relatively low temperatures in which H_2 can be produced (for methanol steam reforming (MSR) below 400 °C and for ethanol steam reforming (ESR) below 550 °C) compared to methane steam reforming (which takes place at above 800 °C). This makes the bio-based alcohols promising raw materials for H_2 production. Most of the recent studies have focused on catalyst and reactor development for methanol and ESR.

In the steam reforming of light alcohols to hydrogen, several reactions take place, the main overall reactions being as follows (see (4.1)) for methanol [7]

$$CH_3OH + H_2O \rightarrow CO_2 + 3H_2 \qquad \Delta H^{\circ}_{298K} = 49.7 \text{ kJ mol}^{-1} \qquad (4.1)$$

and for ethanol (see (4.2)) [8]:

$$C_2H_5OH + 3H_2O \rightarrow 2CO_2 + 6H_2 \qquad \Delta H^{\circ}_{298K} = 174 \text{ kJ mol}^{-1} \qquad (4.2)$$

In addition, the reactions of direct decomposition (see (4.3) and (4.4a)–(4.4c)), the water gas shift (WGS) reaction (see (4.5)), and Boudouard reaction (see (4.6)) as well coke formation via polymerization of ethane (see (4.7)) may be or are plausible to occur at the same time (e.g. [7–14]).

$$CH_3OH \rightleftharpoons CO + 2H_2 \qquad \Delta H^{\circ}_{298K} = 91 \text{ kJ mol}^{-1} \qquad (4.3)$$

$$C_2H_5OH \rightleftharpoons \begin{cases} CO + CH_4 + H_2 & \Delta H^{\circ}_{298K} = 49 \text{ kJ mol}^{-1} & (4.4a) \\ C_2H_4 + H_2O & \Delta H^{\circ}_{298K} = 45 \text{ kJ mol}^{-1} & (4.4b) \\ C_2H_4O + H_2 & \Delta H^{\circ}_{298K} = 68 \text{ kJ mol}^{-1} & (4.4c) \end{cases}$$

$$CO + H_2O \rightleftharpoons CO_2 + H_2 \qquad \Delta H^{\circ}_{298K} = 41 \text{ kJ mol}^{-1} \qquad (4.5)$$

$$2CO \rightarrow CO_2 + C(s) \qquad \Delta H^{\circ}_{298K} = 172 \text{ kJ mol}^{-1} \qquad (4.6)$$

$$nC_2H_4 \rightarrow \text{polymer} \rightarrow C(s) \qquad (4.7)$$

However, the formed coke can also be removed from the catalyst surface through the coke steam reforming (see (4.8)) or the reverse Boudouard reaction (see (4.6)) [7,8,10]:

$$C(s) + H_2O \rightarrow CO + H_2 \qquad \Delta H^{\circ}_{298K} = 131 \text{ kJ mol}^{-1} \qquad (4.8)$$

For methanol conversion, several reaction routes have been reported such as decomposition, steam reforming and catalytic conversion to different fuel compounds (Figure 4.1). In MSR, dehydrogenation, dehydration and decomposition occur (e.g. [15]). Zhang *et al.* [6] have presented the ESR reactions over skeletal Ni catalysts based on their studies as follows (Figure 4.2). The reactions are proposed to occur in steps, the first one being ethanol dehydrogenation followed by decomposition reactions [6].

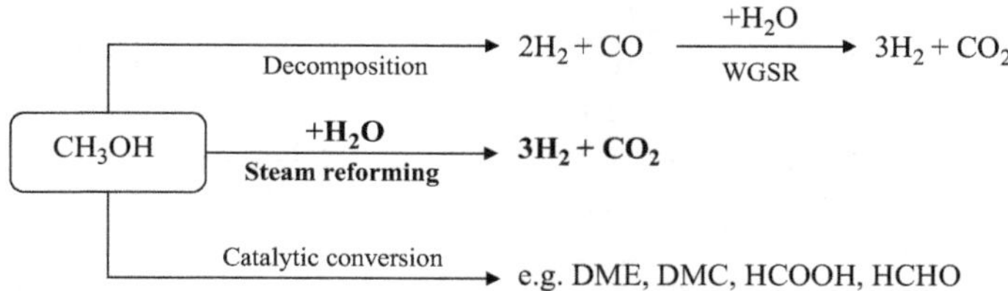

Figure 4.1   Steam reforming and other potential routes for methanol conversion (DME = dimethyl ether, DMC = dimethyl carbonate)

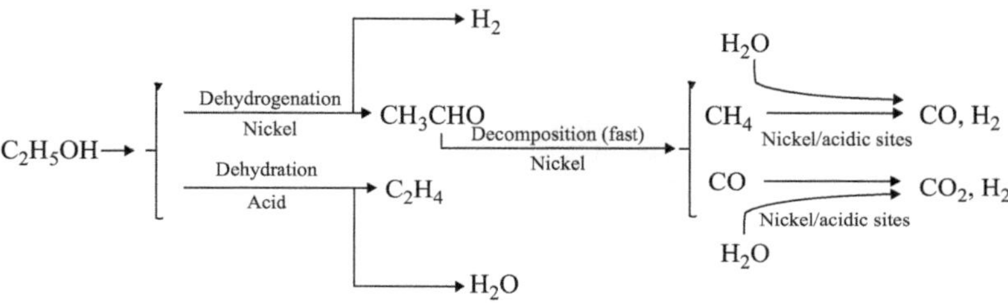

Figure 4.2   Reaction scheme of ESR on Ni-based catalysts [6, reprinted with permission of Wiley]

Hammoud *et al.* [1] have listed, especially in the on-board reforming process, the following superior advantages for methanol in hydrogen production compared to other liquid fuels:

1.   Methanol is a low-cost chemical with a low boiling point.
2.   Reforming can be done at low temperatures and atmospheric pressure.
3.   Methanol is a simple molecule with a high molar ratio of hydrogen to carbon and easy to store.
4.   Low CO concentration formed as CO is poison for catalysts.
5.   No emissions of environmentally harmful compounds, such as NO_x, SO_x.

Ethanol is also an efficient and promising source in hydrogen production as it (e.g. [6,8]):

1.   can be obtained easily from renewable resources,
2.   is non-toxic and free of sulphur or nitrogen (no NO_x and SO_x emissions),
3.   is thermodynamically feasible to decompose, and
4.   is easy to store and transport.

4.2   Catalysts for alcohols steam reforming

An extensive number of papers and articles have been published on methanol and ESR catalysts. Steam reforming of alcohols has been actively studied and reviewed using numerous catalytic materials based on transition and precious metals (e.g. Cu, Ni, Pd and Pt) supported on various metal and mixed oxides (e.g. Al_2O_3, ZnO, TiO_2, ZrO_2, CeO_2, CeO_2–ZrO_2 and $Al_2O_3/CeO_2/ZrO_2$) as well as on carbon supports (e.g. activated carbon (AC), multiwalled carbon nanotubes (MWCNTs)) [1,3,7,16–22]. This chapter summarizes the latest catalyst developments for MSR and ESR reactions, and their respective preparation methods and future directions are as well discussed.

Copper (>580 articles) and nickel (>750 articles) based catalysts for methanol and ESR, respectively, have been at glance of research during the last two decades. Figure 4.3 shows the number of publications dealing with these reactions with Cu and Ni during the years 1996–2015 based on the Web of Science database (adapted in January 2016) [23].

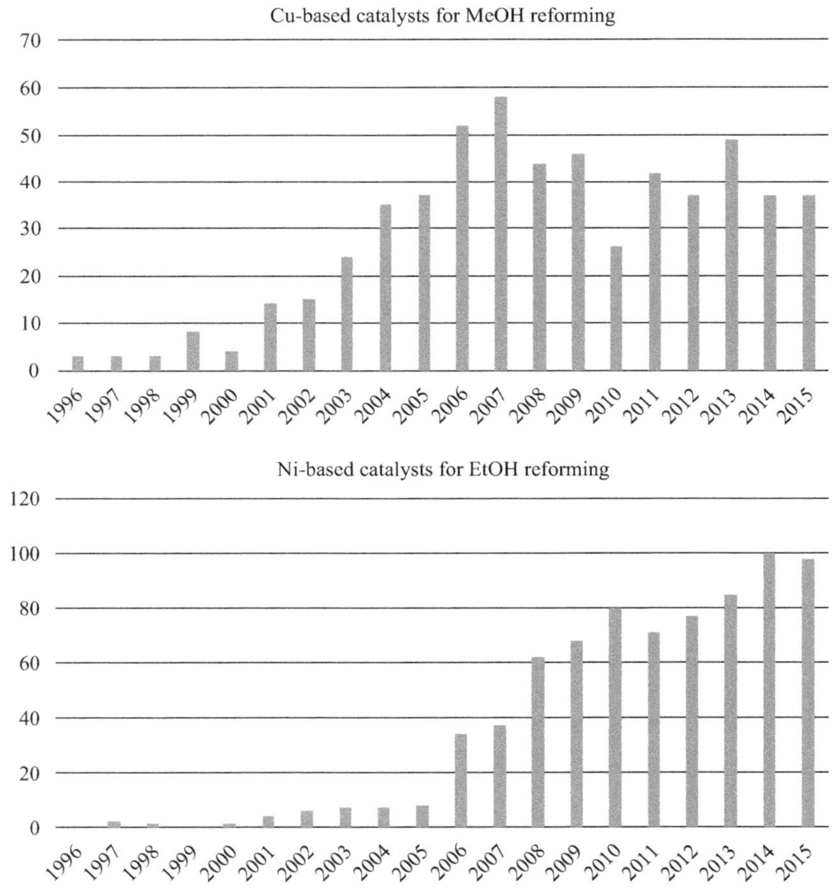

Figure 4.3   Number of publications related to methanol and ethanol steam reforming using Cu- and Ni-based catalysts, respectively [23]

Both copper- and nickel-based catalytic materials on various supports and with several additives including noble metals and oxides are developed and tested in alcohol reforming [24,25]. However, the need of more efficient and stable catalysts is crucial since chemical, mechanical and thermal deactivation phenomena such as coke formation, surface pore blocking and sintering are possible and present in the reforming reactions. In addition, selectivity towards hydrogen without any significant formation of carbon monoxide is essential (e.g. [1,2]).

4.2.1   Catalysts for methanol steam reforming

In MSR, various catalytic materials have been investigated with different compositions under various operating conditions. A wide range of non-noble, especially Cu, and noble metals (Pd and Pt) supported on various carrier materials have been studied in MSR. In Table 4.1, recently studied catalysts in MSR are addressed, the emphasis being on Cu-based catalysts. Most of the studies on MSR catalysts' development are focused on Cu-based materials due to the high activity, low cost and abundant reserves of copper [26]. In this part, we have restricted the review only to Cu-based catalysts used in MSR. Copper is deposited on various carrier materials with different compositions and combination of other metal/metal oxides. Cu is more active at low temperatures ($<$350 °C), but deactivates fast due to the reaction conditions and changes in the active phase [9,27]. The deactivation of Cu can be reduced or eliminated by adding promoters and co-catalysts such as ZnO, GaO$_x$, CeO$_2$–ZrO$_2$ [7,28–31].

In Table 4.1, the recent studies conducted using different operating conditions and catalyst compositions have led to variations in conversions and selectivities. The method of preparation is the most important step to prepare active and selective catalysts. Preparing the support/carrier materials and using metal modification have a significant effect on the overall activity. Moreover, the effect of synthesis conditions, precursors, compositions and pre-treatment steps leads to different physico-chemical properties of the catalyst materials. For example, a traditional method of preparation such as metal impregnation is not effective compared to a microwave assisted combustion method [28].

Novel preparation methods and new materials are designed and developed recently showing promising results in MSR. Even the lab-made catalysts prepared with surfactant-assisted co-precipitation and followed by Cu impregnation are more stable and active when operated at high space velocities compared to the MSR commercial catalysts [9]. MSR is thermodynamically favourable at higher reaction temperatures that is at above 200 °C, and it is an endothermic reaction [25,32]. The main requirements for an MSR catalyst are as follows: it should have high stability (i.e. coke and poison resistance) and high mechanical and thermal stability (no sintering); it should be attrition resistant, and it should have high activity per unit volume in the reactor as well as high selectivity, reproducibility and low cost (preparation and materials costs). The salient features of MSR catalysts possess high basicity (optimal of weak to strong basic sites), reasonable high specific surface area, strong metal–support interactions and optimal catalytic system composition to

Table 4.1 List of Cu-based catalysts in methanol steam reforming

Catalyst and its composition (wt.%)	Preparation method (support + metal)	Reaction conditions	Conversion and selectivity (%)	Salient features	Reference
$CuO/ZnO/Al_2O_3$	Microwave assisted combustion	$T = 240\ °C$, $m_{cat} = 0.4$ g, $R = 1.5$, $GHSV = 10{,}000$ mL g^{-1} h^{-1}	$X_{CH_3OH} = 100$, $S_{H_2} = 80$	Shaped catalyst, higher specific surface area and active phase dispersion, CuO and ZnO are main active sites, optimal fuel to nitrate ratio is needed. TOS = 20 h	[28]
$CuO/ZnO/CeO_2/Al_2O_3$ ($Cu_{50}Zn_{30}Ce_{10}Al_{10}$)	Co-precipitation	$m_{cat} = 0.2$ g, $d_p = 0.34$ mm, $R = 1.2$	$X_{CH_3OH} = 80$, $S_{CO} = 0.055$	Effect of dopants on the activity, over CeO_2 the CO selectivity decreases	[34]
$Cu_{0.1}Fe_{0.9}Al_2O_4$	Single step solution combustion	$T = 300\ °C$, $R = 1.1$, $Q_{tot} = 0.6$ mL h^{-1}, $GHSV = 30{,}000$ h^{-1}	$X_{CH_3OH} =\sim 98$, $S_{CO} = \sim5$	Combustion synthesis is superior than impregnation, Cu^{2+} ion sites are more active than CuO dispersed by impregnation	[4]
$Cu_{0.20}Ce_{0.80}O_2Ga_{0.023}$	Co-precipitation	$T = 400\ °C$, $m_{cat} = 6.73$ g, $d_p = 4\text{–}7$ nm, $R = 2$, $Q_{tot} = 0.15$ mL min^{-1}	$X_{CH_3OH} = 95$, $S_{CO} < 0.05$	CuO being highly dispersed on the fluorite CeO_2 support, high deactivation	[35]
15% Cu-MCM-41	One-pot procedure (solution dissolving)	$T = 300\ °C$, $R = 3$, $GHSV = 2{,}838$ h^{-1}	$X_{CH_3OH} = 89$, $S_{H_2} = 99$, $S_{CO} = 0.8$	Cu loading is the crucial factor, Cu above 15 wt.% decreases the Cu dispersion and leads less resistant to deactivation. High SSA leads to high stability TOS = 48 h	[2]

Catalyst	Synthesis method	Conditions	Performance	Remarks	Ref.
15% Zn–TiO$_2$	Facile one-step synthesis	$T = 350\ ^\circ$C	$X_{CH_3OH} = 88$, $S_{H_2} = \sim 100$, $S_{CO} = 1.3$	TiO$_2$ mesoporous structure hindering the crystal growth, nature of metal, optimal Zn loadings is important parameters	[3]
10% Cu/ZnAl	Wet impregnation memory effect (carbonate solution-aqueous)	$T = 250\ ^\circ$C, $R = 2$, $Q_{tot} = 0.8$ mL h^{-1}	$X_{CH_3OH} = 52$, $S_{H_2} = 75.44$	High activity is based on optimal Cu concentration and catalyst reducibility	[1]
Cu$_{0.06}$Zn$_{0.06}$/γ-Al$_2$O$_3$/Al	Anodic oxidation + electroless deposition	$T = 350\ ^\circ$C, $R = 1,\ 1.5$, $Q_{tot} = 9,000$ mL h^{-1}	$X_{CH_3OH} = 78$	Optimal electro-less deposition time and optimal Cu/Zn concentrations (6 wt.% each)	[29]
Ni$_{0.2}$–Cu$_{0.8}$/a*-ZrO$_2$	Precipitation method + sequential impregnation	$T = 350\ ^\circ$C, $m_{cat} = 0.3$ g, $R = 1$	$X_{CH_3OH} = 100$, $S_{H_2} = 99$	Support structure is the main influence on the activity and degree crystallinity decreases the activity, Cu is more selective than Ni	[7]
45%CuO/20%ZnO/20%CeO$_2$–15% ZrO$_2$	Co-precipitation	$T = 240\ ^\circ$C, GHSV $= 1,200$ h^{-1}, $R = 1.2$	$X_{CH_3OH} = 100$, $S_{H_2} = 75$, $S_{CO} < 0.8$	Precipitant concentrations remarkably influenced the catalyst structure and property, Concentration of precursor (0.1 mol L^{-1}) and precipitant (0.5 mol L^{-1}) exhibited highest activity	[30]
Cu/ZrO$_2$ (ca. 27.3% ZrO$_2$)	Fractionated precipitation	$T = 230\ ^\circ$C, $m_{cat} = 0.5$ g, $R = 1.2$, WSHV $= 4.8$ mL g$_{cat}^{-1}$ h^{-1}	$X_{CH_3OH} = 95$, $S_{H_2} = 74.5$, $S_{CO} = 0.25$	Optimal ZrO$_2$ loadings, synergic effect between copper and zirconia, high Cu SSA, dispersion and also adsorption of water over ZrO$_2$ due to Cu$^+$	[36]

(*Continues*)

Table 4.1   (*Continued*)

Catalyst and its composition (wt.%)	Preparation method (support + metal)	Reaction conditions	Conversion and selectivity (%)	Salient features	Reference
Cu–Ga/ZnO	Incipient wetness impregnation	$T = 320\ °C$, $m_{cat} = 0.1$ g, $R = 1.3$, $Q_{tot} = 1.662$ mL h^{-1}	$X_{CH_3OH} = 96$	Promoter effect of Ga_2O_3, formates formation	[31]
10%Cu/Ce$_{0.6}$Zr$_{0.4}$O$_2$	Surfactant assisted co-precipitation + incipient wetness impregnation	$T = 320\ °C$, $m_{cat} = 0.1$ g, $R = 1$, $Q_{tot} = 0.4$ mL h^{-1}, GHSV $= 40,000$ h^{-1}	$X_{CH_3OH} = 9$, $S_{CO} = 18$	Cu-phases and Cu$^+$/Cu^0 ratio is crucial, 48% of methanol conversion after 50 h TOS	[9]

*a-ZrO_2 = amorphous ZrO_2, T = reaction temperature (°C), m_{cat} = catalyst mass (g), R = molar ratio (H_2O/CH_3OH); d_p = particle diameter; SSA = specific surface area, Q_{tot} = total flow, TOS = time-on-stream (h), X_{CH_3OH} = methanol conversion (%), S_{H_2} = hydrogen selectivity, S_{CO} = CO selectivity, S_{CH_4} = CH$_4$ selectivity, S_{CO_2} = CO_2 selectivity, GHSV = gas hourly space velocity, WHSV = weight hour space velocity.

obtain high performance. The type of carrier/support materials also plays a vital role that of the metal/metal oxide particles [2,3,7,19]. The carrier materials should have high steam and methanol adsorption capacity and avoid the phase transformation of the active phase. In one study, ZnO addition to Cu/CeO_2 was shown to reduce the unwanted CO formation. As reported, ZnO promotes the WGS reaction and increases the oxygen mobility of the CeO_2 [33]. In Table 4.1, the catalysts used in the MSR and their performance are presented.

Several preparation methods including, for example, wet and incipient wetness impregnations, electroless deposition and co-precipitation among other methods are used in Cu insertion on supports as presented in Table 4.1. The MSR reaction mechanism route depends on the type of metal/metal oxides. Over the noble metals, methanol dehydrogenation takes place, whereas over the non-noble metal catalysts (e.g. Cu-based) direct SR of methanol takes place [37]. One of the best results has been obtained by Yu *et al.* [32] at relatively low temperature (150 °C) as complete methanol conversion was observed to take place via direct reforming without CO formation over $Cu/ZnGa_2O_4$ spinel oxide catalysts with a high H_2 production rate. According to the research done by them, Cu with 3–4 nm size particles and clusters stabilized over defective $ZnGa_2O_4$ phase are found to be the most active centres for direct reforming [32]. Bimetallic catalytic systems such as intermetallic compounds, for example Pd–Cu on a spinel structure of $ZnAl_2O_4$ are found to be promising due to their multifunctional behaviour and the H_2 spill-over effect [38].

4.2.2 Catalysts for ethanol steam reforming

In the case of ESR, the nickel-based catalysts are found to be the most active and promising catalysts. Ni activity in ESR on various supports such as SBA-15 [8], montmorillonite [39,40], as well as on numerous oxides (in a single form or in mixed composites) (e.g. [41–44]) besides carbonaceous (AC and CNT supports) (e.g. [20–22]) are recently investigated. Several preparation methods including, for example wet and wetness impregnations, co-impregnation and (co)precipitation are used in metal insertion on supports as presented in Table 4.2.

In our earlier studies, the CNTs and AC as well as graphite carbon black decorated with Ni (5–10 wt.% of nominal loadings) using impregnation have been investigated. It was found that with a low metal loading, the particle size has a greater influence on the catalyst activity [21,45]. Further tests were carried out with Pt and ZnO promoted Ni/CNT catalysts, for enhancing the reforming activity and reducing the CO formation. Significant changes were observed in the textural properties of the pre-treated CNTs and Ni, NiPt and NiZnO decorated CNTs due to the synthesis steps. The promotional effect of Pt in $Ni_{10}Pt_x$/CNT ($x = 1$, 1.5 and 2 wt.%) catalysts was insignificant in ESR. The formation of ethylene over the studied CNT-based catalysts was found to be very low, i.e. <0.3 vol.%. Solid carbon formation on surfaces was, however, detected during the reforming experiments [21,45].

The effect of support has been reported to have an influence on various phenomena during reactions such as dispersion, chemical, poisoning, electronic

Table 4.2 Examples of catalysts for ethanol steam reforming

Catalyst	Preparation method (support + metal modification)	Reaction conditions	Conversion of ethanol and selectivity (%)	Salient features	Reference
Ni/ZnO–Al$_2$O$_3$	Precipitation + wet impregnation	$T = 500\ ^\circ$C, $m_{cat} = 0.02$ g, $R = 3$	$X_{ethanol} = 100$, $S_{H_2} = 72$, $S_{CO} = 18$, $S_{CO_2} = 8$	Highly stable for TOS 25 h, effect of Zn addition is found to more stable, optimal ZnO loadings is crucial	[41]
15%Ni–6%Sr/Al$_2$O$_3$–ZrO$_2$ xerogel	Single epoxide-driven sol–gel + co-impregnation of Ni	$T = 450\ ^\circ$C, $m_{cat} = 0.1$ g, $Q = 1$ mL h^{-1}, $R = 6$, WHSV $= 28\ 280$ mL h^{-1} g$_{cat}^{-1}$	$X_{ethanol} = 100$, $S_{CH_4} = 29.4$, $S_{CO_2} = 70.6$, $S_{H_2} = 85$	Strontium promoter reduce the acidity and enhance the Ni dispersion	[49]
LaNi$_{0.7}$Co$_{0.3}$O$_3$/ZrO$_2$	Precipitation (ZrO$_2$) + one step citrate complexing	$m_{cat} = 0.025$ g, $Q_{tot} = 1.2$ mL h^{-1}, $R = 3$, GHSV $= 264{,}000$ mL h^{-1} g$_{cat}^{-1}$	$X_{ethanol} = 100$, $S_{CO} = 20$, $S_{CO_2} = 10$, $S_{H_2} = 60$	Good activity and stability for TOS 50 h, synergistic effect of the Ni–Co alloy nanoparticles on the ZrO$_2$, La$_2$O$_3$ reduce coke formation	[47]
Ni$_3$Al$_{0.8}$Fe$_{0.2}$	Co-precipitation	$T = 550\ ^\circ$C, $m_{cat} = 0.05$g, $R = 6$, WHSV $= 147$ mol$_{ethanol}$ h^{-1} kg$_{cat}^{-1}$	$X_{ethanol} = 100$, $S_{CO} = 0.15$, $S_{CO_2} = 68$, $S_{H_2} = 70$, $S_{CH_4} = 0.25$	Highly stable for TOS 90 h, synergistic effect between aluminium and iron, limiting C$_2$H$_4$ formation	[50]
8% Ni La$_2$O$_3$–ZrO$_2$ (6 wt.% La$_2$O$_3$)	Wet impregnation	$T = 350\ ^\circ$C, $m_{cat} = 1$ g, $d_p = 100$–200 μm, $R = 30$, $Q_{tot} = 0.1$ mL min^{-1}	$X_{ethanol} = 100$, $S_{CO} = 0$, $S_{CO_2} = 20$, $S_{H_2} = 62$, $S_{CH_4} = 8$	High H$_2$ yield, TOS 24 h, CeO$_2$ and La$_2$O$_3$ lead to better Ni dispersion	[43]

Catalyst	Synthesis method	Operating conditions	Performance	Remarks	Ref.
20 wt.%Ni/Y_2O_3–Al_2O_3 (1:1)	Co-precipitation + impregnation	$T = 500\ ^{\circ}\mathrm{C}$, $m_{cat} = 0.15$ g, $R = 13$, $Q_{tot} = 0.05$ mL min^{-1}	$X_{ethanol} = 100$, $S_{CO} = 0$, $S_{CO_2} = 65$, $S_{CH_4} = 34$	Highly stable for TOS 60 h, Y_2O_3 was found effective on dehydrogenation, Y/Al ratio influence the activity and 1:1 is the best optimum mole ratio	[44]
5 wt.% Ni/MgO	Precipitation + incipient wetness impregnation	$T = 500\ ^{\circ}\mathrm{C}$, $m_{cat} = 0.02$ g, $R = 10$	$X_{ethanol} = 72$, $S_{H_2} = 70$, $S_{CO} = 3$, $S_{CO_2} = 28$, $S_{CH_4} = 2$	Precipitation and aging MgO-based catalyst resulted in high basicity and high activity. Exhibited the highest NiO reduction degree	[42]
$Cu_{0.2}$–Ni_{15}–Al_2O_3–ZrO_2	Single-step epoxide-driven sol–gel method	$T = 450\ ^{\circ}\mathrm{C}$, $m_{cat} = 0.1$ g, $R = 6$, WHSV $= 28{,}280$ mL h^{-1} g^{-1}	$X_{ethanol} = 100$, $S_{CO} = 1.8$, $S_{CO_2} = 62$, $S_{CH_4} = 0$	High H_2 yield of 87% achieved TOS $= 16.7$ h, optimal Cu content is needed to achieve highest performance	[51]
20 wt.% Ni/Zr–0.7%CeO_2	Co-precipitation method + wet impregnation	$T = 700\ ^{\circ}\mathrm{C}$, $m_{cat} = 3$ g, $R = 9$, WHST $= 99{,}609$ kg$_{cat}$ s kmol$_{EtOH}^{-1}$	$X_{ethanol} = 100$, $S_{CO} = 15.1$, $S_{CO_2} = 11$, $S_{CH_4} = 0.2$	Zr in ceria avoids sintering, keeps OSC and facilitates the transition of Ce^{4+} to Ce^{3+}, CeO_2 improves the H_2O dissociation and prevents carbon formation. High H_2 yield of 5.3 mol obtained	[52]

(*Continues*)

Table 4.2 (*Continued*)

Catalyst	Preparation method (support + metal modification)	Reaction conditions	Conversion of ethanol and selectivity (%)	Salient features	Reference
Ni/15%La$_2$O$_3$–10%CeO$_2$–γ-Al$_2$O$_3$	Successive wet impregnation	$T = 500\ °C$, $m_{cat} = 0.1$ g, $d_p = 0.18$–0.36 mm, $R = 3$, $Q_{tot} = 0.09$ mL min^{-1}, GHSV $= 26{,}000$ h^{-1}	$X_{ethanol} = 100$, $S_{CO} = 1.8$ $S_{CO_2} = 7.3$, $S_{CH_4} = 3.6$, $S_{H_2} = 82$	Promoters decrease ethylene formation by reducing the strong Lewis acid sites of γ-Al$_2$O$_3$, TOS $= 24$ h	[53]
6%Ni–1.2%Au/SBA-15	Incipient wetness impregnation	$T = 550\ °C$, $m_{cat} = 0.2$ g, $R = 3$, $Q_{tot} = 0.02$ mL min^{-1}, GHSV $= 10{,}920$ h^{-1}	$X_{ethanol} = 100$, $S_{CO} = 10$ $S_{CO_2} = 20$, $S_{CH_4} = 5$, $S_{H_2} = 80$	TOS $= 25$ h, Au promoter improves the interaction between the SBA-15 and Ni phase and forms a well-dispersed with smaller Ni particles	[40]
10%NiO–10%ZnO/ MWCNTs	Incipient wetness impregnation	$T = 350\ °C$, $m_{cat} = 0.1$ g, $R = 4$, $Q_{tot} = 0.02$ mL min^{-1}, GHSV $= 10{,}920$ h^{-1}	$X_{ethanol} = 100$, $S_{CO} = 0.6$ $S_{CO_2} = 17$, $S_{H_2} = {\sim}76$	ZnO promoted Ni$_{10}$/ MWCNTs catalysts performed better due to enhanced WGS activity	[21]
Ni/Ce$_{0.9}$Zr$_{0.1}$O$_2$	Impregnation	$T = 600\ °C$, $m_{cat} = 0.044$ g, $R = 6$	$X_{ethanol} = 87$, $S_{CO} = 27$, $S_{CO_2} = 61$, $S_{H_2} = 66$		[54]
NiCo/Ce$_{0.9}$Zr$_{0.1}$O$_2$	Impregnation	$T = 600\ °C$, $m_{cat} = 0.044$ g, $R = 6$	$X_{ethanol} = 97$, $S_{CO} = 24$, $S_{CO_2} = 70$, $S_{H_2} = 81$	Addition of Co increased the H$_2$ yield and slightly decreased CO formation	[54]

$T =$ reaction temperature (°C), $m_{cat} =$ catalyst mass (g), $R =$ molar ratio (H$_2$O/C$_2$H$_5$OH); $d_p =$ particle diameter; SSA $=$ specific surface area, $Q_{tot} =$ total feed flow, TOS $=$ time-on-stream (h), $X_{ethanol} =$ ethanol conversion (%), $S_{H_2} =$ hydrogen selectivity (%), $S_{CO} =$ CO selectivity (%), $S_{CH_4} =$ CH$_4$ selectivity (%), $S_{CO_2} =$ CO$_2$ selectivity (%), GHSV $=$ gas hourly space velocity, WHSV $=$ weight hour space velocity, WHST $=$ weight hour space time.

and bifunctional effects. Suitable and usable support materials in ESR are required to have at least the following properties based on the findings by Lin *et al.* [46]: favouring good metal dispersion and stability, hydrophilicity promoting water adsorption and activation, good electronic and chemical efficiency towards the desired reaction products, and tolerance towards coking. Addition of other metals or additives besides Ni/NiO has been found to have an advantageous impact on the hydrogen evolution yields and particle sizes of metals on surfaces [40,47]. Recently a new class of ESR catalysts was reported in Jo *et al.* [48], who introduced a $30Ni_{8.5}Mn_{1.5}/70SiO_2$ core–shell-structured catalyst in which Mn addition improves the redox properties of Ni and enhances the stability by reducing coke and CO formation via the WGS reaction. In Table 4.2, some examples of the catalysts that have recently been developed and used in ESR and their performance in terms of conversion of ethanol and selectivity are presented.

4.3   Advances and drawbacks

One of the main drawbacks in steam reforming of alcohols is the high temperatures needed, and thus the sintering of active metal has to be take into account. Sintering of Cu particles can occur already at relatively low temperatures. The Hüttig and Tamman temperatures of metallic Cu are 134 and 405 °C, respectively, of CuO; these temperatures are 207 and 527 °C and of Cu_2O 179 and 481 °C, respectively. In the case of nickel, the Hüttig and Tamman temperatures for the metallic Ni particles are 245 and 590 °C. However, for NiO, the values are 396 and 841 °C, respectively [18,55]. This means that without stabilizers, phase transformations of these metals and oxides leading to sintering start at quite low temperatures.

Coking of a catalyst is another serious drawback. It is well known that in the case of steam reforming of hydrocarbons (e.g. methane) or using light alcohols (methanol, ethanol), the coke formation deactivates the catalyst by surface fouling. Carbon deposits on the surface can occur via various reaction routes forming different types of carbonaceous compounds causing, for example pore blocking and solid carbon coverage on the catalyst surface (e.g. [56]). It has been reported that with the addition of promoters (such as La, Mg, Sr, K and Ca) into the support materials, the carbon formation reactions (via decomposition) can be hindered, and deposition of solid carbon on surfaces can be avoided. The promoting materials are assumed to have an effect on the acid site strength and number, water dissociation, metal particle size, stability and deactivation rate [43,49,57]. In Table 4.3, some advantages and drawbacks are summarized for ethanol and methanol reforming.

There are some key points in catalyst development for alcohols reforming reactions, which should be taken into account. Li and Gong [58] have presented four 'key learning points': (1) the knowledge of the role of active sites on surfaces, (2) the improvement of anti-sintering properties using synthetic base-metal catalysts, (3) the surface oxygen mobility and its role on catalyst activity and stability and (4) the process intensification applications.

Table 4.3   Some of the benefits and drawbacks in applying methanol and ethanol as reactants in hydrogen production

	Methanol	**Ethanol**
Advantages	Sources from industrial wastes	Non-toxic
	Easy availability	Safe handling
	High H/C ratio	Easy storage
	Easy storage	Renewable source
	Highly miscible with water	Liquid at RT
	No sulphur or nitrogen contents	Availability
	H_2 production at low T (150–350 °C)	No sulphur or nitrogen contents
	Low CO and coke formation at low T	
Disadvantages	Toxic and flammable compound	Highly energy intensive
	Environmentally harmful	Use for edible purposes
	High CO_2 emissions and capture costs	Flammable compound
		Coking on surfaces

4.4 Conclusions

The demand of hydrogen is increasing, and its production is needed to be done in a more environmental and efficient way. Steam reforming of light alcohols such as methanol and ethanol can be one alternative route for sustainable production of hydrogen. The light alcohols are excellent hydrogen sources which can be reformed at relatively low reaction temperatures.

There is a big challenge in performing the steam reforming reaction at low temperatures (<450 °C) using alcohols in order to increase the hydrogen production yield and selectivity. Besides these demands, there is a need to control the CO formation, coking on the surfaces of the catalysts and sintering of active metals and metal oxides. Thus, catalyst supports and metal catalysts have a critical role to produce H_2 at low operating temperatures.

In this part, the copper- and nickel-based catalysts are reviewed in methanol and ESR reactions, respectively. It can be concluded that recently developed catalytic materials are good and their performance in reforming reactions has been improved remarkably. However, there are still open questions in the catalyst development to be answered like preventing the surface carbon formation and selectivity towards hydrogen without any carbon monoxide formation as a by-product.

List of abbreviations

AC	activated carbon
a-ZrO_2	amorphous ZrO_2
DMC	dimethyl carbonate
DME	dimethyl ether

GCB	graphite carbon black
GHSV	gas hourly space velocity
ESR	ethanol steam reforming
MSR	methanol steam reforming
(MW)CNT	(multiwalled) carbon nanotube
SSA	specific surface area (m^2/g)
TOS	time-on-stream (h)
WHST	weight hour space time
WHSV	weight hour space velocity
WGS(R)	water gas shift (reaction)

List of symbols

d_p	particle diameter (nm)
m_{cat}	catalyst mass (g)
Q_{tot}	total flow
R	molar ratio (H_2O/CH_3OH or H_2O/C_2H_5OH)
S_Y	selectivity (%), in where $Y = H_2$, CO, CO_2 or CH_4
T	reaction temperature (°C)
X_{CH_3OH}	methanol conversion (%)
$X_{ethanol}$	ethanol conversion (%)

References

[1] Hammoud D., Gennequin C., Aboukaïs A., Abi Aad E. 'Steam reforming of methanol over $x\%$ Cu/Zn–Al 400 500 based catalysts for production of hydrogen: preparation by adopting memory effect of hydrotalcite and behavior evaluation'. *International Journal of Hydrogen Energy* 2015;**40**: 1283–1297.

[2] Deshmane V.G., Owen S.L., Abrokwah R.Y., Kuila D. 'Mesoporous nano-crystalline TiO_2 supported metal (Cu, Co, Ni, Pd, Zn, and Sn) catalysts: effect of metal-support interactions on steam reforming of methanol'. *Journal of Molecular Catalysis A: Chemical.* 2015;**408**:202–213.

[3] Deshmane V.G., Abrokwah R.Y., Kuila D. 'Synthesis of stable Cu–MCM-41 nanocatalysts for H_2 production with high selectivity via steam reforming of methanol'. *International Journal of Hydrogen Energy.* 2015;**40**(33): 10439–10452.

[4] Maiti S., Llorca J., Dominguez M., *et al.* 'Combustion synthesized copper-ion substituted $FeAl_2O_4$ ($Cu_{0.1}Fe_{0.9}Al_2O_4$): a superior catalyst for methanol steam reforming compared to its impregnated analogue'. *Journal of Power Sources.* 2016;**304**:319–331.

[5]   IEA (2009) World Energy Outlook 2009, 698 p.

[6]   Zhang C., Li S., Wu G., *et al.* 'Steam reforming of ethanol over skeletal Ni-based catalysts: a temperature programmed desorption and kinetic study'. *AIChE Journal.* 2014;**60**(2):635–644.

[7]   Lytkina A.A., Zhilyaeva N.A., Ermilova M.M., Orekhova N.V., Yaroslavtsev A.B. 'Influence of the support structure and composition of Ni–Cu-based catalysts on hydrogen production by methanol steam reforming'. *International Journal of Hydrogen Energy.* 2015;**40**(31):9677–9684.

[8]   Kim D., Kwak B.S., Min B.-K., Kang M. 'Characterization of Ni and W co-loaded SBA-15 catalyst and its hydrogen production catalytic ability on ethanol steam reforming reaction'. *Applied Surface Science.* 2015;**332**: 736–746.

[9]   Das D., Llorca J., Dominguez M., Colussi S., Trovarelli A., Gayen A. 'Methanol steam reforming behaviour of copper impregnated over CeO_2–ZrO_2 derived from a surfactant assisted co-precipitation route'. *International Journal of Hydrogen Energy.* 2015;**40**(33);10463–10479.

[10]   Hou T., Zhang S., Chen Y., Wang D., Cai W. 'Hydrogen production from ethanol reforming: catalysts and reaction mechanism'. *Renewable and Sustainable Energy Reviews.* 2015;**44**:132–148.

[11]   Segal S.R., Carrado K.A., Marshall C.L., Anderson K.B. 'Catalytic decomposition of alcohols, including ethanol, for in situ H_2 generation in a fuel stream using a layered double hydroxide-derived catalyst'. *Applied Catalysis A: General.* 2003;**248**:33–45.

[12]   Bshish A., Yaako Z., Narayanan B., Ramakrishnan R., Ebshish A. 'Steam-reforming of ethanol for hydrogen production'. *Chemical Papers.* 2011;**65**(3): 251–266.

[13]   Ni M., Leung D.Y.C., Leung M.K.H. 'A review on reforming bio-ethanol for hydrogen production'. *International Journal of Hydrogen Energy.* 2007;**32**: 3238–3247.

[14]   Panagiotopoulou P., Verykios X.E. 'Mechanistic aspects of the low temperature steam reforming of ethanol over supported Pt catalysts'. *International Journal of Hydrogen Energy.* 2012;**7**(21):16333–16345.

[15]   Frank B., Jentoft F.C., Soerijanto H., Kröhnert J., Schlögl R., Schomäcker R. 'Steam reforming of methanol over copper-containing catalysts: influence of support material on microkinetics'. *Journal of Catalysis.* 2007;**246**: 177–192.

[16]   Spivey J.J., Agrell J., Lindström B., Pettersson L.J., Järås S.G. 'Catalytic hydrogen generation from methanol', in Spivey J.J., Agarwal S.K. (eds.). *Catalysis.* (UK, The Royal Society of Chemistry, 2002), vol. 16, pp. 67–132.

[17]   Sá S., Silva H., Brandão L., Sousa J.M., Mendes, A. 'Catalysts for methanol steam reforming – a review'. *Applied Catalysis B: Environmental.* 2010; **99**(1–2):43–57.

[18]   Matsumura Y. 'Durable Cu composite catalyst for hydrogen production by high temperature methanol steam reforming'. *Journal of Power Sources.* 2014:**272**:961–969.

[19] Mierczynski P., Vasilev K., Mierczynska A., Maniukiewicz W., Szynkowska M. I., Maniecki T.P. 'Bimetallic Au–Cu, Au–Ni catalysts supported on MWCNTs for oxy-steam reforming of methanol'. *Applied Catalysis B: Environmental*. 2016;**185**:281–294.

[20] Seelam P.K. 'Hydrogen production by steam reforming of bio-alcohols: conventional and membrane-assisted catalytic reactors', Ph.D. Thesis, University Oulu, ACTA series C473: ISBN 978-952-62-0277-8, December 2013.

[21] Seelam P.K., Rautio A.R., Huuhtanen M., Turpeinen E., Kordás K., Keiski R.L. 'Low temperature steam reforming of ethanol over advanced carbon nanotube based catalysts'. *Green Processing Synthesis*. 2015;**4**(5):355–368.

[22] Rautio A.-R., Seelam P.K., Mäki-Arvela P., Huuhtanen M., Keiski R.L., Kordas K. 'Carbon supported catalysts in low temperature steam reforming of ethanol: study of catalyst performance'. *RSC Advances*. 2015;**5**: 49487–49492.

[23] Web of Science. Available from http://www.isiknowledge.com/ [Adapted January 2016].

[24] Wu H.S., Lesmana D. 'Short review: Cu catalyst for autothermal reforming methanol for hydrogen production'. *Bulletin of Chemical Reaction Engineering & Catalysis*. 2012;**7**:27–42.

[25] Contreras J.L., Salmones J., Colín-Luna J.A., *et al.* 'Catalysts for H_2 production using the ethanol steam reforming: a review'. *International Journal of Hydrogen Energy*. 2014;**39**:18835–18853.

[26] Yong S.T., Ooi C.W., Chai S.P., Wu X.S. 'Review of methanol reforming-Cu-based catalysts, surface reaction mechanisms, and reaction schemes'. *International Journal of Hydrogen Energy*. 2013;**38**(22):9541–9552.

[27] Luo X., Hong Y., Wang F., *et al.* 'Development of nano Ni_xMg_yO solid solutions with outstanding anti-carbon deposition capability for the steam reforming of methanol'. *Applied Catalysis B: Environmental*. 2016;**194**: 84–97.

[28] Ajamein H., Haghighi M. 'On the microwave enhanced combustion synthesis of $CuO–ZnO–Al_2O_3$ nanocatalyst used in methanol steam reforming for fuel cell grade hydrogen production: effect of microwave irradiation and fuel ratio'. *Energy Conversion and Management*. 2016;**118**:231–242.

[29] Reddy E.L., Lee H.C., Kim D.H. 'Steam reforming of methanol over structured catalysts prepared by electroless deposition of Cu and Zn on anodically oxidized alumina'. *International Journal of Hydrogen Energy*. 2015; **40**(6):2509–2517.

[30] Zhang L., Lei J.-T., Tian Y., *et al.* 'Effect of precursor and precipitant concentrations on the catalytic properties of $CuO/ZnO/CeO_2–ZrO_2$ for methanol steam reforming'. *Journal of Fuel Chemistry and Technology*. 2015;**43**(11):1366–1374.

[31] Toyir J., Piscina P.R., Homs N. 'Ga-promoted copper-based catalysts highly selective for methanol steam reforming to hydrogen; relation with the hydrogenation of CO_2 to methanol'. *International Journal of Hydrogen Energy*. 2015;**40**(34):11261–11266.

[32] Yu K.M., Tong W., West A., *et al.* 'Non-syngas direct steam reforming of methanol to hydrogen and carbon dioxide at low temperature'. *Nature Communication*. 2012;3:1230.

[33] Tonelli F., Gorriz O., Arrúa L., Abello M.C. 'Methanol steam reforming over Cu/CeO$_2$ catalysts: influence of zinc addition'. *Quimica Nova*. 2011; **34**(8):1334–1338.

[34] Wan Y., Zhou Z., Cheng Z. 'Hydrogen production from steam reforming of methanol over CuO/ZnO/Al$_2$O$_3$ catalysts: catalytic performance and kinetic modelling'. *Chinese Journal of Chemical Engineering*. 2016;24:1186–1194.

[35] Pohar A., Hočevar S., Likozar B., Levec J. 'Synthesis and characterization of gallium-promoted copper–ceria catalyst and its application for methanol steam reforming in a packed bed reactor'. *Catalysis Today*. 2015;**256**:358–364.

[36] Zhou J., Zhang Y., Wu G., Mao D., Lu G. 'Influence of the component interaction over Cu/ZrO$_2$ catalysts induced with fractionated precipitation method on the catalytic performance for methanol steam reforming'. *RSC Advances*. 2016;**6**:30176–30183.

[37] Oar-Arteta L., Remiro A., Epron F., *et al.* 'Comparison of noble metal- and copper-based catalysts for the step of methanol steam reforming in the dimethyl ether steam reforming process'. *Industrial and Engineering Chemistry Research*. 2016;**55**(12):3546–3555.

[38] Mierczynski P., Vasilev K., Mierczynska A., Maniukiewicz W., Maniecki T.P. 'Highly selective Pd–Cu/ZnAl$_2$O$_4$ catalyst for hydrogen production'. *Applied Catalysis A: General*. 2014;**479**:26–34.

[39] Li T., Zhang J., Xie X., Yin X., An X. 'Montmorillonite-supported Ni nanoparticles for efficient hydrogen production from ethanol steam reforming'. *Fuel*. 2015:**143**:55–62.

[40] He S., He S., Zhang L., *et al.* 'Hydrogen production by ethanol steam reforming over Ni/SBA-15 mesoporous catalysts: effect of Au addition'. *Catalysis Today*. 2015;**258**:162–168.

[41] Anjaneyulu C., Costa L.O.O., Ribeiro M.C., *et al.* 'Effect of Zn addition on the performance of Ni/Al$_2$O$_3$ catalyst for steam reforming of ethanol'. *Applied Catalysis A: General*. 2016;**519**:85–98.

[42] Wurzler G.T., Rabelo-Neto R.C., Mattos L.V., Fraga M.A., Noronha F.B. 'Steam reforming of ethanol for hydrogen production over MgO-supported Ni-based catalysts'. *Applied Catalysis A: General*. 2016;**518**:115–128.

[43] Dan M., Mihet M., Tasnadi-Asztalos Z., Imre-Lucaci A., Katona G., Lazar M.D. 'Hydrogen production by ethanol steam reforming on nickel catalysts: effect of support modification by CeO$_2$ and La$_2$O$_3$'. *Fuel*. 2015;**147**: 260–268.

[44] Ma H., Zhang R., Huang S., Chen W., Shi Q. 'Ni/Y$_2$O$_3$–Al$_2$O$_3$ catalysts for hydrogen production from steam reforming of ethanol at low temperature'. *Journal of Rare Earths*. 2012;**30**(7):683–690.

[45] Seelam P.K., Huuhtanen M., Sápi A., *et al.* 'CNT-based catalysts for hydrogen production by ethanol reforming'. *International Journal of Hydrogen Energy*. 2010;**35**:12588–12595.

[46] Lin J., Chen L., Shin Choong C.K., Zhong Z., Huang L. 'Molecular catalysis for the steam reforming of ethanol'. *Science China, Chemistry*. 2015;**58**:60–78.

[47] Zhao L., Han T., Wang H., Zhang L., Liu Y. 'Ni–Co alloy catalyst from $LaNi_{1-x}Co_xO_3$ perovskite supported on zirconia for steam reforming of ethanol'. *Applied Catalysis B: Environmental*. 2016;**187**:19–29.

[48] Jo S.W., Kwak B.S., Kim K.M., *et al.* 'Reasonable harmony of Ni and Mn in core@shell-structured $NiMn@SiO_2$ catalysts prepared for hydrogen production from ethanol steam reforming'. *Chemical Engineering Journal*. 2016;**288**:858–868.

[49] Song J.H., Han S.J., Yoo J., Park S., Kim D.H., Song K. 'Effect of Sr content on hydrogen production by steam reforming of ethanol over $Ni–Sr/Al_2O_3–ZrO_2$ xerogel catalysts'. *Journal of Molecular Catalysis A: Chemical*. 2016;**418–419**:68–77.

[50] Abelló S., Bolshak E., Gispert-Guirado F., Farriol X., Montané D. 'Ternary Ni–Al–Fe catalysts for ethanol steam reforming'. *Catalysis Science and Technology*. 2014;**4**:1111–1122.

[51] Han S.J., Song J.H., Bang Y., *et al.* 'Hydrogen production by steam reforming of ethanol over mesoporous $Cu–Ni–Al_2O_3–ZrO_2$ xerogel catalysts'. *International Journal of Hydrogen Energy*. 2016;**41**(4):2554–2563.

[52] Patel M., Jindal T.K., Pant K.K. 'Kinetic study of steam reforming of ethanol on Ni-based ceria–zirconia catalyst'. *Industrial and Engineering Chemistry Research*. 2013;**52**(45):15763–15771.

[53] Osorio-Vargas P., Flores-González N.A., Navarro R.M., Fierro J.L.G., Campos C.H., Reyes P. 'Improved stability of Ni/Al_2O_3 catalysts by effect of promoters (La_2O_3, CeO_2) for ethanol steam-reforming reaction'. *Catalysis Today*. 2016;**259**:27–38.

[54] Moretti E., Storaro L., Talon A., *et al.* 'Ceria-zirconia based catalysts for ethanol steam reforming'. *Fuel*. 2015;**153**:166–175.

[55] Moulijn J.A., van Diepen A.E., Kapteijn F. 'Catalyst deactivation: is it predictable? What to do?' *Applied Catalysis A. General*. 2001;**212**:3–16.

[56] Argyle M.D., Bartholomew C.H. 'Heterogeneous catalyst deactivation and regeneration: a review'. *Catalysts*. 2015;**5**(1):145–269.

[57] Trane-Restrup R., Dahl S., Jensen A.D. 'Steam reforming of ethanol: effects of support and additives on Ni-based catalysts'. *International Journal of Hydrogen Energy*. 2013;**38**:15105–15118.

[58] Li S., Gong J. 'Strategies for improving the performance and stability of Ni-based catalysts for reforming reactions'. *Chemical Society Reviews*. 2014;**43**:7245–7256.

Chapter 5

Transition metal catalysts for hydrogen production by low-temperature steam reforming of methane

Antonio Vita[1]

Abstract

This chapter discusses different catalytic systems based on transition metals (nickel, rhodium, ruthenium, platinum) for the hydrogen production by steam reforming (SR) of methane at low temperature (≤ 823 K).

The design of robust catalysts for low temperature (≤ 823 K) reforming processes is fundamental for an optimized integration between reforming reactors and concomitant separation/purification steps that usually work at low temperature; therefore, the preparation methods will be also described and compared, especially, considering their contribution to develop catalytic materials with properties as high surface area, high active metal dispersion, low particle size, resistance to carbon formation, opportune metal load and so on.

All these features and others, which will be discussed along this chapter, are very important to overcome deactivation phenomena related and in some cases enhanced by reforming processes conducted at low temperature.

The aim of this chapter is to analyse, from a different point of view, some synthesis routes, generally reported in literature as methods to prepare catalysts for high temperature reforming processes. Thus, correlations between chemical–physical/morphological properties and catalytic activity towards low-temperature SR of methane will be evidenced in order to explore the potential to use the prepared materials for application in integrated processes that combine low-temperature SR reactions with separation/purification step.

5.1 Introduction

The hydrogen consumption is expected to increase dramatically in the near future. In the petroleum refineries, it has already been used to produce clean transportation

[1]Institute for Advanced Energy Technologies (ITAE), "Nicola Giordano", National Research Consilium (CNR), Messina, Sicily, Italy

fuels [1]. In the energy field, the developments in fuel cell (FC) technologies [2] have generated a need to convert the conventional fuels such as natural gas or coal and biofuels as biogas [3] or bioethanol [4] to either pure hydrogen or syngas for efficient power generation. In addition, the decreasing supply of crude oil and rising demand for clean transportation fuels (almost free from sulphur, to meet the stringent environmental regulations imposed in several countries) in recent years led to intensive efforts to the development of alternative sources of fuels through various conversion technologies, including gas-to-liquid [5], coal-to-liquid [6] and biomass-to-liquid (BTL) [7], which involve both hydrogen and syngas as key components.

Industrially, on a large scale, hydrogen production is generally conducted by steam reforming (SR) of methane that theoretically offers the highest H_2/CO ratio (close to 3, since a part of the hydrogen comes from water) respect the other principal reforming processes partial oxidation and autothermal reforming.

The general SR reaction for hydrocarbons and alcohols is represented as follows:

$$C_nH_mO_k + (n - k)H_2O \rightarrow nCO + \left(n - k + \frac{n}{2}\right)H_2, \quad (n > K), \quad \Delta H^{\circ}_{298} > 0$$

$$(5.1)$$

The basic reaction of methane SR is as follows:

$$CH_4 + H_2O \rightarrow 3H_2 + CO, \quad \Delta H^{\circ}_{298} = 205.9 \text{ kJ mol}^{-1} \tag{5.2}$$

In addition, water–gas shift (WGS) represents an important side reaction that must be considered:

$$CO + H_2O \leftrightarrow CO_2 + H_2 \quad \Delta H^{\circ}_{298} = -41.13 \text{ kJ mol}^{-1} \tag{5.3}$$

The methane SR process is normally designed to operate at high pressure ($\sim$25–30 bar), high steam-to-carbon (S/C) molar ratios (typically in the range of 2.5–3) and at the temperatures around 973–1,273 K [8]. About the amount of steam, even if, (5.1) and (5.2) suggest that only 1 mole of H_2O is required for 1 mole of CH_4; the reaction in practice is being performed using a high S/C ratio in order to reduce the risk of carbon deposition on the catalyst surface.

Moreover, since the overall reaction is endothermic, it is necessary to supply the needed heat by some routes. In SR of natural gas, this is accomplished by combustion of a part of the fuel in direct-fired or indirectly fired furnaces. To achieve an almost complete conversion of the fuel to syngas (CO and H_2), a very high temperature and long residence time is necessary, meaning an overall energy loss and a huge size for the reforming reactor.

To obtain H_2 with high pure grade for industrial application ($H_2 = 99.999\%$) or for heat and electricity production by polymer electrolyte membrane–based FC ($H_2 \approx 99\%$) [9], the obtained syngas is subjected to several downstream processes that contribute to increase the size of the complete system. The gas exiting the reformer is cooled to about 623 K and then subjected to the WGS reaction in a high-temperature shift (HTS) converter.

The current process for the industrial production of pure H_2 employs pressure swing adsorption for the purification of H_2 after the shift reaction [10].

Alternatively, after the HTS, catalytic processes can be used to reduce the CO content as low-temperature shift (LTS) reaction, preferential oxidation (PROX, see (5.4)) and/or methanation reaction (see (5.5)) [11].

$$CO + 0.5O_2 \rightarrow CO_2 \quad \Delta H^{\circ}_{298} = -283 \text{ kJ mol}^{-1} \tag{5.4}$$

$$CO + 3H_2 \rightarrow CH_4 + H_2O, \quad \Delta H^{\circ}_{298} = -205.9 \text{ kJ mol}^{-1} \tag{5.5}$$

Typical catalysts for reforming processes are Ni based, since they offer sufficient activity with low cost and high availability [12–14]. Precious metal catalysts such as Rh, Pt and Ru can also be used in SR processes, giving activities higher than Ni-based catalytic materials [15].

However, high cost and low availability are the major drawbacks. Moreover, different support materials and dopants have been reported to improve thermal stability and coke resistance [16,17].

Many of these, studied in investigations at laboratory scale, are prepared with the aim to produce materials which can be used as a stable, active and selective catalysts for processes that require high temperatures to effectively convert methane into H_2-rich gas mixtures. At low reforming temperature, deactivation phenomena, principally due to the carbon deposition, are promoted.

In Figure 5.1, are reported the range conditions in which carbon formation is thermodynamically favoured as a function of *S/C* ratio, temperature and pressure for SR of methane. It is evident that in the temperature range of interest ($\approx$823 K),

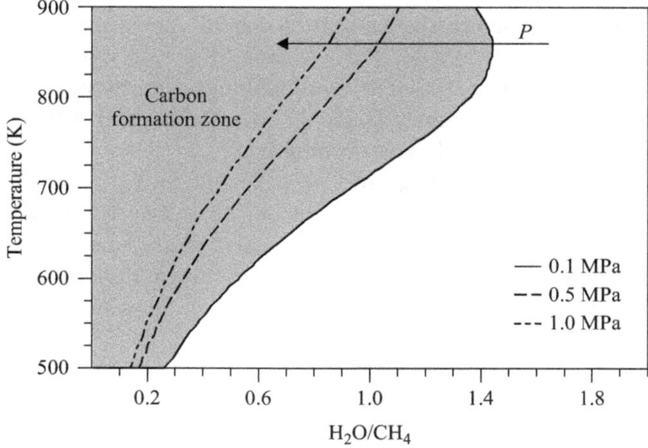

Figure 5.1   Carbon formation conditions for a steam–methane system

the methane–steam system is driven towards carbon formation only at low *S/C* ratios and high pressure.

The possibility to integrate the reforming step with the purification step (based, principally, on H_2-selective membrane technologies) in a single reactor, open new scenery to the development of preparation methodologies able to fabricate materials that can be effectively used in low-temperature reforming processes (≤ 823 K). Methane SR at low temperature results in low conversion of the fuels due to thermodynamic limitation; however, this limit can be surpassed by concomitant separation of H_2 by a selective membrane [18]. The fuel and steam are fed into the reactor under pressure, as the SR reaction proceeds; H_2 is driven across the membrane by the pressure difference, at the same time the shift of the reforming reaction occurs [19]. The fuel can be completely converted, and pure hydrogen is separated from the reaction products. An important advantage of a reforming reactor, equipped with hydrogen-permeable membranes, is the reduction of the number of vessels and consequently of the capital cost as well as simplification of the process and design.

In this regards, the role of the reforming catalyst is fundamental at least as that of the membrane. In order to attain high methane conversion, both a highly active catalyst and a high-permeability membrane are required.

In general, any preparation method involves a sequence of several complex steps that contribute to obtain a solid phase with new properties acquired and stabilized during the various steps. Changes in the preparative details may result in dramatic alterations in the properties of the final catalyst.

The properties of a good catalyst for low reforming temperature processes are not particularly different from those of catalytic materials used in conventional high-temperature reforming. Even in this application area, properties which determine directly catalytic activity and selectivity, here such factors as bulk and surface chemical composition, active phase dispersion, local microstructure and phase composition are important as well as resistance to carbon formation, porosity, high thermal and mechanical stability.

This chapter will summarize shortly some of the preparation procedures reported in the recent literature for the synthesis of solid catalysts based on transition metals (Ni, Rh, Ru and Pt). The reported preparation routes and the properties of the resulted catalytic materials will be correlated with the performance towards the SR of methane at low temperature.

Of course, establish guidelines for the development of a scientific basis for catalyst preparation may be a very ambitious objective, but the aim of this chapter is to identify general correlations between the catalyst properties and the catalytic performance as well as identify correlation between preparation procedures and the introduction or the improvement of these properties in the resulted catalysts.

5.2   Ni-based catalysts

Ni/Al_2O_3-based catalysts are the most common catalytic materials used for the SR of methane due to their high activity and low cost [20] but are also more prone to deactivation for carbon formation or sintering [21]. It was extensively reported that the support can contribute to the characteristic properties of the developed catalyst, especially to limit the carbon formation [22–25]. In a recent review, Angeli *et al.* [26] have evidenced the decisive role of some supports to enhance the catalyst activity for the methane SR process at low temperature. The authors, among the studies reported, have denoted that the most common materials used as supports or support dopants are CeO_2, ZrO_2 and their mixed oxides. High oxygen storage capacity and redox properties lead to efficient coke resistance, making these materials advantageous over conventionally used Al_2O_3 or $MgAl_2O_4$.

Matsumura *et al.* have investigated the performance of nickel-based catalysts prepared by an impregnation technique for SR reaction at low temperatures [27]. Metal oxide supports such as γ-alumina, silica and zirconia were impregnated with nickel nitrate by evaporation of the aqueous solution at 353 K. After drying in air at 383 K overnight, the catalytic materials were heated at 973 K for 3 h. Properties as nickel particle size and Brunauer–Emmett–Teller (BET) surface area have been recorded after the reduction procedure; the data are reported in Table 5.1. The catalyst tested at 773 K with an *S/C* equal to 2 and a low weight hourly space velocity (WHSV) equal to 15,000 mL h^{-1} g_{cat}^{-1} (catalyst = 0.3 g, total flow = 75 mL min^{-1}) have shown different performances (Table 5.2). The reported results are significantly far from the thermodynamic equilibrium calculations; the Ni/SiO_2-based sample loss is activity quickly because of the oxidation of the Ni particles, also the Ni/Al_2O_3-based catalyst loses its initial activity probably due to formation of spinel ($NiAl_2O_4$). Only the Ni/ZrO_2 shows stable performance. The authors suggest that the formation of active hydroxyl groups on the support surface plays an important role in the SR mechanism. In this study, low carbon formation was detected on the used catalysts; this means that it cannot be responsible for the deactivation phenomena.

A Ni/CeO_2–ZrO_2–θ–Al_2O_3 catalyst has been prepared using a double impregnation technique by Liu *et al.* [28]; first, a γ-Al_2O_3 was impregnated with a solution containing Ce–acetate and Zr–nitrate ($CeO_2/ZrO_2 = 0.25$) at room temperature for 12 h. The pre-coated sample was calcined at 1,173 K for 6 h; during the heat treatment, the γ-Al_2O_3 was transformed into θ-Al_2O_3. Nickel was loaded by impregnation using an appropriate amount of $Ni(NO_3)_2 \cdot 6H_2O$ after the drying process at 373 K. The sample was calcined at 773 K for 6 h. The nickel loading was 12 wt%. In general, the catalyst shows very high activity, and CH_4 conversion near the equilibrium at *S/C* equal to 3 is approached at temperatures from 723 to 923 K. In addition, the catalyst shows a high stability without observable deactivation for 200 h during the SR reaction at 923 K and WHSV equal to 1,251 mL CH_4 g_{cat}^{-1} h^{-1}.

Angeli *et al.* [29] prepared, via the wet impregnation method, nickel-based catalysts using as precursors of the active metal $Ni(NO_3)_2 \cdot 6H_2O$ to obtain metal loading of 10 wt%. Lanthanum-doped cerium–zirconium oxide (78 wt% ZrO_2, 17 wt% CeO_2, 5 wt% La_2O_3) and lanthanum-doped zirconium hydroxide (oxide-based composition: 90 wt% ZrO_2, 10 wt% La_2O_3) were used as catalyst supports. The characteristics of the prepared catalysts are reported in Table 5.1 with the principal preparative steps [29]. The data show similar properties between the

Table 5.1 Summary table for the characteristics of the employed methods and the resulting properties of the Ni-based catalysts

Catalysts	Active metal content Ni (wt%)	Preparation method	Support calcination (K/h)	Catalyst drying (K/time)	Catalyst calcination (K/h)	Specific surface area BET ($m^2\ g^{-1}$)	Metal dispersion D (%)[c]	Metal particle size TEM (nm)	Metal particle size XRD (nm)
Ni/SiO$_2$ [27]	20%	Impregnation	–	383/overnight	973/3	237	–	–	15
Ni/TiO$_2$ [36]	10%	Impregnation	773/1	383/overnight	673/3	300	2.6	–	13
Ni/Al$_2$O$_3$ [27]	20%	Impregnation	–	383/overnight	973/3	129	–	–	11
Ni/Al$_2$O$_3$ [34]	13.5%	Impregnation	–	–	1,123/5	106.8	6.4	–	9
Ni/Al$_2$O$_3$[b] [34]	12%	–	–	–	–	12.3	–	–	23.1
Ni/TiO$_2$ [36]	10%	Impregnation	773/1	383/overnight	673/3	40	1.3	–	6
Ni/ZrO$_2$ [27]	20%	Impregnation	–	383/overnight	973/3	0.24	–	–	11
Ni/CeO$_2$–ZrO$_2$–Al$_2$O$_3$ (Ce/Zr = 0.25[a]) [28]	12%	Impregnation	1,173/6	373/–	823/6	–	–	–	–
Ni/90%ZrO$_2$–10% La$_2$O$_3$ [29]	10%	Impregnation	973/2	393/overnight	–	45.07	1.62	–	15
Ni/17%CeO$_2$–78% ZrO$_2$–5%La$_2$O$_3$ [29]	10%	Impregnation	873/2	393/overnight	–	37.17	1.2	–	18
NiAl$_2$O$_4$/Al$_2$O$_3$ (Ni/Al = 0.3)[a] [32]	24%	Impregnation	–	382/overnight	1,123/3	62	18	10.2	<2–5
NiAl$_2$O$_4$/Al$_2$O$_3$ (Ni/Al = 0.2)[a] [32]	17%	Coprecipitation/ deposition	–	383/overnight	1,123/4	84	33	9	<2–5
NiAl$_2$O$_4$/Al$_2$O$_3$ (Ni/Al = 0.3)[a] [32]	24%	Coprecipitation/ deposition	–	383/overnight	1,123/4	57	20	9	<2–5
Ni/MgO$_2$–La$_2$O$_3$– Al$_2$O$_4$ [33]	25%	Coprecipitation/ deposition	–	383/overnight	1,173/5	103.4	–	–	19
Ni$_{0.5}$Mg$_{2.5}$(Al)O (Ni + Mg/Al = 3)[a] [34]	16.5%	Coprecipitation	–	378/–	1,123/5	173.6	11	–	6.8
Ni/TiO$_2$ [36]	5%	Impregnation	773/1	383/overnight	673/3	40	5.2	–	–
Ni/TiO$_2$ [36]	10%	Impregnation	773/1	383/overnight	673/3	42	2.8	–	7
Ni/TiO$_2$ [36]	20%	Impregnation	773/1	383/overnight	673/3	41	0.6	–	13

[a]Atomic ratio.
[b]Commercial catalyst.
[c]H$_2$-chemisorption.

Table 5.2   Summary table for operative conditions and performance of the Ni-based catalysts towards the methane steam reforming

Catalysts	Active metal content Ni (wt%)	T_r (K)	S/C	P (MPa)	WHSV (mL h^{-1} g^{-1})	GHSV (h^{-1})	Conversion[a] X_{CH4} (%) Exp.	Conversion[a] X_{CH4} (%) Eq.	TOF[d] (s^{-1})	Carbon (wt%) on catalyst/TOS[e]
Ni/SiO$_2$ [27]	20%	773	2	0.1	15,000	–	21.8	33.9	–	0.2/5 h
Ni/Al$_2$O$_3$ [27]	20%	773	2	0.1	15,000	–	15.0	33.9	–	0.12/5 h
Ni/Al$_2$O$_3$ [34]	13%	773	2	0.1	1.6×10^6	–	9.35	34	0.27	–
Ni/Al$_2$O$_3$[b] [34]	12%	773	2	0.1	1.6×10^6	–	3.55	33.9	0.03	–
Ni/ZrO$_2$ [27]	20%	773	2	0.1	15,000	–	14.1	33.9	–	–
Ni/CeO$_2$–ZrO$_2$–Al$_2$O$_3$ (Ce/Zr = 0.25[a]) [28]	12%	673	3	0.1	20,000	–	14.6	19.8	–	–
Ni/CeO$_2$–ZrO$_2$–Al$_2$O$_3$ (Ce/Zr = 0.25[a]) [28]	12%	773	3	0.1	20,000	–	45.0	43.6	–	–
Ni/CeO$_2$–ZrO$_2$–Al$_2$O$_3$ (Ce/Zr = 0.25[a]) [28]	12%	823	3	0.1	20,000	–	62.9	59.7	–	–
Ni/90%ZrO$_2$–10%La$_2$O$_3$ [29]	10%	773	3	0.1	–	30,000	28	43.6	–	–
Ni/90%ZrO$_2$–10%La$_2$O$_3$ [29]	10%	673	3	0.1	–	70,000	3.6	19.8	–	–
Ni/90%ZrO$_2$–10%La$_2$O$_3$ [29]	10%	773	3	0.1	–	70,000	20.6	43.6	–	–
Ni/90%ZrO$_2$–10%La$_2$O$_3$ [29]	10%	823	3	0.1	–	70,000	38.1	59.7	–	–
Ni/17%CeO$_2$–78%ZrO$_2$–5%La$_2$O$_3$ [29]	10%	773	3	0.1	–	30,000	41.6	43.6	–	0.05/90 h
Ni/17%CeO$_2$–78%ZrO$_2$–5%La$_2$O$_3$ [29]	10%	673	3	0.1	–	70,000	13.4	19.8	–	–
Ni/17%CeO$_2$–78%ZrO$_2$–5%La$_2$O$_3$ [29]	10%	773	3	0.1	–	70,000	28.9	43.6	–	–
Ni/17%CeO$_2$–78%ZrO$_2$–5%La$_2$O$_3$ [29]	10%	823	3	0.1	–	70,000	42.4	59.7	–	–
NiAl$_2$O$_4$/Al$_2$O$_3$ (Ni/Al = 0.3) [32][c]	24%	723	3	0.1	–	38,400	4	30.2	0.6	–
NiAl$_2$O$_4$/Al$_2$O$_3$ (Ni/Al = 0.3) [32][c]	24%	823	3	0.1	–	38,400	26	59.7	1.6	–
NiAl$_2$O$_4$/Al$_2$O$_3$ (Ni/Al = 0.2) [32]	17%	723	3	0.1	–	38,400	14	30.2	–	–
NiAl$_2$O$_4$/Al$_2$O$_3$ (Ni/Al = 0.2) [32]	17%	823	3	0.1	–	38,400	47	59.7	–	–
NiAl$_2$O$_4$/Al$_2$O$_3$ (Ni/Al = 0.3) [32]	24%	723	3	0.1	–	38,400	17	30.2	–	–
NiAl$_2$O$_4$/Al$_2$O$_3$ (Ni/Al = 0.3) [32]	24%	823	3	0.1	–	38,400	46	59.7	–	–
Ni/MgO$_2$–La$_2$O$_3$–Al$_2$O$_4$ [33]	25%	723	3	0.9	8,000–64,000	–	13.1	13.2	–	–
Ni/MgO$_2$–La$_2$O$_3$–Al$_2$O$_4$ [33]	25%	773	3	0.9	8,000–64,000	–	19.4	19.5	–	–
Ni$_{0.5}$Mg$_{2.5}$(Al)O (Ni + Mg/Al = 3) [34]	16.5%	773	2	0.1	1.6×10^6	–	14.6	33.9	0.35	–
Ni/TiO$_2$ [36]	10%	773	3	0.1	6,000	–	40	43.6	–	–

[a]X_{CH4} (%) = [CH$_4$]$_{in}$ − [CH$_4$]$_{out}$/[CH$_4$]$_{in}$.
[b]Commercial catalyst.
[c]Prepared by impregnation.
[d]Surface atom–based reac. rate (mol/mol Me$_{surf}$s).
[e]Time on stream (TOS).

prepared samples; the impregnation procedure resulted in a decrease of the specific surface area of the pure support for both the samples due to the pore blockage by the metal particles. Moreover, the dispersion on the catalysts was very similar, and, as expected, the crystallite size also shows no substantial differences. Regarding the catalytic activity towards methane SR, the sample with ceria in the support structure shows better performances respect the Ni/Zr–La catalyst, approaching the equilibrium at 773 K with a S/C equal to 3 at relatively high gas hourly space velocity (GHSV $= 30,000$ h^{-1}), see Section 5.2 [29]. As the temperature decreases, the conversion decrease and the differences among the catalysts are more pronounced. The authors ascribe the better catalytic activity of the Ni/Ce–Zr–La catalyst to the active role of reduced ceria in the reforming mechanism through the activation of steam and the increase of the reforming rate by the oxygen back spillover process.

The tendency towards the formation of carbon was assessed using a model biogas (CH$_4 = 50\%$, CO$_2 = 40\%$) as feed type at $S/C = 3$, $T = 773$ K and 0.7 MPa (to simulate the condition of membrane reactor) because the high content of CO$_2$ may contribute to the accumulation of carbonaceous deposits [30,31]. Very stable behaviour, even after 90 h on stream, was observed for Ni/Ce–Zr–La. The stability can be ascribed to the presence of ceria in the support which provides active oxygen for the minimization of coke formation and also to the lanthana dopant, known for its thermal stabilizing effect, and to the low affinity for coking. The high mobility of oxygen formed by the decomposition of H$_2$O on the support surface and its diffusion through ceria facilitates the oxidation of any carbonaceous deposits formed. In addition to effect of ceria, the presence of La$_2$O$_3$ contributes to the resistance in coke accumulation, though the reaction of active La$_2$O$_2$CO$_3$ with the deposited carbon species.

Jiménez-González *et al.* [32] have investigated the possibility to use the nickel aluminate phase (NiAl$_2$O$_4$) as an effective precursor for producing highly active and stable Ni/alumina catalysts for the SR of methane with a S/C ratio of 3 in the 723-923 K temperature range. Two alumina-supported samples with a Ni loading of 17% and 24 wt% were prepared by coprecipitation–deposition method through the drop-by-drop addition of a 0.6 M solution of NH$_4$OH (350-400 cm^3) under constant stirring into an aqueous solution (50 cm^3) of a mixture of nickel acetate and aluminium nitrate (1:2 Ni/Al molar ratio) and γ-Al$_2$O$_3$ alumina (BET $= 133$ m^2 g^{-1}). The amounts of Ni(CH$_3$COO)$_2$·4H$_2$O and Al(NO$_3$)$_3$·9H$_2$O were 7.3 and 21.9 g, respectively, for the CP/A (17 wt%) sample and 18 and 54.7 g, respectively, for the CP/A (24 wt%) sample. The temperature was kept at 298 K. The initial pH was about 3, and it took 2 h to reach a value equal to 8. Afterwards, the precipitates were aged for 30 min at this pH value before being filtered and washed with hot deionized water. On the other hand for comparison, a NiAl$_2$O$_4$/Al$_2$O$_3$ with a 24 wt%Ni content was also synthesized by impregnation. All the samples were dried at 383 K overnight and then calcined at 1,123 K in static air for 4 h at a heating rate of 10 K min^{-1} in order to promote the formation of the NiAl$_2$O$_4$ phase. The textural properties of the prepared catalysts are reported in Table 5.1 [32]. The data evidence similar properties between the coprecipitated catalysts. Both samples are characterized by high dispersion and low metal particle sizes respect the catalysts prepared by impregnation previously described and also respect the catalyst prepared for comparison. Table 5.2 reports the performances of the prepared catalysts [32] under different operating conditions ($T = 723$–823 K, $S/C = 3$, time on stream $= 12.5$ h); previously, the catalysts were in situ activated by reduction with a flow of 5% H_2/N_2 at 1,123 K for 2 h. As expected, methane conversion increased with temperature. The samples prepared by coprecipitation showed a comparable activity with a small difference in CH_4 conversion. Instead, because the dispersion of these coprecipitated catalysts was significantly larger, higher conversion levels were consequently noticed in comparison with the sample prepared by impregnation. The authors ascribe the bad performance of the impregnated catalyst to the presence of larger crystallite size (10.2 nm, Table 5.1), assigned to the significant abundance of NiO species in the calcined catalytic precursor. Instead, the coprecipitation route led to structurally homogeneous $NiAl_2O_4$ catalysts with a relatively reduced presence of NiO species. After high-temperature reduction, which resulted in the formation of small nickel crystallites (9 nm, Table 5.1), a relatively high dispersion and a large metallic surface area were achieved. Also the eventual formation of coke was investigated on the sample with 17 wt% of Ni by temperature-programmed oxidation coupled to mass spectrometry. As the evolution of the CO_2 ($m/z = 44$) signal was almost flat, the presence of carbonaceous deposits was considered negligible.

The coprecipitation–deposition method was used also by Chen *et al.* [33] to prepare catalysts based on Ni/MgO–La_2O_3–Al_2O_3 structure. First, the $Mg_4(OH)_4$ $(CO_3)_2$ of particle sizes less than 100 mm was suspended in deionized water in a beaker, then metal nitrates and ammonia solution with concentrations about 1.0 mol dm^{-3} were simultaneously introduced into a beaker under moderate stirring. After the precipitation process, the resulted mixture was further aged for about 2.0 h, then filtered, washed with ammonia added deionized water, dried at 383 K for 12.0 h, and then calcined at 1,173 K for 4.0 h. The textural properties of the prepared catalyst are reported in Table 5.1. The catalytic performances were tested at 723–823 K, $P = 0.9$ MPa and $S/C = 3.0$; the results are reported in Table 5.2, prior to reaction; the catalyst was activated in a 30.0% (v/v) H_2–Ar flux of 50 mL min^{-1} at 1,023 K for 3.0 h. The obtained data have indicated that, at 723 and 773 K, in the GHSV range of 8,000–64,000 mL g_{cat}^{-1} h^{-1}; the methane conversion was very close to thermodynamic equilibrium. In addition, the performances increase the temperature and slightly decrease the WHSV until it shows 96,000 mL g_{cat}^{-1} h^{-1}. The higher performances of the catalyst may be related to its preparation method. The authors suggest that during the synthesis process, all metallic ions in catalyst precursors could be incorporated into the hydrotalcite structure. Thus, the incorporation of lanthanum and magnesium into the nickel–alumina catalyst can contribute to enhance the catalytic activity as well as the resistance to carbon deposition. More in detail, the high activity may be ascribed to the formation of spinel structure during calcinations and after reduction. The authors supposed that surface or bulk nickel aluminate was related to the hydrotalcite structure formation during coprecipitation and calcination. During reduction, due to the strong interaction of alumina with nickel oxide, isolated (mono-dispersed) nickel atoms may be formed. The interaction of those species with the reactants may be related to the high SR activities at lower temperatures considering that the surface reaction of adsorbed carbon species, and adsorbed steam was the rate-determining step.

Nickel-loaded Mg(Al)O catalysts have been prepared by Takehira *et al.* [34,35]. The catalysts prepared by the coprecipitation method have been tested under different SR conditions; in particular, the $Ni_{0.5}/Mg_{2.5}(Al)O$ catalysts have been tested at low reforming temperature (773 K) for kinetic studies. An aqueous solution containing the nitrates of Mg^{2+}, Ni^{2+} and Al^{3+} was added slowly to an aqueous solution of sodium carbonate; the pH of the solution was adjusted at 10. The precipitate was aged at 363 K for 12 h, then dried in air at 378 K and calcined 1,123 K for 5 h. After the calcination $Mg_{2.5}(Al,Ni_{0.5})O$ periclase was obtained as powders and used as the precursor of $Ni_{0.5}/Mg_{2.5}(Al)O$ catalyst; the Ni loading was found to be 16.5 wt%. For comparison purpose, a 13.5 wt%Ni/γ-Al_2O_3 catalyst was prepared by incipient wetness impregnation method. In addition, a commercial 12 wt%Ni/γ-Al_2O_3 catalyst was also used as control. The principal catalyst properties, obtained by several characterization techniques, are reported in Table 5.1 [34,35]. The characterization results evidence that the catalyst prepared by precipitation has a higher specific surface area and Ni metal dispersion than catalysts prepared by incipient wetness impregnation instead; as expected, the particle size follows an opposite trend. The activity of the catalysts was evaluated in conventional SR condition at high temperature (973 K) after reduction pre-reduction under H_2/N_2 (5/25 mL min^{-1}) at 1,173 K for 30 min. However, the turnover frequency (TOF) was evaluated in SR of methane with a small amount of catalyst (10 mg) at low temperature (773 K) with a WHSV equal to 1.6×10^6 mL g^{-1} cat h^{-1}. As reported in Table 5.2 [34,35], the order of activity of the catalysts follows the order of Ni dispersion as well as the order of BET surface area, implying significant effects of those parameters on the catalyst performances. The activity gap between the samples further increases after the steaming treatment, which was performed in a $H_2/H_2O/N_2$ (20/100/25 mL min^{-1}) gas mixture for 10 h at 1,173 K to evaluate the catalysts deactivation. After steaming, the deactivation was the most enhanced for the commercial catalyst ($CH_4^{Conv} = 0.43\%$), followed by 13.5 wt%Ni/γ-Al_2O_3 ($CH_4^{Conv} = 3.75\%$) and $Ni_{0.5}/Mg_{2.5}(Al)O$ ($CH_4^{Conv} = 5.42\%$) catalysts. The analysis of the TOF values of the catalysts, calculated on the basis of the total Ni amount, allows a more precise evaluation of catalyst deactivation. The TOF values of commercial catalyst were one unit smaller compared with those of the $Ni_{0.5}/Mg_{2.5}(Al)O$ and 13.5 wt%Ni/γ-Al_2O_3 samples. Moreover, heavy deactivation took place on this commercial catalyst after steaming. In general, the deactivation for all the catalysts is due to phenomena related with the Ni particle sintering and the Ni oxidation. The potential for using Ni/TiO_2 as a catalyst for the SR reaction at temperatures less than or equal to 773 K was investigated by Kho *et al.* [36]. The nickel-based catalysts were prepared via wet impregnation using Aeroxide P25 TiO_2 as the support material. The calcined (773 K) TiO_2 powder was suspended in 5 mL of ultra-pure water containing a defined amount of dissolved $Ni(NO_3)_2 \cdot 6H_2O$. The resulting paste was dried at 383 K overnight and then calcined at 673 K for 3 h.

Catalysts with nominal nickel loadings of 5, 10 and 20 wt% were prepared. The results of the characterization tests, reported in Table 5.1, indicate that nickel impregnation and calcination had a negligible impact on the specific surface area of the TiO_2. Instead, the dispersion of nickel metal on the oxide was observed to decrease with increasing nickel loading ranging from 5.2% to 0.6% as the nickel loading increased from 5 to 20 wt%. As expected, the particle size increases with the increase of Ni loading. The catalytic activity was evaluated after the activation under 10 vol% H_2/Ar (50 mL min^{-1}) at 673 K for 1 h. CH_4 gas was introduced into the catalyst bed with a WHSV of 6,000 mL CH_4 h^{-1} g$_{cat}^{-1}$. Different temperature (673–773 K) and H_2O/CH_4 (1–3) ratios were investigated; the catalytic activity was compared with nickel (10 wt% nominal) supported on silica and gamma-alumina prepared using the procedure described above. The best performances, reported in Table 5.2 [36], were obtained at 773 K and $H_2O/CH_4 = 3$ with the sample containing the 10 wt% of Ni amount. The catalyst has shown a stable (96 h) methane conversion ($\approx$40%) near the equilibrium. Comparing the performance of the prepared catalyst, it can be argued that the difference in average nickel crystal size may contribute to the difference in methane activation at these temperatures. In addition, the TiO_2-based catalysts show another distinct difference, respect the inert oxide supports such as Al_2O_3 and SiO_2, relative to the strength of interaction of the nickel deposits with the metal oxide supports. The authors suggest that stronger interacting nickel species are the principal participants in low-temperature SR reactions. The formation of these species occurs during the calcination/reduction steps through the dissolution of nickel into the TiO_2 matrix. Finally, a H_2O/CH_4 molar ratio equal to 3 can be effective in reducing the extent of solid carbon formation.

5.3   Noble metal–based catalysts (Rh, Ru and Pt)

Noble metal–based catalysts are more active and stable than Ni catalysts in reforming reactions showing high performances at very low concentrations and resulting more resistant to coke formation [37–42]. From the other side, noble metal–based catalysts are much more expensive than nickel-based catalysts. Several scientists involved in this field have investigated on the order of catalytic activity and selectivity towards hydrogen production via methane SR reaction, including also other metals of the group VIII. Kikuchi *et al.* [43] studied SR of methane at atmospheric pressure in the temperature range 623–873 K; the relative activity was found to be Rh, Ru > Ni > Ir > Pd $\sim$ Pt $\gg$ Co, Fe. Similar results have been obtained by Rostrup-Nielsen and Hansen [44] after a series of experiments with Ru, Rh, Pd, Ir, Ni and Pt on MgO supports, measuring SR activity at 823 K and atmospheric pressure. The relative activities were reported to be Ru, Rh > Ir > Ni > Pt, Pd. Almost identical activity was obtained by Qin and Lapszewicz [45] for noble metals supported on MgO under similar reaction conditions with a temperature range of 873–117 K and atmospheric pressure: Ru > Rh > Ir > Pt > Pd. Based on the studies described above, there is a good agreement on the activity sequence for the noble metals. From the opposite side, the study of Wei and Iglesia [46] showed that Pt performances in terms of C–H bond activation are higher than Ir, Rh and Ru. Instead, recently, Jones *et al.* [47] confirmed that at 773 K Rh and Ru are the most active pure transition metals for methane SR, whereas Ni, Ir, Pt and Pd are significantly less active. The different order of activity obtained can be explained with the fact that different studies, carried out under different conditions, find different results. At low temperatures for the noble metals, the CO formation step is kinetically the most important reaction step [47]. However, as the temperature and reactivity of the metal increase, the most kinetically relevant step switches from being CO formation to dissociative methane adsorption [46,47]. Summarizing, even if the activity of noble metals towards the SR reaction has been the subject of numerous interpretations and discussions, there is a reasonably general consensus regarding the trend in the order of reactivity, among the noble metals. Rh and Ru are the most active especially at low SR temperature; experimental results obtained over Rh- and Ru-based catalysts are summarized in Tables 5.3 and 5.4.

Rh is one of the most active and stable VIII metals (Ni, Ru, Rh, Pt, Pd and Ir) which can catalyse the CH_4 reforming with steam or CO_2 [38,48–50]. Several studies have been investigating the activity of Rh-based catalysts for CH_4 SR over different supports. Most of these studies suggest that the support affects the catalytic activity in steam methane reforming indirectly by influencing the dispersion and the reduction degree of the metal phase [26]. Wei and Iglesia [46] investigated the catalytic activity of Rh/Al_2O_3 catalysts with different Rh content (0.1, 0.2, 0.4, 0.8, 1.6 wt%) prepared by incipient wetness impregnation of Al_2O_3 with an aqueous solution of $Rh(NH_4)_3Cl_6$; the resulting paste was dried at 393 K in ambient air for 12 h and calcined at 1,123 K for 5 h. The results of the metal dispersion analysis show that the dispersion decrease (50.1%–25.1%) with increasing Rh content (0.1–1.6 wt%). The features of the samples with Rh content equal to 0.1 and 1.6 wt% are reported in Table 5.3 [46]. The analysis of the effects of Rh dispersion on turnover rates (normalized by Rh surface atoms measured by H_2 chemisorption) suggests that small clusters are more active than larger Rh crystallites, the turnover rates, calculated at 873 K, increased monotonically with increasing Rh dispersion as shown in Table 5.4. The influence of Rh nanoparticle size, type of support and synthesis procedures on the catalytic performance in steam methane reforming has been investigated by Ligthart *et al.* [51] to clarify the nature of the rate-controlling step. A set of Rh catalysts was prepared using ZrO_2, CeO_2, $CeZrO_2$ and SiO_2 supports by pore volume impregnation, using aqueous solutions of $Rh(NO_3)_3 \cdot nH_2O$ of appropriate concentration. Prior to impregnation, the supports were calcined in a mixture of 20 vol.% O_2 in N_2 at different temperatures (623–1,173 K) for 4 h. The impregnated supports were dried at 383 K overnight and finally calcined at different temperatures (823–1,273). Table 5.3 [51] reports the principal textural properties of some representative samples. The general trend of the dispersion results confirms the strong influence of Rh loading and calcination temperatures; the dispersion increases with the decreasing of Rh amount; as expected, the catalysts with lower Rh loading also show a smaller particle size. Moreover, the dispersion of the $CeZrO_2$-catalysts is typically somewhat higher than that of the

Table 5.3   Summary table for the characteristics of the employed methods and the resulting properties of the noble metals–based catalysts

Catalysts	Active metal content (wt%)	Preparation method	Support calcination (K/h)	Catalyst drying (K/h)	Catalyst calcination (K/h)	Specific surface area BET ($m^2\ g^{-1}$)	Metal dispersion (%)[a]	Metal particle size TEM (nm)	Metal particle size Chems. (nm)[a]
Rh/Al_2O_3 [46]	Rh 0.1%	Impregnation	–	393/–	1,123/5	–	50.1	–	1.4
Rh/Al_2O_3 [46]	Rh 1.6%	Impregnation	–	393/–	1,123/5	–	25.1	–	4
Rh/ZrO_2 [51]	Rh 0.1%	Impregnation	873/–	383/3	873/4	–	79	–	1.4
Rh/ZrO_2 [51]	Rh 0.8%	Impregnation	873/–	384/3	873/4	–	69	–	1.6
Rh/ZrO_2 [51]	Rh 1.6%	Impregnation	873/–	385/3	1,173/4	–	24	4.5	4.7
Rh/CeO_2 [51]	Rh 0.1%	Impregnation	823/–	386/3	823/4	–	83	–	1.3
Rh/CeO_2 [51]	Rh 0.8%	Impregnation	823/–	387/3	823/5	60	51	–	2.1
Rh/CeO_2 [51]	Rh 1.6%	Impregnation	823/–	388/3	1,173/4	–	21	–	3.8
$Rh/Ce_{0.25}Zr_{0.75}O_2$ [51]	Rh 1.6%	Impregnation	873/–	383/3	873/4	84	69	–	1.6
Rh/SiO_2 [51]	Rh 1.6%	Impregnation	873/–	384/3	1,173/4	149	33	3.3	–
$Rh/Ce_{0.6}Zr_{0.4}O_2$ [38]	Rh 0.8%	Impregnation	1,173/6	353/–	673/5	38.8	27.9	–	3
$Rh/Ce_{0.15}Zr_{0.85}O_2$ [37]	Rh 3%	Impregnation	773/4	323/2	773/5	–	–	–	–
$Ru/Ce_{0.15}Zr_{0.85}O_2$ [37]	Ru 3%	Impregnation	773/4	323/2	773/5	–	–	–	–
$Pt/Ce_{0.15}Zr_{0.85}O_2$ [37]	Pt 3%	Impregnation	773/4	323/2	773/5	–	–	–	–
Ru/ZrO_2 [41]	Ru 1%	Impregnation	–	353/–	1,023[b]	–	25	4.9	–
Ru/MgO [53]	Ru 1.5%	Impregnation	923/3	–	673/3	6.7	44.7[c]	2.9 (XRD)	–
Ru/Nb_2O_5 [53]	Ru 1.5%	Impregnation	–	–	–	114.7	86.1[c]	1.5 (XRD)	–
Ru/Nb_2O_5 [53]	Ru 1.5%	Impregnation	773/6	–	673/3	–	–	–	–
Ru/Al_2O_3[40]	Ru 1.5%	Impregnation	923/3	403	1,073/3	191.5	0.8[c]	5–50	–
Rh/CeO_2 [40]	Rh 1.5%	Impregnation	923/3	403	1,073/3	13.9	22.5[c]	5–10	–
Pt/CeO_2 [40]	Pt 1.1%	SCS[d]	–	–	1,073/3	14	12.7[c]	3–7	–

[a]H_2-Chemisorption.
[b]Aging for 336 h in a H_2O/H_2 (ratio $\approx$ 1) at 3.1 MPa.
[c]CO-Chemisorption.
[d]Solution combustion synthesis (SCS).

Table 5.4   Summary table for operative conditions and performance of the noble metals–based catalysts towards the methane steam reforming

Catalysts	Active metal content (wt%)	T_r (K)	S/C (molar)	P (MPa)	WHSV (N mL h^{-1} g^{-1})	GHSV (h^{-1})	Conversion[a] Exp. X_{CH4} (%)	Conversion[a] Eq. X_{CH4} (%)	TOF[f] (s^{-1})	TOF[g] (h^{-1})	Carbon deposition
Rh/Al$_2$O$_3$ [46]	Rh 0.1%	873	1.25	0.1	–	–	–	85	5.7	–	–
Rh/Al$_2$O$_3$ [46]	Rh 1.6%	873	1.25	0.1	–	–	–	86	2.2	–	–
Rh/ZrO$_2$ [51]	Rh 0.1%	773	3	0.12	–	–	–	43.6	14[b]–4.7[c]	0.43[b]–0.14[c]	67.4[d]–0.57[e]
Rh/ZrO$_2$ [51]	Rh 0.8%	773	3	0.13	–	–	–	43.6	12.9[b]–11.2[c]	2.5[b]–2.21[c]	–
Rh/ZrO$_2$ [51]	Rh 1.6%	773	3	0.14	–	–	–	43.6	5.4[b]–5.3[c]	0.71[b]–0.69[c]	9.5[d]–0.36[e]
Rh/CeO$_2$ [51]	Rh 0.1%	773	3	0.15	–	–	–	43.6	9.3[b]–3[c]	0.41[b]–0.13[c]	49.9[d]–0.6[e]
Rh/CeO$_2$ [51]	Rh 0.8%	773	3	0.16	–	–	–	43.6	11.3[b]–9.1[c]	1.97[b]–1.58[c]	7.2[d]–0.33[e]
Rh/CeO$_2$ [51]	Rh 1.6%	773	3	0.17	–	–	–	43.6	5.5[b]–5.2[c]	0.63[b]–0.6[c]	6.1[d]–0.2[e]
Rh/Ce$_{0.25}$Zr$_{0.75}$O$_2$ [51]	Rh 1.6%	773	3	0.18	–	–	–	43.6	11.5[b]–10.8[c]	2.3[b]–2.17[c]	–
Rh/SiO$_2$ [51]	Rh 1.6%	773	3	0.19	–	–	–	43.6	6.8[b]–5.7[c]	1.24[b]–1.03[c]	–
Rh/Ce$_{0.6}$Zr$_{0.4}$O$_2$ [26,38]	Rh 0.8%	823	4	0.15	13,440[j,k]	–	48–27	61.7	–	0.12[h]–0.036[i]	–
Rh/Ce$_{0.15}$Zr$_{0.85}$O$_2$ [37]	Rh 3%	773	2	0.1	9,000[k]	–	28.1	33.9	–	–	–
Ru/Ce$_{0.15}$Zr$_{0.85}$O$_2$ [37]	Ru 3%	773	2	0.1	9,000[k]	–	21.4	33.9	–	–	–
Pt/Ce$_{0.15}$Zr$_{0.85}$O$_2$ [37]	Pt 3%	773	2	0.1	9,000[k]	–	20.5	33.9	–	–	–
Ru/ZrO$_2$ [41]	Ru 1%	823	3.98	0.13	40,000	–	22	62.7	–	–	–
Ru/ZrO$_2$ [41]	Ru 1%	773	3.98	0.13	40,000	–	7	52	11.1	–	–
Ru/MgO [53]	Ru 1.5%	773	4	0.1	1,980	–	45	52.1	–	–	–
Ru/Nb$_2$O$_5$ [53][l]	Ru 1.5%	773	4	0.1	1,980	–	52	52.1	–	–	–
Ru/Nb$_2$O$_5$ [53][m]	Ru 1.5%	773	4	0.1	1,980	–	5	52.2	–	–	–
Ru/Al$_2$O$_3$[40]	Ru 1.5%	773	3	0.2	20,000	–	42	43.6	–	–	–
Ru/Al$_2$O$_3$[40]	Ru 1.5%	673	3	0.2	20,000	–	11.5	19.8	–	–	–
Rh/CeO$_2$ [40]	Rh 1.5%	773	3	0.3	20,000	–	42	43.6	–	–	–
Rh/CeO$_2$ [40]	Rh 1.5%	673	3	0.3	20,000	–	19	19.8	–	–	–
Pt/CeO$_2$ [40]	Pt 1.1%	773	3	0.4	20,000	–	20	43.6	–	–	–
Pt/CeO$_2$ [40]	Pt 1.1%	673	3	0.4	20,000	–	3.3	19.8	–	–	–

[a]X_{CH4} (%) = [CH$_4$]$_{in}$ − [CH$_4$]$_{out}$/[CH$_4$]$_{in}$.
[b]Initial average rate between 0.5 and 2.5 h.
[c]Final average rate between 15 and 17 h.
[d]Coke mmol mmol^{-1} Rh$_{surf}$.
[e]Coke mmol g$_{cat}$$^{-1}$.
[f]Surface atom–based reac. rate (mol mol^{-1} Me$_{surf}$s).
[g]Weight-based reac. rate (mol g$_{cat}$$^{-1}$ h^{-1}).
[h]Initial rate after 1 h.
[i]Final rate after 25 h.
[j]CH$_4$ inlet flow/mass of catalyst.
[k]CH$_4$/H$_2$ = 1.5.
[l]Thermally untreated Nb$_2$O$_5$.
[m]Nb$_2$O$_5$ thermally treated (773k for 6 h).

ZrO_2- and CeO_2-supported catalysts. The evaluation of SR activity and turnover rates was carried out in a fixed-bed reactor at 773 K with $S/C = 3$ and $P = 0.12$ MPa; the total gas flow was 200 mL min^{-1}; the data relative to the most representative sample are reported in Table 5.4 [51]. The results show that the intrinsic activity does not depend on the type of support. The support only affects the catalytic activity in steam methane reforming indirectly by influencing the dispersion. As shown in Table 5.4 [51], the initial intrinsic activity decreases with increasing Rh particle size, but for the catalyst with the highest Rh dispersion, deactivation is much more pronounced than for the samples with low dispersion. It is also noteworthy that the initial intrinsic reaction rates do not vary significantly between catalysts based on different supports and with similar dispersion. Another important conclusion is that all catalysts containing Rh particles smaller than about 3 nm deactivate; instead, catalysts containing larger nanoparticles exhibit a quite stable SR activity. The analysis on the spent catalysts suggests that the very small particles have not sintered during the SR reaction but instead have oxidized. Halabi *et al.* [38] investigated the catalytic SR of methane over Rh/$Ce_{0.6}Zr_{0.4}O_2$ catalyst in the relatively low-temperature range of 748–973 K. The Rh is loaded on the support, previously prepared by a calcination (1,173 K for 6 h) of a mixture of commercially available ceria and zirconia supports with a composition ratio of (60:40 wt.%), by dry impregnation with a $Rh(NO_3)_3$ solution to obtain a metal loading of 0.8%. The principal catalyst features are reported in Table 5.3 [38]. Comparing the dispersion (27.9%) and the particle size (3 nm) values of this catalyst with that reported for the 1.6 wt%Rh/$Ce_{0.25}Zr_{0.75}O_2$ ($D = 69\%$, p.s. = 1.6 nm [51]), is evident the effect of the surface area (85.7% m^2 g^{-1}) of the used support ($Ce_{0.6}Zr_{0.4}O_2$) and the calcination temperature (1,173 K) on final catalyst morphological features. In the case of the previous described sample [51], the high surface area (271 m^2 g^{-1}) of the $Ce_{0.25}Zr_{0.75}O_2$ support and the low calcination temperature (873 K) contributes to obtain a high dispersion and also a small particle size, despite the higher Rh amount. The catalytic performance tests were carried under $S/C = 4$, at 823 K. The catalyst has shown a good initial conversion of about 48% after 1 h coupled with high reaction rates of about 0.12 mol g$_{cat}$$^{-1}$ h^{-1}. After the first 12 h, the catalyst reaches a stable activity of 27% CH_4 conversion with slow deactivation and a final reaction rate equal to 0.036 mol g$_{cat}$$^{-1}$ h^{-1}. The deactivation may be attributed to the surface reconstruction during the first 5 h of test and secondarily to a small amount of produced carbon. At a low space velocity, the catalyst has shown a low deactivation with an insignificant loss of surface area and activity over 100 h time on stream even at a low S/C ratio of 1. The authors suggest that the use of Rh can substantially suppress carbon formation due to a smaller dissolution of carbon into these metals. Moreover, the addition of zirconia to CeO_2 enhances the oxygen storage capacity of ceria and its redox properties; thus, the possible carbon producing reactions can also be diminished due to the redox reaction of the carbonaceous species formed on the Rh surface from CH_4 dissociation with oxygen spillover from ceria. The performance of different noble metals–based catalysts (3 wt%Me/$Ce_{0.15}Zr_{0.85}O_2$ Me = Rh, Ru, Pt) was investigated by Kusakabe *et al.* [37] under SR condition. The catalytic activity of

10 wt%Ni/$Ce_{0.15}Zr_{0.85}O_2$ was also determined for comparison. Platinum, rhodium and ruthenium and nickel were loaded on the $Ce_{0.15}Zr_{0.85}O_2$ (previously prepared by urea hydrolysis method [52] and calcined at 773 K for 4 h, BET = 63.6 m^2 g^{-1}) by incipient wetness impregnation with H_2PtCl_6, $RhCl_3$, $RuCl_3$ and $Ni(NO_3)_2$ solutions, respectively, followed by drying in a vacuum at 323 K for 2 h. After drying, the sample was calcined in air at 773 K for 5 h. The influence of metal type was investigated at low reaction temperatures (773 K) with a *S/C* molar ratio equal to 2. The data reported in Table 5.4 [37] show that among the investigated samples, the highest activity was obtained for the 3 wt%Rh/$Ce_{0.15}Zr_{0.85}O_2$ catalyst for which the methane conversion was 28.1% at 773 K; instead the catalytic activity of 10 wt %Ni/$Ce_{0.15}Zr_{0.85}O_2$ (CH_4 conversion = 9.9%) is lower than those of noble metal catalysts in spite of its higher metal loading. In addition, the authors ascribe the absence of carbon deposits (after 5 h of reaction) to the positive influence of the support; particularly, the reducibility of Ce–ZrO_2 makes the production of highly mobile oxygen species possible through a redox cycle of Ce^{4+}/Ce^{3+}. The carbonaceous materials on the Me sites can react with the oxygen spillover from the support, as a result, carbon deposition on the surface of the catalyst can be limited. Jakobsen *et al.* [41] prepared a 1 wt%Ru/ZrO_2 catalysts for use in the kinetic measurements by an incipient wetness impregnation of a ZrO_2 support with an aqueous solution of $Ru(NO)(NO_3)_3$. The sample was dried at 353 K in ambient air and subsequently reduced at 873 K in H_2 for 4 h. To achieve a stable catalyst, the samples were aged at 1,023 K for 336 h in a H_2O/H_2 (ratio ~1) gas mixture at 3.1 MPa total pressure. The kinetic studies have revealed that the methane dissociative adsorption is again the rate determining step. At high temperature, the active surface is free, but at lower temperatures, CO and hydrogen are present at the surface and reduce the activity due to blockage of the active sites. The results obtained by Jakobsen *et al.* are in agreement with the kinetic model of Rostrup-Nielsen [44] along with the data and model presented by Wei and Iglesia [46]. In addition, a TOF of 11.1 s^{-1} registered at 773 K confirms that at low temperature Rh and Ru show similar activity. A comparative analysis of Ru-based catalysts on different supports (MgO and Nb_2O_5) have been carried out by Amjad *et al.* [53] to identify highly active and selective catalysts for the SR process (T = 673–102, *S/C* = 4, WHSV = 1,980 N mL h^{-1} g^{-1}). The catalysts were prepared by incipient wetness impregnation method using an aqueous solution of a $Ru(NO)(NO_3)_3$ or $RuCl_3$ to obtain a nominal 1.5 wt.% of Ru as active element. The supports and the resulted catalysts were calcined at different temperatures with different procedures. In Tables 5.3 and 5.4 [53] are reported the data related to the preparation procedure as well as the chemical–physical characteristics of some representative samples. In general, among all of the catalysts prepared, the calcination treatment on the support, or on the Ru-impregnated catalysts, greatly affected the dispersion and the overall performance in particular at low reaction temperature (773 K). The Ru-based catalysts obtained by Ru deposition on thermally untreated Nb_2O_5, Table 5.3 [53] performed slightly better compared to the MgO-based catalysts. The worst performances, between Nb_2O_5 based catalysts, belong to the sample prepared by the double calcination treatment, at 773 K on the Nb_2O_5 support, then

at 673 K for 3 h after the Ru deposition. On the contrary, the authors report that the calcination treatment after the Ru impregnation favoured the dispersion of Ru on the uncalcined support. The elevate catalytic activity of these samples is also ascribed to the structural and surface interaction between the Ru and the support. These catalysts presented a starting amorphous structure with Ru present mainly as Ru^{4+}, which changed to tetragonal niobia with metallic Ru during the SR reaction. For all of the catalysts tested, no carbon formation was noticed at the end of the test campaign. Amjad *et al.* [40] have synthesized different noble metals (Rh, Ru and Pt) deposited on two different oxide carriers (CeO_2 and Al_2O_3). The catalysts were also tested towards methane SR at low reaction temperature (673–773 K) with a $S/C = 3$. The Rh- and Ru-based catalysts were synthesized in two different steps, by employing first a solution combustion synthesis (SCS) for preparing the carriers (CeO_2 and Al_2O_3) and then the incipient wetness impregnation for depositing the noble metals. The oxides were prepared starting from a solution of the ceria precursor (cerium nitrate $Ce(NO_3)_3 \cdot 6H_2O$, or aluminium nitrate $Al(NO_3)_3 \cdot 9H_2O$) and urea (CH_4N_2O) as fuel. The aqueous solution was heated up to 873 K in a furnace to form the desired oxides. The so-synthesized powders were then calcined in static air for 3 h at 650 °C. Active metals were deposited on the carriers by incipient wetness impregnation after drying step at 404 K. The samples were calcined at 1,073 K for 3 h. The noble metal content of Ru- and Rh-based catalysts were 1.5% in weight. Supported 1.1 wt.% Pt/CeO_2 catalysts were prepared by one-shot oxalyldihydrazide-nitrate self-combustion synthesis [54]. The choice of the oxide carrier (Al_2O_3, CeO_2) strongly influences the chemical–physical properties of the resulted catalysts. The highest BET value of the Ru-based sample was due to the typically high BET of the starting alumina, instead the noble metal dispersion, despite the highest BET displayed by the Ru-based catalyst, apparently is promoted by the CeO_2. Among the catalysts investigated, the 1.5 wt%Rh/CeO_2 has shown the best catalytic performances for methane SR reaction at low temperature as it gave a methane conversion near the calculated thermodynamic equilibrium at both the temperature investigated, 42% and 19% at 673 and 773 K, respectively. The authors suggest that the highest dispersion value of the Rh could be responsible of the highest catalytic activity recorded during the catalysts screening towards the methane SR reaction. Considering the results obtained at the 673–773 K temperature range, the relative catalytic activity was found to be $Rh \geq Ru > Pt$. This trend is in agreement with the sequences reported by several authors [43–45,47].

5.4 Conclusions and future trends

Large-scale SR process for H_2 production has been well established in the chemical industry, but to meet some requirements for distributed production, it should be modified in order to reach high-energy efficiency and ability to respond to the diversification of practical applications at smaller scales.

Thus, the modification of conventional methane SR with selective membranes for hydrogen production represents an alternative way to improve the efficiency of this energivorous process. Coupling low temperature (≤ 823 K) SR with membrane technology, high CH_4 conversion and H_2 yield can be achieved as well as decrease in energy demands. A membrane reformer shows some potential advantages respect the conventional SR units, is more compact, simpler, more active and more efficient, since the reaction and hydrogen separation are performed simultaneously in a single reactor.

Referring to the low-temperature SR process, the design of opportune catalysts is a key aspect to improve performance, durability, cost effectiveness and sustainability of membrane reactor technology.

Based on the studies described in this chapter, some general considerations can be carried out irrespective of the type of active metal or support used for the catalysts preparation. High surface area, high active metal dispersion, low particle size, resistance to carbon formation and opportune metal loading are some of the basic properties for a catalytic material to be employed in the methane SR at low temperature.

Ni-based catalysts have been widely investigated in SR of methane. For reforming processes at high temperature, nickel has shown a comparable activity to noble metal, but at low temperature Ni is inferior to noble metals in terms of fuel conversion, H_2 production, catalyst oxidation and coke resistance. On the other hand, noble metals, which are also highly active in methane SR, are more expensive but less prone to coke formation and oxidation deactivation. In addition, the high activity at relatively low loading (0.1–1.6 wt%) makes them competitive and interesting candidates for low temperature reaction. Even if, at low temperature, the Ni-based catalysts show a lower catalytic activity respect catalysts with noble metals; their use is possible if we consider the enhanced contribution of the membrane reactor in terms of CH_4 conversion and H_2 production. Moreover, an opportune design of the structure of catalysts allows their utilization. Generally, a nickel loading in the 10–25 wt% range can be considered suitable for reforming at low temperature (≤ 823 K). The low cost of Ni metal justifies the higher loading amount used in order to increase the activity per catalyst volume. On the other hand, high Ni amounts contribute to the formation of particles with large sizes (7–23 nm) influencing the dispersion values that generally are in the 1%–18% range for the catalysts prepared by impregnation. In addition, nickel shows lower activity when supported on inert oxides. Among the studies reported in this chapter, the effect of the support on the performance of the catalysts can be considered very important as it can not only improve the stability of the catalyst but the performance of the catalyst as well. The support can indirectly affect the activity by changing the dispersion of the metal, but there are also cases in which the support can directly participate to the reaction steps facilitating the adsorption of reactants. The CeO_2, ZrO_2 TiO_2 and mixed oxides ($CeO_2/ZrO_2/La_2O_3$, $CeO_2/ZrO_2/Al_2O_3$, ZrO_2/Al_2O_3) are principally used as supports or support dopants. High oxygen storage capacity and redox properties lead to efficient coke resistance, making these materials advantageous over conventionally used Al_2O_3, $MgAl_2O_4$ or SiO_2. Regarding the preparation procedure, the impregnation is the most used technique to deposit the Ni on the support generally previously prepared by precipitation rules. Also the coprecipitation and the coprecipitation/deposition methods are used, generally, the obtained catalysts show a higher dispersion ($D = 11\%$–33%) respect the catalysts prepared by impregnation. The catalysts ($NiAl_2O_4/Al_2O_3$, $Ni/MgO_2/La_2O_3/Al_2O_3$) prepared by these routes show also promising catalytic performances (near the thermodynamic equilibrium).

Compared to Ni-based catalysts, noble metals–based catalysts exhibit superior catalytic performance in the methane SR at low temperature, that is high fuel conversion, high resistance to carbon deposition sintering and oxidation of metal particles. In addition, the experimental reaction rates are almost higher respect the values reported by for the Ni-based catalysts. The improved catalytic performance of this type of catalysts can be attributed to the high dispersion ($D = 12\%$–83%) of the active metal particles. The presence of nanoparticles (1–5 nm) well dispersed on the support surface provided a higher number of active sites that contribute to enhance the catalytic activity of the catalyst. Among noble metals, Rh and Ru show the most promising results in low temperature operation. Also in this case, the use of supports with redox properties (CeO_2, ZrO_2 and CeO_2/ZrO_2) affects positively the catalytic performances of the prepared catalysts. Overall, considering the results obtained in the examined 673–773 K temperature range, the relative catalytic activity was found to be Rh $\geq$ Ru $>$ Pt. This trend is in agreement with the sequences reported by several authors. Also in the case of noble metals, the impregnation of supports prepared by precipitation is the most used method for the dispersion of the metal on the carrier surface. Alternatively, promising performances at low temperature (673–773 K) have been obtained with catalysts prepared by combining the SCS with the impregnation technique.

Based on the studies reported in this chapter, appears evident the importance to control the design of the catalytic materials that together with the experimental conditions (*S/C* molar ratio, reaction temperature, etc.) significantly affect the lifetime of a reforming system based on low temperature SR process. Many aspects as specific surface area, metal dispersion, particle size are very important to obtain high performances overcoming the principal deactivation phenomena (coke deposition, metal oxidation, particle sintering) related to reforming processes conducted at low temperature.

Low-cost nickel-based catalyst can be used, but the studies reported in literature cannot be considered conclusive, especially for reforming processes conducted at low reaction temperature and further efforts are needed to solve the deactivation problems.

From the other side, the utilization of small amounts of noble metals should be considered for the design of efficient catalysts. Precious metals (especially Rh and Ru) have high activity per gram, and thus only small amounts are required to obtain a final catalytic material with high activity per unit volume. In addition, processes for recycling precious metals are already part of the business loop for automotive, petroleum and chemical applications. Thus, their robustness (especially durability), respect the Ni-based catalysts could have an important economic impact on the efficiency of small-scale processes in which they are used. Nevertheless, advances in research and optimization of catalysts and reactor designs in order to reduce the amount of noble metals are fundamental for their definitive application.

Abbreviations

ATR	autothermal reforming
BoP	balance of plant
BET	Brunauer–Emmett–Teller
BTL	biomass-to-liquid
CTL	coal-to-liquid
D	dispersion
GHSV	gas hourly space velocity
GTL	gas-to-liquid
HTS	high temperature shift
LTS	low temperature shift
FC	fuel cell
PEM	poly electrolyte membrane
P	pressure
PROX	preferential oxidation
POX	partial oxidation
PSA	pressure swing absorption
S/C	steam to carbon molar ratio: H_2O/CH_4
SR	steam reforming
TEM	transmission electron microscopy
TOF	turnover frequency
TOS	time on stream
Tr	reaction temperature
WGS	water gas shift
WHSV	weight hourly space velocity
XRD	X-ray diffraction

References

[1]　Elsherif M., Manan Z.A., Kamsah M.Z., 'State-of-the-art of hydrogen management in refinery and industrial process plants', *Journal of Natural Gas Science and Engineering*, 2015;**24**:346–356.

[2]　Behling N., Williams M.C., Managi S., 'Fuel cells and the hydrogen revolution: analysis of a strategic plan in Japan', *Economic Analysis and Policy*, 2015;**48**:204–221.

[3]　Budzianowski W.M., 'A review of potential innovations for production, conditioning and utilization of biogas with multiple-criteria assessment', *Renewable and Sustainable Energy Reviews*, 2016;**54**:1148–1171.

[4]   Du C.M., Mo J.M., Li H., 'Renewable hydrogen production by alcohols reforming using plasma and plasma-catalytic technologies: challenges and opportunities', *Chemicals Review*, 2015;**115**:1503–1542.

[5]   Arutyunov V.S., Savchenko V.I., Sedov I.V., Fokin I.G., Nikitin A.V., Strekova L.N., 'New concept for small-scale GTL', *Chemical Engineering Journal*, 2015;**282**:206–212.

[6]   Jina E., Zhang Y., Hea L., Harris H.G., Tenga B., Fan M., 'Indirect coal to liquid technologies', *Applied Catalysis A: General*, 2014;**476**:158–174.

[7]   Sunde K., Brekke A., Solberg B., 'Environmental impacts and costs of woody biomass-to-liquid (BTL) production and use – a review', *Forest Policy and Economics*, 2011;**13**:591–602.

[8]   Zamaniyan A., Behroozsarand A., Ebrahimi H., 'Modeling and simulation of large scale hydrogen production', *Journal of Natural Gas Science and Engineering*, 2010;**2**:293–301.

[9]   Jordal K., Anantharaman R., Peters T.A., *et al.*, 'High-purity H_2 production with CO_2 capture based on coal gasification', *Energy*, 2015;**88**:9–17.

[10] Ribeiro A.M., Grande C.A., Lopes F.V.S., Loureiro J.M., Rodrigues A.E., 'A parametric study of layered bed PSA for hydrogen purification', *Chemical Engineering Science*, 2008;**63**:5258–5273.

[11] Gosavi P.V., Biniwale R.B., 'Effective cleanup of CO in hydrogen by PROX over perovskite and mixed oxides', *International Journal of Hydrogen Energy*, 2012;**37**:3958–3963.

[12] Seoa J.G., Youna M.H., Jung J.C., Songa I.K., 'Hydrogen production by steam reforming of liquefied natural gas (LNG) over mesoporous nickel–alumina aerogel catalyst', *International Journal of Hydrogen Energy*, 2010;**35**:6738–6746.

[13] Sehested J., 'Four challenges for nickel steam-reforming catalysts', *Catalysis Today*, 2006;**111**:103–110.

[14] Zeppieri M., Villa P.L., Verdone N., Scarsella M., Filippis P.D., 'Kinetic of methane steam reforming reaction over nickel- and rhodium-based catalysts', *Applied Catalysis A: General*, 2010;**387**:147–154.

[15] Vita A., Cristiano G., Italiano C., Pino L., Specchia S., 'Syngas production by methane oxy-steam reforming on Me/CeO_2 (Me = Rh, Pt, Ni) catalyst lined on cordierite monoliths', *Applied Catalysis B: Environmental*, 2015;**162**:551–563.

[16] Lisboa J.S., Santos D.C.R.M., Passos F.B., Noronha F.B., 'Influence of the addition of promoters to steam reforming catalysts'. *Catalysis Today*, 2005;**101**:15–21.

[17] Harshini D., Lee D.H., Jeonga J., *et al.*, 'Enhanced oxygen storage capacity of $Ce_{0.65}Hf_{0.25}M_{0.1}O_{2-\delta}$ (M = rare earth elements): applications to methane steam reforming with high coking resistance', *Applied Catalysis B: Environmental*, 2014;**148**:415–423.

[18] Gallucci F., Basile A., 'Pd–Ag membrane reactor for steam reforming reactions: a comparison between different fuels', *International Journal of Hydrogen Energy*, 2008;**33**:1671–1687.

[19]   Tong J., Matsumura Y., 'Pure hydrogen production by methane steam reforming with hydrogen-permeable membrane reactor', *Catalysis Today*, 2006;**111**:147–152.

[20]   Rostrup-Nielsen J.R., Sehested J., Nørskov J.K., 'Hydrogen and synthesis gas by steam and CO_2 reforming', *Advances in Catalysis*, 2002;**47**:65–139.

[21]   Rostrup-Nielsen J.R., 'Catalytic steam reforming'. In Anderson J.R., Boudart M., Eds., *Catalysis: Science and Technology*, vol. 5, 1984, pp. 1–117, Springer-Verlag, New York.

[22]   Ashrafi M., Proll T., Pfeifer C., Hofbauer H., 'Experimental study of model biogas catalytic steam reforming: thermodynamic optimization', *Energy & Fuels*, 2008;**22**:4182–4189.

[23]   Tomishige K., Chen Y.G., Kujimoto K., 'Studies on carbon deposition in CO_2 reforming of CH_4 over nickel–magnesia solid solution catalysts', *Journal of Catalysis*, 1999;**181**:91–103.

[24]   Tsipouriari V.A., Verykios X., 'Carbon and oxygen reaction pathways of CO_2 reforming of methane over Ni/La_2O_3 and Ni/Al_2O_3 catalysts studied by isotopic tracing techniques', *Journal of Catalysis*, 1999;**187**:85–94.

[25]   Wang J., Tai Y., Dow W., Huang T., 'Study of ceria-supported nickel catalyst and effect of yttria doping on carbon dioxide reforming of methane', *Applied Catalysis A: General*, 2001;**218**:69–79.

[26]   Angeli S.D., Monteleone G., Giaconia A, Lemonidou A.A., 'State-of-the-art catalysts for CH_4 steam reforming at low temperature', *International Journal of Hydrogen Energy*, 2014;**39**:1979–1997.

[27]   Matsumura Y., Nakamori T. 'Steam reforming of methane over nickel catalysts at low reaction temperature', *Applied Catalysis A: General*, 2004;**258**:107–114.

[28]   Liu Z.-W., Jun K.-W., Roh H.-S., Park S.-E., 'Hydrogen production for fuel cells through methane reforming at low temperatures', *Journal of Power Sources*, 2002;**111**:283–287.

[29]   Angeli S.D., Turchetti L., Monteleone G., Lemonidou A.A., 'Catalyst development for steam reforming of methane and model biogas at low temperature', *Applied Catalysis B: Environmental*, 2016;**181**:34–46.

[30]   Zhao J., Zhou W., Ma J., 'Pretreatment as the crucial step for biogas reforming over Ni–Co bimetallic catalyst a mechanistic study of CO_2 pretreatment', *International Journal of Hydrogen Energy*, 2014;**39**:13429–13436.

[31]   Zhang W.D., Liu B.S., Zhu C., Tian Y.L., 'Preparation of La_2NiO_4/ZSM-5 catalyst and catalytic performance in CO_2/CH_4 reforming to syngas', *Applied Catalysis A: General*, 2005;**292**:138–143.

[32]   Jiménez-González C., Boukha Z., De Rivas B., González-Velasco J.R., Gutiérrez-Ortiz J.I., López-Fonseca R., 'Behavior of Coprecipitated $NiAl_2O_4$/Al_2O_3 Catalysts for Low-Temperature Methane Steam Reforming', *Energy Fuels*, 2014;**28**:7109–7121.

[33]   Chen Y., Wang Y., Xu H., Xiong G., 'Efficient production of hydrogen from natural gas steam reforming in palladium membrane reactor', *Applied Catalysis B: Environmental*, 2008;**80**:283–294.

[34] Takehira K., '"Intelligent" reforming catalysts: trace noble metal-doped Ni/ Mg(Al)O derived from hydrotalcites', *Journal of Natural Gas Chemistry*, 2009;**18**:237–259.

[35] Ohi T., Miyata T., Li D., *et al.*, 'Sustainability of Ni loaded Mg–Al mixed oxide catalyst in daily startup and shutdown operations of CH_4 steam reforming', *Applied Catalysis A: General*, 2006;**308**:194–203.

[36] Kho E.T., Scott J., Amal R., 'Ni/TiO_2 for low temperature steam reforming of methane', *Chemical Engineering*, 2016;**140**:161–170.

[37] Kusakabe K., Sotowa K.I., Eda T., Iwamoto Y., 'Methane steam reforming over $Ce-ZrO_2$-supported noble metal catalysts at low temperature', *Fuel Processing Technology*, 2004;**86**:319–326.

[38] Halabi M.H., De Croon M.H.J.M., Van der Schaaf J., Cobden P.D., Schouten J.C., 'Low temperature catalytic methane steam reforming over ceria–zirconia supported rhodium', *Applied Catalysis A: General*, 2010;**389**:68–79.

[39] Schädel B.T., Duisberg M., Deutschmann O., 'Steam reforming of methane, ethane, propane, butane, and natural gas over a rhodium-based catalyst', *Catalysis Today*, 2009;**142**:42–51.

[40] Amjad U.E.-S., Vita A., Galletti C., Pino L., Specchia S., 'Comparative study on steam and oxidative steam reforming of methane with noble metal catalysts', *Industrial Engineering Chemistry Research*, 2013;**52**:15428–15436.

[41] Jakobsen J.G., Jørgensen T.L., Chorkendorff I., Sehested J., 'Steam and CO_2 reforming of methane over a Ru/ZrO_2 catalyst', *Applied Catalysis A: General*, 2010;**377**:158–166.

[42] Carvalho L.S., Martins A.R., Reyes P., *et al.*, 'Preparation and characterization of $Ru/MgO-Al_2O_3$ catalysts for methane steam reforming', *Catalysis Today*, 2009;**142**:52–60.

[43] Kikuchi E., Tanoka S., Yamazaki Y., Morila Y., 'Steam reforming of noble metal catalysts (part 1)', *Bulletin of the Japan Petroleum Institute*, 1974;**16**:95–98.

[44] Rostrup-Nielsen J.R., Hansen J.-H.B., 'CO_2-reforming of methane over transition metals', *Journal of Catalysis*, 1993;**144**:38–49.

[45] Qin D., Lapszewicz J., 'Study of mixed steam and CO_2 reforming of CH_4 to syngas on MgO supported metals', *Catalysis Today*, 1994;**21**:551–560.

[46] Wei J., Iglesia E., 'Mechanism and site requirements for activation and chemical conversion of methane on supported Pt clusters and turnover rate comparisons among noble metals', *Journal of Physical Chemistry B*, 2004;**108**:4094–4103.

[47] Jones G., Jakobsen J.G., Shim S.S., *et al.*, 'First principles calculations and experimental insight into methane steam reforming over transition metal catalysts', *Journal of Catalysis*, 2008;**259**:147–160.

[48] Kurungot S., Yamaguchi T., 'Stability improvement of $Rh/\gamma-Al_2O_3$ catalyst layer by ceria doping for steam reforming in an integrated catalytic membrane reactor system', *Catalysis Letters*, 2004;**92**:181–187.

[49] Wei J., Iglesia E., 'Structural requirements and reaction pathways in methane activation and chemical conversion catalyzed by rhodium', *Journal of Catalysis*, 2004;**225**:116–127.

[50] Wang H.Y., Ruckenstein E., 'Carbon dioxide reforming of methane to synthesis gas over supported rhodium catalysts: the effect of support', *Applied Catalysis A: General*, 2000;**204**:143–152.

[51] Ligthart D.A.J.M., Van Santen R.A., Hensen E.J.M., 'Influence of particle size on the activity and stability in steam methane reforming of supported Rh nanoparticles', *Journal of Catalysis*, 2011;**280**:206–220.

[52] Thammachart M., Meeyoo V., Risksomboon T., Osuwan S., 'Catalytic activity of CeO_2–ZrO_2 mixed oxide catalysts prepared via sol–gel technique: CO oxidation', *Catalysis Today*, 2001;**68**:53–61.

[53] Amjad U., Lenzi G.G., Fernandes-Machado N.R.C., Specchia S., 'MgO and Nb_2O_5 oxides used as supports for Ru-based catalysts for the methane steam reforming reaction', *Catalysis Today*, 2015;**257**:122–130.

[54] Pino L., Vita A., Cipitì F., Lagana M., Recupero V., 'Performance of Pt/CeO_2 catalyst for propane oxidative steam reforming', *Applied Catalysis A: General*, 2006;**306**:68–77.

Chapter 6

Supercritical water gasification of biomass to produce hydrogen

C. Cannilla[1], G. Bonura[1] and F. Frusteri[1]

Abstract

Supercritical water gasification (SCWG) processes have recently received significant attention as a sustainable technology for the production of hydrogen starting from wet biomass. In this chapter, the influence of biomass composition for the production of hydrogen under SCWG is evaluated. The influence of the main reaction conditions as temperature, pressure and feed concentration on hydrogen yield are discussed together with a critical rationalization of data.

6.1   Introduction

Among different hydrogen production methods, gasification is an effective route to convert carbon-containing feeds into syngas constituted of carbon monoxide (CO), hydrogen (H_2), methane (CH_4) and small quantities of other light hydrocarbons (C_nH_m), carbon dioxide (CO_2) and steam (H_2O) (through partial oxidation at elevated temperatures) [1]. The syngas could be further processed to produce more hydrogen by water–gas shift reaction. However, the energy efficiency of biomass conversion technologies decreases by increasing moisture content of feed. The direct combustion for example requires biomass drying, but the energy spent in evaporating water from biomass is so large that the net energy production results negative [2]. Moreover, the moisture affects both the operation of the gasifier and the composition of syngas. Low-biomass density is another shortcoming which negatively influences the transportation and handling. For these reasons and for the high cost of drying steps, although wet biomass grows rapidly and abundantly around the world, it is not considered as a promising feedstock for conventional gasification processes. Thus, the attention has been recently addressed towards the possibility to treat the wet biomass in supercritical phase exploiting the water as

[1]CNR-ITAE "Nicola Giordano", Messina, Italy

reaction medium without the need of expensive drying pre-treatment with the aim to produce hydrogen or methane.

Supercritical water (SCW) – defined as water above its critical point (374 °C, 22.1 MPa) – could be considered an environmentally benign reaction medium being inexpensive, no toxic and no flammable [3]. SCW is characterized by particular physical and transport properties (as dielectric constant and ion product) which can be varied significantly by manipulating temperature and density with dramatic consequences for its solvent behaviour. At high temperatures in fact water loses the solvent properties towards polar and ionic compounds and, in some range of temperature and pressure, non-polar compounds become highly miscible. Moreover, at 400 °C, all inorganic gases and simple organic compounds are completely miscible with water thus opening a window for generating aqueous solutions with high solutes concentrations of largely different polarity which is impossible at ambient conditions.

To evaluate the H_2 production by biomass supercritical water gasification (SCWG), extensive investigations have been conducted in recent years [4–14]. Model compounds such as cellulose, lignin, glucose and glycerol have been widely tested to get information on the chemistry of biomass gasification in SCW. On the other hand, gasification of the real biomass such as sawdust and starches, clover grass and corns silage, baby food and zoo mass in SCW was investigated too [15]. The need to produce a tar-free product gas from the gasification of biomass, the removal of tars and the reduction of the methane have been the main topics of several studies.

In this chapter, the influence of biomass composition and type for the production of H_2 under SCWG is reviewed, along with the influence of process conditions on the H_2 yield and products composition.

6.2   Gasification under supercritical water

Gasification is the only commercial, large-scale route for converting solids to gases and one of the cleanest technologies for solid conversion [16]. In particular, the gasification of carbonaceous, hydrogen containing feeds is an effective method for thermal H_2 production and it is considered a key technology in the transition to a hydrogen economy [17]. Gasification under SCW differs from the conventional dry process since the water acts as a reactant, solvent, catalyst and hydrogen donor via various reactions. Gaseous products are mainly CH_4, H_2, CO, CO_2 and C_1–C_4 whereas, as side products, some bio-oils, char and tar could be formed too. Compared with thermal gasification at ambient pressure, SCW gasification occurs at temperature below 700 °C, thus making also possible to use waste heat from other processes [13,18].

Pioneering researches were carried out by Modell *et al.* [19] with maple wood sawdust in SCW obtaining a rapid, direct route to gases without char formation, although these experiments mainly produced low CH_4 yields. Afterwards, several other research groups developed processes for biomass gasification in SCW.

Table 6.1   Some of reactions involved in biomass SCWG

Hydrolysis cellulose	$(C_6H_{10}O_5)_n + nH_2O \rightarrow nC_6H_{12}O_6$	(6.1)
Glucose decomposition	$C_6H_{12}O_6 \rightarrow C_xH_yO_z$	(6.2)
Steam reforming	$C_xH_yO_z + (2x - z)H_2O \rightarrow xCO + (2x - z + y/2)H_2$	(6.3)
	$C_xH_yO_z + (x - z)H_2O \rightarrow xCO + (2x - z + y/2)H_2$	
Hydrothermal pyrolysis	$C_xH_yO_z \rightarrow CO(CH_4, H_2, CO_2) + CH_xO_y$	(6.4)
Char formation	$C_xH_yO_z \rightarrow wC + C_{x - w}H_yO_z$	(6.5)
Water gas shift reaction	$CO + H_2O \rightarrow CO_2 + H_2$	(6.6)
Methanation	$CO + 3H_2 \rightarrow CH_4 + H_2O$	(6.7)

Elliott *et al.* [20] converted biomass to CH_4-rich gas using reduced metal catalysts based on ruthenium, rhodium, osmium, iridium or their mixtures in a temperature range of 300–450 °C and 13 MPa. Minowa and Inoue [21] found that H_2-rich gas could be obtained in hot-compressed water at 350 °C and 18 MPa from biomass with reduced nickel catalyst and sodium carbonate. Then, Antal and its research group reached complete glucose gasification at 600 °C in 30 s of residence time [22].

Despite the efforts and achievements of many research groups, the chemistry of biomass gasification is not fully understood yet, since the reaction pathways consists of many steps, involving a huge number of compounds [23], above all when lignocelluloses wastes, characterized by a complex structure and chemistry, are used. Many solid and liquid phase reactions in fact are involved in the formation of gas: biomass *depolymerization* and *hydrolysis* to oligomers or monomers, *decomposition* to monomers, *steam reforming, pyrolysis, char formation* though intermediates and char from pyrolysis of the feedstock, *water gas shift* and *methanation*. All these reactions play a significant role in the gasification chemistry (see Table 6.1) [7,12].

SCWG of biomass is an endothermic process, if H_2 is the desired product, water gas shift reaction (WGSR, see (6.6)) should be dominant and methanation (6.7) should be restrained. Several intermediate reactions – sometimes competing ones – are involved too and some of them are described in Table 6.2 [24].

Kruse *et al.* [25,26] for example identified the key compounds in biomass conversion as phenols, furfural, acids and aldehydes (such as lactic and levulinic acids or acetic and formic acids and aldehydes). Such intermediates then converted by steam reforming into syngas containing hydrogen [10]. Kinetic models of gasification proposed by Resende *et al.* showed that the prevalence of a reaction and a particular compound formation were strongly dependent both on temperature and reaction time [27]. Reactions responsible for gas formation from intermediate (water-soluble products) are most significant at short reaction times, whereas a reaction that redistributes the different gases (e.g. WGSR) becomes the most important at longer residence time [12]. So, H_2 was primarily produced via steam reforming reaction (see (6.3)) at short residence time and via WGSR (see (6.6)) at longer reaction time and higher temperature.

Table 6.2   Formation of some intermediates and H_2 in biomass SCWG

$C_6H_{12}O_6 + 6H_2O \rightarrow 6HCOOH + H_2$	(6.8)
$C_6H_{12}O_6 + 6H_2O \rightarrow 4HCOOH + CO_2 + H_2$	(6.9)
$C_6H_{12}O_6 + 4H_2O \rightarrow 4HCOOH + CH_3COOH + 4H_2$	(6.10)
$C_6H_{12}O_6 + 2H_2O \rightarrow 2HCOOH + 2CH_3COOH + 2H_2$	(6.11)
$C_6H_{12}O_6 + 2H_2O \rightarrow 2CH_3COOH + 2CO_2 + 4H_2$	(6.12)
$HCOOH \rightarrow CO + H_2O$	(6.13)
$CH_3COOH \rightarrow CH_4 + CO_2$	(6.14)
$CH_4 + H_2O \rightarrow CO + 3H_2$	(6.15)

Biomass also contains small concentrations of various inorganic salts, the solubility of which in supercritical media could be very low, causing salt deposition on reactor walls and other parts of equipment. Moreover, the SCW conditions (acidic and oxidizing conditions, extreme pH values, pressure changes, high temperature, etc.) favour the corrosion process, which represents a serious problem for reactor design and safety. On this account, high temperature and pressure resistant materials – which are also resistant to corrosion in SCW – should be used for equipment construction [12]. As regards the char/coke formation, they may originate from biomass resistant to decomposition. Anyhow, also some parts of aromatics compounds such as lignin or other unsaturated species could eventually polymerize to tar and char materials. Therefore, reactor plugging represents also a critical technological problem for the SCW gasification of biomass not easy to be solved. Tubular reactors for example are susceptible to system shut-downs due to reactor plugging resulting from the formation of char at the heating section and the buildup of ash inside the reactor.

Temperature, heating rate, pressure, residence time, feedstock concentration and pre-treatment strongly influence the H_2 yield in SCWG but, among them, the reaction temperature appears the most important parameter [28]. Below the critical point, in fact, the higher the density is, the higher the ionic product (K_W) is and the ionic reaction mechanism is prevalent. On the other hand, at supercritical state, the decrease of water density causes the drop of K_W so, at higher temperature and low-density conditions, free radical reactions are dominant and formation of gases is favoured [12,13]. In particular, H_2 and CO_2 are the dominant gases at temperatures greater than 420 °C and the conversion rate are high even without the use of catalysts, whereas CH_4 and CO_2 are the main products at temperature below 420 °C [29].

Under the hydrothermal condition of SCW, water itself can act like an acid or base catalyst and many organic compounds – that usually do not react in water without the presence of strong acid or base catalysts – may readily react [2]. Nevertheless, it is impossible to achieve complete biomass conversion by gasification at low-moderate temperatures, under subcritical water gasification (SubCW) conditions ($T < T_c$ at a pressure above its saturation pressure), and at short reaction times without a proper catalyst [30]. For this reason, the design of stable catalysts tolerant towards dissolved inorganics compounds is one of the main challenges in hydrothermal gasification of wet biomass [31]. Moreover, since the water density could affect the interaction between catalysts and reactants, also rates and equilibrium of the reaction should be further controlled [32,33]. In general, the catalyst lowers the reaction temperature for biomass degradation and accelerates the reaction with technological and economic benefits [10]. They could enhance the process efficiency by fast gasification of reactive intermediates produced by hydrolysis or dehydration, depressing the re-polymerization and formation of char/tar and conducting the reaction towards the desired product. This is especially important for aromatic (phenol) intermediates, for which a good catalyst must achieve a fast cleavage of the C–C bond in aromatic rings.

Both homogeneous and heterogeneous catalysis has received much attention for the catalytic gasification of biomass under SCW conditions. The effect of alkali salts on gas yield is well known, so many studies have been carried out in presence of such kind of catalysts. Jin *et al.* [10] for example reported that the order of the catalytic effect on H_2 yield of biomass gasification is: $KOH > Ca(OH)_2 > K_2CO_3 > LiOH > NaOH > Na_2CO_3$. Then, comparison between LiOH, NaOH and KOH showed that the stronger the alkalinity is, the weaker the catalytic effect is for hydrogen production. Muangrat *et al.* [34] investigated the gasification by partial oxidation of glucose under SubCW conditions and found this order for H_2 yield: $NaOH > KOH > Ca(OH)_2 > K_2CO_3 > Na_2CO_3 > NaHCO_3$, thus suggesting that the metal hydroxides could produce higher H_2 yield than the carbonates or bicarbonate.

Representative food processing wastes (molasses and rice bran) were decomposed under hydrothermal condition by using NaOH, KOH and $Ca(OH)_2$ and the H_2 yield was improved via the WGSR by intermediate formation of formate salts. Moreover, NaOH, KOH and $Ca(OH)_2$ inhibited and suppressed tar and char formation.

By adding alkaline salts, a formate salt ($HCOO^-K^+$) is formed which reacts with water to produce hydrogen [23]:

$$K_2CO_3 + H_2O \rightarrow KHCO_3 + KOH \tag{6.16}$$

$$KOH + CO \rightarrow HCOOK \tag{6.17}$$

$$HCOOK + H_2O \rightarrow H_2 + KHCO_3 \tag{6.18}$$

Instead, the CO_2 is produced by the reaction of $KHCO_3$:

$$2KHCO_3 \rightarrow H_2O + K_2CO_3 + CO_2 \tag{6.19}$$

$$H_2O + CO \leftrightarrow HCOOK \leftrightarrow H_2 + CO_2 \tag{6.20}$$

As regards heterogeneous catalysts, many studies have been addressed towards the individuation of a solid system characterized by high activity, hydrothermal stability and resistance to carbon deposition [35]. To enhance the H_2 production, activated carbon (AC) and supported transition metal catalysts are typically used for biomass SCWG. Compared to homogeneous alkali catalysts, they could exhibit

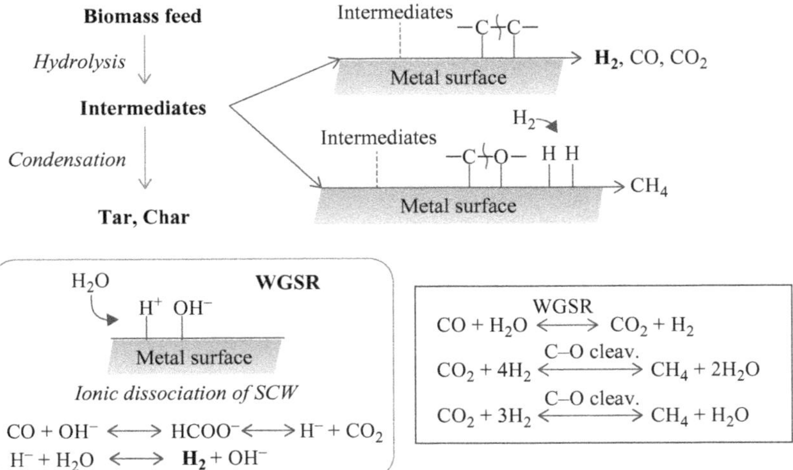

Figure 6.1   Reaction pathways for the catalytic decomposition of organic compounds in SCWG

higher catalytic activity and are easier to be recycled. Nickel-based catalysts are the most investigated systems due to relatively low cost of nickel and high activity towards WGSR, methanation and H_2 reactions and also towards tar crack [36,37]. Nevertheless, hydrothermal instability and carbon deposition are still the main problems to overcome for the Ni-based catalysts development [10].

In Figure 6.1, some different reaction pathways for the catalytic decomposition of organic compounds are reported. The biomass hydrolysis leads to intermediates which can be decomposed on the surface of the metal catalyst through C–C cleavage in conjunction with the WGSR to give a gas rich in H_2 and CO_2 [38]. The shift reaction could be initiated through interaction of CO with OH^- formed by ionic dissociation of SCW on the metal surface and forming the formate ion which decomposes into CO_2 and H^-. The hydride anion further interacts with water, forming H_2 and OH^- by electron transfer [39]. In case of oxygenated biomass compounds, CH_4 could be formed both as primary (cleavage of the C–O bond) or secondary product (metal-catalysed CO_2 hydrogenation).

Particular attention has been devoted to evaluate the synergic effect using a solid catalyst in the presence of acid or alkaline homogeneous medium. Jin *et al.* [10], for example, by investigating the peanut shell gasification in a stainless steel autoclave, investigated the effect of Raney-Ni and NaOH: NaOH forms a slurry, reacts with biomass and leads to the formation of formates compounds, which then degrade to H_2 and CO_2; it accelerates the gasification of phenols to form benzene and cyclohexane for H_2 production. At the same time, Raney Ni reacts with liquid and gas intermediate, favours C–C bond and C–H bond cleavage to obtain gaseous and liquid products enhancing the carbon gasification efficiency (CGE) at relatively low temperature. The positive synergic effect of homogeneous and heterogeneous catalysts had been found also by Minowa and Ogi [40] by using Ni and Na_2CO_3 in autoclave (350 °C; 18–20 MPa) for the thermochemical conversion of cellulose.

6.3   Feedstock influence on SCWG

6.3.1   *Lignocellulosic biomass and sewage sludge*

Lignocellulosic biomass – the most abundant type of biomass on earth – is mainly composed of carbohydrate polymers (cellulose, hemicellulose) and an aromatic polymer (lignin) tightly bound among them. Moreover, other substances including minerals and organic molecules (tannins, waxes terpene, fatty acids and proteins) could be present in some percentages and could vary considerably in their compositions. Lignocellulosic biomass can be broadly classified into *virgin biomass*, all terrestrial plants such as trees, bushes and grass, *waste biomass*, produced as a low value by-product of various industrial sectors such as agricultural and forestry, and *energy crops*, or rather specific crops produced as a raw material for production of second generation biofuels. Recent investigations of high-temperature gasification with food and agriculture waste, for example, have pointed out several critical issues that affect the gasification efficiency (GE), H_2 yield and technological and engineering obstacles in relation with proper and stable reactor design [41]. The most important challenge is the plugging of reactor caused by inorganic salts precipitating from biomass and carbonaceous product (char, coke). In fact, inorganic salt solubility is significantly lower under SCW than under ambient condition, and the eutectic melting of inorganic salt could cause trouble in the continuous operation of the gasification systems.

In this context, it is interesting to mention the sewage-sludge exploitation too considering that the compost produced from its use is only a small per cent of the total amount of waste produced and that the amount of water can be greater than 90% on a wet mass basis [13,15,29]. In the past two decades, in fact, also the wastewater treatment has gained significant interest [42] since the hydrogen production from sewage sludge may be a solution both for cleaner fuel as well as sewage-sludge disposal problems. Differently from lignocellulosic biomass, sewage sludge typically consists of 41 wt% protein, 25 wt% lipid, 14 wt% carbohydrate and the rest is constituted by ash and biodegradable and recalcitrant organic compounds, as well as pathogens and heavy metals [43]. In the treatment of activated sludge using the SCWG technology, the sludge is first fed into a non-catalytic pre-treatment vessel operated at a temperature range of 250–400 °C to hydrolyse the biopolymers. Due to the low polarity of water near its critical point, the inorganic ash precipitates and can be easily removed from the solution. Finally, the hydrolysed feed is injected into a catalytic SCW reactor to convert the dissolved organics to a gas mixture [38].

An overview of studies regarding real biomass gasification under SCW is reported in Table 6.3. Many of experimental data showed that the GE and H_2 yield increase with temperature. Lu *et al.* [44], for example, have obtained higher H_2 yield increasing the temperature from 600 to 650 °C in the SCWG of wood

Table 6.3   Gasification under SCW of different real biomasses

Reactant	Reactor	Operation conditions			Catalyst	Experimental results and comment	Reference
		T (°C)	P (MPa)	RT			
Wood sawdust, rice straw, rice shell, wheat, corn stalk, peanut shell, stalk, corn cob sorghum stalk (CMC)	SS tubing (6 or 9 mm i.d.)	600–650	30	27 s	–	All the biomasses have been gasified with 40% H_2 molar fraction. Reactor temperature and inner diameter influenced the gas yield. High reactor residence time and heating rate, low biomass content and small biomass particles favoured high H_2 yield	[44]
Rice husk	SS 316 tube (270 mm, 1.65 mm)	400–680	32	60 min	–	CGE was ~60%–70% and increased linearly with T. A 20% drop in CE was due to increase in biomass concentration from 2 to 14 wt%	[2]
Cassava biomass	SS batch reactor (500 mL)	350–380		120 min		Char yield was similar to that formed with starch, but lower H_2 yield was obtained due to low H_2 content of the original biomass sample	[45]
Sawdust, straw, sewage sludge, lignin	Batch reactors (100–1,000 mL)	400–600	31–35	120 min	K_2CO_3	At $T > 550$ °C, complete biomass gasification to a H_2 rich product was reached also in absence of catalyst	[47]
Tobacco, corn, cotton, sunflower, oreganum stalk, corncob, Cr-tanned and vegetable-tanned waste	Inconel 625-lined tumbling batch autoclave (1 L)	500	24–33	60 min	–	Yields and gas composition depended on the organic materials other than cellulose and lignin amount in lignocellulosic materials. The presence of Cr negatively affected tanned waste gasification	[15]

Feedstock	Reactor	Temperature	Pressure	Time	Catalyst/additive	Remarks	Ref.
Sunflower oil, corn, carrot, bean, beef, mayonnaise, cooked beef, tropical fruit salad, chicken soup, cat food, molasses and glucose	SS 316 reactor (500 mL)	330	13.5	120 min	NaOH, H_2O_2	>H_2 production for carbohydrate-rich food waste, specifically glucose, molasses, whey powder, tropical fruit mixture > glutamic acid > dried mixed food waste > sunflower oil rice bran, chicken soup and cat food	[41]
Subbituminous, bituminous, lignite	Autoclave (20 mL)	600–700	12–105	10 min	NaOH, $Ca(OH)_2$ for capturing the CO_2	Water–carbon reaction, SCWR, CO_2 and other pollutants removal, all processes could be conducted in a single reactor under optimal conditions	[50]
Depithed bagasse liquid extract and sewage sludge (22 wt%)	Flow tubular reactor (Inconel 625 tubing 9.53–4.75 mm i.d.)	600	35.5	4–6 h	Activate carbon as spruce wood charcoal and macadamia shell charcoal	Almost 100% conversion to high H_2 yield. Deactivation of catalyst after <4 h without a swirl generator in the entrance of the reactor	[22]
Corn and potato starch	3 tubular flow reactors (Hastelloy C-276)	650	22		Carbon catalysts	>2 L g^{-1} of gas with high content of H_2 (57 mol%) were realized at the highest temperatures. Problems of plugging and reactor corrosion	[51]
Sawdust, cellulose (with CMC)	Batch reactor (140 mL)	450–500	27	20 min	Ru/C, Pd/C, CeO_2, nano-CeO_2, nano$(CeZr)xO_2$	Catalytic activities were Ru/C > Pd/C > nano$(CeZr)xO_2$ > nano-CeO_2 > CeO_2. The addition of CMC favoured the gasification efficiency	[52]
Beechwood sawdust	Incoloy 825 high-pressure tubing (5.4 mm i.d.)	600–650	28		NaOH	Biomass flash pyrolysis pre-treatment was exploited to concentrate the minerals in the char and not fed them to the reactor	[53]

(*Continues*)

Table 6.3 (*Continued*)

Reactant	Reactor	Operation conditions			Catalyst	Experimental results and comment	Reference
		T (°C)	P (MPa)	RT			
Corncob combined with CMC	Alloy tube and Hastelloy C-276 tube	650–775	25	40 s	25	$T > P >$ conc. $>$ r.t. effect. Diluted acid hydrolysis pre-treatment increased H_2 yield from 15.23 to 19.6 mol_{H2} kg^{-1} (feed: 3 wt%)	[28]
Rice straw, sawdust	SS tubing (i.d. 6.53 mm)	400	25	20	Ni-5132P	The gasification ratios were lower than expected from their components: interactions between each component occurred affecting the GE	[70]
Sunflower stalk corn-cob; vegetable-tanned leather	Inconel 625-lined tumbling batch autoclave (1 L)				K_2CO_3 Trona	Trona showed gasification activity similar to that of K_2CO^3. The use of this cheap material instead of commercially produced alkali materials can be preferable in the gasification of biomass in SCW	[74]
Peanut shell	SS 316 autoclave	400	24–28	20 min	NaOH, Raney Nickel	The synergic effect of NaOH and Raney Ni increased the H_2 yield	[10]

sawdust, rice straw, rice shell, wheat stalk, peanut shell and sorghum stalk. They used 2–3 wt% of sodium carboxymethylcellulose (CMC) to form a uniform and stable viscous paste thus aiding continuous feeding of multiphase mixture and favouring the GE. Moreover, by increasing the pressure, the H_2 yield increased along with a decrease in CH_4 and CO yields. High reactor residence time and heating rate favoured the gas yield along with smaller biomass particles which were more easily gasifiable. High content of biomass caused reactor plugging problems. Basu and Mettanant [2] gasified rice husks and an increasing in more than 50% in H_2 yield was obtained raising the temperature from 650 to 700 °C (32 MPa and 60 min). The CGE (~60%–70%) also increased linearly with the temperature, but a 20% drop in this value was due to the biomass concentration increase from 2% to 14%.

William and Onwudili [45] for cassava biomass similarly reported that higher H_2 yields were obtained at higher reaction temperature and that cassava waste produced a similar level of char than cellulose. Venkitasamy *et al.* [46] also investigated the influence of temperature for the SCW of sawdust and rice straw in a closed batch reactor obtaining an increase in gas and H_2 yield raising the temperature from 500 to 750 °C.

Schmieder *et al.* [47] reached the complete gasification of real biomass (sawdust, straw) and wastes (sewage sludge and lignin) at temperature higher than 550 °C and no difference was found between experiments with and without addition of K-based catalysts. This was due to the high potassium content in the studied biomass, for example for straw (ash: 4.6 wt% with ~15 wt% K).

D'Jesùs *et al.* [48] examined the SCWG of clover grass and corn silage by using a continuous flow reactor system. Pressure had no significant effect, but temperature, residence time and biomass particle size strongly influenced the biomass conversion. The addition of potassium significantly affected the corn starch GE, but, also in this case, it had no significant effect on the gasification of the K-containing natural products.

Yanik *et al.* [15] studied lignocellulosic and tannery wastes gasification at 500 °C, confirming that the H_2 yields (8.1–9.3 g_{H2} $kg^{-1}_{biomass}$) and gases composition depend also on the organic material other than cellulose and lignin contents of lignocellulosic material. Moreover, by using biomass with similar lignin content, the coke formation was different (five times higher with oreganum stalk than with sunflower stalk) thus suggesting that coke formation strongly depends on the structure and interactions between other components in biomass than lignin.

Muangrat *et al.* [41], by studying reactions of various food classes in SubCW (330 °C and 13.5 MPa) with sub-stoichiometric amount of H_2O_2 for partial oxidation and NaOH as catalyst, suggested that the potential for CGE and H_2 production depended on the class of food wastes, specifically not on the carbon percentage present in a material but on the chemical nature of carbon atom. In general, carbohydrate-rich biomass (molasses, tropical fruit mixture, whey powder) have a greater potential to form H_2 gas compared to other types as proteins and lipids. On this account, Kruse *et al.* [49] reported that phytomass (plant biomass) produced more products than zoomass (animal biomass).

NaOH and – in general – alkali can also adsorb CO_2 in the form of carbonate or bicarbonate, thus improving the H_2 purity in the effluent and shifting the WGSR in the forward direction. For this reason, Lin *et al.* [50] proposed a H_2 production process from lignite, bituminous and organic wastes, HyPr-RING, using CaO or/and $Ca(OH)_2$ as the adsorbent of CO_2, thus integrating WGSR and CO_2 adsorption reaction in a single reactor. For example, 170 cm^3 of gas with 80% of H_2 was produced from 0.1 g of the subbituminous Taiheiyo coal at 700 °C, by converting 90% of the carbon. Moreover, some organic materials containing chlorine and sulphur also produced gases (H_2, CH_4) and no chlorine or sulphur gases since they were captured by additives as Ca and Na.

Xu *et al.* [22] studied SCWG of some depithed bagasse and sewage sludge (22 wt%) using flow-type tubular reactor. Almost 100% conversion with high H_2 yield gases was observed for all feedstocks at 600 °C and 34.5 MPa in the presence of carbon catalysts, but deactivation of catalysts was observed. A problem with plugging of the reactor due to char was observed by Antal *et al.* [51] during corn and potato-starch biomass gasification at 650 °C. Metals present in the Hastelloy reactor tube catalysed the gasification and reforming reactions and some biomass feedstocks deactivated the catalyst by coke deposition on the reactor wall. However, they suggested that the coke could be easily and quickly removed from the reactor by combustion in flowing air.

Hao *et al.* [52] obtained almost complete gasification of 10 wt% sawdust or cellulose with Ru/C catalysts producing 20–40 g_{H2} per kg feedstock at 500 °C and 27 MPa. They observed such catalytic activities order: Ru/C > Pd/C > nano(CeZr)xO$_2$ > nano-CeO_2 > CeO_2.

To overcome some of the feeding issues in SCWG, Penninger and Rep [53] pre-treated beech wood sawdust by flash pyrolysis to concentrate the minerals in the char produced and not fed them to the reactor. GE values of ~60%–80% were obtained with H_2, CH_4, CO and CO_2 as their major gas. Low concentration of soda in the feed promoted the reaction. However, problems encountered in their experiments were associated with tar build-up in the preheater. Then, to further improve the biomass GE, a two-step H_2 production with the hydrolysis pre-treatment was also explored by Lu *et al.* [28].

6.3.2   Cellulose and hemicellulose

Cellulose is one of the most common substances used in predicting the behaviour of agricultural and food processing waste biomass under the hydrothermal gasification. It is formed by long linear chain from glucose molecules linked in the form of D-anhydroglucopyranose units with (1 → 4)-β-D-glycosidic ether bridges and the repeating unit is cellobiose. Similar to cellulose, hemicellulose is a complex macromolecular component of biomass built from sugar units. The backbone chain frequently consists of pentoses (e.g. xylans) or alternating units of mannose and glucose or galactose units. Hemicelluloses possess side chains linked to the main chain including acetic acid, pentoses acids and deoxyhexoses which are responsible for the solubility of the hemicelluloses in water and/or alkali. This solubility occurs

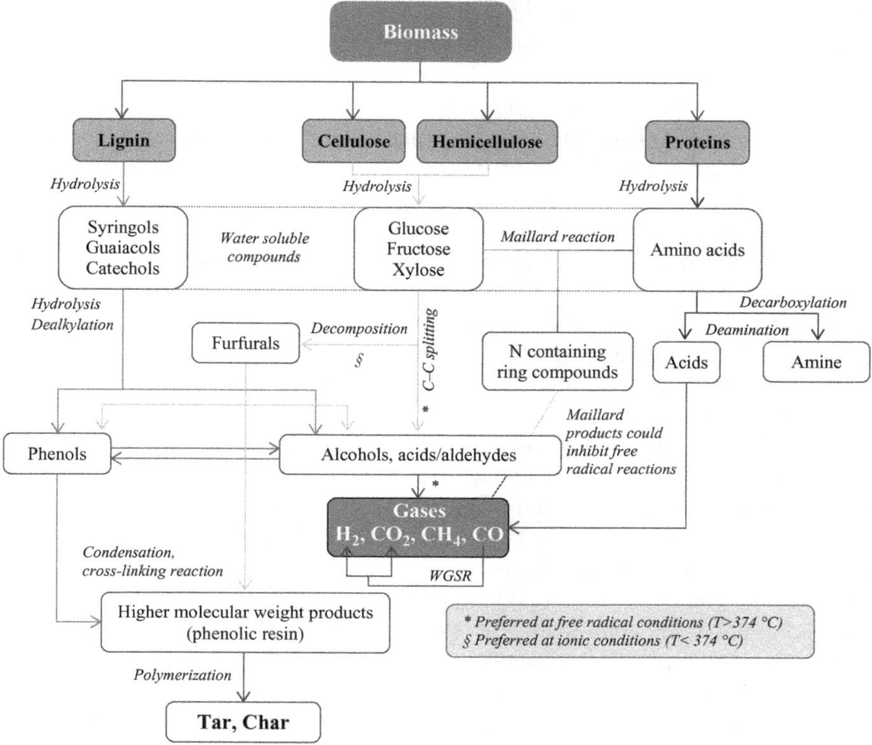

Figure 6.2   Simplified pathways of biomass decomposition under SCWG

only if the hemicelluloses are separated from other plant matter. Instead within the plant, they are mostly connected to lignin by covalent links and are thus fixed in the fibre structure [54]. The overall idealized reaction using cellulose as the model compound can be represented by:

$$C_6H_{10}O_5 + 7H_2O \rightarrow 12H_2 + 6CO_2 \qquad (6.21)$$

Several researchers suggested a reaction mechanism for the decomposition of cellulose and hemicellulose in SubCW and SCW as a way of understanding biomass gasification which is schematized in Figure 6.2 along with the simplified pathways of lignin. The effect of protein compounds under SCWG is depicted too.

Cellulose decomposition proceeds mainly through hydrolysis, dehydration and retro-aldol condensation [24]. Rapid hydrolysis can take place (at around 200–250 °C) at the glycosidic bond of cellulose to form water soluble sugars, both oligomers (cellobiose, cellotriose, cellotetraose, etc.) and monomers (glucose, fructose), without no production of gas, oil or char under 240 °C. The formation of furfural from fructose is due to a multiple water elimination. Dehydration and retro-aldol condensation can occur at the reducing end of cellulose to form levo-glucosan, glycolaldehyde and erythrose [11]. Often the char is formed during the decomposition of cellulose, by dehydration reaction from low molecular weight compounds and ring closure to furfural derivatives and phenols, which in fact represent the tar [55]. This reaction path is preferred in SubCW at ionic conditions. At free radical conditions ($T > 374\ °C$) instead, gas is produced directly from the decomposition of the water soluble products; in particular, glucose and fructose further undergo to fast decomposition to various compounds such as carboxylic acids, alcohols, aldehydes and ketones [25,56] which are all highly reactive compounds and easily decompose to gases via decarbonylation and decarboxylation. Furfural and phenols could also decompose into gases in SCW but this reaction is slower than the decomposition of the glucose.

In Table 6.4, some results obtained in SCWG of cellulose using different reactors and catalysts are compared.

The chemical nature of the carbon atoms may refer to both the type of bond linkages the carbon atoms have along with the nature and type of the other element(s) sharing the linkages. William and Onwudili [45], for example, suggested that the type of polymer linkages might be responsible for the difference in gasification products from carbohydrates such as cellulose and starch characterized by $(1 \rightarrow 4)$-β-D glycosidic and $(1 \rightarrow 4)$-α-D glycosidic bonds, respectively. Specifically, starch produced a higher H_2 yield than cellulose under identical reaction conditions, the cellulose bonds being stronger than those of the starch polymer. Resende *et al.* [27] investigated a no-catalytic gasification of cellulose for a range of parameters, confirming that at higher temperature the rate of formation of all gases increased.

In 1985, Modell *et al.* [57] reported the gasification of cellulose over a Ni-based catalyst with a batch reactor at 374 °C and 22 MPa obtaining CO, CO_2 and H_2 without char formation. Then Minowa *et al.* [40] investigated cellulose gasification at 350 °C and 16.5 MPa with a reduced nickel catalyst and reported that 70% of the carbon could be gasified. The catalyst with different supports had different activity not only due to the properties of materials, but also to the overall catalyst size; authors indicated that only the nickel particles present on the external surface contribute to the gasification. Usui *et al.* [58] presented Pd/Al_2O_3 with highest catalytic activity for cellulose gasification among supported Ni, Pd or Pt catalysts. Watanabe *et al.* [59] conducted the batch experiments for H_2 production from cellulose and glucose in SCWG (400–440 °C) and observed that the H_2 yield with ZrO_2 was almost twice than that without catalyst. Moreover, with NaOH, the GE was about five times higher and the CO yield was negligibly small. Guan *et al.* [60] used a batch reactor with K_2CO_3 and $Ca(OH)_2$ as catalysts: the H_2 and CH_4 yields increased by 70% and 40% as the temperature raised from 500 to 550 °C at 26 MPa ($12.6\ mol_{H2}\ kg^{-1}$ and $4.1\ mol_{CH4}\ kg^{-1}$). As expected, by increasing the temperature, CH_4 decreases reacting with water to form H_2 and CO_2 and, by increasing the K_2CO_3 amount, WGSR enhanced with correspondent higher production of both H_2 and CO_2.

Park and Tomiyasu [61] reported cellulose gasification in autoclave over RuO_2 and obtained CH_4, CO_2 and H_2 at 400 °C and 44 MPa: all hydrogen atoms in the

Table 6.4 Gasification under SCW of cellulose as model compounds

Reactant	Reactor	Operation conditions			Catalyst	Experimental results and comment	Reference
		T (°C)	P (MPa)	RT (min)			
Cellulose, starch, glucose	Batch reactor	300–380	9.5–22.5		–	$>H_2$ yield for glucose, followed by starch and cellulose; $>$chars, CO, C_1–C_4 for cellulose	[45]
Cellulose	Quartz capillary tubes (2 mm i.d., 0.58 cm^3)	400–600	~22	60	–	Gas yields and H_2 mol fraction are lower in quartz reactors than in SS reactors	[27]
Cellulose	Batch reactor	374	22		Ni-based catalyst	No char formation	[57]
Cellulose	SS autoclave (120 mL)	350	16	30–60	50 wt%Ni–SiO_2/ Al_2O_3	70% of the carbon could be gasified. Nickel particles presented on the external surface could only contribute to the gasification	[40]
Cellulose	Autoclave	350	16–20		Ni–Pd–Pt catalysts	Pd/Al_2O_3 showed highest activity	[58]
Cellulose, glucose	SS 316 tube bomb reactor with inner volume of 6 cm^3	400–440	30–35		NaOH, ZrO_2	H_2 yield with ZrO_2 was twice than that obtained without catalyst	[59]
Cellulose	Batch reactor (316L SS, 140 mL)	450–500	24–26	20	K_2CO_3, $Ca(OH)_2$	Good catalytic effect. CH_4 was dominant at relatively low temperature. The combination of both K_2CO_3 and $Ca(OH)_2$ catalysts increased the H_2 yield	[60]
Cellulose	Autoclave	450	44	120	RuO_2	Higher CH_4 yield was obtained than H_2 yield	[61]
Cellobiose	Flow tubular reactor (Inconel 625 tubing with 9.53 mm o.d. and 4.75 mm i.d.)	600	35.5	60	Coconut shell activated carbon	Completely biomass gasification	[51]
Cellulose	Batch reactor	500	500	20	CeO_2, $nCeO_2$, $n(CeZr)_xO_2$, Pd/C, Ru/C	Maximal values of H_2 yield, GE and CE were obtained with Ru/C catalyst	[44]
Cellulose	SS tubing (i.d. 6.53 mm)	400	25		Ni-5132P	H_2 yield was almost twice as much as that without catalyst for all the feedstocks	[76]

gaseous products were originated from water molecules and the catalytic effect of ruthenium oxide resulted from a redox couple of Ru^{IV}/Ru^{II}. In the presence of coconut shell AC, cellobiose was almost completely gasified at 600 °C and 34.5 MPa, with low CO yield and high H_2 and CO_2 yield [51]. Gasification of cellulose in the presence of metal catalysts, including CeO_2, $nCeO_2$, $n(CeZr)_xO_2$, Pd/C, Ru/C was examined by Lu *et al.* [44] in a batch reactor at 500 °C and an initial pressure of 4.0 MPa. After 20 min of reaction, the maximal values of H_2 yield (~17 mol kg^{-1}), GE (~115%) and CGE (~100%) were obtained with Ru/C.

6.3.3   Glucose

SCWG with glucose as model compounds has been investigated by many researchers, however, due to very fast hydrolysis of the cellulose, the gasification of glucose and cellulose leads to identical gas yields. Some literature data are reported in Table 6.5.

Although glucose in SCW is expected to be gasified through a variety of reaction pathways (see Figure 6.2), glucose steam reforming (see (6.22)) and WGSR (see (6.6)) reactions have received particular attention because of the importance of their role in determining the degree of gasification and composition of gaseous products:

$$C_6H_{12}O_6 + 6H_2O \rightarrow 6CO_2 + 12H_2 \tag{6.22}$$

Already in 1975, Amin *et al.* [62] obtained a H_2-rich gas from the catalytic gasification of glucose in water at 374 °C and 22.1 MPa, mainly through WGSR with a low efficiency (20%) of carbon gasification. Antal *et al.* [63] reported that without catalysts, low concentrations of glucose (0.1 M) and various wet biomass species can be completely gasified in SCW to a H_2 rich syngas containing almost no CO (at 600 °C and 34.5 MPa) already after a residence time of 34 s whereas higher concentration of glucose evidenced incomplete gasification [51]. Heterogeneous catalysis was thus employed to increase the GE of concentrated feeds. Reactors were properly fabricated to accommodate the catalyst. It was demonstrated that concentrated glucose solution could be completely gasified in SCW at temperature higher than 600 °C with the help of carbon based catalysts [51,52,63]. Experiments were carried out by employing a variety of activated carbon (AC) (spruce wood charcoal, macadamia shell charcoal, coals AC and coconut shell AC) at high temperatures (600–650 °C) under supercritical pressures (22–34.5 MPa) and concentrated glucose feeds (22 wt%) at a weight hourly space velocity (WHSV) of 22.2 h^{-1} [22]. CGE near 100% were easily achieved and the extension of surface area of the carbon does not greatly affect the catalytic performance. The amount of CO produced varied according to the type of AC used. However, although complete glucose gasification to high H_2 yields was achieved, carbon catalysts deactivated progressively.

As regards the effect of the temperature, the H_2 yield by gasification of a glucose solution (0.6 M) increased sharply with temperature over 660 °C; CO yield increased with temperature at lower temperature but, after the reaching of a

Table 6.5   Gasification under SCW of some model compounds

Reactant	Reactor	Operation conditions			Catalyst	Experimental results and comment	Reference
		T (°C)	P (MPa)	RT			
Glucose, glycerol	Flow tubular reactor (Inconel 625 tubing with 9.53 mm o.d. and 4.75 mm i.d.)	600	34.5	44 s	Spruce wood, charcoal, macadamia or coconut shell charcoal, AC	Spruce wood charcoal allowed reaching 99% CGR of a solution of 22 wt% glucose with production of 21.4 mol_{H2} kg^{-1} and 7.5 mol_{CH4} kg^{-1}. Glycerol completely decomposes without catalyst after 44 s	[22]
Glucose	Continuous Tubular	650	25	3.6 min	–	0.1 M glucose solution was completely gasified without char formation	[65]
Glucose (0.6 M)	Tubular-flow reactor Hastelloy C-276 tube (9.53 mm o.d. 6.22 mm i.d., 670 mm)	480–750	28	10–50 s		H_2 yield increased with temperature over 660 °C. Carbon efficiency was 100% at 700 °C for 10–50 s	[85]
Glucose Catechol Vanillin Glycine	Batch reactors (100 and 1,000 mL) Tubular flow reactor	600	250	30, 60, 120 s	KOH, K_2CO_3	At $T > 550$ °C, complete gasification of glucose was possible with trace of solid and oily by-products. By addition of KOH, a H_2 rich gas was obtained with low CO, CH_4 an C_2–C_4 concentrations. At lower feed concentrations (≤ 0.2 M), residence times of ~30 s were required. At higher feed concentration (≥ 0.6 M) and constant K concentrations soot and tar formation appeared	[47]
Glucose	Union tee reactor (316 SS, 24 mL)	310–350	10–21		Raney Ni; Ni(acac)₂, Co(acac)₂, Fe(acac)₃	Raney Ni was a more effective catalyst compared to homogeneous catalysts	[66]
Glucose Glycine Glycerol Lauric acid Humic acid	Non stirred 316 SS batch reactor (50 mL)	380	23	15 min	Raney Ni, Ni/α-Al₂O₃, Ru/C and Ru/γ-Al₂O₃	Carbon conversion on Raney nickel: glycerol > glucose > glycine > lauric acid > humic acid. Carbon conversion on Ru catalyst: glycerol > glycine > glucose > lauric acid > humic acid. It catalysed also methanation	[38]

(*Continues*)

Table 6.5 (*Continued*)

Reactant	Reactor	Operation conditions			Catalyst	Experimental results and comment	Reference
		T (°C)	P (MPa)	RT			
Glucose	Bench-scale continuous down-flow tubular reactor	700	24		Ru-modified Ni/γ-Al$_2$O$_3$	H$_2$ yield ~50 mol kg^{-1} glucose over 33 h on stream Ru$_{0.1}$Ni$_{10}$/γ-Al$_2$O$_3$ exhibited higher activity and stability	[35]
Glucose, glycerol, pinewood	Quartz capillary reactor	600–700	25–30	140 s	KOH, NaOH	Complete gasification reached	[68]
Glucose (17 wt%)	Quartz capillary reactor	450–700	5–50		Ru/TiO$_2$	Complete gasification with 3 wt% Ru/TiO$_2$	[68]
Glucose	Continuous tubular reactor	700–800	24–25	2 s	Ru/Al$_2$O$_3$	12 mol$_{H2}$ mol^{-1}$_{glucose}$ reached (the stoichiometric limit). At high glucose concentration (>5 wt%) tar formed too	[39]
Glucose (17 wt%)	SS316 tube bomb reactor (6 cm^3)	400		15 min	CeO$_2$, MoO$_3$, TiO$_2$, ZrO$_2$	Gasification efficiency (=CO + CO$_2$): MoO$_3$ > ZrO$_2$ > CeO$_2$ > TiO$_2$; H$_2$ yield: ZrO$_2$ > CeO$_2$ > MoO$_3$ > TiO$_2$	[69]
Glucose	Hastelloy C reactor (75 mL)	330	13.5	60–120	Alkaline catalysts	NaOH > KOH > Ca(OH)$_2$ > K$_2$CO$_3$ > Na$_2$CO$_3$ > NaHCO$_3$	[34]
Glucose	Autoclave (190 mL)	400–500	30–50	1.8–16.3 min	K$_2$CO$_3$	Key compounds identified: furfural, phenol, phenols, acids	[23]
Glucose, glycerol	Quartz capillary tubes	400–600	30	60 s	NaOH, Ru/TiO$_2$	Complete conversion was achieved only for very diluted solutions (1 wt%)	[29]
Glucose	Batch autoclave (Inconel 625, 1 L)	500	30	60 min	K$_2$CO$_3$, Raney nickel	A decrease in gas yield at slow heating showed that in the technical process the biomass should be heated as fast as possible	[67]

maximum it dropped rapidly; CO_2 yield increased with temperature over the temperature range considered. CGE reached 100% at 700 °C indicating complete conversion of glucose to gaseous products. The H_2 gasification efficiency higher than 100% indicated that SCW contributed some of the hydrogen in the product gas, confirming that the water under SC conditions act as both a H_2 source as well as a solvent for glucose gasification [64]. A very strong effect of temperature was observed by Hao *et al.* [65] in the glucose SCWG at 25 MPa: an increase in reaction temperature from 500 to 650 °C resulted in 167% increase in the CGE and more than 300% increase in the GE. At the same time, by increasing glucose concentration, the GE decreased.

However, by comparing glucose and cellulose SCWG, the highest amount of H_2 was obtained with glucose, whereas cellulose was found to produce the greatest amount of chars, CO and C_1–C_4. Moving from SubCW to SCW resulted in a decrease in the oil and char yield with a corresponding increase in the gas yield, mainly CO_2 [45].

Schmieder *et al.* [47] conducted the gasification of glucose by using two-batch reactors and a flow type apparatus (600–700 °C; 25–30 MPa). The use of K_2CO_3 or KOH (contained in real biomass as an ash) allowed total gasification to H_2 and CO_2 within 140 s with low concentrations of CO, CH_4 and C_2–C_4 hydrocarbons in the product gas (<1, ~3 and <1 vol.%, respectively). Sinağ *et al.* [23] indicated that K_2CO_3 favours the glucose decomposition to formic acid, an intermediate of gas formation, and restrains the furfural formation which instead can be converted to phenols and then to tar and coke (see Figure 6.2). Muangrat *et al.* [41] reached a very high H_2 gas production from glucose at a low temperature of 330 °C and a pressure of 13.5 MPa thanks to the synergic role of H_2O_2 and NaOH.

As already said, Ni-based catalysts are efficient materials for SCWG [36,37]. On this account, Azadi *et al.* [66] studied hydrothermal gasification of glucose solutions (0.06–0.65 M) in presence of three organometallic salts, $Ni(acac)_2$, $Co(acac)_2$ and $Fe(acac)_3$, under SubSCW conditions. At 350 °C, in contrast to homogeneous catalysts, Raney nickel was a more effective catalyst producing five folds more H_2 in a shorter period of time with a high heating rates. The values obtained with Raney nickel were comparable with that obtained at 500 °C with longer reaction time [29,67]. Then, Azadi *et al.* [38] investigated the SCWG of model compounds of activated sludge. Among them, glucose, glycine, glycerol, lauric acid and humic acid were used as model compounds for carbohydrates, proteins, alcohols and glycerolipids, fatty acids and humic substances, respectively, at 380 °C with Raney Ni, Ni/α-Al_2O_3, Ru/C and Ru/γ-Al_2O_3 catalysts. By using Raney nickel, the carbon conversion generally increased with reduction in the number of C–C bonds present per unit mass of the molecules, except for the glycine which contains a nitrogen atom in its structure. The presence of inorganic ash decreased the gasification yield of glucose and glycerol and increased the gasification yield of glycine. Zhang *et al.* [35] demonstrated that the addition of small amounts of Ru in Ni-based catalyst could improve Ni dispersion thus accounting for the enhanced activity and higher stability of $Ru_{0.1}Ni_{10}/\gamma$-Al_2O_3 catalyst. In addition, a small amount of Ru could also enhance the reducibility of NiO.

As regards the resistance of reactor material, in general, the alkali may dissolve the protective metal oxide on the reactor walls causing important problems of corrosion. If the outer metal oxide layer dissolves, it exposes fresh, temporary reduced metal to SCW. The metal can quickly oxidize in SCW further producing hydrogen. For this reason, Kersten *et al.* [68] conducted over 700 experiments in small sealed quarts capillary tubes to investigate the SCWG in the absence of metal reactor surfaces: NaOH increased the H_2 yield from 9.9 to 17–21 vol.% following the gasification of 17 wt% glucose at 600 °C, 30 MPa, and 60 s, whereas, in the absence of catalyst, it was confirmed that complete gasification was only possible at very low concentrations, below 2 wt%. The addition of Ru/TiO_2 catalyst allowed glucose solutions of up to 17 wt% to be gasified. According to Byrd *et al.* [39], glucose was gasified in SCW in a continuous tubular reactor at short residence times. The addition of Ru/Al_2O_3 catalyst enhanced the conversion and H_2 yield by reducing CH_4 formation. At high glucose concentration (>5 wt%), the formation of tar was observed.

Watanabe *et al.* [69] investigated the acidity and basicity of metal oxide catalysts in glucose SCWG at 400 °C and found this GE order: $ZrO_2 > CeO_2 > MoO_3 > TiO_2$ (anatase) $> TiO_2$ (rutile). With CeO_2 and ZrO_2, the H_2 yields was higher than that without the catalyst but by adding MoO_3 and TiO_2, the H_2 formation was suppressed. Then, the H_2 yield was enhanced in the presence of NaOH but it was inhibited by H_2SO_4.

6.3.4 Lignin

Lignin is a complex, stable highly aromatic biopolymer available in different composition and molecular weight, characterized by a chemically and physically heterogeneous branched structure mostly built from three phenyl propane (C_6–C_3) subunits. *Softwood* lignin consists almost exclusively of *trans-p*-coniferyl alcohol (con); in contrast, *hardwood* lignin is composed of *trans-p*-sinapyl alcohol (sin) and coniferyl alcohol units in varying ratios. *Grass* lignin has a higher content of *trans-p*-coumaryl alcohol (cou) than other types of lignin [70]. Isolation of lignin from biomass causes structural change and these differences can affect the gasification characteristics of biomass. Indeed, as structure of model lignin samples were altered by isolation method, Madenoğlu *et al.* [71] concluded that model lignin samples could not truly represent lignin structure in biomass. The highly cross-linked phenol alcohol structure – bonded together with strong ether bonds (C–O–C) – makes lignin the most resistant component of lignocelluloses and, in general, its efficient decomposition to gases is more difficult than that of cellulose and hemicellulose. In literature, many studies about the lignin gasification under SubCW or SCW are reported and some of them are summarized in Table 6.6.

Lignin decomposition starts with hydrolysis and dealkylation to low molecular weight fragments having reactive functional groups and compounds (see Figure 6.2). The high reactivity of low molecular weight fragments (formaldehyde, syringol, guaiacol, catechol, etc.) could cause their re-polymerization and formation of char and tar (solid residue) [55,72]. A cross-linking reaction among these

Table 6.6   Gasification under SCW of lignin and some model compounds

Reactant	Reactor	Operation conditions			Catalyst	Experimental results and comment	Reference
		T (°C)	P (MPa)	RT			
Lignin/phenol	SS SUS316 tube bomb reactor (10 cm^3)	400		10–64 min	–	The increase in phenol/lignin ratio allowed the obtainment of lower TIS* yield, lighter TIS production and prevented polymerization of the TIS products	[73]
Lignin, cellulose, xylan	Different reactors	500–775	27		Ca(OH)$_2$, K$_2$CO$_3$	Systematic experimental and analytical study using different reaction and biomass compounds	[29]
Lignin	Quartz capillary tubes (2 mm. i.d., 0.58 cm^3)	375–725		60	–	Manipulating lignin loading provided an efficient means to control the CH$_4$/H$_2$ molar ratio. The highest H$_2$ yield was 7.1 mol kg^{-1}, obtained at 725 °C and 60 min	[74]
Lignin, cellulose mixture	316SS steel tubing batch-type	350	25	20 min	– Ni-based catalyst	Lignin content affected the amount and composition of gas. Cellulose and xylan are hydrogen donor to lignin. Cellulose-lignin mixture required a larger amount of Ni catalyst compared to cellulose alone	[75]
Lignin	SS tubing (i.d. 6.53 mm)	400	25	20 min	Ni-5132P	Different lignin reagents showed different lignin gasification characteristics	[70]
Lignin	Continuous flow reactor	390–450	25	0.5–10 s		Complete depolymerization could be achieved within a 5 s residence time. Gas formation arose from lignin during the early period of lignin depolymerization	[72]
Lignin	SS tube bomb reactors (6 cm^2)	250–400		360	Ni/MgO	The amount of gases produced increased with an increase in Ni loading on magnesium oxide; MgO decomposed lignin to reactive intermediates and Ni promoted reaction between intermediates and water to form gases	[33]

(Continues)

Table 6.6 (*Continued*)

Reactant	Reactor	Operation conditions			Catalyst	Experimental results and comment	Reference
		T (°C)	P (MPa)	RT			
Lignin	SUS 316 tube reactor (6 cm^3)	400	37.1		Ruthenium trivalent salts	Gasification activity order: Ru/C $\approx$ Ru(NO)(NO$_3$)$_3$/C $\approx$ Ru(NO)(NO$_3$)$_3$/TiO$_2$ > RuCl$_3$/C $\approx$ RuCl$_3$/TiO$_2$	[77]
Lignin	SUS 316 SS bomb reactor (6 cm^3)	400	30		NaOH–ZrO$_2$	Zirconia allowed to obtain a H$_2$ yield twice than that obtained without catalyst; base catalysts gave 2 times higher H$_2$ yield than zirconia	[78]
Catechol	Tubular flow (i.d. 8 mm) and tumbling reactors (1,000 mL)	600–700	20–40	1–2 min	KOH, LiOH	More than 99% gasification was achieved at 600 °C	[79]
Alkylphenols	SS 316 tube bomb reactor (6 cm^3)	400	28.8	15 min	Ru/γ-Al$_2$O$_3$; Pt/γ-Al$_2$O$_3$; Pd/γ-Al$_2$O$_3$; Ru/C; Rh/C; Pd/C	Ru > Rh > Pt > Pd, the reactivity of *o*- and *p*-alkyl phenols were higher than those of *m*-alkylphenol	[32]
Lignin 4-propylphenol	316 stainless steel bomb reactor	400			Ru/TiO$_2$, Ru/C, Ru/Al$_2$O$_3$, Rh/C, Pt/C, Pt/Al$_2$O$_3$, Pd/C, Pd/Al$_2$O$_3$, Ni/Al$_2$O$_3$	Decomposition to low molecular weight compounds was enhanced by increasing the water density. Gasification of the low molecular weight compounds was accelerated over metal catalysts	[80]

*TIS = tetrahydrofuran-insoluble compounds.

reactive degradation fragments and residual lignin could also give higher molecular weight fragments [71,73].

Guo *et al.* [29] suggested that a temperature of 700 °C or higher is necessary for complete gasification of lignin; without catalysts, the CE increased from 41% for 1.5 wt% lignin at 500 °C to 90% for 3 wt% lignin above 700 °C. The positive effect of temperature to maximizing H_2 yield was also discussed by Resende *et al.* [74] in lignin gasification from 350 to 725 °C.

Effect of lignin content on gaseous product composition and interaction between lignin and cellulose and xylan (as model of hemicellulose) was investigated by Yoshida and Matsumura [75] at 350 °C and 25 MPa in a batch reactor in 20 min. Cellulose gives the highest H_2 yield whereas a decrease in H_2 yield was recorded for the mixtures containing lignin. Therefore, lignin acted as an inhibitor to syngas production thus the intermediate products from cellulose and hemicelluloses likely could react with lignin reducing the H_2 formation. On the contrary, cellulose or hemicelluloses act as H_2 donor to lignin. Also Karagöz *et al.* [55] reported that – without the catalysts – reactivity of lignin was lower than that of cellulose and real biomass, sawdust and rice risk at 280 °C.

Afterwards, Yoshida *et al.* gasified lignin, cellulose and their mixture with commercial Ni-based catalysts at 400 °C and 25 MPa. As expected, SCWG of cellulose-lignin mixture requires more Ni-catalyst compared to cellulose alone. The mixture with *hardwood* and *grass lignin* was gasified much more easily than *softwood lignin* [70]. Poisoning of catalyst was observed due to carbon production, sulphur addition to the lignin structure during sulphite pulping or Kraft pulping and tarry products formation by the reaction between cellulose and lignin. Ando *et al.* [76] found that – although lignin content of Japan cedar was similar to that of chinquapin – the residue yield obtained with such species were different. Specifically Japan cedar, containing softwood lignin showed higher resistance to degradation than chinquapin and bamboo containing hardwood lignin thus confirming the different behaviour of various lignin species.

Yong and Matsumura [72] investigated the lignin decomposition under different heating rate and temperature (390–450 °C) and short residence times (0.5–10 s) at 25 MPa. Char was formed at both short and long residence times, suggesting that this cross-linking process occurs instantaneously. The presence of phenolic compounds at short residence times also indicated that ether bonds in the lignin are easily degraded and that gas formation mainly comes from lignin depolymerization.

Madenoğlu *et al.* [71] investigated five biomass samples with different amount of lignin and thus different cellulose, hemicellulose and lignin ratio in a continuous flow reactor at 600 °C and 35 MPa. As expected, CGE changes with the biomass type, i.e. acorn has lower CGE and higher residue yield in spite of having relatively low lignin content (12.5 wt%); at the same time, CGE of extracted acorn (40.0 wt% lignin) was higher than acorn. This could be due to their different composition: acorn in fact contains nearly 75% of tannin which is a polycondensation product of glucose and gallic acid, but tannin (in respect to lignin) contains less aliphatic groups which could increase gasification yield.

Therefore, a significant relationship between the lignin content and the product yield and the gas composition was not distinguishable for all the selected biomass samples, likely due to the differences in the structure and composition of lignin in the feeds. Sato *et al.* [33] investigated the lignin gasification in the presence of Ni/MgO from 250 to 400 °C. The metal and the support play different roles in gasification: MgO decomposed lignin to reactive intermediates, whereas Ni promoted reaction between intermediates and water to form gases. By increasing the Ni amount up to 20 wt% Ni, higher gases amount was produced, reaching the highest yield of 78% at 400 °C and 0.3 g cm^{-3} for a 360 min reaction time.

Yamaguchi *et al.* [77] evaluated lignin gasification over ruthenium salts in SCW observing this gasification activity order: Ru/C $\approx$ Ru(NO)(NO$_3$)$_3$/C $\approx$ Ru(NO)(NO$_3$)$_3$/TiO$_2$ > RuCl$_3$/C $\approx$ RuCl$_3$/TiO$_2$. Activities of RuCl$_3$/C and RuCl$_3$/TiO$_2$ catalysts were low likely because large Ru metal particles were formed having fewer active sites and chloride ions adsorbed on the Ru metal particles poisoned the catalyst.

Watanabe *et al.* [78] studied the effect of both homogeneous NaOH solution and metal heterogeneous (ZrO$_2$) catalysts on the lignin gasification (400 °C – 30 MPa, 0.5 M; 15–60 min) and confirmed a higher H$_2$ yield with NaOH (two times higher) than ZrO$_2$ and that WGSR was promoted lowering the CO yield at all reaction times.

Studies carried out by Schmieder *et al.* [47] by using catechol and vanillin as lignin model compounds showed that at 600 °C, after 30 s of residence time (20–30 MPa, 0.2 M), 10.5 mol H$_2$ per mole of catechol (82% of theoretical H$_2$ formation) was obtained (see Table 6.5). Vanillin gasification was easier than catechol, more than 99% destruction efficiency being reported even without the use of KOH as catalyst. Also Kruse *et al.* [79] studied the gasification of pyrocatechol using both a batch autoclave and a tubular flow reactor. More than 99% of biomass was already gasified at 600 °C; at 500 °C and 25 MPa, the increase in KOH amount from 0 to 5 wt% enhanced the H$_2$ yield by 40 vol% and depressed the CO yield from 40 to 0.7 vol%. Also in this case, this result was attributed to formation of formates by the addition of alkali salts.

Sato *et al.* [32] conducted the gasification of alkylphenols over supported noble metal catalysts and reported this order of activity: Ru/γ-Al$_2$O$_3$ > Ru/C, Rh/C > Pt/γ-Al$_2$O$_3$ > Pd/C and Pd/γ-Al$_2$O$_3$. The noble metal catalysts were effective for decomposition of the benzene ring of alkylphenols for the gasification in the presence of water. As regards the isomers of propylphenols, the reactivities of *o*- and *p*-propylphenols were relatively higher than those of *m*-propylphenols. Later Osada *et al.* [80] studied the gasification of lignin and 4-propylphenol (as a model of low molecular weight compound from lignin) over supported metal catalysts in SCW at 400 °C. They confirmed that the gasification of lignin proceeded through two steps: (i) decomposition of lignin to low molecular weight compounds and (ii) gasification of the lower molecular weight compounds over metal catalysts. Moreover, if lignin gasification rate was enhanced by the increase of water density, this was not true for 4-propylphenol gasification. This result indicates that the water density has a major effect on the first step of lignin gasification, or rather decomposition to low molecular weight compounds.

Considering that a single type of model compound cannot completely simulate the complexity of biomass composition, as already discussed, some researchers studied gasification of mixed model compounds [81]. Goodwin and Rorrer [82] for example used xylose and phenol as model compounds for hemicellulose and lignin (750 °C, 25 MPa) and found that xylose promotes the apparent conversion rate of phenol. Weiss-Hortala *et al.* [83] also confirmed that if phenol was present, the GE of glucose was dramatically reduced. Moreover, Castello *et al.* [84] found that during SCWG of glucose/phenol mixtures, phenol behaved as an inert component in terms of gas production, but as inhibitor towards H_2.

6.3.5   Glycerol

The interest in SCWG of glycerol was increasing in recent years and most of published works focused on the effects of operating parameters and development of related catalysts [85]. Considering the high production of glycerol as a by-product of biodiesel plants and the decrease in its price, the individuation of new routes to convert it into high value added products is very attractive. The maximum theoretical H_2 yield obtainable from glycerol is 7 mol according to the following equation:

$$C_3H_8O_3 + 3H_2O \rightarrow 7H_2 + 3CO_2 \tag{6.23}$$

Xu *et al.* [22] proposed the decomposition of glycerol in SCW without catalyst to a H_2 rich syngas containing no CO (44 s at 600 °C and 34.5 MPa). Antal *et al.* [86] studied the glycerol SCWG at 500 °C and 34.5 MPa, describing the free radical chemistry for its decomposition, whereas Bühler *et al.* [87] offered a detailed reaction pathway and a kinetics study proposing both ionic and free radical reactions if glycerol is treated at low temperature (349–475 °C), with very low gases production. Although without a proper catalyst near-theoretical H_2 yields were obtained for dilute glycerol concentrations at 800 °C [68], feed containing up to 40 wt% glycerol were completely gasified in the presence of 5 wt% Ru/Al_2O_3 (700–800 °C, 25 MPa, 1–6 s) thanks to the high activity of Ru-based catalysts towards C–C bond scission [88]. Onwudili and Williams [89] reported the alkaline hydrothermal gasification of a typical biodiesel plant waste containing a mixture of glycerol and unrecovered fatty acid methyl esters (FAME). Crude glycerol was easily converted to H_2 without significant formation of solid reside, whereas FAME showed high stability towards decomposition except for some hydrolysis reactions to the corresponding fatty acids with significant soap formation. Guo *et al.* [90] identified the intermediates of glycerol under SCWG (487–600 °C, 25 MPa) and developed the first quantitative kinetics model describing the individual gaseous products. The reaction rates analysis based on the model showed that the main sources of H_2 production were glycerol pyrolysis and steam reforming of intermediate products, whereas the rate of WGSR was very low, indicating that the WGSR was not the main source of H_2. The temperature estimated by kinetics model for completely SCWG of 10 wt% glycerol solution was 600 °C with 7 s as residence time.

6.4   Critical rationalization of data

Among many hydrogen production methods, the gasification under SCW represents a promising technology for the conversion of biomass with high moisture content, with the main advantages to achieve high solid conversion and both high reaction efficiency and H_2 selectivity. As previously reported, the amount of key compounds formed under SCWG and their formation pathways strongly depend on the experimental conditions such as temperature, reaction time and the catalyst used. In addition, the potential for carbon gasification and H_2 production strongly depends on the composition and concentration of biomass used too. Furthermore, also the structure of biomass in terms of carbon percentage and chemical nature is a decisive factor influencing the conversion rate and the extent of intermediates relevant for H_2 production.

To study the complex mechanisms involved into gasification of biomass, many efforts have been done by many research groups using both real biomass and model compounds. Nevertheless, the chemistry of biomass degradation is not fully understood yet, since the reaction pathways consist of many steps, involving a huge number of compounds. The gaseous yield depends on chemical reactions involved and their reaction rate while the product gas composition would be governed by the chemical equilibrium of the reaction. The main key points conditions are: (i) the rate of biomass decomposition to intermediates that form H_2 has to be much greater than the rate of formation of polymeric unreactive substances and (ii) the rate of the water gas shift reaction must be promoted, whereas (iii) the methanation of either CO or CO_2 has to be inhibited [59].

Some of the most interesting results obtained in biomass gasification under SCW and discussed in this chapter have been rationalized in Figure 6.3 and main considerations are here reported.

Among different reaction parameters, the temperature appears the most important one to take into consideration to obtain high GE and high H_2 yield. At higher temperature and low-density conditions, formation of gases is favoured thanks to the prevalence of free radical reactions [12,13]. Considering that high GE and H_2 yield are the results of promoted free radical mechanism reactions and WGSR, at temperature higher than 600 °C, a H_2-rich product gas can be formed from a variety of biomass sources with near complete conversion and low char and tar formation also without catalyst. For example, 23.4 g_{H2} kg^{-1} $C_{in\ hazelnut\ shell}$, 35.8 g_{H2} kg^{-1} $C_{in\ tomatoes\ residue}$, 40.4 g_{H2} kg^{-1} $C_{in\ cauliflower\ residue}$ had been obtained in a continuous flow reactor at 600 °C [71] and 44.0 g_{H2} $kg^{-1}_{glucose}$ at 650 °C [65], but 52.5 g_{H2} $kg^{-1}_{corncob}$ at 750 °C [28] and 56.0 g_{H2} kg^{-1}_{lignin} at 775 °C [29]. A temperature of 700 °C or higher, in fact, is necessary for lignin complete gasification, since, among different components of biomass, it is one of most difficult to decompose to gases due to the existence of very stable (*p*-hydroxymethoxyphenyl) propane units.

By comparing biomasses with similar K content, cellulose, hemicellulose and lignin content, the yields and composition of gases obtained from them are significantly different, suggesting that also the structure and the interaction between other organic components play an important role. For example, at 500 °C,

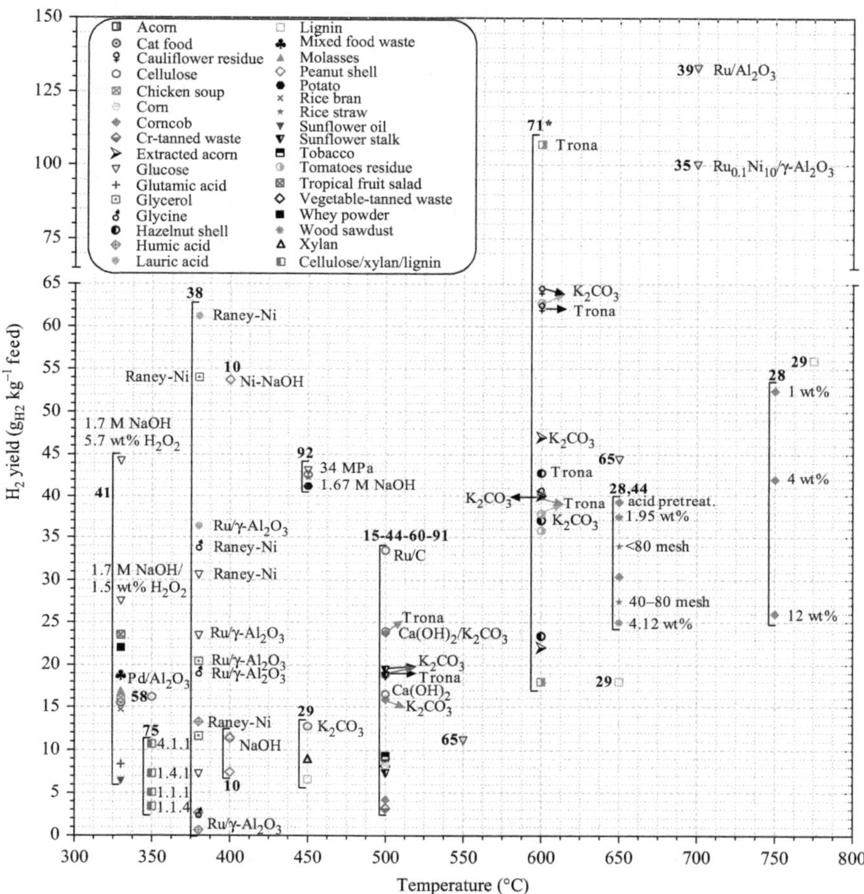

Figure 6.3   *Rationalization of data in terms of H_2 production yield as a function of biomass feed and experimental conditions. If not specified, catalyst was not employed. For Reference 71, the H_2 yield is referred as $g_{H2}\ kg_C^{-1}$.*

gasification of corncob (with high amount of cellulose and hemicellulose) resulted in the very low amount of H_2 production [15,44], whereas the gas obtained from tobacco stalks contained the highest H_2 content (39.5%). In case of tannery wastes, vegetable-tanned wastes produced an amount of gas and coke similar with the lignocellulosic materials, higher than Cr-tanned waste. Therefore, the presence of chromium negatively affects the gasification process.

When biomass mixtures are gasified, lignin acted as an inhibitor to syngas production: specifically, by using a cellulose/xylan/lignin mixture of 1/1/4, the results obtained at 350 °C confirmed the negative influence of lignin with a low H_2 yield (3.4 $g_{H2}\ kg^{-1}_{reactant}$). The use of equal amounts of cellulose/xylan/lignin mixture increased the H_2 yield by 50% (5.04 g_{H2} $kg^{-1}_{reactant}$). At last, the highest amount of gas, 17 mol of H_2, CO_2, CH_4 and C_2H_6 per gram of reactant was obtained with a cellulose/xylan/lignin mixture of 4/1/1, corresponding to 10.6 g_{H2} $kg^{-1}_{reactant}$ [75].

In order to promote the lignocellulosic separation from whole biomass thus making it more accessible for the conversion to sugar under SCWG, an acid pretreatment has been suggested. Under the optimal operation conditions (650 °C, 25 MPa, 40 s, 2 wt% corn cob + 1 wt% CMC), without acid hydrolysis pretreatment, the H_2 yield by corncob SCWG was 30.4 g_{H2} kg^{-1}, whereas, after diluted acid hydrolysis pre-treatment, the H_2 yield increased up to 39.2 g_{H2} kg^{-1}. Feedstock including an high content of biomass is more difficult to gasify in SCW resulting in lower H_2 yields, as showed for corncob gasification at 750 °C and for wood sawdust at 650 °C. Specifically, as the corncob concentration increases from 1 wt% up to 12 wt%, the H_2 yield decreases sharply from 52.4 to 26 g_{H2} kg^{-1} [28]. At the same time, smaller biomass particles are easier to degrade: if rice straw particles are lower than 80 mesh, a yield of 34 g_{H2} kg^{-1} is obtained at 650 °C, this value decreases to 27.5 g_{H2} kg^{-1} when the particle size is in the range of 40–80 mesh [44].

SCW itself could act as an acid/base catalyst but, to enhance the CGE and increase in H_2 yield under low temperature and at short time, the design of a suitable catalyst represents one of the most challenging field of research.

In particular homogeneous alkali catalysts promote the biomass gasification reactions and adsorb CO_2 thus increasing the H_2 yield but their recycling is very difficult. Metals present in biomass such as the alkali metals (i.e. sodium and potassium) and alkaline earth metals (i.e. calcium) may catalyse the production of hydrogen too. For example, the CGE was improved by addition of K_2CO_3 and Na_2CO_3 * $NaHCO_3$ * H_2O (Trona) catalysts thanks to the enhancement of the WGSR by intermediate formation of formate. Specifically, at 500 °C and 26 MPa, the H_2 yield from cellulose gasification increased from 8.8 g_{H2} $kg^{-1}_{cellulose}$ (without catalyst) to 16.6 g_{H2} $kg^{-1}_{cellulose}$ and 18.8 g_{H2} kg^{-1} with 1.6 g of $Ca(OH)_2$ and 0.2 g of K_2CO_3, respectively. In addition, the use of both catalysts together allowed obtaining a H_2 yield of 23.9 g_{H2} $kg^{-1}_{cellulose}$ [60]. It is interesting to observe also that the H_2 yield of sunflower stalk – containing 9.7 wt% of lignin – was improved from 7.3 to 19.4 and 18.7 by using K_2CO_3 and Trona, respectively, whereas the H_2 yield of corncob (3.4 wt% of lignin) was increased from 4.2 to 15.9 and 23.7 with the same catalysts [91]. Moreover, the H_2 yield of acorn was doubled by the use of K_2CO_3 and then it was seven times higher than that obtained without catalyst (53.5 mol_{H2} kg_C^{-1} in feed) by using Trona catalyst, likely thanks to the presence of SiO_2 which improves the degradation of tannin [71]. The use of Trona which is a natural mineral in SCWG is economically advantageous. Such results indicate that the effect of catalyst on gasification varies according to the type of biomass, due to the composition of the lignocellulosic materials or rather that the chemical nature of carbon atom strongly influenced the CGE and hydrogen gasification efficiency (HGE) also in the presence of catalyst.

By using both NaOH as catalyst and H_2O_2 as oxidant at 330 °C, rice bran containing ~44% of C produced a lower H_2 yield, 14.7 g_{H2} $kg^{-1}_{rice\ bran}$, than whey powder and molasses (~37.5% and 30.9% of C content respectively), 22 g_{H2} $kg^{-1}_{whey\ powder}$ and 17 g_{H2} $kg^{-1}_{molasse}$. In general lipids-rich samples are the most difficult to decompose into water soluble products and gasifiable intermediates, and therefore they produced the lowest amount of H_2. Under the same reaction conditions, for example, sunflower oil, although constituted by 77.5 wt% of carbon, produced an even lower H_2 yield (6.4 g_{H2} $kg^{-1}_{sunflower\ oil}$) than glucose or glutamic acid [41]. Proteins and lipid from animal sources (chicken soup and cat food) produce low H_2 too. However, this could be due to the reactivity of proteins which could inhibit free-radical reactions needed for gas formation.

As regards the simultaneous use of NaOH and H_2O_2, their synergy contributes to enhance H_2 yield and GE compared to reaction in which either NaOH or H_2O_2 alone are used and also to decrease char and tar/oil formation. A very high H_2 yield at low temperature and pressure (330 °C, 13.5 MPa), in fact was obtained by optimizing the H_2O_2 amount (5.7 wt%), 44.2 g_{H2} $kg^{-1}_{glucose}$ [41] a value comparable to that obtained at 450 °C and 34 MPa, by using NaOH alone [92]. H_2O_2 promoted decarbonylation reactions and production of more CO than by NaOH alone. Then, the consumption of CO in the WGSR to produce CO_2 and H_2 was enhanced.

Instead, regarding solid catalysts, AC and supported transition metal catalysts are typically used, but hydrothermal instability and carbon deposition are the main problems to be solved. Ni-based catalysts have been widely used due to its low cost and high activity but crystallite sintering, support's breakdown and carbon deposition cause their deactivation. For example, with Raney nickel, at 380 °C, the H_2 yield (g_{H2} kg^{-1}_{feed}) follows this trend: lauric acid (61.2) > glycerol (54.0) > glycine (34.0) > glucose (30.6) > humic acid (13.2). Nobel metals have a great activity in SCWG too, in particular Ru, but high cost and limited availability prevent their further development. For example, Ru/γ-Al_2O_3 catalyses the SCWG allowing to obtain this H_2 yield (g_{H2} kg^{-1}_{feed}) trend: lauric acid (36.6) > glucose (23.4) > glycerol (21.0) > glycine (19.2) > humic acid (0.3) [38].

A very interesting result was obtained by adding small amounts of Ru in Ni-based catalysts ($Ru_{0.1}Ni_{10}$/γ-Al_2O_3). A H_2 yield of ~100 g_{H2} $kg^{-1}_{glucose}$ was maintained for 33 h on stream at 700 °C and 24 MPa in a bench-scale continuous down-flow tubular reactor [35]. The higher yield of 12 mol_{H2} $mol^{-1}_{glucose}$ (stoichiometric limit, 133 g_{H2} $kg^{-1}_{glucose}$) was reached by using Ru/Al_2O_3 but at high temperature (700–800 °C).

At last, also the synergic role of homogeneous and heterogeneous catalysts (i.e. Raney Ni and NaOH) was exploited to increase GE, CGE and HGE. If the H_2 yield by peanut shell gasification without catalyst is 7.44 $g\ kg^{-1}$, in fact, it became 11.34 $g\ kg^{-1}$ with NaOH, 11.43 $g\ kg^{-1}$ with Raney Ni catalyst and 53.71 $g\ kg^{-1}$ when both catalysts were used together at a relatively low temperature (400 °C, 22–28 MPa, 20 min) [10].

6.5   Conclusions

Gasification under supercritical water represents a promising technology for the conversion of biomass with high moisture content, with the main advantages to achieve high solid conversion and both high reaction efficiency and H_2 selectivity. Among different reaction parameters, the temperature appears to be the most important one to take into consideration for the obtainment of high gasification efficiency and high H_2 yield. Moreover, biomass chemical structure strongly influences the conversion rate and the intermediates which then lead to H_2 production, so smaller molecules, like glucose or glycerol, allow feeding higher concentration to the reactor without coke formation, whereas biomass containing lignin could be very hard to gasify. In fact, a temperature of 700 °C or higher is necessary for complete gasification of lignin, since, among different components of biomass, it is one of more difficult to decompose to gases due to the existence of very stable propane units. Moreover, by comparing biomasses with similar K content, cellulose, hemicellulose and lignin content, the yields and composition of gases obtained from them are significantly different, suggesting that also the interaction between the organic components play an important role.

It is interesting to highlight that the simultaneous use of NaOH and an oxidant like H_2O_2 further contributes to significantly enhance the H_2 yield. Such synergic contribution allows decreasing in char and tar/oil formation.

From a technological point of view, the need to produce a tar-free product gas from the gasification of biomass, the removal of tars and the reduction of the methane represent the main challenges for the SCWG development. In particular, some of most important factors are the plugging of the reactor caused by the precipitation of inorganic salts (char, coke, tar) and the reactor corrosion due to the harsh SCW environment and the corrosive nature of inorganic compounds contained in different waste biomass. On this account, high temperature and pressure resistant materials which are also resistant to corrosion in SCW should be used for equipment construction.

Concluding, despite the technical obstacles to overcome, SCWG is recognized as a prospective technology for agriculture and food industry waste reuse and it is under development and optimization on various demonstration pilot plants. Further studies are needed to make the process cost competitive with petroleum-based fuels and more efforts are necessary to better highlight the chemistry involved and to develop efficient and stable catalysts.

Abbreviations

SCW	supercritical water
SCWG	supercritical water gasification
SubCW	subcritical water gasification
CMC	sodium carboxymethylcellulose
AC	activated carbon
GE	gasification efficiency
CGE	carbon gasification efficiency
HGE	hydrogen gasification efficiency
WGSR	water gas shift reaction
WHSV	weight hourly space velocity
FAME	fatty acid methyl esters

References

[1] Pereira E.G., da Silva J.N., de Oliveira J.L., Machado C.S. 'Sustainable energy: a review of gasification technologies'. *Renew. Sustainable Energy Rev.* 2012, vol. 16, pp. 4753–4762.

[2] Basu P., Mettanant V. 'Biomass gasification in supercritical water – a review'. *Int. J. Chem. React. Eng.* 2009, vol. 7, pp. 1–61.

[3] Akizuki M., Fujii T., Hayashi R., Oshima Y. 'Effect of water on reactions for waste treatment organic synthesis and biorefinery in sub- and supercritical water'. *J. Biosci. Bioeng.* 2014, vol. 117, pp. 10–18.

[4] Azadi P., Farnood R. 'Review of heterogeneous catalysts for sub- and supercritical water gasification of biomass and waste'. *Int. J. Hydrogen Energy* 2011, vol. 36, pp. 9529–9541.

[5] Tekin K., Karagöz S. 'Non-catalytic and catalytic hydrothermal liquefaction of biomass'. *Res. Chem. Intermed.* 2013, vol. 39, pp. 485–498.

[6] Reddy S.N., Nanda S., Dalai A.K., Kozinski J.A. 'Supercritical water gasification of biomass for hydrogen production'. *Int. J. Hydrogen Energy* 2014, vol. 39, pp. 6912–6926.

[7] Guo Y., Wang S.Z., Xu D.H., Gong Y.M, Ma H.H., Tang X.Y. 'Review of catalytic supercritical water gasification for hydrogen production from biomass'. *Renew. Sustainable Energy Rev.* 2010, vol. 14, pp. 334–343.

[8] Youssef E.A.E.A., Nakhla G., Charpentier P. 'Hydrogen production by supercritical water gasification', in S.A. Sherif, D.Y. Goswami, E.K. (Lee) Stefanakos, A. Steinfeld (eds). *Handbook of Hydrogen Energy*. New York, NY: CRC Press Taylor & Francis Group; 2014. pp. 139–178.

[9] Yakaboylu O., Harinck J., Smit K.G., de Jong W. 'Supercritical water gasification of biomass: a literature and technology overview'. *Energies* 2015, vol. 8, pp. 859–894.

[10] Jin H., Lu Y., Guo L., Zhang X., Pei A. 'Hydrogen production by supercritical water gasification of biomass with homogenous and heterogeneous catalyst'. *Adv. Cond. Matter Phys.* 2014, vol. 2014, pp. 1–9.

[11] Matsumura Y., Minowa T., Potic B., *et al.* 'Biomass gasification in near and supercritical water: status and prospects'. *Biomass Bioenergy* 2005, vol. 29, pp. 269–292.

[12] Pavlovič I., Knez Ž., Škerget M. 'Hydrothermal reactions of agricultural and food processing wastes in sub-and supercritical water: a review of fundamentals, mechanisms and state of research'. *J. Agric. Food Chem.* 2013, vol. 61, pp. 8003–8025.

[13] Osada M., Takafumi S., Watanabe M., Shirai M., Arai K. 'Catalytic gasification of wood biomass in subcritical and supercritical water'. *Combust. Sci. Technol.* 2006, vol. 178, pp. 537–552.

[14] Peterson A.A., Vogel F., Lachance R.P., Fröling M., Antal M.J., Tester J.W. 'Thermochemical biofuel production in hydrothermal media: a review of sub- and supercritical water technologies'. *Energy Environ. Sci.* 2008, vol. 1, pp. 32–65.

[15] Yanik J., Ebale S., Kruse A., Saglam M., Yuksel M., 'Biomass gasification in supercritical water: Part 1. Effect of the nature of biomass'. *Fuel* 2007, vol. 86, pp. 2410–2415.

[16] Rezaiyan N., Cheremisinoff N.P. *Gasification Technologies. A Primer for Engineers and Scientists.* New York, NY: Taylor & Francis Group; 2005, p. 360.

[17] Collot A.G. 'Matching gasification technologies to coal properties'. *Int. J. Coal Geol.* 2006, vol. 65, pp. 191–212.

[18] Calzavara Y., Joussot-Dubien C., Boissonnet G., Sarrade S. 'Evaluation of biomass gasification in supercritical water process for hydrogen production'. *Energy Convers. Manage.* 2005, vol. 46, pp. 615–631.

[19] Modell M., 'Reforming of glucose and wood at critical conditions of water'. *Am. Soc. Mech. Eng.* Paper no. 77-ENAs, 1977, vol. 99, pp. 108–115.

[20] Elliott D.C., Sealock L.J., Baker E.G. 'Method for the catalytic conversion of organic materials into a product gas', US 5616154 A, 1997.

[21] Minowa T., Inoue S. 'Hydrogen production from biomass by catalytic gasification in hot compressed water'. *Renew. Energy* 1996, vol. 16, pp. 1114–1117.

[22] Xu, X., Matsumura Y., Stenberg, J., Antal, M. Jr. 'Carbon catalyzed gasification of organic feedstocks in supercritical water'. *Ind. Eng. Chem. Res.* 1996, vol. 35, pp. 2522–2530.

[23] Sinağ A., Kruse A., Schwarzkopf V. 'Key compounds of the hydropyrolysis of glucose in supercritical water in the presence of K_2CO_3'. *Ind. Chem. Eng. Res.* 2003, vol. 42, pp. 3516–3521.

[24] Onwudili J.A., Williams P.T., 'Production of hydrogen from biomass via supercritical water gasification', in Z. Fang, C. Xu (eds). *Near-critical and Supercritical Water and Their Applications for Biorefineries, Biofuels and Biorefineries 2.* Dordrecht: Springer Science+Business Media 2014.

[25] Kruse, A., Gawlik A. 'Biomass conversion in water at 330–410 °C and 30–50 MPa. Identification of key compounds for indicating different chemical reaction pathways'. *Ind. Eng. Chem. Res.* 2003, vol. 42, pp. 267–279.

[26] Kruse, A., Henningsen T., Sinağ A., Pfeiffer J. 'Biomass gasification in supercritical water: influence of the dry matter content and the formation of phenols'. *Ind. Eng. Chem. Res.* 2003, vol. 42, pp. 3711–3717.

[27] Resende F.L.P, Neff M.E, Savage P.E. 'Noncatalytic gasification of cellulose in supercritical water'. *Energy Fuel* 2007, vol. 21, pp. 3637–3643.

[28] Lu Y., Guo L., Zhang X., Ji C. 'Hydrogen production by supercritical water gasification of biomass: explore the way to maximum hydrogen yield and high carbon gasification efficiency'. *Int. J. Hydrogen Energy* 2012, vol. 37, pp. 3177–3185.

[29] Guo L.J., Lu Y.J., Zhang X.M., Ji C.M., Guan Y., Pei A.X. 'Hydrogen production by biomass gasification in supercritical water: a systematic experimental and analytical study'. *Catal. Today* 2007, vol. 129, pp. 275–286.

[30]   Savage, P.E. 'A perspective on catalysis in sub- and supercritical water'. *J. Supercrit. Fluids* 2009, vol. 47, pp. 407–414.

[31]   Waldner M.H., Krumeich F., Vogel F. 'Synthetic natural gas by hydrothermal gasification of biomass. Selection procedure towards a stable catalyst and its sodium sulfate tolerance'. *J. Supercrit. Fluids* 2007, vol. 43, pp. 91–105.

[32]   Sato T., Osada M., Watanabe M., Shirai M., Arai K. 'Gasification of alkylphenols with supported noble metal catalysts in supercritical water'. *Ind. Eng. Chem. Res.* 2003, vol. 42, pp. 4277–4282.

[33]   Sato T., Furusawa T., Ishiyama Y., *et al.* 'Effect of water density on the gasification of lignin with magnesium oxide supported nickel catalysts in supercritical water'. *Ind. Eng. Chem. Res.* 2006, vol. 45, pp. 615–622.

[34]   Muangrat R., Onwudili J.A., Williams P.T., 'Influence of alkali catalysts on the production of hydrogen-rich gas from the hydrothermal gasification of food processing waste'. *Appl. Catal. B: Environ.* 2010, vol. 100, pp. 440–449.

[35]   Zhang L.H., Xu C.B., Champagne P. 'Activity and stability of a novel Ru modified Ni catalyst for hydrogen generation by supercritical water gasification of glucose', *Fuel* 2012, vol. 96, pp. 541–545.

[36]   Elliott D.C., Neuenschwander G.G., Phelps M.R., Hart T.R., Zacher A.H., Silva L.J. 'Chemical processing in high pressure aqueous environments. 6. Demonstration of catalytic gasification for chemical manufacturing wastewater cleanup in industrial plants'. *Ind. Eng. Chem. Res.* 1999, vol. 38, pp. 879–883.

[37]   Wang W., Padban N., Ye Z., Olofsson G., Andersson A., Bjerle I. 'Catalytic hot gas cleaning of fuel gas from an air-blown pressurized fluidized-bed gasifier'. *Ind. Eng. Chem. Res.* 2000, vol. 39, pp. 4075–4081.

[38]   Azadi P., Afif E., Foroughi H., Dai T.S., Azadi F., Farnood R. 'Catalytic reforming of activated sludge model compounds in supercritical water using nickel and ruthenium catalysts'. *Appl. Catal. B: Environ.* 2013, vol. 134–135, pp. 265–273.

[39]   Byrd A.J., Pant K.K., Gupta R.B. 'Hydrogen production from glucose using Ru/Al_2O_3 catalyst in supercritical water'. *Ind. Eng. Chem. Res.* 2007, vol. 46, pp. 3574–3579.

[40]   Minowa T., Zhen F., Ogi T. 'Hydrogen production from cellulose using a reduced nickel catalyst'. *Catal. Today* 1998, vol. 45, pp. 411–416.

[41]   Muangrat R., Onwudili J.A., Williams P.T. 'Reaction of different food classes during subcritical water gasification for hydrogen gas production'. *Int. J. Hydrogen Energy* 2012, vol. 37, pp. 2248–2259.

[42]   Rulkens W. 'Sewage sludge as a biomass resource for the production of energy: overview and assessment of the various options'. *Energy Fuel* 2008, vol. 22, pp. 9–15.

[43]   Goto M., Nada T., Kodama A., Hirose T. 'Kinetic analysis for destruction of municipal sewage sludge and alcohol distillery wastewater by

supercritical water oxidation'. *Ind. Eng. Chem. Res.* 1999, vol. 38, pp. 1863–1865.

[44] Lu Y.J., Guo L.J., Ji C.M., Zhang X.M., Hao X.H., Yan Q.H. 'Hydrogen production by biomass gasification in supercritical water: a parametric study'. *Int. J. Hydrogen Energy* 2006, vol. 31, pp. 822–831.

[45] Williams P.T., Onwudili J.A. 'Subcritical and supercritical water gasification of cellulose, starch, glucose and biomass waste'. *Energy Fuel* 2006, vol. 20, pp. 1259–1265.

[46] Venkitasamy C., Hendy D., Wilkinson N., Fernando L., Jacoby W. 'Investigation of thermochemical conversion of biomass in supercritical water using a batch reactor'. *Fuel* 2011, vol. 90, pp. 2662–2670.

[47] Schmieder H., Abeln J., Boukis N., *et al.* 'Hydrothermal gasification of biomass and organic wastes'. *J. Supercrit. Fluids* 2000, vol. 17, pp. 145–153.

[48] D'Jesùs P., Boukis N., Kraushaar-Czarnetzki B., Dinjus E. 'Gasification of corn and clover grass in supercritical water'. *Fuel* 2006, vol. 85, pp. 1032–1038.

[49] Kruse A., Maniam P., Spieler F. 'Influence of proteins on the hydrothermal gasification and liquefaction of biomass. 2. Model compounds'. *Ind. Eng. Chem. Res.* 2007, vol. 46, pp. 87–96.

[50] Lin S.Y., Suzuki Y., Hatano H., Harada M. 'Hydrogen production from hydrocarbon by integration of water–carbon reaction and carbon dioxide removal (HyPr-RING Method)'. *Energy Fuel* 2001, vol. 15, pp. 339–343.

[51] Antal M. Jr., Allen S., Schulman D., Xu X., Divilio R. 'Biomass gasification in supercritical water'. *Ind. Eng. Chem. Res.* 2000, vol. 39, pp. 4040–4053.

[52] Hao X., Guo L., Zhang X., Guan Y. 'Hydrogen production from catalytic gasification of cellulose in supercritical water'. *Chem. Eng. J.* 2005, vol. 110, pp. 57–65.

[53] Penninger J.M.L., Rep M. 'Reforming of aqueous wood pyrolysis condensate in supercritical water'. *Int. J. Hydrogen Energy* 2006, vol. 31, pp. 1597–1606.

[54] Monica E.K., Gellerstedt G., Henriksson G. (eds.). *Wood Chemistry and Wood Biotechnology*. Berlin: Walter de Gruyter GmbH & Co., KG, 10785; 2009. p. 308.

[55] Karagöz S., Bhaskar T., Muto A., Sakata Y. 'Comparative studies of oil compositions produced from sawdust, rice husk, lignin and cellulose by hydrothermal treatment'. *Fuel* 2005, vol. 84, pp. 875–884.

[56] Kabyemela B.M., Adschiri T., Malaluan R.M., Arai K. 'Glucose and fructose decomposition in subcritical and supercritical water: detailed reaction pathway, mechanism, and kinetics'. *Ind. Eng. Chem. Res.* 1999, vol. 38, pp. 2888–2895.

[57] Modell M. 'Gasification and liquefaction of forest products in supercritical water', in R.P. Overend, T. Milne, L. Mudge (eds). *Fundamentals of Thermochemical Biomass Conversion*. Netherlands: Springer; 1985. pp. 95–119.

[58] Usui Y., Minowa T., Inoue S., Ogi T. 'Selective hydrogen production from cellulose at low temperature catalyzed by supported group 10 metal'. *Chem. Lett.* 2000, vol. 29, pp. 1166–1167.

[59] Watanabe M., Inomata H., Arai K. 'Catalytic hydrogen generation from biomass (glucose and cellulose) with ZrO_2 in supercritical water'. *Biomass Bioenergy* 2002, vol. 22, pp. 405–410.

[60] Guan Y., Pei A., Guo L. 'Hydrogen production by catalytic gasification of cellulose in supercritical water'. *Front. Chem. Eng. Chin.* 2008, vol. 2, no. 2, pp. 176–180.

[61] Park K.C., Tomiyasu H. 'Gasification reaction of organic compounds catalyzed by RuO_2 in supercritical water'. *Chem. Commun.* 2003, vol. 6, pp. 694–695.

[62] Amin S., Reid R.C., Modell M. 'Reforming and decomposition of glucose in an aqueous phase'. *Intersociety Conference on Environmental System*, San Francisco, CA, July, 21–24, 1975. ASME 75-ENAs-21.

[63] Yu D. Aihara M., Antal M.J. 'Hydrogen production by steam reforming glucose in supercritical water'. *Energy Fuel* 1993, vol. 7, pp. 574–577.

[64] Lee I.G., Kim M.S., Ihm S.K. 'Gasification of glucose in supercritical water'. *Ind. Eng. Chem. Res.* 2002, vol. 41, pp. 1182–1188.

[65] Hao X.H., Guo L.J., Mao X., Zhang X.M, Chen X.J. 'Hydrogen production from glucose used as a model compound of biomass gasified in supercritical water'. *Int. J. Hydrogen Energy* 2003, vol. 28, pp. 55–64.

[66] Azadi P., Khodadai A.A., Mortzavi Y., Farnood R. 'Hydrothermal gasification of glucose using Raney nickel and homogeneous organometallic catalysts'. *Fuel Process. Technol.* 2009, vol. 90, pp. 145–151.

[67] Sinağ A., Kruse A., Rathert J. 'Influence of the heating rate and the type of catalyst on the formation of key intermediates and on the generation of gases during hydropyrolysis of glucose in supercritical water in a batch reactor'. *Ind. Eng. Chem. Res.* 2004, vol. 43, pp. 502–508.

[68] Kersten S., Potic B., Prins W., Van Swaaij W. 'Gasification of model compounds and wood in hot compressed water'. *Ind. Eng. Chem. Res.* 2006, vol. 45, pp. 4169–4177.

[69] Watanabe M., Iida T., Aizawa Y., Ura H., Inomata H., Arai K. 'Conversions of some small organic compounds with metal oxides in supercritical water at 673 K'. *Green Chem.* 2003, vol. 5, pp. 539–544.

[70] Yoshida T., Oshima Y., Matsumura Y. 'Gasification of biomass model compounds and real biomass in supercritical water'. *Biomass Bioenergy* 2004, vol. 26, pp. 71–78.

[71] Madenoğlu T.G., Boukis N., Sağlam M., Yuksel M. 'Supercritical water gasification of real biomass feedstocks in continuous flow system'. *Int. J. Hydrogen Energy* 2011, vol. 36, pp. 14408–14415.

[72] Yong T.L.K., Matsumura Y. 'Reaction kinetics of the lignin conversion in supercritical water'. *Ind. Eng. Chem. Res.* 2012, vol. 51, pp. 11975–11988.

[73] Saisu M., Sato T., Watanabe M., Adschiri T., Arai K. 'Conversion of lignin with supercritical water-phenol mixtures'. *Energy Fuel* 2003, vol. 17, pp. 922–928.

[74] Resende F.L.P., Fraley S.A., Berger M.J., Savage P.E. 'Noncatalytic gasification of Lignin in supercritical water'. *Energy Fuel* 2008, vol. 22, pp. 1328–1334.

[75] Yoshida T., Matsumura Y. 'Gasification of cellulose, xylan and lignin mixture in supercritical water'. *Ind. Eng. Chem. Res.* 2001, vol. 40, pp. 5469–5474.

[76] Ando H., Sakaki T., Kokusho T., Shibata M., Uemura Y., Hatate Y. 'Decomposition behavior of plant biomass in hot-compressed water'. *Ind. Eng. Chem. Res.* 2000, vol. 39, pp. 3688–3693.

[77] Yamaguchi A., Hiyoshi N., Sato O., Osada M., Shirai M. 'Lignin gasification over supported ruthenium trivalent salts in supercritical water'. *Energy Fuel* 2008, vol. 22, pp. 1485–1492.

[78] Watanabe M., Inomata H., Osada M., Sato T., Adschiri T., Arai K. 'Catalytic effect of NaOH and ZrO_2 for partial oxidative gasification of n-hexadecane and lignin in supercritical water'. *Fuel* 2003, vol. 82, pp. 545–552.

[79] Kruse A., Meier D., Rimbrecht P., Shacht M. 'Gasification of pyrocatechol in supercritical water in the presence of potassium hydroxide'. *Ind. Eng. Chem. Res.* 2000, vol. 39, pp. 4842–4848.

[80] Osada M., Sato O, Watanabe M., Arai K., Shirai M. 'Water density effect on lignin gasification over supported noble metal catalysts in supercritical water'. *Energy Fuel* 2006, vol. 20, pp. 930–935.

[81] Su Y., Zhu W., Gong M., Zhou H., Fan Y., Amuzu-Sefordzi B. 'Interaction between sewage sludge components lignin (phenol) and proteins (alanine) in supercritical water gasification'. *Int. J. Hydrogen Energy* 2015, vol. 40, pp. 9125–9136.

[82] Goodwin A.K., Rorrer G.L. 'Conversion of xylose and xylose–phenol mixtures to hydrogen-rich gas by supercritical water in an isothermal microtube flow reactor'. *Energy Fuel* 2009, vol. 23, pp. 3818–3825.

[83] Weiss-Hortala E., Kruse A., Ceccarelli C., Barna R. 'Influence of phenol on glucose degradation during supercritical water gasification'. *J. Supercrit. Fluids* 2010, vol. 53, pp. 42–47.

[84] Castello D., Kruse A., Fiori L. 'Low temperature supercritical water gasification of biomass constituents: glucose/phenol mixtures'. *Biomass Bioenergy* 2015, vol. 73, pp. 84–94.

[85] Tapah B.F., Santos R.C.D., Leeke G.A. 'Processing of glycerol under sub and supercritical water conditions'. *Renew. Energy* 2014, vol. 62, pp. 353–361.

[86] Antal M.J., Mok W.S.L., Roy J.C., Raissi A.T., Anderson D.G.M. 'Pyrolytic sources of hydrocarbons from biomass'. *J. Anal. Appl. Pyrolysis* 1985, vol. 8, pp. 291–303.

[87] Bühler W., Dinjus E., Ederer H.J., Kruse A., Mas C. 'Ionic reactions and pyrolysis of glycerol as competing reaction pathways in near- and supercritical water'. *J. Supercrit. Fluids* 2002, vol. 22, pp. 37–53.

[88] Byrd A.J., Pant K.K., Gupta R.B. 'Hydrogen production from glycerol by reforming in supercritical water over Ru/Al$_2$O$_3$ catalyst'. *Fuel* 2008, vol. 87, pp. 2956–2960.

[89] Onwudili J.A., Williams P.T. 'Hydrothermal reforming of biodiesel plant waste: product distribution and characterization'. *Fuel* 2010, vol. 89, pp. 501–509.

[90] Guo S., Guo L, Yin J, Jin H. 'Supercritical water gasification of glycerol: intermediates and kinetics'. *J. Supercrit. Fluids* 2013, vol. 73, pp. 95–102.

[91] Yanik J., Ebale S., Kruse A., Saglam M., Yuksel M. 'Biomass gasification in supercritical water: part II. Effect of catalyst'. *Int. J. Hydrogen Energy* 2008, vol. 33, pp. 4520–4526.

[92] Onwudili J.A., Williams P.T. 'Role of sodium hydroxide in the production of hydrogen gas from the hydrothermal gasification of biomass'. *Int. J. Hydrogen Energy* 2009, vol. 34, pp. 5645–5656.

Biofuels starting materials for hydrogen production

S. Abramov[1], M. Shalygin[2], V. Teplyakov[2] and A. Netrusov[1]

Abstract

Future developments in energy-efficient processes and potential solutions for the energy-related environmental tasks are coupled with hydrogen-based technologies. Introductory parts of this chapter are focused on the specifics of H_2 generation from biomass. Within the framework of this topic, three platforms are compared: conversion of simple sugars, cellulose, and thermochemical conversion of biomass to hydrogen-containing gaseous mixtures. Three approaches for generation of biofuels starting materials for hydrogen production are considered: the first one includes sugars and organic acids; the second one includes lignocellulose, wood-chips, etc.; finally, the third approach considers the possible routes of biomass gasification. In all cases, the hydrogen needs to be separated (to be recovered) from the hydrogen-containing multicomponent gaseous mixtures of biogenic origin. Membrane-based gas separation processes are considered for H_2 recovery from gaseous sources, including (1) estimation of commercial and lab-scale polymeric membranes for recovery of H_2 from gaseous mixtures, containing additionally CO_2, CO, N_2, CH_4, H_2S, with calculation of standard membrane process itself; (2) membrane contactors for hydrogen recovery from H_2/CO_2 mixtures; (3) combined membrane/pressure-swing adsorption (PSA) systems for hydrogen recovery from gaseous mixtures of biogenic origin. It is shown that H_2 recovery can be successfully realized as a combination of standard membrane method (H_2 preconcentrating) and PSA (H_2 conditioning). Potential of whole process (biomass treatment and H_2 recovery as a fuel) requires the active generation of knowledge for development of the desired bioprocesses and highly selective membranes.

[1]Microbiology Department, Moscow Lomonosov State University, Moscow 119992, Russian Federation
[2]Topchiev Institute for Petrochemical Synthesis RAS, Moscow 119991, Russian Federation

7.1 Introduction

One challenge of scientific research in the early twenty-first century is the development of new technologies that allow the progressive transition from the current energy model, based on fossil fuels, to a more sustainable energy model, based on renewable and carbon-neutral fuels. To date, many alternatives have been explored experimentally to achieve this goal, for example generating electricity from renewable sources and catalytic or bio-catalytic transformation of renewable feedstocks into fuels. Nowadays, the dominating role of hydrocarbon resources in the modern fuel-and-power sector structure is considered by experts to be a potential threat to energetic and economical security of countries. That is why the development and implementation of alternative energy resources is of great importance.

Hydrogen is the key element in many processes of organic synthesis and can be considered as the key energy carrier. In petroleum refining (hydrocracking, hydrorefining), up to 37% of obtained H_2 is utilized for the purpose of quality improvement of hydrocarbon fuels with an enhanced calorific value and reduced the quantity of harmful impurities [1–3]. Hydrogen is widely used (up to 2%) in powder metallurgy, metalworking, production of glass, and synthetic diamonds. Hydrogen is applied as a rocket fuel, the combination of liquid hydrogen with liquid oxygen provides maximal energy per weight unit [120.6 MJ/kg(H_2)]. In the last decades, vehicles with H_2-powered internal combustion engines were developed. Hydrogen–oxygen steam generators for electricity production during peak periods were developed as well. Other perspective areas of hydrogen utilization are production of fats and oils, oxoproducts, synthetic fuels, and semiconductors [4–6]. At the same time, up to 40% of H_2 is losing in waste streams or burned in technological processes in installations for heat production. At present time, the most part of H_2 (58%–80%) is produced by steam conversion of methane. It is important to note that only 62% of hydrogen is produced as target product; the rest 38% of hydrogen is obtained as by-product of other productions.

Current world hydrogen consumption in chemical, petrochemical, and petroleum refining industry is around 45 Mt/year. The prediction of hydrogen consumption and structure of market in the twenty-first century foresees increasing of hydrogen consumption in 16–20 times to the year 2100 and 80% of this increasing is related to hydrogen utilization as an energy carrier. In accordance to estimation [2], if hydrogen content in waste stream is higher than 50%, the price of H_2 recovered by membrane, adsorption, or cryogenic method is 1.5–2 times lower than the price of H_2 obtained by steam conversion of natural gas. Therefore, perspective sources of hydrogen can be such waste gas mixtures as blow-down gases of ammonia and methanol production, gases of catalytic reforming processes, cracking, dehydrogenation, operating of coke ovens and installations of olefins, acetylene, butadiene production as well as biohydrogen produced by bacteria and biosyngas produced by pyrolysis of solid biomass waste and wood. As a result, the consideration of available hydrogen sources of biogenic origin that are as rule multicomponent gas mixtures with considerable amount of hydrogen seems to be of great importance. Depending on the composition of hydrogen-containing gas mixture, the optimal technology for hydrogen recovery can be standard membrane separation or hybrid membrane systems and processes without phase transitions that significantly reduce energy consumption. This chapter represents results of critical analysis of published and own data on application and potential of membrane technologies for hydrogen recovery from biomass treatment products (renewable sources). Introductory parts of review are focused on particularity of H_2 generation from biomass (microbiological routes and pyrolysis). Membrane gas separation processes for H_2 recovery from gaseous sources are considered in this chapter as well. Critical aspects include (1) estimation of commercially and lab-scale polymeric membranes for separation of H_2 from mixtures with CO_2, CO, N_2, CH_4, H_2S and calculation of standard membrane processes itself; (2) membrane contactors for hydrogen recovery from H_2/CO_2 mixtures; (3) combined membrane/pressure-swing adsorption (PSA) systems for hydrogen recovery from gaseous mixtures of biogenic origin. It is shown that H_2 recovery can be successfully realized as a combination of standard membrane method (H_2 pre-concentrating) and PSA (H_2 conditioning). Improving the whole process requires the development of high selective membranes. The formulated problems of H_2 recovery demand multidisciplinary interaction of specialists in the field of biomass treatment, chemistry, membrane technology, and energy production.

7.2   Hydrogen from biomass

As the main criteria for choosing of feedstock that can be used for hydrogen production are considered price, hydrocarbon content, and biodegradability. Simple sugars are the preferred substrate for hydrogen production because they can be easily and quickly decomposed by hydrogen-producing microorganisms. However, from the economical viewpoint, the feedstock that contained pure sugars is comparably expensive. Therefore, lignocellulosic biomass is the most profitable source for hydrogen production. It is largely due to the fact that lignocellulosic biomass is very common an agricultural waste. However, it should be noted that for more than half a century, there were various materials suggested as feedstock for the hydrogen production. As the whole, these materials can be divided into three generations that will be discussed below.

7.2.1   *First generation of starting materials for biohydrogen production*

First generation of starting materials for biohydrogen production consists of simple sugars (saccharose of sugar cane or beet, molasses) or more complex sugars as corn or potato starch. Biological route of hydrogen production from this foodstuff was to use acetone–butanol–ethanol (ABE) fermentation, widely applied in different countries in the world for acetone and butanol production. This fermentation was the main source for the butanol for synthetic rubber production during the 1930s and 1940s, starting as early as in the First World War in Britain and then in

Germany. The first commercial plant for the sterile fermentation of potato to ABE started in USSR in 1935 in Grozny (Chechen Republic, Russia). The by-product of ABE process was a mixture of H_2 and CO_2 in almost equal concentrations, and then chemical scrubbers were used to purify hydrogen from CO_2. Pure hydrogen was used in chemical industry as a reducing agent (ammonia production). The set of reactions for this synthesis can be described as follows: starch $\rightarrow$ glucose $\rightarrow$ ABE fermentation $\rightarrow$ [acetone + butanol + ethanol + H_2 + CO_2 + some minor products (2-propanol, lactate, acetate, and butyrate)].

This fermentation was widely used in the world until the petrochemical routes for ABE obtaining were developed, and in the time of low oil prices, all the plants were closed due to economic inefficiency. The last plant was closed in South Africa in 2002.

Gaseous products of ABE fermentation are H_2 and CO_2, produced in almost equal molar amounts and can be separated by traditional or membrane absorption technique with membrane contactors. Resulted mixture of residual organic acids can be converted into hydrogen and carbon dioxide via light-depending bioprocess by phototrophic bacteria.

First generation of the starting materials for hydrogen production has main pitfall: the feed is foodstuff; therefore, alternative processes based on nonfood organic substrates must be applied for noncompetitive way of biohydrogen production. This comes with the second and the third generations of starting materials.

The present world's population is around 7 billion people, about 1 billion are undernourished, and it is annually increasing by 1%, at the same time the primary-energy consumption is continuously going up with rate of around 5% per year. Meeting the food demands of the world's growing population and providing them with European standard of living has been predicted to require a 100% increase in global food production until 2050. At the same time, it is estimated that the increase in arable land between 2005 and 2050 will be just 5% (FAO Expert meeting 2009).

Production of bioethanol and hydrogen from sugars and starch, and of biodiesel from vegetable oils, globally competes with the production of food and animal feed. These processes are hard to justify in densely populated areas such as central Europe or China—the reason why China already prohibits production of renewable energy from sugar and starch-containing edible parts of plants. Even ethanol production from sugar cane in the tropics (Brazil) with an energy returned on energy invested (EROI) of eight appears to be only a temporally option. This high EROI is reached only when bagasse (the residue from sugar cane left after it has been crushed to extract the juice) is used as the main energy source for ethanol distillation rather than plowed back to soil, which is not sustainable option because of the resulting loss in soil carbon.

7.2.2 *Second generation of starting materials for biohydrogen production (lignocellulose, biomass, algae, etc.)*

When compared with the conventional substrates maize starch and molasses, lignocellulosic biomass is regarded as the most promising substrate for next-generation ABE production for a number of reasons: it is readily available, it is sustainable, and it does not compete with food crops [7,8]. For instance, in China nearly 20 million tons of corncob are produced annually that can be used for the conversion to hydrogen, ethanol, butanol, and other chemical compounds [9].

Second generation of the starting materials for hydrogen production comes to live in the late 1970s. For a long time, it was known that some fungi can degrade lignocelluloses with high efficiency (*Trichoderma reesei*). This is due to synthesis of extracellular cellulases by fungi that degrade cellulose with high rate. The sugars released from this hydrolysis (mainly cellobiose and glucose) are used by fungal cells for growth aerobically. This process of cellulose biodegradation in aerobic zones is widely spread on the Earth, and this is the main route for the conversion of cellulose to CO_2 in global carbon–oxygen cycle.

Lignocellulose (wood and straw) is mainly composed of cellulose, hemicellulose, and lignin. Lignin and cellulose are very difficult components to degrade, although both are rather heterogeneous polymers and differ considerably depending on their origin. Hemicelluloses are composed of pentoses and hexoses and relatively easy to hydrolyze, but in raw material it is protected from hydrolysis by a complex linkage with lignin and cellulose. Lignin is a polymer of phenolic constituents that can be degraded only aerobically, mainly by aerobic fungi. The hydrolysis of cellulose to fermentable sugars is catalyzed by cellulases, which are produced by microorganisms, but not by most animals. Cellulase-catalyzed cellulose hydrolysis is slow relative to amylase-catalyzed starch hydrolysis, but it is even slower if the cellulose is in a complex linkage with lignin. The formation of biofuels from lignocellulose is therefore dependent on a lignocellulose pretreatment in order to make all the celluloses and hemicelluloses accessible to cellulases and hemicellulases in a reasonable time. If such pretreatment is not possible, biofuel formation from these compounds will be very slow. Pretreatment of lignocelluloses is an energy-intensive process; it involves mechanical steps followed by the extraction of the celluloses and hemicelluloses with acid or ammonia.

Use of cellulose and lignocellulose constituents of plant material (wood, straw, etc.) for hydrogen, bioethanol, or biobutanol production is limited by the high stability of lignocelluloses. Mechanical and thermochemical treatments help to overcome this limitation, but these treatments in turn are highly energy intensive. Pretreatment of biomass with specific enzymes is an important field of biotechnical development.

Lignocellulosic biomass is a widely available and sustainable source of fermentable sugars for the biofuels production [10]. It consists of cellulose, hemicellulose, and lignin that strongly intermeshed and chemically bonded by non-covalent forces and by covalent cross-linkages [10,11]. This structure provides plants with structural support, impermeability, resistance to oxidative stress, and microbial attack [12]. Therefore, conversion of lignocellulose to monomeric fermentable sugars in the nature is quite prolonged process. In order to receive enough amounts of fermentable sugars, it is necessary to use pretreatment methods for destruction of interconnections in the lignocellulosic biomass and cellulose and hemicellulose hydrolysis [11].

All pretreatment methods can be classified into mechanical (chipping, milling, and grinding), physical–chemical (thermal, hot water, steam explosion, and ammonia fiber expansion), chemical (pretreatment with solutions of acids, alkalizes, ionic liquids, pretreatment with ozone, and oxidants), and biological methods (enzymes, cellulolytic microorganisms) [11,13]. The choice of pretreatment method mainly depends on physical–chemical properties of lignocellulosic biomass (Table 7.1).

7.2.2.1   Mechanical pretreatment

The goal of mechanical pretreatment consists in crushing of lignocellulose and correspondingly in decreasing of particle size. The size of pretreated biomass can be reduced considerably (from meters to less than micrometers) using different kinds of techniques (cutting or crushing, coarse milling, intermediate microniza-tions, fine grinding, ultra-fine grinding, nano-grinding, etc. [14,15]). This leads to increasing of specific surface area of substrate and decreasing its degree of poly-merization. The smaller particle of biomass the more exposed reactive chemical bonds (such as glycosidic and ester bonds) are subject to attack by catalyst such as protons and enzymes from a liquid phase [15]. Therefore, smaller particle sizes always favor the conversion process, ensures fast reaction and more uniform conversion.

Economy of the mechanical pretreatment process depends on the specific energy requirement (SER) that is usually determined by the type of biomass (its structural heterogeneity, complexity of cell wall constituents, and association of tissues), moisture content, and final particle size [14,16]. Cellulose content, crys-tallinity, *p*-coumaric acids, high level of humidity of raw materials, and small particle size will affect the SER values negatively [14,17], whereas the arabinose/ xylose ratio and accessible surface area positively affect the SER values [14].

The mechanical size reduction of lignocelluloses does not require chemical catalysts [18] and is not accompanied either by the formation of inhibitors (i.e., furfural, 5-(hydroxymethyl)furfural, and phenolic compounds) of fermentative process [14].

7.2.2.2   Physical–chemical pretreatment

Physical–chemical pretreatment methods are widely used, for example hot steam and hot-water pretreatment, steam explosion and ammonia fiber explosion, pre-treatment with organic solvents. These methods are based on the joint variation of process parameters and use of chemical compounds for effects on both physical and chemical properties of lignocellulose [15,19,20].

The conventional initial acid-hydrolysis/extraction step in degradation of lignocellulose is the most wasteful step in the process. A recent paper [21] has reported the use of ball milling as an effective means to induce "mechano-catalysis" between cellulose and clay-based catalysts with layered structures.

The hydrothermal methods are most spread in this category. At temperatures above 150–180 °C, the part of lignocellulosic biomass (firstly hemicellulose, than other polymers) begins to dissolve [22]. Acids that are produced during hemi-cellulose hydrolysis participate in its further dissolution [23,24].

Phenolic and heterocyclic compounds (vanillin, vanillic alcohol, furfural, and hydroxymethylfurfural) almost always present among the by-products of lignin

Table 7.1 Application of different pretreatment methods for cellulose-containing biomass

Pretreatment methods	Substrate	Mechanical pretreatment	Pretreatment conditions	Polymers or solid dissolution (%)	Cellulose conversion yield (%)	Enzymatic hydrolysis
Alkali pretreatment						
Alkaline (NaOH or Ca(OH)$_2$)	Corn stover [93]	Washed, dried, and milled	55 °C, 7.3 wt.%, Ca(OH)$_2$, 4 weeks	NA	98 in 96 h	15 FPU/g for cellulase (Spezyme CP) and 40 CBU/g for β-glucosidase (Novozyme 188)
Alkaline (NaOH)	Kapok fiber (*Ceiba pentandra*) [32]	Not conducted	120 °C, 8.3 wt.%, 2% v/v NaOH	SD 63 in 60 min	>99.9 (glucose yield) in 48 h	70 FPU/g for cellulose (Celluclast 1.5 L, Novozymes A/S, Denmark)
Alkaline (NaOH)	Corn stover (*Zea mays* L. var. *ceratina*) [32]	Biomass was dried (final humidity 5%) and milled (average particle size smaller than 1 mm)	60 °C, 10 wt.%, 0.25 M NaOH	SD 39.01 in 60 min	NA	310 EGU/g for cellulase (Celluclast Conc BG, Daejung Company, Korea) and 70/g CBU for cellobiase (Novozyme 188, Sigma, USA)
Acid pretreatment						
Acid (H$_2$SO$_4$)	Kapok fiber (*Ceiba pentandra*) [32]	Not conducted	140 °C, 8.3 wt.%, 1% v/v H$_2$SO$_4$	SD 46.8 in 45 min	81 (glucose yield) in 48 h	70 FPU/g for cellulose (Celluclast 1.5 L, Novozymes A/S, Denmark)
Acid (H$_2$SO$_4$)	Stem wood of loblolly pine	Biomass was milled (average size 2 mm)	180 °C, 12.5 wt.%, 1% H$_2$SO$_4$, 30 min	NA	35 in 72 h	20 FPU/g for cellulose (Celluclast 1.5 L) and 40 IU/g for β-glucosidase (Novozyme 188)
Hot-water pretreatment						
Water	Kapok fiber (*Ceiba pentandra*) [32]	Not conducted	170 °C, 8.3 wt.%	SD 50 in 45 min	94.6 (glucose yield) in 48 h	70 FPU/g for cellulose (Celluclast 1.5 L, Novozymes A/S, Denmark)
Ammonia fiber expansion						
Ammonia	Poplar [94]	Not conducted	180 °C, 2:1 ammonia to biomass loading, 233% moisture content, 30 min	NA	70 in 72 h	15 FPU/g for cellulase (Celluclast 1.5 L) and 26.25 CBU/g for β-glucosidase (Novozyme 188)

(Continues)

Table 7.1 (*Continued*)

Pretreatment methods	Substrate	Mechanical pretreatment	Pretreatment conditions	Polymers or solid dissolution (%)	Cellulose conversion yield (%)	Enzymatic hydrolysis
			Ionic liquids			
Hot steam and ionic liquid (NaCl–H_2O)	Pubescent (*Quercus pubescens*) [29]	Before use, all samples were ground into a powder of 40–80 mesh and dried in an oven at 110 °C overnight	220 °C, 4 wt.% with 20 wt.% NaCl	More than 99% of cellulose in 2 h	NA	Not conducted
			Hybrid pretreatment			
Acid (H_2SO_4) and alkaline (NaOH) pretreatments	Napier grass (*Pennisetum purpureum*) [46]	Biomass was dried (final humidity 4%) and ground (average particle size 0.3 mm)	50 °C, 10 wt.%, addition of PEG 6000	NA	45 (Cellulose hydrolysis) in 94 h	5 or 25 FPU/g for cellulase (NS50013, Novozymes, Brazil) and 10 CBU/g β-glucosidase (NS50010, Novozymes, Brazil)
HCl and moist heat	Napier grass (*Pennisetum purpureum*) [95]	Harvested leaves were dried at 60 °C for 72 h and milled (average size 1 mm or less)	93.07 °C, 10 wt.%, 4.39% HCl	PD cellulose 19, hemi-cellulose 83.3, lignin 8.4 in 180 min	NA	Not conducted
Hydroxyl radicals pretreatment combined with hot-water treatment	Macroalgae (*Macrocystis pyrifera*) [85]	Biomass dried at 65 °C, and sifted (average size 0.2 mm)	100 °C, 5 wt.%, 0.018% H_2O_2 and 11.9 mM $FeSO_4$ added after cooling of biomass	NA	88.1 (cellulose conversion)	15 FPU/g cellulose (Sigma–Aldrich Co. USA)

Notes: SD, solid dissolution; PD, polymers dissolution; NA, data are not available.

dissolution that can impact inhibitory or toxically on microorganisms during fermentation stage. The formation of these compounds is especially expressed in acidic conditions [25]. The temperature of thermal pretreatment is limited by 250 °C due to the starting of pyrolysis of biomass [26].

7.2.2.3 Hot steam and steam explosion pretreatment

The main goal of the treatment with hot steam as well as treatment by steam explosion is dissolution of hemicelluloses, thereby making the cellulose more accessible to enzymatic hydrolysis. It also provides particle size reduction [15,19,27].

For steam explosion pretreatment, biomass is firstly immersed in steam, then pressure raises and sudden release of pressure causing the biomass to disintegrate. The steam explosion method requires considerably less energy than necessary for mechanical pretreatment for achieving the same increase in specific surface area [15,19,27].

The lignocellulosic biomass can be also pretreated with hot steam [28,29]. For example, in the case of straw treatment, the biomass placed in the reactor and then treated with high-temperature (up to 240 °C) steam under overpressure during few minutes. Later steam is released and quickly cools the biomass. Two factions of products formed in this process: solid, containing mainly cellulose, and a liquid containing pentoses (xylose and arabinose) and a small amount of glucose [30]. During the pretreatment with hot steam, part of hemicellulose is hydrolyzed. It provides formation of acids that provide the further hemicellulose hydrolysis probably contributing to their dissolution [31].

Thus, the difference between steaming and treatment by steam explosion concludes in the fast decrease of pressure and cooling of biomass [26].

7.2.2.4 Hot-water pretreatment

The main task of this method concludes in dissolution of hemicellulose, which makes the cellulose more accessible for enzymatic hydrolysis [32]. For this purpose, the pH value should be maintained between 4 and 7 [33]. During hot-water extraction biomass immerses in liquid water than the temperature of water increases above the normal water boiling temperature. The high temperature inside of the reactor contributes simultaneously existing of liquid water with vapor. At the final stage of hot-water extraction, pressure releases, causing the biomass to expand and disintegrate into fibers. This method has used as one of the steps of commercial technology of furfural production from woody biomass [34]. The biomass after hot-water extraction has improved properties for further processing due to removal of extractives such as carboxylic acids from woody biomass [15].

7.2.2.5 Chemical pretreatment

Most of the leading pretreatment technologies are represented in this category since chemical compounds, such as acids, alkalis, organic solvents, and ionic liquids, have a significant impact on the conversion of original structure of lignocellulosic biomass [13,29,32,35].

Acid pretreatment

The main task of the low-temperature acid pretreatment of lignocellulose is to dissolve hemicellulose and accordingly to increase the availability of cellulose for next step. Due to the process of acid pretreatment, lignin partly dissolves, condenses, and precipitates. The dissolved hemicellulose oligomers can be hydrolyzed with the formation of monomers, furfural, hydroxymethylfurfural, and other volatile products in acidic conditions [25]. Pretreatment with dilute solutions of sulfuric, nitric, and hydrochloric acids is a robust and relatively low-cost method that has shown a promising application on a broad spectrum of lignocellulosic feedstock including herbaceous, softwoods, and hardwoods [32,36–43]. However, this method does not allow us to remove enough amount of lignin from lignocellulose and also many inhibitory compounds (such as acetic acid, phenolic compounds) produced during the hydrolytic reaction [44].

Pretreatment with concentrated solutions of acids largely affects hemicellulose dissolution and dissolved lignin precipitation. The yield of available sugars increases, whereas the time of hydrolysis reduces because of pretreatment with concentrated solutions of acids. However, this type of pretreatment contributes to increased production of inhibitors and also significantly increases the cost of the process.

Alkaline pretreatment

Alkaline pretreatment with the use of sodium hydroxide, ammonia, or calcium hydroxide (lime) is effective for treatment of different types of biomass [13,45,46]. Alkalis specifically target hemicellulose acetyl groups and lignin–carbohydrate ester linkages [11]. These reactions help solubilize and extract lignin from the biomass, reducing nonspecific binding during enzymatic hydrolysis [11,45,46].

NaOH treatment is very effective in increasing digestibility of hardwood and agricultural residues with low lignin content [47]. NaOH pretreatment softens and ruptures the cell wall and removes lignin partially from the biomass by fracturing the ester bonds, thereby decreasing the crystallinity of cellulose [13].

Another effective alkaline process is pretreatment with lime. Lime pretreatment removes lignin, which improves the enzymes effectiveness since elimination of nonproductive adsorption sites and increases access to cellulose and hemicellulose [47].

The majority of studies with NaOH pretreatment were conducted under high temperatures (above 100 °C) or at high concentrations of NaOH (above 1.25 mol/L) [32,45]. An intensive formation of monomeric sugars and low molecular weight compounds, which occurs at temperatures above 100 °C, is increasing the risk of destruction and loss of carbon in the form of carbon dioxide from solution. Therefore, high intensity of the monomeric fractions formation leads to the low level of hemicellulose total recovery value [38]. In addition, monomeric forms of hemicelluloses are easily degraded into volatile compounds such as furfural. This reaction also decreases a final concentration of monomeric sugars [48].

In order to prevent the negative consequences of alkaline pretreatment, the extraction should be carried out at temperature less than 100 °C [13,47].

The processes of dissolution, redistribution and condensation of lignin, as well as the modification of the crystalline state of cellulose, can occur during the alkaline pretreatment. These processes can significantly reduce the benefits of alkaline pretreatment (extraction of lignin and hydration of cellulose [49]). Another significant disadvantage of the alkaline pretreatment is the transformation of the structure of native cellulose into more dense and thermodynamically stable form [50].

Ammonium pretreatment

Ammonia pretreatment is an alternative alkaline pretreatment process; it involves the use of an ammonia solution either at high pressures (ammonia fiber expansion [51]) or in ambient conditions by soaking biomass in aqueous ammonia (e.g., ammonia recycle percolation, soaking in aqueous ammonia, supercritical ammonia, and ammonia–hydrogen peroxide pretreatments [52,53]).

During ammonia fiber expansion, concentrated ammonia is added to dry biomass in a high pressure reactor. After boiling process (approximately 5–45 min), the pressure is rapidly released. Due to such a technique, reactions do not produce significant amount of inhibitors for enzymes and microbes (e.g., organic acids), it is possible to ferment and hydrolyze this substrate without detoxification [11].

Oxidative pretreatment

Oxidants can be used to remove lignin and hemicellulose from biomass in order to increase enzymatic digestibility of cellulose. These compounds can react selectively with lignin aromatics and alkyl/aryl ether linkages or also target hemicellulose and cellulose, which will contribute negatively to the final sugar hydrolysis yields [11].

Solutions of H_2O_2 and peracetic acid are usually used as oxidants. Oxidants catalyze reactions of electrophilic substitution, displacement of the side chains, cleavage of bond alkyl(aryl) ether, or oxidative cleavage of aromatic nuclei [54–56].

It was also demonstrated that H_2O_2 can be used for delignification of biomass at pH level of 11.5. Delignification occurs due to activity of the hydroxyl ion that is the by-product of H_2O_2 degradation at this pH level [57].

There is also high risk of inhibitors creation as lignin oxidation produces soluble compounds (primarily aliphatic aldehydes and aliphatic organic acids) that have been found to inhibit enzymatic hydrolysis of cellulose [11]. At the same time, it was reported that peracetic acid is quite selective to lignin and is using for pretreatment of wood [58].

Pretreatment with ionic liquids

Ionic liquids are alternative solvents that can dissolve cellulose under mild conditions and depolymerize it into oligomers and also provided the decomposition of hydroxymethylfurfural to yield levulinic and formic acids. The anions (e.g., OAc^-, $HCOO^-$, and Cl^-) of ionic liquids participate in the destruction of the inter- and intramolecular networks to facilitate cellulose dissolution. Solutions of chloride salts (e.g., $CrCl_3$, $AlCl_3$, $FeCl_3$, and $ZnCl_2$) can be used as ionic liquids under certain hydrothermal conditions [59]. These metal chlorides tend to be expensive or environment unfriendly.

More promising solvent, $NaCl–H_2O$, can solubilize cellulose, pubescent, corn stover, corncob residue, and mulberry wood. Cl^- played a major role to destroy the hydrogen-bond network through the interaction with a,b-OH-9 of cellobiose in a ratio of 1:1. For cellulose in raw biomass, Cl^- can interact strongly with the end – OH group (OH-3) of a glucose unit in cellulose, which results in the enhanced breaking of both inter- and intramolecular hydrogen bonds in cellulose [59].

However, ionic liquids are ineffective for cellulose depolymerization, susceptible to water, and more expensive than water [59]. As a result of its low cost, safety, and abundance, H_2O is generally regarded as the optimal solvent.

Alternative approaches make use of ionic liquids that can dissolve cellulose and, if coupled with acidic reagents, also generate selected platform chemicals [60,61].

Recently, this approach has been applied with solid catalysts [62] offering both the ease of separation of a solid catalyst with the dissolving power of the ionic liquid. This combination generates exciting prospects for cleaner conversion of cellulose to chemicals. The dissolution of cellulose can be accelerated by a combined application of strong acids with milling [63].

7.2.2.6   Biological pretreatment

Different biological approaches have been developed as environment-friendly alternatives to physical–chemical and chemical pretreatment methods [64]. Microbial pretreatment consists of a solid-state fermentation process in which microorganisms grow on the lignocellulosic biomass selectively degrading lignin (and in some cases hemicellulose), whereas cellulose is expected to remain intact [64].

Another attractive alternative for the pretreatment of lignocellulose consists in using of ligninolytic enzymes. This strategy is substrate specific and offers the possibility to increase reaction rates and delignification efficiency reducing the process time from weeks to hours with no carbohydrate consumption [64].

Microbial pretreatment

Microorganisms that degrade lignocellulose in the nature can be also used for pretreatment stage [65,66]. They attract an attention of industry because lignin is the main barrier to cellulose hydrolysis. Fungi and bacteria depolymerize lignin by secreting extracellular enzymes such as lignin peroxidase. Lignin peroxidase of white rot fungi generates free radical species that attack aromatic rings in the lignin polymer to effectively degrade it [66]. This biological pretreatment step can be performed before biomass saccharification. Delignification activity was demonstrated mainly for "white-rot" (*Basidiomycetes*); however, some *Ascomycetes* can also colonize lignocellulosic biomass (for instance, *Trichoderma reesei* and *Aspergillus terreus*). Besides fungi, certain bacterial strains such as *Bacillus macerans*, *Cellulomonas cartae*, *Cellulomonas uda*, and *Zymomonas mobilis* also show delignification abilities [64]. There are different factors to be considered for an efficient solid-state fermentation such as nutrient addition, moisture content, aeration, pH, temperature, inoculum size, or the microorganism strain.

Enzymatic hydrolysis
In contrast to the chemical hydrolysis, enzymatic hydrolysis is highly substrate specific achieving high yields of glucose [67,68]. It should be noted that the enzymatic hydrolysis is often considered as a necessary final stage of any pretreatment types. Enzymatic delignification can be performed either by using a culture supernatant with different ligninolytic activities or with a prepared solution containing a single purified and concentrated enzyme or their combination (cellulose, glucosidase, xylanase, and cellobiase [45,69]).

The size of pores on the cellulose surface is an essential factor for the process of enzymatic hydrolysis [70]. Thus, the pretreatment of lignocellulose with other techniques can lead to pore size increasing and in turn increasing of efficiency of enzymatic hydrolysis [71].

Enzymatic hydrolysis of lignocellulose is limited by such factors as the degree of cellulose crystallinity, degree of polymerization, moisture content, available surface area, and lignin content [72–75].

Commercially available cellulase preparations proved to be very effective in the industrial hydrolysis of cellulose [76,77]. However, this method has some disadvantages. The cellulases can carry out a nonspecific ability for binding to lignin. Therefore, some alternative techniques like the utilization of surfactants, which have the ability to reduce these bindings, can improve the enzymatic yields [68]. Another issue to be considered in enzymatic hydrolysis is the inhibition caused by the substrate. This problem can be overcome by using techniques of fermentation in simultaneous with the enzymatic hydrolysis step (simultaneous saccharification and fermentation process [68,76]). However, one major disadvantage of this method is the cost of enzymatic preparations. At the same time, some techniques for enhancement of enzymatic hydrolysis yield were suggested: among them, development of more optimal reactor design enabling more efficient extraction of the reaction products to avoid inhibiting of the process [78–80], adding of surfactants to the reaction mixture for washing from substrate, subsequent concentration and reusing of enzyme preparations [81], and optimization of the reaction mode by the sequential adding of substrate portions into reactor [82].

7.2.2.7 Combination of pretreatment methods

It should also be noted that in some cases, it is very effective to combine different methods to improve the efficiency of the pretreatment process [38,39].

A vast number of researches include a combination of mechanical pretreatment with physical–chemical or chemical types of pretreatment. Typically, the particle size of the lignocellulose after pretreatment decreased significantly from centimeters [19,27,83] to millimeters [29] or a few tenths of a millimeters [41,84,85]. This pretreatment considerably increases the reactivity of lignocellulose and enhances efficiency of further steps of pretreatment [15].

Hydrothermal pretreatment can be enhanced by addition of alkali solution [86]. The treatment with $Ca(OH)_2$ under heating (at 100–150 °C) significantly increases availability of biomass containing a low concentration of lignin, but not effective for the treatment of a biomass containing a high concentration of lignin.

Hot steam or hot-water pretreatment can be enhanced by addition of acid solution. Such combination catalyzes the process of hemicellulose dissolution. It also allows decreasing of the operational temperature of pretreatment [23,26].

Enzymatic hydrolysis is also integral part of numbers of research [13,46,87]. This is a final step that allows releasing of fermentable sugars from pretreated biomass (mainly from cellulose or hemicellulose). Combinations of microwave-heated acid pretreatment or steam-heated acid pretreatment with enzymatic hydrolysis of cellulose are also possible. Microwave-heated acid pretreatment generated numerous regular micropores and steam-heated acid pretreatment generated many irregular fragments in destroyed lignocellulose matrix [88].

Other methods of combined pretreatment are also possible: among them, thermal pretreatment combined with oxidative pretreatment and thermal pretreatment combined with alkali or acid pretreatment [86]. Acid pretreatment as a first step can be combined with sulfomethylation treatment as a second step [83].

Hybrid pretreatment methods, such as different types of hydrothermal pretreatment combined with alkali or acid pretreatment, considerably improve the biomass digestibility. However, the extra equipment and operation cost will reduce the advantages of combined methods.

7.2.2.8 Detoxification of pretreated biomass

Except of acids, alkalis, and other chemical reagents that are used for treatment of lignocelluloses, the process of enzymatic hydrolysis, or further fermentation step can be inhibited by reaction products of lignocellulose hydrolysis.

All inhibitors produced during pretreatment could be divided into three groups: (1) organic acids (acetic, formic, etc., their undissociated forms penetrate into the cell, in which they dissociate and decrease the pH value); (2) furfural, 5-furfural, levulinate acid, and humic substances, which are by-products of the degradation of sugars; (3) a wide range of aromatic and polyaromatic compounds produced during degradation of lignin. Different inhibitors in varying degrees can impact on the subsequent process of fermentation from a slight decrease in efficiency to a complete stoppage of the process [89]. For example, two-stage pretreatment process with sulfuric acid and enzyme preparation caused a number of potential inhibitors formation such as furfural, acetic, levulinic, and formic acid [90]. Therefore, in some cases, the detoxification step can increase efficiency of the lignocellulose conversion. For example, biomass after acid pretreatment can be treated by NaOH or $Ca(OH)_2$. NaOH is very efficient in terms of removing toxic compounds such as furfural, phenols, acetol, and hydroxymethylfurfural [68]. Treatment with the $Ca(OH)_2$ is based on the precipitation of the toxic components as well as instability of some inhibitors at high pH. The weakness of this method is the carbohydrate degradation due to high temperature and pH [91,92].

Enhancement of lignocellulose digestibility will contribute to significant development of technologies based on renewable biomass as a feedstock. Therefore, the number of researches in the area of development of pretreatment methods increases each year. In favor of this indirectly implies a significant increase of scientific publications on this subject in the last 10 years (Figure 7.1). Some

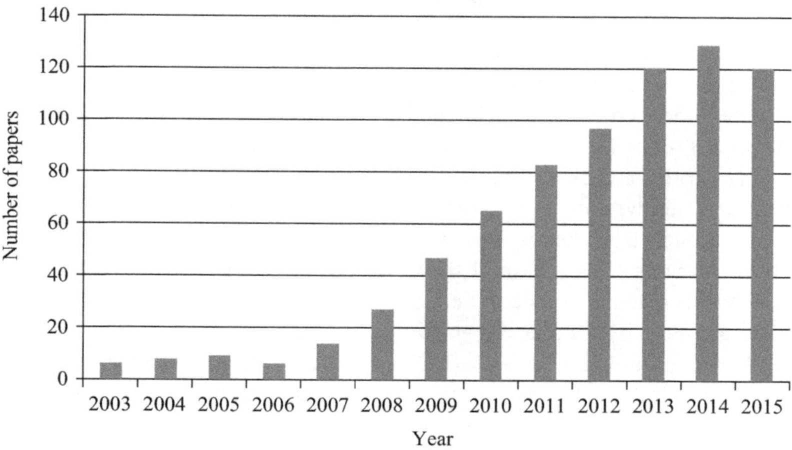

Figure 7.1   *Number of papers in PubMed Data Base found with using of key words combination—"lignocellulose pretreatment" [http://www.ncbi.nlm .nih.gov/pubmed/]*

examples of pretreatments with the main operational characteristics are shown in Table 7.1. Although the range of pretreatment methods is quite wide, it should be noticed that large amount of them is a combination of mechanical pretreatment as the first step and enzymatic hydrolysis as the final step (Table 7.1). In addition, the main pretreatment step can include several types of pretreatment. Although hybrid technologies of pretreatment are often significantly effective than single techniques, the combination of several methods will obviously affect the costs of complex process.

7.2.3   Third generation of starting materials for biohydrogen production

Bacterial hydrogen fermentation is a family of processes that can be roughly divided into three groups: dark fermentation, photofermentation and two-stage processes combining dark fermentation either with photofermentation [96–98]. Dark fermentation consists in the conversion of sugars into H_2, CO_2, and organic acids by microorganisms. Photofermentation is based on the uptake of CO_2 or organic acids by photosynthetic microorganisms and production of hydrogen. Theoretically, any sugar-containing biomass can be used as a feedstock for dark fermentation. Organic acids produced during dark fermentation can be used as a substrate for hydrogen producing photosynthetic microorganisms in order to increase efficiency of process of biomass conversion to hydrogen [96–98]. Thermochemical treatment of biomass, or pyrolysis, leads to production of a mixture of carbon monoxide and hydrogen (synthesis gas) and it is a valuable substrate for large-scale pure hydrogen generation after membrane separation or for microbial production of ethanol and other combustible compounds.

7.2.3.1 Anaerobic dark fermentation for hydrogen production

The temperature level during the dark fermentation is in the mesophilic, thermophilic or extreme thermophilic range depending on the type of microorganism. Dark fermentation converts simple sugars or disaccharides into hydrogen, carbon dioxide, and organic acids. Suitable heterotrophic bacteria include strict anaerobes (such as clostridia and thermophiles), facultative anaerobes (like of genus *Enterobacter*), and aerobes (e.g., *Alcaligenes* and *Bacillus* genera). Anaerobic bacteria produce hydrogen from hexoses in such processes as acetic acid, butyric acid and ABE fermentations. The maximal theoretical value of 4 mol of H_2 per 1 mol of glucose can be reached in acetic acid fermentations. Production of other organic acids and alcohols by facultative anaerobic bacteria as by-products decreases the yield of hydrogen down to 2 mol of H_2 per 1 mol of hexose. The hydrogen yields and production rates of thermophilic and extreme thermophilic bacteria (e.g., *Thermoanaerobacterium thermosaccharolyticum, Caldicellulosiruptor saccharolyticus*, and *Thermotoga neapolitana*) growing at temperatures above 60 °C are often higher than those of mesophilic bacteria growing at ambient temperature because thermophilic bacteria produce acetic acid as the main fermentation by-product.

In view of the approaching industrial-scale biohydrogen production it becomes increasingly important to consider the selection of microbial consortia aiming at increased H_2 yield and improved process robustness under conditions of fluctuating feedstock composition and stochastic process disturbances. A method was developed using the co-culture of *C. thermocellum* and *Clostridium thermosaccharolyticum* to improve hydrogen production via direct thermophilic fermentation of cornstalk waste (CW) [99]. Hydrogen yield with this co-culture fermentation process in 125 mL laboratory anaerobic bottles reached 68.2 mL per 1 g of CW that was 94% higher than that obtained with mono-culture.

Fermentative H_2 production is limited by thermodynamics and only one-third (usually substantially less) of the electrons available in biomass can be finally recovered as molecular hydrogen. The major part of the electrons (at least two-thirds) is bound into the acetate formed as the most important product of biomass fermentation and cannot be released as H_2 for thermodynamic reasons. Even the optimal fermentation of hexoses (as important representatives of lignocellulose) to 2 acetate, 2 CO_2, and 4 H_2 is reached only at enhanced temperatures (above 60 °C [100]); otherwise, butyrate or ethanol are formed as by-products with substantially lower H_2 yields. Thus, H_2 production by biological processes will gain importance only if coupled to photosynthetic reactions.

7.2.3.2 Anaerobic photofermentation for hydrogen production

After anaerobic dark fermentation processes performed in mesophilic or thermophilic conditions resulted mixture of residual organic acids (acetate, lactate, formate, and butyrate) can be converted into hydrogen and carbon dioxide via light-depending process by purple phototrophic bacteria in free or immobilized stage with high efficiency output (up to 95% for lactic acid) [101–103]. This process needs constant gassing of the solar fermentor with inert gas like argon (nitrogen and methane are options too) to avoid self-inhibition by hydrogen produced. Released three gas mixture has to be separated for argon recycling as carrier gas into the solar bioreactor. Purified hydrogen can be used in microbial or enzymatic fuel cells for the electricity generation [104]. Although biological processes and, in particular, dark fermentation have been recognized as a promising approach to hydrogen production, the next key issue appearing for moving this technology to commercialization is high production cost. Biomass residues of agricultural origin have a role to play in this regard as these materials generally are cheap, and many of them are abundant offering the opportunity for utilizing the economy of large-scale processing. However, similar to other potential substrates, each particular material requires individual selection of optimum processing conditions, and this usually includes optimization of parameters of pretreatment methods and fermentation processes. Despite the diversity of lignocellulosic agro-residues, problematic areas of their bioconversion to hydrogen are well covered by research efforts taking place around the world and employing most advanced research methodologies and techniques. In coming years, these research efforts will be followed by pilot- and demonstration-scale setups for hydrogen production from biomass at a larger scale in order to make the technology ready for commercialization. It is expected that the involvement of industry will gradually increase in time as well.

7.2.3.3   Thermochemical treatment of biomass

Gasification is a promising technology for the conversion of biomass resources into biofuels and biochemicals. Gases generated by this process can be utilized for power generation and production of synthetic petroleum and chemicals through Fischer–Tropsch synthesis. Thermochemical conversion of biomass (pyrolysis, analogous to coal gasification) leads to a mixture of carbon monoxide and hydrogen (synthesis gas) as a valuable substrate for hydrogen obtaining after separation or for microbial production of ethanol and other combustible compounds. Synthesis gas can also be used for chemical production of methanol and of long-chain hydrocarbons (Fischer–Tropsch synthesis) to replace petrochemicals. Fischer–Tropsch synthesis is a large-scale industrial process developed before 1925 by Franz Fischer and Hans Tropsch in Mülheim an der Ruhr, Germany. In this process, gas mixture containing mainly carbon monoxide and hydrogen is converted into liquid hydrocarbons. Thermochemical pyrolysis can be recommended as a strategy for energetic utilization of lignocellulosic and other similarly stable organic matter with low water content. Using this process for the biomass conversion into energy, one can avoid release of wastes because all the carbon content of biomass is converted into CO without releasing of lignin as in the case of lignocellulosic biomass conversion by microbial cultures. It should be noticed that utilization of lignin takes place only under the aerobic conditions by white-rot fungi (see above), and this is a slow environmental process. From the other hand, this process of slow releasing of organic matter in the soil maintains its fertility over the years. If this process could be fast, the organic content of the soil could be consumed very quickly that could convert fertile soil into life-less substance like in desert.

One notable advantage of thermochemical conversion is the possibility of valorization of all the biomass lignocellulosic polymers constituting the cell walls (i.e., cellulose, hemicelluloses, and lignin). However, the development of such processes is limited due to the complexity of biomass thermochemical conversion and the fact that products are typically obtained with too low yield and purity. Hence additional upgrading treatments that significantly enhance the conversion efficiency of the process are required.

Pyrolysis is the thermochemical decomposition of dry organic matter (moisture content below 10 wt.%) in the absence of oxygen at moderate temperatures (350–550 °C) and atmospheric pressure [105].

Three types of pyrolysis have been described [105,106]:

- slow pyrolysis at low temperature (300 °C) and low heating rate (<1 K/s) gives 35% of oil, 35% of char, and 30% of gas;
- intermediate pyrolysis at moderate temperature (500 °C) and moderate heating rate (1–10 K/s) gives 50% of oil, 20% of char, and 30% of gas;
- fast pyrolysis at moderate temperature (500 °C) and fast heating rate (>10 K/s) gives 75% of oil, 12% of char, and 13% of gas.

Biomass is usually considered as raw material with three components: cellulose, hemi-cellulose, and lignin. In general, cellulose and hemicellulose have a lower pyrolysis temperature. Compared with cellulose and hemicellulose, the pyrolysis temperature range of lignin is higher. Cellulose, hemicellulose, and lignin are mainly pyrolyzed at 300–390, 200–350, and 200–450 °C, respectively [107].

One possible alternative is to use a hybrid process that involves the simultaneous conversion of organic matter (from coal and biomass) into synthesis gas (syngas) via gasification followed by biological conversion utilizing microorganisms that are able to convert major components of syngas (CO and H_2) into multicarbon compounds. Agricultural residues and woody biomass are not applicable for direct microbial conversion due to poor biodegradability of the lignin moiety called "recalcitrance"; however, lignin is one of the most energy-rich components and creates 10%–20% of the entire biomass present in nature [108]. Lignin content of biomass can be utilized effectively in syngas-based biorefineries. Moreover, after catalytic reforming, the pyrolysis gas is suitable for hydrogen production or directly utilization as the feedstock of fuel cell [109].

Thermochemical reactions like pyrolysis, combustion and gasification is performed at high temperatures using dry lignocellulosic biomass of straw or wood. The technology assumes the transformation of biomass into H_2 and CO, various useful chemicals or fuel. It took all the best what had been developed for technology of char gasification.

The ratio of C and H_2O in the dry and ash-free lignocellulose can be regarded as approximately 1:1 (by weight); however, content of the releasing heat approximately 20% greater due to containing of C–H and C–OH bonds. As a rule, the hydrogen production from biomass is considered as three stages process: (1) pyrolysis of raw biomass to produce the main hydrogen containing gas products, volatile tar, and solid char; (2) cracking and reforming of liquid oil and gases; and (3) char gasification [110]. Pyrolysis uses heat to degrade a feedstock in an oxygen absence conditions into combustible mixture of gases along with liquid oil and char fractions. The yields of end products of pyrolysis and the composition of gas mixtures depend on following parameters: temperature, biomass species, particle size, heating rate, operating pressure, and reactor configuration, as well as the type of append catalysts [111]. Depending on mentioned parameters raw biosyngas contains around 10–50 vol.% of H_2, 5–56 vol.% CO_2, and 17–45 vol.% of CO and other impurities such as H_2S, N_2, CH_4, and H_2O (Table 7.2). CH_4 can be further converted into H_2 and CO by air reforming, and CO_2 can be separated using conventional processes [112].

Among many types of biomass (Table 7.2) that can be pyrolyzed to produce H_2, lignocellulosic biomass has received more attention because it does not require the diversion of feed resources. As an example, the wood is composed of cellulose (35%–50%), hemicelluloses (20%–30%), and lignin (20%–30%) as well as inorganic salts and extractives (low-molecular-weight organics). Thus, the biomass pyrolysis is in fact the pyrolysis of a mixture of cellulose, hemicellulose, and lignin in the presence of various minor compounds. Lignocellulosic biomass (plant biomass) conversion can be achieved from various resources including (1) forestry wastes such as logging wastes, sawmill wood waste and residues of the trees and shrubs; (2) agricultural residues like animal and crop wastes (e.g., corn stover); and (3) energy crops like corn, sugarcane, grasses, and aquatic plants like water hyacinth.

Although various research groups have investigated suitable technology and processes for bio-hydrogen and biosyngas production, the issue of the gas treatment for H_2 recovery has not received much attention.

Table 7.2   Hydrogen-containing gas mixtures of biogenic origin. Reprinted, with permission, from Reference 104

Source	Gas composition (vol.%)							Reference
	H_2	CH_4	C_2+	CO	CO_2	N_2	H_2S	
Biohydrogen	1–80	0–2	–	–	6–12	1–80	0–12	[113,114]
Biohydrogen	57–60	–	–	–	39–43	1–5	–	[115]
Biosyngas	25–42	1	–	25–42	10–35	2–5	1	[116]
Biosyngas, model	33	–	–	17	–	50	–	[112]
Solid waste pyrolysis	31–32	25–31	–	20–27	7–13	–	–	[116]
Catalytic pyrolysis of pine tree	49.7–52.8	3.8–5.3	1.1–1.6	33.7–34.5	8.6–8.7	–	–	[117]
Pyrolysis of wet sewage sludge	36.7	14.7	7.1	21.8	13.6	6.3	–	[118]
Pyrolysis of coffee hulls	9.3–40.1	0–11.3	0–2.5	20.6–32.7	17.7–56.6	–	–	[119]
Gasification of Siberian elm	44.3	5.9–7.9	–	17.1	26.8–31.1	–	–	[112]

7.3   Membrane recovery of hydrogen: basic regularities for polymeric membranes; classification of membranes

7.3.1   *Application and potential of standard membrane technology for hydrogen recovery from gaseous mixtures*

Membrane, adsorption, absorption, and cryogenic technologies are applied in industry for the recovery and purification of H_2 [120–126]. Utilization of particular technology is determined by desired productivity and target purity of the product. Technological zones for gas separation processes with low-energy consumption (membrane, short cycle adsorption, ultra short cycle adsorption) are shown in Figure 7.2. As it can be seen from Figure 7.2, membrane processes occupy reasonably broad range: from local (mobile) membrane equipment till high capacity installations.

Membrane methods take particular place due to the absence of energy consumption for phase transitions. Membrane technologies for hydrogen recovery appeared in the market for the first time in 1980: Separator PRIZM produced by Monsanto for H_2 recovery from blow-down gas of ammonia production. It should be noted that the first gas separating flat-sheet membranes under industrial scale were developed in USSR in 1977 [127]. At present time, there are several hundreds of membrane installations for H_2 recovery in chemical, petrochemical, and petroleum refining industrial plants all around the world. The number of companies that produces membrane installations and utilizes membrane modules of own or off-site production is quite high. Leading positions are occupied by US companies: IGS GLOBAL, Air Products, NATCO, CMS, Newpoint Gas, Parker NNI, On Site Gas Systems Inc., Holtec Gas Systems, ProSep Technologies Inc., MTR, and PRAXAIR. Big companies from other countries are UBE (Japan), Air Liquide

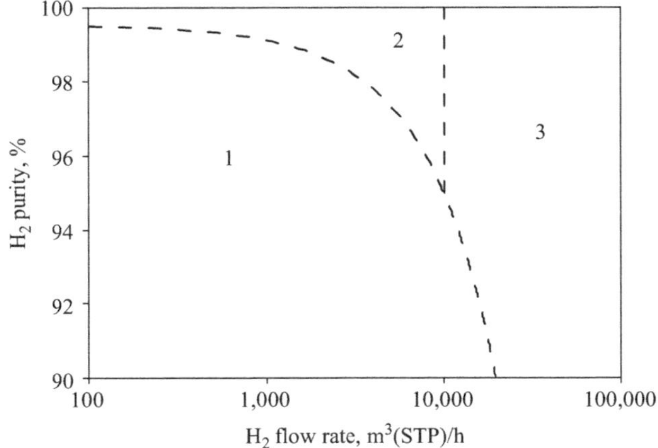

Figure 7.2   Technological zones of H_2 purification processes: 1—membranes, 2—ultra short cycle adsorption, 3—short cycle adsorption. Reprinted, with permission, from Reference 104

(France), INMATEC, BORSIG Membrane technology (Germany), and CAN Gas Systems Company Limited (China). Russia is represented by such companies as GRASIS, KRIOGENMASH, VLADIPOR. Installations for H_2 production of technical purity are produced and supplied by Air Liquide, UBE, PRAXAIR, UOP, and GRASIS. At present time, membrane recovery of hydrogen from waste H_2-containing gases is approved and fast developing technology.

The separation of gases can be carried out by two types of membranes that have different mass-transfer mechanisms: porous and non-porous [6,128]. More than 500 patents on H_2-selective membranes were published during 2000–11. More than 50% of patents are devoted to the synthesis of organic (polymeric) membranes, around 25% of patents are devoted to porous inorganic membranes and around 25% of patents are devoted to non-porous inorganic membranes based on palladium and its alloys. Systematic review of patent literature, investigation groups, and recommendations for development of certain membranes (polymeric, ceramic, palladium based) for recovery/purification of H_2 is represented in References 4, 129, and 130. Published data show that high selective membranes for single-stage H_2 recovery from gas mixtures containing CO_2, CO, N_2, CH_4, and other components still not exist. Nevertheless in Reference 131, it is mentioned that industrial membrane module based on palladium alloy (PdInRu) for H_2 purification is already developed. This module has following characteristics: operating temperature 250–800 °C, feed pressure 20 MPa, initial H_2 concentration 30%–98%, H_2 concentration in product stream 99.9999% and higher, product stream pressure 0.1–5 MPa, lifetime more than 2 years. It is important to note that specific productivity of palladium alloys is strongly dependent on high temperature and therefore significant energy consumption is necessary to provide high productivity process.

7.3.1.1 Gas separation by nonporous polymeric membranes

Polymeric membranes are considered as the most advanced development in gas separation processes. The driving force of gas transfer through polymeric non-porous membrane is the difference of chemical potentials of components. Mass transfer through polymeric membrane is a complex process including sorption and dissolution of a substance on one surface, diffusion through the volume of polymer to the second surface and desorption from it. Table 7.3 represents characteristics of polymers used for industrial gas separation membranes production.

One can see that for the most of polymers used in industry H_2/CO_2 selectivity varies in very narrow range from 0.2 to 3.9. At the same time, polymers with higher selectivity demonstrate lower permeability coefficient. Therefore, it is quite difficult to select membrane for hydrogen recovery from multicomponent gas mixtures which contain CO_2. On the other hand, these polymeric membranes can be used for preconcentration of H_2 for its following recovery by other technique.

7.3.1.2 Commercially available polymers and membrane modules

More than 2000 new polymers for gas separation were synthesized during the last 30 years. Nevertheless, only a few polymers that demonstrate acceptable permeability and selectivity are used for industrial production of membranes and

Table 7.3   Characteristics of polymers for production of commercially available gas separation membranes. Reprinted, with permission, from Reference 104

Polymer	Permeability coefficient, Barrer (10^{-10} cm^3(STP) cm/(cm^2 s cmHg))					Selectivity			Reference
	CO_2	N_2	CH_4	H_2	CO	H_2/CO_2	H_2/N_2	H_2/CH_4	
	Glassy polymers								
Cellulose acetate	10	0.33	0.36	24	–	2.4	72.7	66.7	[6]
Polysulfone	5.6	0.25	0.25	14	–	2.5	56	56	[6]
Polyimide	13	0.6	0.4	50	–	3.9	83.3	125	[6]
Tetrabromopolycarbonate	4.5	0.18	0.12	16*	0.16*	3.6	88.8	133.3	[132]
Polyvinyltrimethylsilane	190	11	22	200	15*	1.1	16.7	9.1	[127]
	Rubbery polymers								
Silicon rubber	2700	250	800	550	–	0.20	2.2	0.69	[6]

*Estimation of permeability by method described in References 70 and 127.

membrane modules: for example, polysulfone, polyimide, cellulose acetate, tetra-bromopolycarbonate, and silicon rubber. The reason of selection of a certain polymer for membrane production is the importance of other criteria except permeability and selectivity. Industrial application of membrane material is determined by such important parameters as ability of formation of thin and mechanically strong film, durability, chemical stability, price, etc. Effective thickness of selective layer of industrial membranes is often in the range of 0.05–0.2 μm. This layer is formed on a porous nonselective support layer. Membranes can be flat or hollow fiber geometry, packing density of membrane module (specific membrane area) being higher in the case of hollow fiber membranes. Realization of membrane process at industrial scale usually demands high membrane area; therefore, membrane modules with hollow fibers are used predominantly (around 75% of cases), membrane modules with flat membranes are used more often in the form of spiral wound (around 15% of cases) than in the form of flat-sheets (usually disk-type) that is caused as by technical as by economic reasons [6]. Selectivity of some industrial gas separation membranes is presented in Table 7.4.

It can be seen from Table 7.4 that H_2/CO_2 selectivity of membranes is quite low. Therefore, it is difficult to design effective membrane separation process for hydrogen recovery from multicomponent gas mixtures that contain CO_2. On the other hand, polymeric membranes can be very effective for separation of H_2/CO or H_2/N_2 pairs. Main developers and producers of membranes and technological processes are presented in Table 7.5.

Variety of H_2-containing gas mixtures of biogenic and technogenic origin demands the estimation of membrane properties of polymeric membranes for a set of mixture components. On the contrary, permeability data of many desired components are not available for the most of commercial membranes. Nevertheless, it is possible to estimate necessary values using theoretical method

Table 7.4   Selectivity of commercially available polymeric membranes. Reprinted, with permission, from Reference 104

Selectivity	Membrane polymer					
	Polyaramide (Medal)	Cellulose acetate (Separex)	Polysulfone (Permea)	Polyimide (Ube)	Tetrabromo polycarbonate (MG)	Silicon rubber
H_2/CO_2	3.0	2.4	2.5	3.8	3.5	0.2
H_2/N_2	>200	72–80	56–80	88–200	90	22
H_2/CO	100	30–66	40–56	50–125	100–123	0.69
H_2/CH_4	>200	60–80	80	100–200	120	0.8

Table 7.5   Main developers and producers of membranes and membrane systems. Reprinted, with permission, from Reference 104

Producer	Process	Polymer	Type of module*
Permea (air products)	Gas separation	Polysulfone	HF
Medal (Air Liquide)	Air separation	Polyimide, polyamide	HF
IMS (Praxair)	Air separation, recovery of H_2	Polyimide	HF
GENERON (MG)	Air separation, recovery of H_2	Tetrabromopolycarbonate	HF
Separex (UOP)	CO_2/CH_4 separation	Cellulose acetate	SW
Kvaerner	CO_2/CH_4 separation	Cellulose acetate	SW
Cynara (Natco)	CO_2/hydrocarbons separation	Cellulose acetate	HF
Aquilo	Air separation	Polyphenylenoxide	HF
Parker-Hannifin	Air separation, recovery of H_2	Polyamide, polyphenylenoxide	HF
UBE	Vapor/gas separation, air separation	Polyimide	HF
GKSS Licensees	Hydrocarbons recovery	Siloxane-containing copolymers	DT
MTR	C_4H_{10}/N_2 separation, Hydrocarbons recovery	Polyalkylsiloxane	SW

*HF, hollow fiber membrane module; SW, spiral wound membrane module; DT, disk-type membrane module.

suggested in Reference 133. In the present work, we used such approach in order to predict the permeability of membranes for certain gases. Details of this approach were described earlier in Reference 127 and recent achievements in Reference 134.

Comparison of estimated values with available ones and own experimental data were found to be in satisfactory agreement for all considered membranes. This approach was also validated by comparison of results of calculation with experimental data reported in the literature for large number of other polymers.

Table 7.6   Permeance of commercial and lab-scale gas separation membranes with estimated values for a set of gases. Reprinted, with permission, from Reference 104

Membrane or membrane module	Gas permeance, $L(STP)/(m^2\ h\ atm)$									
	H_2	He	CO_2	O_2	SO_2	H_2S	N_2	CO	CH_4	C_2H_6
Hollow fiber GENERON®	160*	180	45	13.6	10.3*	4*	1.8	1.6*	1.3	0.56*
Hollow fiber AIR PRODUCTS®	151	151	104	22.7	47.5*	14.3*	3.8	6.6	6.3	3.79*
Spiral wound Silar®	440	250	2,000	400	2,570	1,195	190	270	545	1,644*
Flat-sheet membrane polyvinyltrimethylsilane [79]	2,000	1,800	1,600	450	1,000*	350*	120	150*	220	142*
Fluorinated composite polyvinyltrimethylsilane flat-sheet membrane [79,135]	355*	342	249	29.6*	77.7*	23.9*	10.6	9.46*	6.2	4.81*
Matrimid 5218® hollow fiber membrane [79,135]	214*	238.6	70.6	3.76*	21.4*	2.84*	0.258*	0.57*	1.23	0.21*
Fluorinated composite Matrimid 5218® hollow fiber membrane [79,135]	145*	207.9	18.87	0.69*	2.99*	0.35*	0.03*	0.07*	0.14	0.01*

*Estimated values.

Obtained data (Tables 7.6 and 7.7) confirm that recovery of H_2 from multicomponent mixtures containing CO, CO_2, N_2 by membranes is difficult task. Therefore, it is necessary to increase polymer selectivity keeping good film-formation properties, high mechanical characteristics, and permeability for effective separation of such gas mixtures as H_2/CO_2, He/CO_2, and $He/CO_2/CH_4$. This purpose can be achieved, for example, by fluorination of membrane [135].

On the other side, these polymeric membranes can be used for preconcentrating of H_2 for its following recovery by other technique as mentioned above. Some examples of H_2 membrane separation processes calculation are shown below.

7.3.1.3   Calculation of H_2 concentrating from multicomponent mixtures by membrane method

Optimization methods and general approach for membrane module simulation are well known from literature, for example [136,137]. These approaches use several models of membrane unit and calculation of process mainly performed for binary gas mixtures separation. Calculation of multicomponent gas mixtures membrane separation in this work was carried out using computer program developed at TIPS RAS [138]. Calculations were performed for biosyngas separation as an example. The program allows calculating of gas separation parameters under different flow conditions in membrane module: counter-current, co-current, cross-flow, and complete mixing. Based on initial conditions of calculated system such as feed and

Table 7.7   Selectivity of commercial and lab-scale gas separation membranes with estimated values for a set of gases [6,127,132,133]. Reprinted, with permission, from Reference 104

Membrane or membrane module	Ideal selectivity							
	H_2/CO_2	H_2/N_2	H_2/CH_4	H_2/CO	He/CO_2	He/N_2	He/CH_4	He/CO
Hollow fibers GENERON®	3.6	88.9	123	100	4.0	100	139	113
Hollow fibers AIR PRODUCTS®	1.5	39.7	24.0	22.9	1.5	39.7	24.0	22.9
Spiral wound, Silar®	0.2	2.3	0.8	1.6	0.1	1.3	0.5	0.9
Flat sheet membrane polyvinyltrimethylsilane (PVTMS) [135]	1.3	16.7	9.1	13.3	1.1	15.0	8.2	12.0
Fluorinated composite polyvinyltrimethylsilane flat sheet membrane (PVTMS-F2) [135]	1.4	33.5	57.3	37.5	1.4	32.3	55.2	36.2
Matrimid 5218® hollow fibers membrane [135]	3.0	829	174	372	3.4	925	194	414
Fluorinated composite Matrimid 5218® hollow fibers membrane (Matrimid 5218®-F2) [135]	7.7	4,830	1,035	2,070	11.0	6930	1,485	2,970

permeate pressure, flow rate, membrane surface area, and membrane separation properties, the concentration of gases in permeate and retentate was obtained depending on stage-cut that determines as the ratio of permeate flux to feed flux. The corresponding properties of membranes for calculation are given in Tables 7.6 and 7.7. The choice of model for comparative calculations was done experimentally by using industrial hollow fiber membrane module GENERON® and laboratory scale disk-type module with PVTMS membranes. $He/H_2/CO_2/N_2$ and $He/CO_2/O_2$ mixtures were used in comparative experiments. The best agreement between calculation and experimental data was found for counter-current flow for the case of hollow fiber module and cross-flow for the case of disk-type membrane module.

In order to show how separation properties of membrane affect characteristics of process calculations were performed for several membranes. Results of membrane separation of wet sewage sludge pyrolysis gas (obtained from urban waste water and treatment plants) (*36.7% H_2, 13.6% CO_2, 14.7% CH_4, 6.3% N_2, 21.8% CO, 7.1% C_2H_6, and 0.8% O_2*) and pyrolysis of pine tree (*49.7% H_2, 8.7% CO_2, 5.5% CH_4, 34.5% CO, and 1.6% C_2H_6*), and biosyngas (*42% H_2, 24% CO, 30% CO_2, 1% CH_4, 2% N_2, and 1% H_2S*) are represented in Figures 7.3 and 7.4 and Tables 7.8 and 7.9. The following operating conditions were considered: feed gas flow rate 0.1–60 m³(STP)/h, pressures in feed/permeate sides 10/1 atm, membrane area 10 m². Figures 7.3 and 7.4 demonstrate the concentration of H_2 in permeate and retentate of respective membrane module with variation of stage-cut (abbreviation of membranes can be found in Table 7.7).

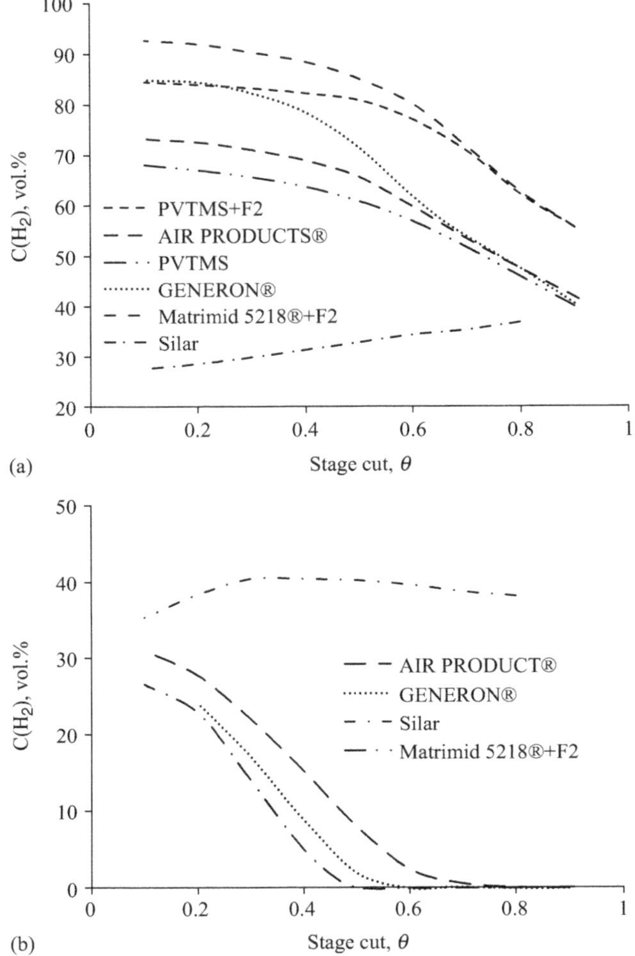

Figure 7.3 *The dependence of H₂ concentration in permeate (a) and retentate (b) on stage cut for wet sewage sludge pyrolysis gas membrane separation. Reprinted, with permission, from Reference 104*

Obtained results show that by application of single-membrane separation stage with membranes based on high selective polyimide polymer, especially fluorinated Matrimid 5218®, allow to recover from both considered gas mixtures more than 95% of hydrogen with concentration in permeate more than 90 vol.%. All considered membranes allow increasing of hydrogen concentration from 36 to 49 up to 70 vol.% that is enough for the subsequent effective H_2 purification by PSA method. It should be noted that membranes based on tetrabromopolycarbonate, polysulfone, and polyimide provide recovery degree in the range of 92%–97%. Tables 7.8 and 7.9 show obtained gas compositions after membrane separation with the mentioned characteristics.

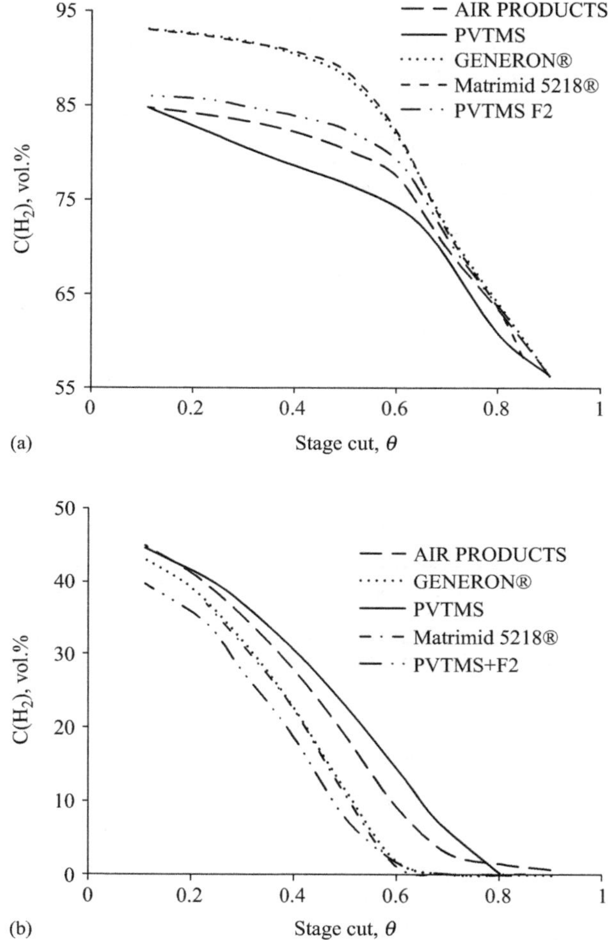

Figure 7.4   The dependence of H$_2$ concentration in permeate (a) and retentate (b) on stage cut for pyrolysis gas of pine-tree membrane separation. Reprinted, with permission, from Reference 104

As was expected, high H$_2$ purity can be achieved by membrane methods only at low recovery degrees (see Table 7.10). On the other hand, as it is mentioned above, membrane methods of gas separation can be used for the preconcentration of hydrogen at high H$_2$ recovery degrees for its further purification by PSA method. Thereby there are polymeric membranes and modules that can solve the problem of hydrogen preconcentration of mixtures containing 35–49 vol.% of hydrogen. In any way characteristics of membrane separation strongly depend on feed mixture composition; therefore, in each case corresponding calculation is required. As opposed to the separation of the bulk species Table 7.10 also shows that the residual carbon monoxide might be an issue.

Table 7.8   Calculation of pine-tree pyrolysis gas membrane separation (stage cut corresponds to preconcentration of H_2 up to 70 vol.%). Reprinted, with permission, from Reference 104

Component	GENERON®, $\theta = 0.71$		Air Product®, $\theta = 0.70$	
	Concentration in permeate (%)	Concentration in retentate (%)	Concentration in permeate (%)	Concentration in retentate (%)
H_2	70.0	0.1	70.0	2.9
CO_2	12.8	0.5	11.7	1.0
CO	14.9	80.2	15.6	78.9
CH_4	2.0	13.9	2.3	12.8
C_2H_6	0.3	5.3	0.4	4.4

Table 7.9   Calculation of pyrolysis pine-tree gas membrane separation (stage cut correspond to preconcentration of H_2 up to 70 vol.%). Reprinted, with permission, from Reference 104

Component	PVTMS, $\theta = 0.68$		Matrimid 5218®, $\theta = 0.72$	
	Concentration in permeate (%)	Concentration in retentate (%)	Concentration in permeate (%)	Concentration in retentate (%)
H_2	70.0	5.8	70.0	0.0
CO_2	11.9	1.8	12.1	0.0
CO	14.3	78.0	13.4	85.0
CH_4	3.2	10.7	3.9	10.3
C_2H_6	0.6	3.7	0.6	4.7

Table 7.10   Calculation of one-stage gas separation by membrane at maximum preconcentration of H_2 (pressure drop 10/0.1 atm, feed flow rate 10 m^3(STP)/h). Reprinted, with permission, from Reference 104

Component	GENERON®, $\theta = 0.08$, wet sewage sludge pyrolysis gas		PVTMS, $\theta = 0.07$, biosyngas	
	Concentration in permeate (%)	Concentration in retentate (%)	Concentration in permeate (%)	Concentration in retentate (%)
H_2	90.0	42.0	84.0	35.3
CO_2	8.1	12.1	12.7	14.9
CH_4	0.01	1.2	0.3	1.1
N_2	0.1	3.6	0.1	5.5
H_2S	–	–	0.08	1.1
CO	0.7	22.2	1.5	42.1
C_2H_6	0.1	12	–	–
O_2	1	0.5	–	–

Although a decreasing of the H_2S fraction in permeate down to 0.3 vol.% is considered feasible, it can be recovered completely at subsequent PSA stage. Also it is not sufficient to recover only the hydrogen from feed gas as permeate, CO concentration in retentate increases as it is shown in Tables 7.8 and 7.9, so it can be used after preliminary treatment in chemical synthesis such as Fischer–Tropsch process or syngas production.

Nevertheless, presented calculations predict that H_2 concentration can be increased from 35 up to 70 vol.% by application of commercially available membranes.

7.3.2   Membrane contactors for hydrogen recovery from gaseous mixtures of bio-origin

As was mentioned above, the problem of H_2/CO_2 separation (purification of bio-hydrogen) is difficult to solve by standard membrane technology due to the low selectivity of membranes for this pair of gases. Nevertheless, application of gas–liquid membrane contactors (GLMCs) allows one to overcome this problem [139–141]. GLMC is a device in which mass exchange between gas and liquid phases takes place via a membrane. Such a combination of absorption and membrane separation techniques unites advantages of both methods and provides following features: possibility of application of wide range of industrial CO_2 absorbents (as physical as chemical); high selectivity of CO_2-containing gas mixtures separation; determined and constant geometry of mass exchange area; very wide range of possible velocities of gas and liquid streams; independence of gas and liquid streams from each other; high specific area of mass exchange; independence of gravity force direction (free orientation of module). Main drawback of GLMCs is additional mass-transfer resistance due to the presence of a membrane. Application of porous membranes that demonstrate low mass-transfer resistance in GLMCs is possible only in some particular cases. These GLMCs require accurate control of trans-membrane pressure drop in order to avoid formation of bubbles in liquid or penetration of liquid into membrane pores that leads to dramatic rising of mass-transfer resistance in membrane. Some studies have shown that liquid penetrates into membrane pores during the time even if proper trans-membrane pressure is maintained. Therefore, application of nonporous membranes seems to be much more prospective.

Nonporous membranes demonstrate much higher mass-transfer resistance compared to porous membranes. Nevertheless, starting from a certain level of membrane permeance, the dominant contribution of mass-transfer resistance becomes related to liquid phase. Therefore, in practical cases permeance of nonporous membranes could be enough high for its application. Wide investigation and application of GLMCs based on nonporous membranes is limited at present time due to the absence of commercially available high permeable nonporous membranes that additionally demonstrate long-time stability in contact with industrial CO_2 absorbents.

The application of GLMCs with nonporous membranes based on poly(vinyltrimethylsilane) (produced by Kuskovo Chemical Factory, Russia) and aqueous

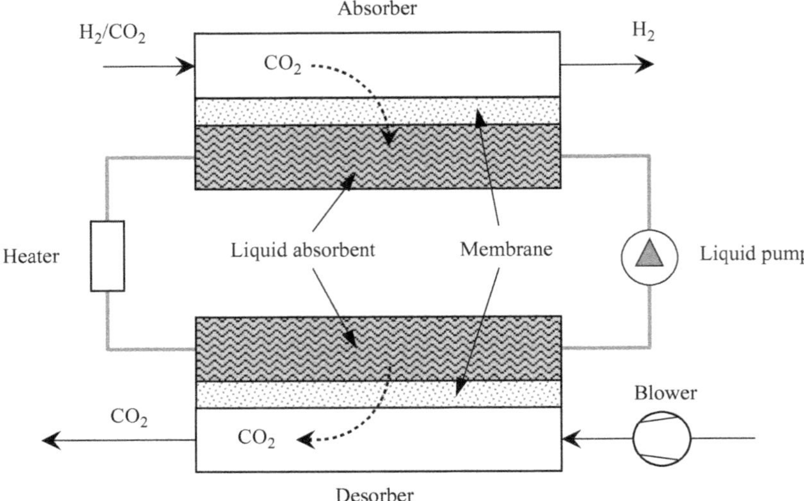

Figure 7.5   Biohydrogen separation by gas–liquid membrane contactor system. Reprinted, with permission, from Reference 104

potassium carbonate as CO_2 absorbent was demonstrated in pilot-scale biogas separation process [88] and lab-scale biohydrogen separation [142]. Mathematical model of mass-transfer in GLMC with chemical absorbent of CO_2 was developed in Reference 143; obtained results of theoretical calculations were in good agreement with experimental data.

Purification of biohydrogen by separation system based on GLMCs (Figure 7.5) with nonporous membranes and chemical CO_2 absorbent in liquid phase (aqueous potassium carbonate) was studied theoretically using the model developed in Reference 143.

Separation characteristics of the system were investigated by variation of such operating parameters as H_2 concentration in feed (60–80 vol.%), relative velocity of feed stream, relative velocity of liquid absorbent, permeance of membrane (0.34–3.6 m^3(STP)/(m^2 h atm)). Results of theoretical study of H_2/CO_2 gas mixture separation show the possibility of effective hydrogen purification; such system provides high separation factors and low losses of hydrogen.

Obtained dependencies of hydrogen concentration in product stream on velocity of liquid absorbent are nonlinear and have extrema (Figure 7.6). Such behavior provides easy optimization of separation process for particular composition of feed mixture. Possibility of the system characteristics adjustment is especially important for the application in biohydrogen purification since bioreactor productivity and composition of feed gas may vary during the time.

The influence of membrane permeance is shown in Figure 7.7. Concentration of H_2 in product stream increases strongly when CO_2 membrane permeance rises from 0.34 up to approximately 1.5 m^3(STP)/(m^2 h atm); after this value weak

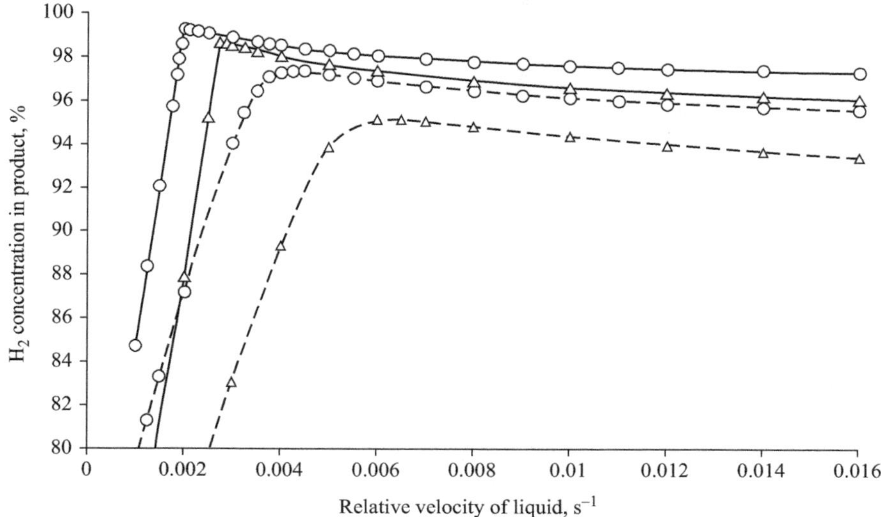

*Figure 7.6 Concentration of H_2 in product stream at membrane permeance
2.7 m^3(STP)/(m^2 h atm) depending on velocity of liquid, initial
concentration of H_2 in feed ($\circ$, 70%; $\triangle$, 60%), and relative
velocity of feed stream (——, 0.1 s^{-1}; - - -, 0.15 s^{-1}). Reprinted,
with permission, from Reference 104*

increase of H_2 concentration is observed. It means that mass-transfer resistance in membrane is dominant in the initial region and becomes comparable to resistance in liquid phase in the second region. These results allow us to establish the value of 1.5 m^3(STP)/(m^2 h atm) as sufficient level of membrane permeance for realization of biohydrogen purification process.

Industrial nonporous membranes based on polyvinyltrimethylsilane (permeance around 2.7 m^3(STP)/(m^2 h atm) for H_2 and CO_2) were used for assembling of laboratory scale GLMCs. Mass-transfer characteristics of the modules were experimentally investigated and found to be in good agreement with results of modeling.

7.3.3   Combined membrane systems for hydrogen recovery from gaseous mixtures of technogenic and bio-origin

Combined membrane systems for hydrogen recovery from gaseous mixtures is known enough long time ago [144], but for variety of hydrogen-containing mixtures, it continues to upgrowth. Combined unit including membrane and PSA blocks for separation of multicomponent gaseous mixture separation is described in References 125 and 126. This unit consists of membrane modules (hollow fiber type and disk-type) and lab-scale PSA block (volume of adsorbent about 1 dm^3) that assigned for study of membrane and PSA combination for separation aims. Simple scheme of this unit with example of combination for biosyngas separation is presented in Figure 7.8.

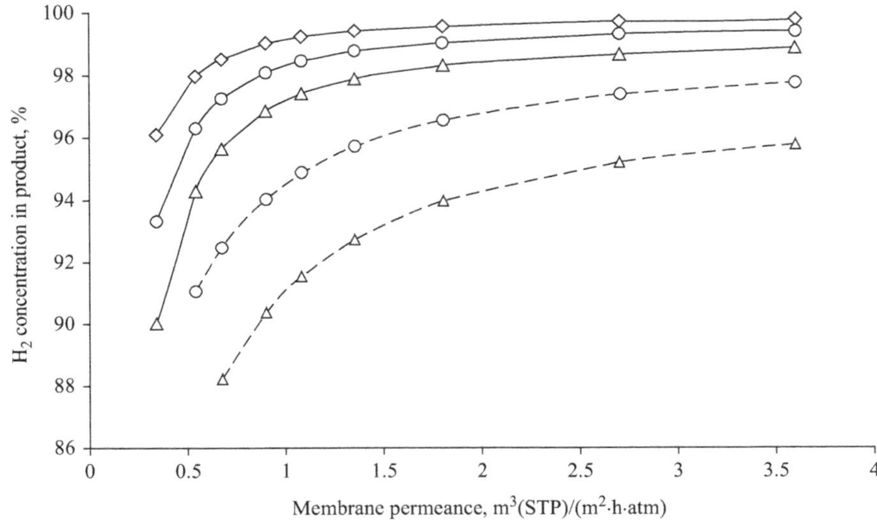

Figure 7.7   Concentration of H_2 in product stream at optimal velocity of liquid depending on CO_2 membrane permeance, initial concentration of H_2 in feed (◊, 80%; ○, 70%; Δ, 60%), and relative velocity of feed stream (——, $0.1\ s^{-1}$; - - -, $0.15\ s^{-1}$). Reprinted, with permission, from Reference 104

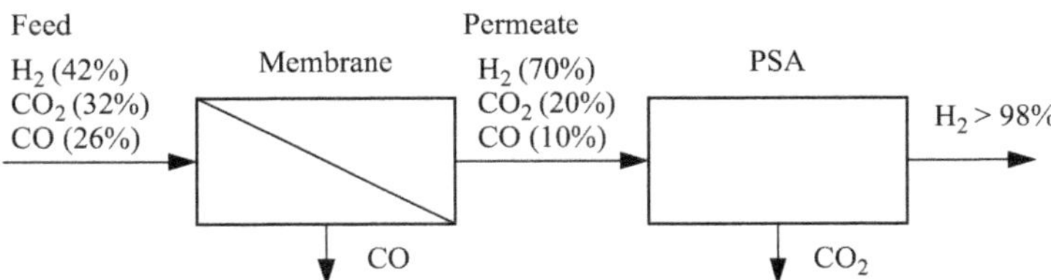

Figure 7.8   Scheme of combined membrane/PSA method for H_2 recovery with calculated values of streams composition. Reprinted, with permission, from Reference 104

Scheme in Figure 7.8 was tested by using model mixture $He/CO_2/O_2$—42.5%/32%/25.5%, in which He was used instead of H_2 for safety reasons and O_2 instead of CO. Product gas flow rate was 45.6 L/h. Gas pressure was varied in the range of 4–10 bar in membrane block and 4–7 bar in PSA block, comparative data are presented in Table 7.11.

As can be seen from Table 7.11, the combined membrane/PSA system provides high quality product in comparison with separation by single processes (PSA or membrane). It was estimated that hydrogen recovery degree from biosyngas may achieve 80%–97% in the case of hydrogen preconcentration up to 70 vol.% by

Table 7.11　Separation of model gas mixture by different methods. Reprinted, with permission, from Reference 104

Stream	Composition of gas mixture (vol.%)		
	O_2	CO_2	He
Feed	25	32	43
Product gas after membrane separation alone	10	29	61
	10	21	69
Product gas after PSA separation alone (without membrane separation)	11	0.07	89
Product gas after combined membrane/PSA separation	1.8	~0	98

membrane stage. As a result, 99.9 vol.% purity hydrogen can be obtained after PSA stage. Obtained results demonstrate that the investigation of effective recovery of H_2 needs to be focused on improving membrane selectivity and optimization of PSA step. It is known that at present time, the industrial PSA contains 4–16 adsorbed columns with 2–4 layers of different adsorbents. Proposed scheme can help to simplify these processes by introducing membrane block for preliminary concentrating of desired product and decreasing of undesirable impurities to provide the durability of process itself and additionally decrease the pressure down to 10 bar and less. It should be noted that successful results demonstrated by using commercial membranes can be improved by development of membranes with higher selectivity and more efficient adsorbents.

7.4　Prospects of commercial membranes application for biohydrogen recovery

The authors made an attempt to outline main activity in development of energy saving technologies for recovery of hydrogen from gaseous mixtures of bio-origin. It is shown that in this case different hydrogen-containing gaseous compositions are obtaining under low pressure and "soft" temperature region. Membrane technology gives the chance to solve problems of hydrogen production from biomass as energy carrier. Specific requirements to membranes are traditional [145]: high productivity and high selectivity in relation to H_2. It is estimated that the prospective polymers for this aim need to have hydrogen permeability >1,000 Barrer and selectivity for the pair H_2/CO more than 100 and for the pair H_2/CO_2 more than 10 at least. It can be done, for example, by using super-permeable glassy-like di-substituted polyacetylenes with modification providing the strong increasing of required separation selectivity. Additionally, further development of hybrid technology (combination of standard membrane processes with membrane contactors and/or PSA) can provide realization of biomass treatment as source of hydrogen as energy carrier.

7.5   Conclusion and summary

The authors tried to consider the potential of hydrogen production from biomass as renewable source. Three platforms of biofuels starting materials for hydrogen production are considered: the first one includes sugars and organic acids; the second one includes lignocellulose, woodchips, etc.; the third approach concerns the possible routes of thermochemical biomass conversion. In all cases, hydrogen needs to be separated (recovered) from the hydrogen-containing multicomponent gaseous mixtures. Membrane gas separation processes as lower energy consuming ones for H_2 recovery from gaseous sources demonstrate positive application of (1) commercial and lab-scale polymeric membranes for recovery of H_2 from gaseous mixtures containing additionally CO_2, CO, N_2, and CH_4; (2) membrane contactors for H_2/CO_2 gas mixtures separation; (3) combined membrane/PSA systems for effective hydrogen recovery from multicomponent gaseous mixtures. Combination of membrane (for H_2 preconcentrating) and PSA (for H_2 purification) separation systems demonstrates possibility of achieving high hydrogen recovery degree (up to 80%–97%) from biosyngas and production of 99.9 vol.% purity hydrogen. Obtained data correspond to well-known membranes and adsorbents and can be considerably improved by using of new membrane materials and adsorbents.

The presented paper shows the need in multidisciplinary studies combining the specialists in the field of microbiology, membrane science, and catalytic transformations of gaseous mixtures into valuable products.

Acknowledgments

The authors thank Dr. Olga Amosova for her help in preparation of this chapter. This work was carried out in the frames of State Program. The biotechnological part of this work was supported in part by grant No. 16-14-00098 of Russian Science Foundation.

List of abbreviations

5-HMF	5-(hydroxymethyl)furfural
ABE	acetone–butanol–ethanol fermentation
CBU	cellobiase unit
CW	cornstalk waste
DT	disk-type membrane module
EGU	endoglucanase unit
EROI	energy returned on energy invested
FPU	filter paper unit
GLMC	gas–liquid membrane contactors

HF	hollow fiber membrane module
IU	international unit
NA	data are not available
PD	polymers dissolution
PSA	pressure-swing adsorption
PVTMS	polyvinyltrimethylsilane
SD	solid dissolution
SER	specific energy requirement
SSF	simultaneous saccharification and fermentation
SW	spiral wound membrane module
TIPS RAS	A.V. Topchiev Institute of Petrochemical Synthesis
USSR	The Union of Soviet Socialist Republics

References

[1] Hoffman P. *Tomorrow's Energy: Hydrogen, Fuel Cells, and the Prospects for a Cleaner Planet*. Cambridge, MA: Massachusetts Institute of Technology; 2012, 355 p.

[2] Tarasov B.P., Lototskii M.V. 'Hydrogen for energy production: problems and perspectives'. *International Scientific Journal for Alternative Energy and Ecology*. 2006;**8**(40):72–90 (in Russian).

[3] Separation Technology R&D Needs for Hydrogen Production in the Chemical and Petrochemical Industries. Report of Chemical Industry Vision2020 Technology Partnership. 2005. 74 p.

[4] Ritter J.A., Ebner A.D. 'State-of-the-art adsorption and membrane separation processes for hydrogen production in the chemical and petrochemical industries'. *Separation Science and Technology*. 2007;**42**(6):1123–1193.

[5] Tarasov B.P., Lototskii M.V. 'Hydrogen energetics: past, present, view to the future'. *Russian Chemical Journal*. 2006;**L**(6):5–18 (in Russian).

[6] Baker R.W. *Membrane Technology and Application*, 2nd ed. New York, NY: John Wiley & Sons Ltd.; 2004, 538 p.

[7] Jones D.T., Woods D.R. 'Acetone–butanol fermentation revisited'. *Microbiological Reviews*. 1986;**50**(4):484–524.

[8] Wang L., Chen H.Z. 'Increased fermentability of enzymatically hydrolyzed steam-exploded corn stover for butanol production by removal of fermentation inhibitors'. *Process Biochemistry*. 2011;**46**:604–607.

[9] Bai D., Li S., Liu Z.C., Cui Z. 'Enhanced L-(+)-lactic acid production by an adapted strain of *Rhizopus oryzae* using corncob hydrolysate'. *Applied Biochemistry and Biotechnology*. 2008;**144**(1):79–85.

[10] Perez J., Munoz-Dorado J., de la Rubia T., Martinez J. 'Biodegradation and biological treatments of cellulose, hemicellulose and lignin: an overview'. *International Microbiology*. 2002;**5**:53–63.

[11]   da Costa Sousa L., Chundawat S.P., Balan V., Dale B.E. '"Cradle-to-grave" assessment of existing lignocellulose pretreatment technologies'. *Current Opinion Biotechnology*. 2009;**20**(3):339–347.

[12]   Mafuleka S., Kana E.B.G. 'Modelling and optimization of xylose and glucose production from Napier grass using hybrid pre-treatment techniques'. *Biomass and Bioenergy*. 2015;**20**:200–208.

[13]   Hong E., Kim D., Kim J., *et al.* 'Optimization of alkaline pretreatment on corn stover for enhanced production of 1,3-propanediol and 2,3-butanediol by *Klebsiella pneumoniae* AJ4'. *Biomass and Bioenergy*. 2015;**77**:177–185.

[14]   Barakat A., Monlau F., Solhy A., Carrere H. 'Mechanical dissociation and fragmentation of lignocellulosic biomass: effect of initial moisture, biochemical and structural proprieties on energy requirement'. *Applied Energy*. 2015;**142**:240–246.

[15]   Liu S. 'A synergetic pretreatment technology for woody biomass conversion'. *Applied Energy*. 2015;**144**:114–128.

[16]   Motte J.-C., Sambusiti C., Dumas C., Barakat A. 'Combination of dry dark fermentation and mechanical pretreatment for lignocellulosic deconstruction: an innovative strategy for biofuels and volatile fatty acids recovery'. *Applied Energy*. 2015;**147**:67–73.

[17]   Mani S., Tabil L.G., Sokhansanj S. 'Grinding performance and physical properties of wheat and barley straws, corn stover and switchgrass'. *Biomass and Bioenergy*. 2004;**27**:339–352.

[18]   Barakat A., Mayer-Laigle C., Solhy A., Arancon R.A.D., de Vries H., Luque R. 'Mechanical pretreatments of lignocellulosic biomass: towards facile and environmentally sound technologies for biofuels production'. *RSC Advances*. 2014;**4**:48109–48127.

[19]   Theuretzbacher F., Lizasoain J., Lefever C., *et al.* 'Steam explosion pretreatment of wheat straw to improve methane yields: investigation of the degradation kinetics of structural compounds during anaerobic digestion'. *Bioresource Technology*. 2015;**179**:299–305.

[20]   Nakasaki K., Mimoto H., Tran Q.N.M., Oinuma A. 'Composting of food waste subjected to hydrothermal pretreatment and inoculated with *Paecilomyces* sp. FA13'. *Bioresource Technology*. 2015;**180**:40–46.

[21]   Hick S.M., Griebel C., Restrepo D.T., *et al.* 'Mechanocatalysis for biomass-derived chemicals and fuels'. *Green Chemistry*. 2010;**12**(3):468–474.

[22]   Bobleter O. 'Hydrothermal degradation of polymers derived from plants'. *Progress in Polymer Science*. 1994;**19**:797–841.

[23]   Gregg D.J., Saddler J.N. 'Factors affecting cellulose hydrolysis and the potential of enzyme recycle to enhance the efficiency of an integrated wood to ethanol process'. *Biotechnology and Bioengineering*. 1996;**51**(4):375–383.

[24]   Zhu Y., Lee Y.Y., Elander R.T. 'Optimization of dilute-acid pretreatment of corn stover using a high-solids percolation reactor'. *Applied Biochemistry and Biotechnology*. 2005;**121–124**:1045–1054.

[25]   Ramos L.P. 'The chemistry involved in the steam treatment of lignocellulosic materials'. *Química Nova*. 2003;**26**(6):863–871.

[26] Brownell H.H., Yu E.K.C., Saddler J.N. 'Steam-explosion pretreatment of wood: effect of chip size, acid, moisture content and pressure drop'. *Biotechnology Bioengineering*. 1986;**28**:792–801.

[27] Kärcher M.A., Iqbal Y., Lewandowski I., Senn T. 'Comparing the performance of *Miscanthus* × *giganteus* and wheat straw biomass in sulfuric acid based pretreatment'. *Bioresource Technology*. 2015;**180**:360–364.

[28] Angelidaki I., Ellegaard L., Ahring B. 'Applications of the anaerobic digestion process'. *Advances in Biochemical Engineering/Biotechnology*. 2003;**82**:1–34.

[29] Jiang H., Han B., Ge J. 'Enhancement in the enzymatic digestibility of hybrid poplar with poor residual hemicelluloses after Na_2SO_3 pretreatment'. *Bioresource Technology*. 2015;**180**:338–344.

[30] Thomsen M., Thygesen A., Thomsen A. 'Hydrothermal treatment of wheat straw at pilot plant scale using a three-step reactor system aiming at high hemicellulose recovery, high cellulose digestibility and low lignin hydrolysis'. *Bioresource Technology*. 2008;**99**(10):4221–4228.

[31] Laser M., Schulman D., Allen S.G., Lichwa J., Antal M.J., Lynd L.R. 'A comparison of liquid hot water and steam pretreatments of sugar cane bagasse for bioconversion to ethanol'. *Bioresource Technology*. 2002;**81**(1): 33–44.

[32] Tye Y.Y., Lee K.N., Abdullah W.N.W., Leh C.P. 'Effects of process parameters of various pretreatments on enzymatic hydrolysability of *Ceiba pentandra* (L.) Gaertn. (Kapok) fibre: a response surface methodology study'. *Biomass and Bioenergy*. 2015;**75**:301–313.

[33] Mosier N., Hendrickson R., Ho N., Sedlak M., Ladisch M.R. 'Optimization of pH controlled liquid hot water pretreatment of corn stover'. *Bioresource Technology*. 2005;**18**:1986–1993.

[34] Chheda J.N., Román-Leshkov Y., Dumesic J.A. 'Production of 5-hydroxymethylfurfural and furfural by dehydration of biomass-derived mono- and poly-saccharides'. *Green Chemistry*. 2007;**9**:342–350.

[35] Gandolfi S., Ottolina G., Consonni R., Riva S., Patel I. 'Fractionation of hemp hurds by organosolv pretreatment and its effect on production of lignin and sugars'. *ChemSusChem*. 2014;**7**(7):1991–1999.

[36] Nguyen Q.A., Tucker M.P., Boynton B.L., Keller F.A., Schell D.J. 'Dilute acid pretreatment of softwoods'. *Applied Biochemistry and Biotechnology*. 1998;**70–72**(1):77–87.

[37] Wyman C.E., Dale B.E., Elander R.T., *et al.* 'Comparative sugar recovery and fermentation data following pretreatment of poplar wood by leading technologies'. *Biotechnology Progress*. 2009;**25**(2):333–339.

[38] Panagiotopoulos I.A., Bakker R.R., Budde M.A.W., Vrije T., Claassen P.A. M., Koukios E.G. 'Fermentative hydrogen production from pretreated biomass: a comparative study'. *Bioresource Technology*. 2009;**100**: 6331–6338.

[39] Vrije T., Bakker R.R., Budde M.A.W, Lai M.H., Mars A.E., Claassen P.A.M. 'Efficient hydrogen production from the lignocellulosic energy crop

Miscanthus by the extreme thermophilic bacteria *Caldicellulosiruptor saccharolyticus* and *Thermotoga neapolitana*'. *Biotechnology for Biofuels.* 2009;**2**:12.

[40] Jensen J.R., Morinelly J.E., Gossen K.R., Brodeur-Campbell M.J., Shonnard D.R. 'Effects of dilute acid pretreatment conditions on enzymatic hydrolysis monomer and oligomer sugar yields for aspen, balsam, and switch grass'. *Bioresource Technology.* 2010;**101**(7):2317–2325.

[41] Shi J., Ebrik M.A., Wyman C.E. 'Sugar yields from dilute sulfuric acid and sulfur dioxide pretreatments and subsequent enzymatic hydrolysis of switchgrass'. *Bioresource Technology.* 2011;**102**(19):8930–8938.

[42] Wei W., Wu S., Liu L. 'Enzymatic saccharification of dilute acid pretreated eucalyptus chips for fermentable sugar production'. *Bioresource Technology.* 2012;**110**:302–307.

[43] Abramov S.M., Sadraddinova E.R., Shestakov A.I., *et al.* 'Turning cellulose waste into electricity: hydrogen conversion by a hydrogenase electrode'. *PLoS ONE.* 2013;**8**(11):e83004.

[44] Kim Y., Ximenes E., Mosier N.S., Ladisch M.R. 'Soluble inhibitors deactivators of cellulase enzymes from lignocellulosic biomass'. *Enzyme and Microbial Technology.* 2011;**48**:408–415.

[45] Michalska K., Bizukoj M., Ledakowicz S. 'Pretreatment of energy crops with sodium hydroxide and cellulolytic enzymes to increase biogas production'. *Biomass and Bioenergy.* 2015;**80**:213–221.

[46] Camesasca L., Ramírez M.B., Mairan G., Ferrari M.D., Lareo C. 'Evaluation of dilute acid and alkaline pretreatments, enzymatic hydrolysis and fermentation of napiergrass for fuel ethanol production'. *Biomass and Bioenergy.* 2015;**74**:193–201.

[47] Bali G., Meng X., Deneff J.I., Qining S., Ragauskas A.J. 'The effect of alkaline pretreatment methods on cellulose structure and accessibility'. *ChemSusChem.* 2015;**8**:275–279.

[48] Bobleter O. 'Hydrothermal degradation of polymers derived from plants'. *Progress in Polymer Science.* 1994;**19**:797–841.

[49] Gregg D.J., Saddler J.N. 'Factors affecting cellulose hydrolysis and the potential of enzyme recycle to enhance the efficiency of an integrated wood to ethanol process'. *Biotechnology and Bioengineering.* 1996;**51**(4):375–383.

[50] Pettersen R.C. 'The chemical composition of wood' (chapter 2), in: R.M. Rowell (Ed.), *The Chemistry of Solid Wood*, Advances in Chemistry Series. Washington, DC: American Chemical Society; 1984. vol. 207, pp. 57–126.

[51] Sendich E., Laser M., Kim S., *et al.* 'Recent process improvements for the ammonia fiber expansion (AFEX) process and resulting reductions in minimum ethanol selling price'. *Bioresource Technology.* 2008;**99**:8429–8435.

[52] Weimer P.J., Chou Y.-C.T., Weston W.M., Chase D.B. (Eds.). 'Effect of supercritical ammonia on the physical and chemical structure of ground wood'. *Biotechnology and Bioengineering Symposium*; Gatlinburg, TN, USA, Jan 1986 (vol. **17**, pp. 5–18, no. CONF-860508; Journal ID: CODEN: BIBSB).

[53]   Kim S.B., Lee Y.Y. 'Fractionation of herbaceous biomass by ammonia–hydrogen peroxide percolation treatment'. *Applied Biochemistry and Biotechnology*. 1996;**57/58**:147–156.

[54]   Chen Y., Cheng J.J., Creamer K.S. 'Inhibition of anaerobic digestion process: a review'. *Bioresource Technology*. 2008;**99**:4044–4064.

[55]   Rabelo S.C., Filho R.M., Costa A.C. 'A comparison between lime and alkaline hydrogen peroxide pretreatments of sugarcane bagasse for ethanol production'. *Applied Biochemistry and Biotechnology*. 2008;**144**:87–100.

[56]   Erden G., Filibeli A. 'Improving anaerobic biodegradability of biological sludges by Fenton pre-treatment: effects on single stage and two-stage anaerobic digestion'. *Desalination*. 2010;**251**:58–63.

[57]   Gould J.M., Freer S.N. 'High-efficiency ethanol production from lignocellulosic residues pretreated with alkaline H_2O_2'. *Biotechnology and Bioengineering*. 1984;**26**(6):628–631.

[58]   Yin D.T., Jing Q., AlDajani W.W., *et al*. 'Improved pretreatment of lignocellulosic biomass using enzymatically-generated peracetic acid'. *Bioresource Technology*. 2011;**102**(8):5183–5192.

[59]   Swatloski R.P., Spear S.K., Holbrey J.D., Rogers R.D. 'Dissolution of cellulose with ionic liquids'. *Journal of the American Chemical Society*. 2002;**124**:4974–4975.

[60]   Zhao H., Holladay J.E., Brown H., Zhang Z.C. 'Metal chlorides in ionic liquid solvents convert sugars to 5-hydroxymethylfurfural'. *Science*. 2007;**316**(5831):1597–1600.

[61]   Binder J.B., Raines R.T. 'Simple chemical transformation of lignocellulosic biomass into furans for fuels and chemicals'. *Journal of American Chemical Society*. 2009;**131**(5):1979–1985.

[62]   Villandier N., Corma A. 'One pot catalytic conversion of cellulose into bio-degradable surfactants'. *Chemical Communications*. 2010;**46**(24):4408–4410.

[63]   Meine N., Rinaldi R., Schüth F. 'Solvent-free, catalytic depolymerisation of cellulose to water-soluble oligosaccharides'. *ChemSusChem*. 2012;**5**(8): 1449–1454.

[64]   Moreno A.D., Ibarra D., Alvira P., Tomas-Pejo E., Ballesteros M. 'A review of biological delignification and detoxification methods for lignocellulosic bioethanol production'. *Critical Reviews in Biotechnology*. 2015;**35**(3): 342–354.

[65]   Taylor C.R., Hardiman E.M., Ahmad M., Sainsbury P.D., Norris P.R., Bugg T.D. 'Isolation of bacterial strains able to metabolize lignin from screening of environmental samples'. *Journal of Applied Microbiology*. 2012;**113**(3): 521–530.

[66]   Cook C., Francocci F., Cervone F., *et al*. 'Combination of pretreatment with white rot fungi and modification of primary and secondary cell walls improves saccharification'. *Bioenergy Research*. 2015;**8**:175–186.

[67]   Jørgensen H., Kristensen J.B., Felby C. 'Enzymatic conversion of lignocellulose into fermentable sugars: challenges and opportunities'. *Biofuels Bioproducts and Biorefining*. 2007;**1**:119–134.

[68] Van Dyk J., Pletschke B. 'A review of lignocellulose bioconversion using enzymatic hydrolysis and synergistic cooperation between enzymes: factors affecting enzymes, conversion and synergy'. *Biotechnology Advances.* 2012;**30**(6):1458–1480.

[69] Obata O., Akunna J., Walker G. 'Hydrolytic effects of acid and enzymatic pre-treatment on the anaerobic biodegradability of *Ascophyllum nodosum* and *Laminaria digitata* species of brown seaweed'. *Biomass and Bioenergy.* 2015;**80**:140–146.

[70] Luterbacher J.S., Moran-Mirabal J.M., Burkholder E.W., Walker L.P. 'Modeling enzymatic hydrolysis of lignocellulosic substrates using confocal fluorescence microscopy I: filter paper cellulose'. *Biotechnology and Bioengineering.* 2015;**112**(1):21–31.

[71] Palonen H., Thomsen A.B., Tenkanen M., Schmidt A.S., Viikari L. 'Evaluation of wet oxidation pretreatment for enzymatic hydrolysis of softwood'. *Applied Biochemistry and Biotechnology.* 2004;**117**(1):1–17.

[72] Puri V.P. 'Effect of crystallinity and degree of polymerization of cellulose on enzymatic saccharification'. *Biotechnology and Bioengineering.* 1984;**26**(10):1219–1222.

[73] Koullas D.P., Christakopoulos P., Kekos D., Macris B.J., Koukios E.G. 'Correlating the effect of pretreatment on the enzymatic hydrolysis of straw'. *Biotechnology and Bioengineering.* 1992;**39**(1):113–116.

[74] Chang V.S., Holtzapple M.T. 'Fundamental factors affecting biomass enzymatic reactivity'. *Applied Biochemistry and Biotechnology.* 2000;**84–86**: 5–37.

[75] Laureano-Perez L., Teymouri F., Alizadeh H., Dale B.E. 'Understanding factors that limit enzymatic hydrolysis of biomass'. *Applied Biochemistry and Biotechnology.* 2005;**121–124**:1081–1099.

[76] Shinozaki Y., Kitamoto H.K. 'Ethanol production from ensiled rice straw and whole-crop silage by the simultaneous enzymatic saccharification and fermentation process'. *Journal of Bioscience and Bioengineering.* 2011;**111**(3): 320–325.

[77] Morales-Rodriguez R., Meyer A.S., Gernaey K.V., Sin G. 'A framework for model-based optimization of bioprocesses under uncertainty: lignocellulosic ethanol production case'. *Computers and Chemical Engineering.* 2012; **42**:115–129.

[78] Knutsen J.S., Davis R.H. 'Cellulase retention and sugar removal by membrane ultrafiltration during lignocellulosic biomass hydrolysis'. *Applied Biochemistry and Biotechnology.* 2004;**113–116**:585–599.

[79] Andrić P., Meyer A.S., Jensen P.A., Dam-Johansen K. 'Reactor design for minimizing product inhibition during enzymatic lignocellulose hydrolysis II. Quantification of inhibition and suitability of membrane reactors.' *Biotechnology Advances.* 2010;**28**:407–425.

[80] Yang J., Zhang X., Yong Q., Yu S. 'Three-stage hydrolysis to enhance enzymatic saccharification of steam-exploded corn stover'. *Bioresource Technology* 2010;**101**:4930–4935.

[81] Tu M., Zhang X., Paice M., MacFarlane P., Saddler J.N. 'The potential of enzyme recycling during the hydrolysis of a mixed softwood feedstock'. *Bioresource Technology*. 2009; **100**(24):6407–6415.

[82] Rosgaard L., Andric P., Dam-Johansen K., Pedersen S., Meyer A.S. 'Effects of substrate loading on enzymatic hydrolysis and viscosity of pretreated barley straw'. *Applied Biochemistry and Biotechnology*. 2007;**143**: 27–40.

[83] Zhu S., Huang W., Huang W., Wang K., Chen Q., Wu Y. 'Pretreatment of rice straw for ethanol production by a two-step process using dilute sulfuric acid and sulfomethylation reagent'. *Applied Energy*. 2015;**154**:190–196.

[84] Jung Y.H., Kim H.K., Park H.M., *et al.* 'Mimicking the Fenton reaction-induced wood decay by fungi for pretreatment of lignocellulose'. *Bioresource Technology*. 2015;**179**:467–472.

[85] Gao F., Gao L., Zhang D., Ye N., Chen S., Li D. 'Enhanced hydrolysis of *Macrocystis pyrifera* by integrated hydroxyl radicals and hot water pretreatment'. *Bioresource Technology*. 2015;**179**:490–496.

[86] Chang V.S., Nagwani M., Kim C.H., Holtzapple M.T. 'Oxidative lime pretreatment of high-lignin biomass: poplar wood and newspaper'. *Applied Biochemistry and Biotechnology*. 2001;**94**(1):1–28.

[87] Vasco C.A., Guo M., Zhang X. 'Dilute acid pretreatment of Douglas fir forest residues: pretreatment yield, hemicellulose degradation, and enzymatic hydrolysability'. *Bioenergy Research*. 2015;**8**:42–52.

[88] Cheng J., Lin R., Ding L., *et al.* 'Fermentative hydrogen and methane cogeneration from cassava residues: effect of pretreatment on structural characterization and fermentation performance'. *Bioresource Technology*. 2015;**179**:407–413.

[89] Chen Y., Stevens M.A., Zhu Y., Holmes J., Moxley G., Xu H. 'Reducing acid in dilute acid pretreatment and the impact on enzymatic saccharification'. *Journal of Industrial Microbiology and Biotechnology*. 2012;**39**(5): 691–700.

[90] Panagiotopoulos I.A., Bakker R.R., de Vrije T., Koukios E.G. 'Effect of pretreatment severity on the conversion of barley straw to fermentable substrates and the release of inhibitory compounds'. *Bioresource Technology*. 2011;**102**(24):11204–11211.

[91] Palmqvist E., Hahn-Hagerdal B. 'Fermentation of lignocellulosic hydrolysates. I: inhibition and detoxification'. *Bioresource Technology*. 2000;**74**:17–24.

[92] Purwadi R., Niklasson C., Taherzadeh M.J. 'Kinetic study of detoxification of dilute-acid hydrolyzates by $Ca(OH)_2$'. *Journal of Biotechnology*. 2004;**114**:187–198.

[93] Kim S., Holtzapple M.T. 'Lime pretreatment and enzymatic hydrolysis of corn stover'. *Bioresource Technology*. 2005;**96**(18):1994–2006.

[94] Kumar R., Wyman C.E. 'Effects of cellulase and xylanase enzymes on the deconstruction of solids from pretreatment of poplar by leading technologies'. *Biotechnology Progress*. 2009;**25**:302–314.

[95]   Mafuleka S., Kana E.B.G. 'Modelling and optimization of xylose and glucose production from Napier grass using hybrid pre-treatment techniques'. *Biomass and Bioenergy* 2015;**20**:200–208.

[96]   Claassen P.A.M., de Vrije T. 'Non-thermal production of pure hydrogen from biomass: HYVOLUTION'. *International Journal of Hydrogen Energy*. 2006;**31**(11):1416–1423.

[97]   Hallenbeck P.C. 'Fermentative hydrogen production: principles, progress, and prognosis'. *International Journal of Hydrogen Energy*. 2009;**34**(17): 7379–7389.

[98]   Azbar N., Levin D.B. *State of the Art and Progress in Production of Bio-hydrogen (e-book)*. Sharjah: Bentham Science Publishers; 2012. Available from benthamscience.com/ebooks

[99]   Li Q., Liu C.-Z. 'Co-culture of *Clostridium thermocellum* and *Clostridium thermosaccharolyticum* for enhancing hydrogen production via thermophilic fermentation of cornstalk waste'. *International Journal of Hydrogen Energy*. 2012;**37**(14):10648–10654.

[100] Verhaart M.R.A., Bielen A.A.M., van der Oost J., Stams A.J.M., Kengen S.W.M. 'Hydrogen production by hyperthermophilic and extremely thermophilic bacteria and archaea: mechanisms for reductant disposal'. *Environmental Technology*. 2010;**31**(8–9):993–1003.

[101] Modigell M., Holle N. 'Reactor development for a biosolar hydrogen production process'. *Renewable Energy*. 1998;**14**(1/4):421–426.

[102] Netrusov A., Abramov S., Sadraddinova E., Shestakov A., Shalygin M., Teplyakov V. 'Membrane-assisted separation of microbial gaseous fuels from renewable sources'. *Desalination and Water Treatment*. 2010;**14**(1–3): 252–258.

[103] Gebicki J., Modigell M., Schumacher M., van der Burg J., Roebroeck E. 'Comparison of two reactor concepts for anoxygenic H_2 production by *Rhodobacter capsulatus*'. *Journal of Cleaner Production*. 2010;**18**(1):36–42.

[104] Teplyakov V.V., Shalygin M.G., Abramov S.M., Netrusov A.I. 'Membrane recovery of hydrogen from gaseous mixtures of biogenic and technogenic origin'. *International Journal of Hydrogen Energy*. 2015;**40**(8):3438–3451.

[105] Knoef H. *Handbook on Biomass Gasification*. Enschede: BTG Biomass Technology Group; 2012.

[106] International Energy Outlook: 2010. Washington, DC; 2010 July. 327 p. Report No.: DOE/EIA-0484(2010).

[107] Powell E.E., Hill G.A. 'Economic assessment of an integrated bioethanol-biodiesel-microbial fuel cell facility utilizing yeast and photosynthetic algae'. *Chemical Engineering Research and Design*. 2009;**87**(9):1340–1348.

[108] Curtiss, P.S., Kreider, J.F. 'Algaculture as a feedstock source for biodiesel fuel—a life cycle analysis'. In: *ES2009: Proceedings of the ASME Third International Conference on Energy Sustainability*, vol. 1. San Francisco, CA: ASME; 2009; pp. 171–179.

[109] Greenwell H.C., Laurens L.M.L., Shields R.J., Lovitt R.W., Flynn K.J. 'Placing microalgae on the biofuels priority list: a review of the

technological challenges'. *Journal of Royal Society Interface*. 2010;7(46): 703–726.

[110] Wijffels R.H., Barbosa M.J. 'An outlook on microalgal biofuels'. *Science*. 2010;**329**(5993):796–799.

[111] Stephenson A.L., Kazamia E., Dennis J.S., Howe C.J., Scott S.A., Smith A.G. 'Life-cycle assessment of potential algal biodiesel production in the United Kingdom: a comparison of raceways and air-lift tubular bioreactors'. *Energy & Fuels*. 2010;**24**(7):4062–4077.

[112] Greenwell H.C., Laurens L.M.L., Shields R.J., Lovitt R.W., Flynn K.J. 'Placing microalgae on the biofuels priority list: a review of the technological challenges'. *Journal of Royal Society Interface*. 2010;**7**(46):703–726.

[113] Chisti Y. 'Biodiesel from microalgae'. *Biotechnology Advances*. 2007;**25**(3): 294–306.

[114] Falkowski P., Scholes R.J., Boyle E., *et al*. 'The global carbon cycle: a test of our knowledge of earth as a system'. *Science*. 2000;**290**(5490):291–296.

[115] Bridgwater A.V. 'Renewable fuels and chemicals by thermal processing of biomass'. *Chemical Engineering Journal*. 2003;**91**(2–3):87–102.

[116] Bridgwater A.V. 'Review of fast pyrolysis of biomass and product upgrading'. *Biomass and Bioenergy*. 2012;**38**(3):68–94.

[117] Collard F., Blin J. 'A review on pyrolysis of biomass constituents: mechanisms and composition of the products obtained from the conversion of cellulose, hemicelluloses and lignin'. *Renewable and Sustainable Energy Reviews*. 2014;**38**(1):594–608.

[118] Daniell J., Köpke M., Simpson S. 'Commercial biomass syngas fermentation'. *Energies*. 2012;**5**(12):5372–5417.

[119] Xu X., Jiang E., Wang M., Xu Y. 'Dry and steam reforming of biomass pyrolysis gas for rich hydrogen gas'. *Biomass and Bioenergy*. 2015;**78**(1): 6–16.

[120] Cendrowska A. 'Hydrolysis kinetics of cellulose of forest and agricultural biomass'. *European Journal of Wood and Wood Products*. 1997;**55**(3): 195–196.

[121] Nickerson T.A., Hathaway B.J., Smith T.M., Davidson J.H. 'Economic assessment of solar and conventional biomass gasification technologies: financial and policy implications under feedstock and product gas price uncertainty'. *Biomass Bioenergy*. 2015;**74**:47–57.

[122] Zhang Y.H. 'Reviving the carbohydrate economy via multiproduct lignocellulose biorefineries'. *Journal of Industrial Microbiology and Biotechnology*. 2008;**35**(5):367–375.

[123] Baker Richard W., Lokhandwala Kaafid A. 'Process, including PSA and membrane separation, for separating hydrogen from hydrocarbons'. US Pat. 6,183,628 (2002).

[124] Rao Madhukar B., Sircar Shivaji, Abrardo Joseph M., Baade William F. 'Hydrogen recovery by adsorbent membranes'. US Pat. 5,447,559 (1994).

[125] Amosova O.L., Malykh O.V., Teplyakov V.V. 'Membrane-adsorption methods of recovery of hydrogen from multicomponent gas mixtures of

biotechnology and pertochemistry'. Membranes 2008; **38**(2):26–40 (in Russian).

[126] Amosova O.L., Malykh O.V., Teplyakov V.V. 'Integrated membrane/PSA systems for hydrogen recovery from gas mixtures'. *Desalination and water treatment*. 2010; **14**(1–3):119–125.

[127] Teplyakov V.V., Meares P. 'Correlation aspects of the selective gas permeabilities of polymeric materials and membranes'. *Gas Separation & Purification*, 1990; **4**(2):68–72.

[128] Mulder M. 'Basic principles of membrane technology'. Kluwer Academic Publishers: Dodrecht/Boston/London. 1996; p. 513.

[129] Lu G.Q., Diniz da Costa J.C., Duke M., Giessler S., Sokolow R., Williams R.H., Kreutz T. 'Inorganic membranes for hydrogen production and purification: A critical review and perspective'. *Journal of Colloid and Interface Science*. 2007; **314**:89–603.

[130] Ockwig N.W., Netoff T.M. 'Membranes for hydrogen separation'. *Chem. Rev.* 2007; **107**:4078–4110.

[131] Slovetskii D.I., Chistov E.M., Roshan N.R. 'Production of pure hydrogen'. *International Scientific Journal for Alternative Energy and Ecology*. 2004; **1**(9):43–46.

[132] Hellums M.W., Koros W.J., Husk G.R., Paul D.R. 'Gas transport in halogen-containing aromatic polycarbonates'. *J. Appl. Polym. Sci.* 1991; **43**:1977.

[133] Teplyakov V.V., Malykh O.V., Amosova O.L., Golub A.Yu. Yastrebov R.A. 'Data base "Functional Data Base on permeability parameters of permanent and acid gases, lower hydrocarbons, toxic gas impurities through polymeric materials and membranes with function of calculative estimation of values which data are absent"'. Certificate No 2011620549, dated 28 Jul 2011 (in Russian).

[134] Teplyakov V., Golub A., Malykh O. 'Polymeric membrane materials: new aspects of empirical approaches to prediction of gas permeability parameters in relation to permanent gases, linear lower hydrocarbons and some toxic gases'. HYPERLINK "http://www.sciencedirect.com/science/journal/00018686" Advances in Colloid and Interface Science. 2011; **164**(1–2): 89–99.

[135] Syrtsova D.A., Kharitonov A.P., Teplyakov V.V., Koops G.-H. 'Improving gas separation properties of polymeric membranes based on glassy polymers by gas phase fluorination'. *Desalination*. 2004; **163**:273–279.

[136] Hinchliffe A.B., Porter K.E. 'Gas Separation Using Membranes: 1. Optimization of the Separation Process Using New Cost Parameters'. *Ind. Eng. Chem. Res.* 1997; **36**(3):821–829.

[137] Marriott J., Sorensen E. 'A general approach to modeling membrane modules'. *Chemical Engineering Science*. 2003; **58**(22):4975–4990.

[138] Teplyakov V.V., Malykh O.V., Amosova O.L., Yastrebov R.A., 'Software: Estimation of membrane separation of multicomponents gaseous mixtures by using Data Base on membranes with function of prediction unknown

experimental data', Certificate No 2011615930, dated 28 Jul 2011 (in Russian).

[139] Teplyakov, V., Sostina, E., Beckman, I., Netrusov, A. 'Integrated membrane system for gas separation in biotechnology potential and prospects'. *World Journal of Microbiology and Biotechnology*. 1996; **12**:477–485.

[140] Gabelman, A., Hwang, S.-T. 'Hollow fiber membrane contractors'. *Journal of Membrane Science*. 1999; **159**(1):61–106.

[141] Schumacher, M., Modigell, M., Teplyakov, V.V., Zenkevich, V.B. 'A membrane contactor for efficient CO2 removal in biohydrogen production'. *Desalination*. 2008; **224**(1–3):186–190.

[142] Beggel F., Nowik I., Modigell M., Shalygin M., Teplyakov V., Zenkevitch V. 'A novel gas purification system for biologically produced gases'. *Journal of Cleaner Production*. 2010; **18**:S43–S50.

[143] Shalygin M.G., Yakovlev A.V., Khotimskii V.S., Gasanova L.G., Teplyakov V.V. 'Membrane contactors for biogas conditioning'. *Petroleum Chemistry*. 2011; **51**(8):601–609.

[144] Feng H., Pan C.Y., Ivory J., Ghosh D. 'Integrated membrane/adsorption process for gas separation'. *Chemical Engineering Science*. 1998; **53**(9): 1689–1698.

[145] Baker R.W. 'Future direction of membrane gas separation technology'. *Ind. Eng. Chem. Res*. 2002; **41**:1393–1411.

Chapter 8

Fixed bed membrane reactors for ultrapure hydrogen production: modeling approach

Marjan Alavi[1,2], Adolfo Iulianelli[2],
Mohammad Reza Rahimpour[1,3], Reza Eslamloueyan[1],
Marcello De Falco[4], Giuseppe Bagnato[2]
and Angelo Basile[2]

Abstract

This chapter deals with the modeling approach toward membrane reactors, making a short overview on the most significative findings in the specialized literature. In detail, 1-D, 2-D, and 3-D models are analyzed, pointing out the role of such parameters as the membrane permeability mechanism and hydrogen flux, reaction kinetics, and heat and mass transport inside the reactor and within the catalyst pellets, able of influencing the accuracy of the model.

8.1 Introduction

Mathematical modeling of a reactor is an essential tool in process engineering. Basically, reactor modeling is done to predict the behavior of the system in dynamic/steady state mode and optimize the operating conditions by a suitable optimizing algorithm. This task can be done before running the experiments to avoid longer operating time and effort in the experimental phase.

The perm-selectivity properties of a membrane such as permeability and selectivity are able to enhance the performance of a catalytic reaction. The membrane can act as an extractor and it facilitates the selective removal of one of the products. The equilibrium can be shifted to the chosen direction according to the

[1]Department of Chemical Engineering, School of Chemical and Petroleum Engineering, Shiraz University, Mollasadra Street, Shiraz 71345, Iran

[2]ITM-CNR c/o University of Calabria, via P. Bucci Cubo 17/C, 87036 Rende (CS) 87036, Italy

[3]Department of Chemical Engineering and Materials Science, University of California, Davis, 1 Shields Avenue, Davis, California 95616, United States

[4]Faculty of Engineering, University of Rome "Campus Bio-medico", Via Alvaro del Portillo 21, 00128 Rome, Italy

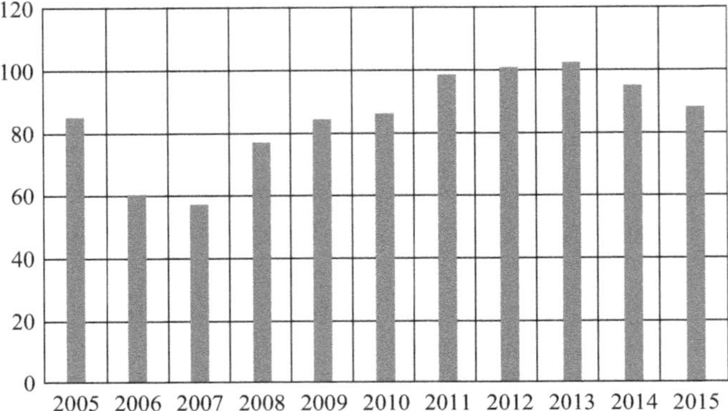

Figure 8.1   Recent number of publications on MR modeling [Scopus Web site, the searched keywords: Membrane Reactor Modeling]

Le Chatelier principle. This task can be performed by selectively removing one of the reaction products (i.e. hydrogen) using membranes. The utilization of membrane reactors (MRs) make possible the reaction rate increase, reducing the by-product formation and lowering the energy requirements, driving consequently to a much flexible process [1].

In a typical packed bed MR, the catalyst is packed in the tube/shell side of the membrane while produced hydrogen permeates through the membrane and is recovered in the shell/tube side. In this case, the driving force is related to the hydrogen partial pressure difference between the reaction and permeation side [2,3].

Utilizing fluidized bed reactors shows major advantages over packed beds such as: improved heat and mass transfer, bed uniformity, and elimination of external mass-transfer resistances [4–9].

However, some issues such as difficulties in the reactor manufacturing and employing membrane technology as well as possibility of catalyst erosion are involved in using this configuration [10].

The number of recent publications on MR modeling is illustrated in Figure 8.1.

In this chapter, the packed bed MR modeling, as the most common configuration, are reviewed and discussed from the modeling point of view, giving details regarding the most recent findings in this field.

8.2   Modeling an MR

Models are categorized into three groups based on the ways to be formed:

- White box or theoretical models that are developed by applying physical and chemical principles such as conservation laws of mass, energy and momentum, kinetic and transport expressions, and the physical characteristics of the reactor [11]. Once validated by experimental data, they are applicable within a wide range of operation. These models give a complete insight through the system but developing and solving such models require significant time and the computers with a high-level of computational capacity.

- Black box or empirical models that are obtained through fitting experimental data (e.g., Artificial Neural Network Models) [12]. These models are easy to derive but less helpful due to their lack of validity further than the conditions of the experimental data.
- Gray box or semi empirical models that are theoretical models in which some parameters (e.g., reaction-kinetics rate coefficients, catalyst adsorption coefficients) are calculated using data fitting [13,14]. These models provide a good understanding of the system accompanied by a good generalization over a wide range of data, and they need less effort than theoretical models.

Another categorization of reactor modeling is based on the heterogeneous or pseudo-homogeneous assumption that defines the complexity and accuracy of the model. In a heterogeneous model, the fluid and the catalyst particles are considered as two different phases, and the balance equations are imposed for both phases, whereas in a pseudo-homogeneous model, they are considered as a single pseudo-phase and the balances are imposed for only one phase [15].

White box modeling of a traditional reactor (TR) is basically applying the mass, energy, and momentum balances, accompanied by the needed kinetics and transport equations. As a result, a set of differential equations (ODEs or PDEs) are formed. By employing a suitable solving method, they result in the concentration profiles of each component, temperature, and pressure in the reactor. For MRs, other parameters have to be included, such as the hydrogen flux through the membrane, heat transport inside the reactor and from the external to the reactor.

An MR consists of reaction and permeation zones as illustrated in Figure 8.2.

A hydrogen production reaction takes place in the reaction side while the produced hydrogen is recovered from the permeation side where an inert gas is

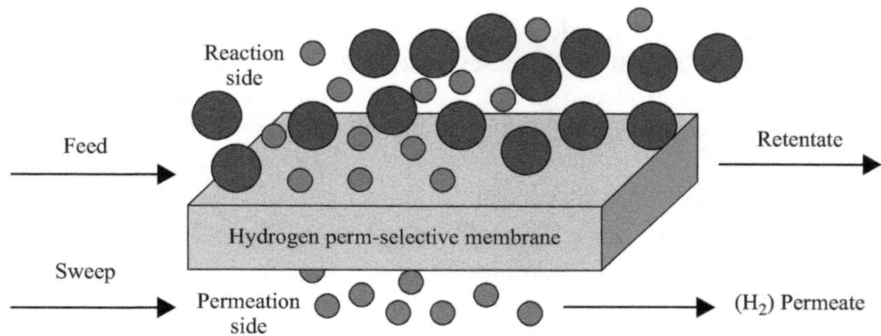

Figure 8.2   Scheme of a H$_2$ perm-selective MR

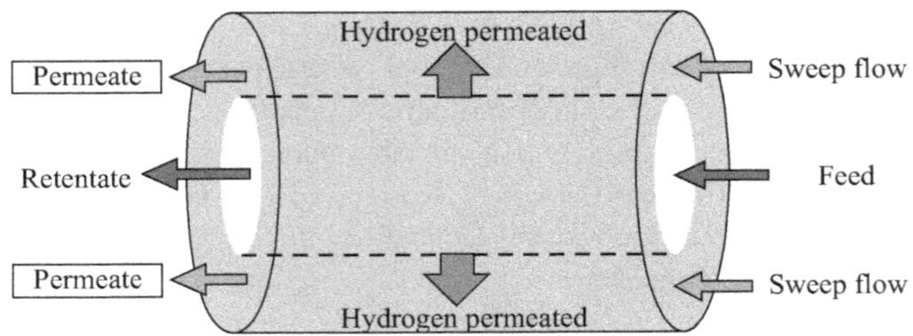

Figure 8.3   Co-current flow in an MR

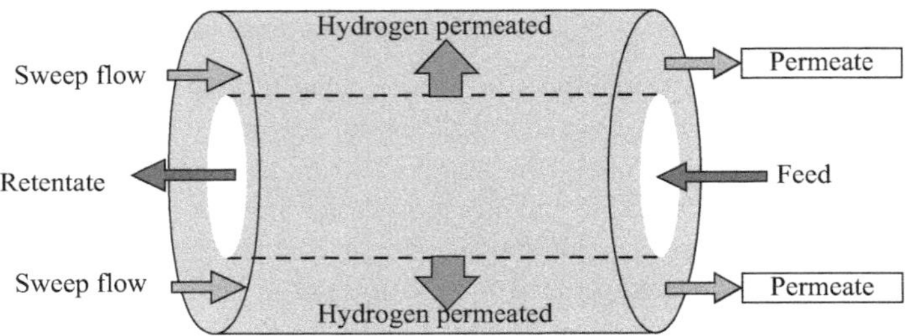

Figure 8.4   Counter-current flow in an MR

flowing to sweep the hydrogen and keep the hydrogen partial pressure as low as possible. Therefore, there are four different streams, which are as follows:

- Feed: the inlet stream of the reaction side.
- Retentate: the outlet of the reaction side.
- Sweep: the inlet of the permeation side.
- Permeate: the outlet of the permeation side.

So, the balance equations should be applied for both the reaction and the permeation sides, and the accurate flux of hydrogen permeation must be utilized.

Depending on the type of reaction and the process necessities, the catalyst is packed in the tube or shell sides. Two configurations are possible for MRs, co-current and counter-current. In the co-current mode, the feed in reaction zone and the sweep flow are introduced in the same direction (Figure 8.3), whereas in a counter-current mode, they are flowing in opposite directions (Figure 8.4).

8.2.1   Tubular reactor modeling

The majority modeling of the reactors belongs to the tubular ones due to their abundant uses in industrial applications. The mass balance for each membrane side is written for a differential volume of length dz (Figure 8.5).

By supposing that the compositions, temperature and pressure only change in the axial direction, and neglecting axial diffusion, a plug flow assumption is acceptable.

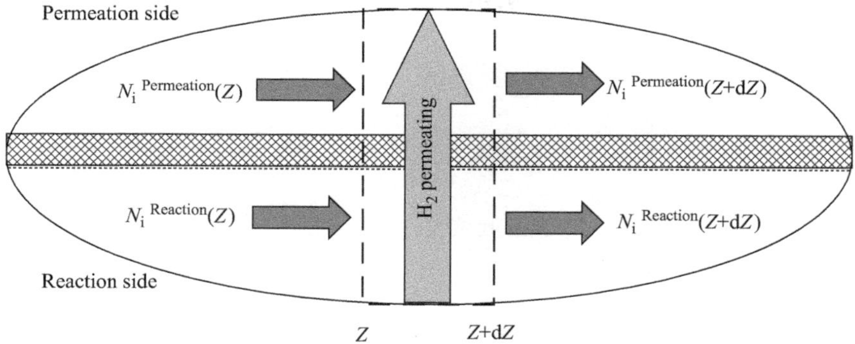

Figure 8.5   A schematic of Pd-based tubular MR

The number of space dimensions is an important parameter to select the type of the model. One-dimensional modeling is the most common type of MR modeling. The properties variations in only one dimension (axial coordinate) are considered in this type, so their derivations and solutions are quite simple.

Because of the hydrogen permeation and heat transfer through the membrane, sometimes it is important to consider radial concentrations and temperature gradients as well. So, it is essential to include the radial profiles and apply a different differential volume for two-dimensional modeling. In this situation, a new term is involved in the balances as the radial diffusion, and the term of hydrogen permeation is omitted from the balance equations and is used as a boundary condition for both the permeation and reaction zones.

Tri-dimensional modeling is also possible, even though these models show the highest complexity as they consider the whole geometry and take into account the profiles in angular directions as well. So their applications are limited unless a non-symmetrical reactor is used.

As a result, four different modeling strategies for MRs are possible:

1.   One-dimensional model, plug-flow.
2.   One-dimensional model, with axial diffusion.
3.   Two-dimensional model, with axial and radial diffusion.
4.   Tri-dimensional model, with axial, radial and angular diffusion.

8.2.1.1   Mass balance

Mass balances are necessary to calculate the concentration gradient in a reactor. When a component balance is applied to a control volume of a dynamic reactor system, the main terms of the mass balance equation will be: the input and output flows of ith component through the control volume, the rate of formation, the rate of permeation and the rate of accumulation of ith component in the control volume.

The time-dependent term is zero in steady state mode, and the reaction term only exists in reaction zone modeling. The rate of hydrogen permeation is positive for the permeation side and negative for the reaction side. This rate depends on the

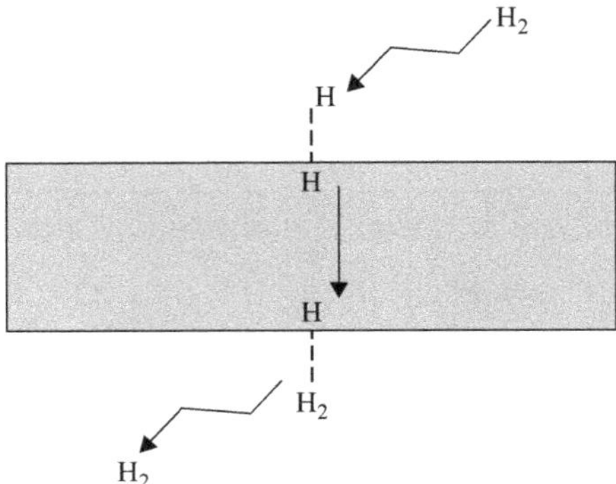

*Figure 8.6   Solution–diffusion mechanism of hydrogen permeation through a
dense metal membrane*

type of membrane and the permeation mechanism. Hydrogen transport in dense
metals such as Pd-based membranes occurs via a solution/diffusion mechanism
[16]. This mechanism follows six activated steps as illustrated in Figure 8.6:

- Diffusion of molecular hydrogen from the bulk to the surface of the metal
  membrane;
- Reversible dissociation adsorption of molecular hydrogen at the gas/metal
  interface;
- Dissolution of atomic hydrogen into the palladium matrix;
- Association of hydrogen atom on the metal surface;
- Desorption of molecular hydrogen from the surface;
- Diffusion of molecular hydrogen away from the surface to the bulk.

Each step may control the hydrogen permeation through the membrane depending
on temperature, pressure, composition of the gas mixture, and the membrane
thickness. The hydrogen permeation flux can be expressed by (8.1) as:

$$J_{H_2}^{\text{Permeating}} = \frac{B_{H_2}}{\delta}\left(p_{H_2,\text{retentate}}^n - p_{H_2,\text{permeate}}^n\right) \tag{8.1}$$

where J_{H_2} is the hydrogen flux permeating through the membrane, B_{H_2} is the
hydrogen permeability, δ is the membrane thickness, $p_{H_2,\text{retentate}}$ and $p_{H_2,\text{permeate}}$ are
the hydrogen partial pressures in the retentate and permeate sides. n is the variable
in the range of 0.5–1 depending on the rate-limiting step of hydrogen diffusion and
can be calculated through experiments [17].

When the hydrogen diffusion into the bulk is the rate-limiting step and,
meanwhile, no other gas permeation is noticed, then the hydrogen permeating flux
can be calculated by Sievert's law as (8.2).

$$J_{H_2}^{\text{Permeating}} = \frac{B_{H_2}}{\delta}\left(p_{H_2,\text{retentate}}^{0.5} - p_{H_2,\text{permeate}}^{0.5}\right) \tag{8.2}$$

Equation (8.2) indicates that the hydrogen partial pressure difference between the retentate and the permeate sides of the membrane acts as the driving force of the separation process. It also shows that the hydrogen flux is inversely proportional to the membrane thickness. Thinner membranes display better fluxes but, at the same time, their mechanical resistance and strength are reduced.

To consider the temperature effect on the hydrogen permeability of the membrane, the relationship between the hydrogen permeability and the temperature as an Arrhenius law is:

$$B_{H_2} = B_{H_2}{}^0 \cdot \exp\left(-\frac{E_a}{R \cdot T}\right) \tag{8.3}$$

where $B_{H_2}{}^0$ is the pre-exponential factor, E_a is the apparent activation energy, R is the universal gas constant and T is the temperature. $B_{H_2}{}^0$ and E_a are evaluated by experimental procedure.

So, considering a Pd-based membrane which is fully perm-selective to hydrogen, the permeation rate of i is:

$$J_i^{\text{Permeating}} = \begin{cases} 0 & \text{if } i \neq H_2 \\ \text{Equation (8.1)} & \text{if } i = H_2 \end{cases} \tag{8.4}$$

where $i = 1, 2, 3, \ldots, N_{\text{species}}$, the components in the system.

1D models
The component mass balance equation for a 1D model, considering transient state is available in Table 8.1 and the different terms of these equations are expressed in Table 8.2.

2D models
Table 8.3 presents the terms of component mass balance for a 2D model, considering the transient state.

8.2.1.2   Concentration polarization

The real driving force for hydrogen permeation can be less than the one described in (8.1). When hydrogen permeates through the membrane, a layer of non-permeating components accumulates on the surface of the membrane and acts as a mass transfer resistance to the permeation. This effect, called concentration polarization, can decrease the separation efficiency of the system since it reduces the permeation of hydrogen. This is a main design parameter in membrane processes.

According to the experimental efforts of Mori *et al.* [18], to reduce the negative effect of this phenomena, it is possible to use the reactors with narrower diameters to increase the linear velocity of the reactants or install baffle plates inside the reactors. Another option is using the fluidized bed reactors that minimizes the concentration polarization [19].

Using smaller diameters to avoid the concentration polarization causes higher pressure drop in the reactor. Tiemersma *et al.* [20] observed that at higher membrane permeability concentration, polarization effect was observed which could be

Table 8.1   Component mass balance of a tubular Pd-alloy MR with cylindrical symmetry (Transient, 1D with axial dispersion)

<table>
<tr><td rowspan="4">Reaction side</td><td>Equation</td><td>$\dfrac{1}{RT^{\text{Reaction}}}\dfrac{\partial p_i^{\text{Reaction}}}{\partial t} = \dfrac{1}{RT^{\text{Reaction}}}D_i\dfrac{\partial^2 p_i^{\text{Reaction}}}{\partial z^2} - \dfrac{\partial N_i^{\text{Reaction}}}{\partial z}$

$+ \displaystyle\sum_{j=1}^{N_{\text{Reaction}}} v_{i,j}r_j - \dfrac{A^{\text{Membrane}}}{V^{\text{Reaction}}}J_i^{\text{Permeating}}$</td></tr>
<tr><td>Initial condition</td><td>$p_i^{\text{Reaction}}(z) = p_i^{\text{Reaction, initial}}$</td></tr>
<tr><td rowspan="2">Boundary condition</td><td>$p_i^{\text{Reaction}}\big|_{Z=0} = p_i^{\text{Feed}}$</td></tr>
<tr><td>$\dfrac{\partial p_i^{\text{Reaction}}}{\partial z}\bigg|_{Z=L} = 0$</td></tr>
<tr><td rowspan="4">Permeation side</td><td>Equation</td><td>$\dfrac{1}{RT^{\text{Permeation}}}\dfrac{\partial p_i^{\text{Permeation}}}{\partial t} = \dfrac{1}{RT^{\text{Permeation}}}D_i\dfrac{\partial^2 p_i^{\text{Permeation}}}{\partial z^2} - \dfrac{\partial N_i^{\text{Permeation}}}{\partial z}$

$+ \dfrac{A^{\text{Membrane}}}{V^{\text{Permeation}}}J_i^{\text{Permeating}}$</td></tr>
<tr><td>Initial condition</td><td>$p_i^{\text{Permeation}}(z) = p_i^{\text{Permeation, initial}}$</td></tr>
<tr><td rowspan="2">Boundary condition</td><td>$p_i^{\text{Permeation}}\big|_{Z=0} = p_i^{\text{Sweep}}$ (Co-current flow)

$p_i^{\text{Permeation}}\big|_{Z=L} = p_i^{\text{Sweep}}$ (Counter-current flow)</td></tr>
<tr><td>$\dfrac{\partial p_i^{\text{Permeation}}}{\partial z}\bigg|_{Z=L} = 0$</td></tr>
</table>

Table 8.2   Terms of the component mass balance equation

Term	**Expression**
$\dfrac{1}{RT^{\text{Reaction}}}\dfrac{\partial p_i^{\text{Reaction}}}{\partial t}$	Accumulation term, existing in transient state and zero at steady state
$-\dfrac{\partial N_i^{\text{Reaction}}}{\partial z}$	Convective flux variation of the ith species along the reaction side
$+\displaystyle\sum_{j=1}^{N_{\text{Reaction}}} v_{i,j}r_j$	Reaction term involving ith species in all the reactions
$-\dfrac{A^{\text{Membrane}}}{V^{\text{Reaction}}}J_i^{\text{Permeating}}$	Permeation term of the ith species through the membrane
$\dfrac{1}{RT^{\text{Reaction}}}D_i\dfrac{\partial^2 p_i^{\text{Reaction}}}{\partial z^2}$	Axial dispersion term, present in second-order model and is zero if Peclet number (Pe) is less than 1. $Pe_{AB} = \dfrac{(\text{Volumetric flow rate/cross sectional area})L}{D_{AB}} \leq 1$

Table 8.3   Component mass balance of a tubular Pd-alloy MR with cylindrical symmetry (Transient, 2D with axial and radial dispersion)

<table>
<tr><td rowspan="4">Reaction side</td><td>Equation</td><td>$\dfrac{\partial p_i^{\text{Reaction}}}{\partial t} = D_{i,z}\dfrac{\partial^2 p_i^{\text{Reaction}}}{\partial z^2} + D_{i,r}\dfrac{\partial^2 p_i^{\text{Reaction}}}{\partial r^2}$
 $+ RT^{\text{Reaction}}\left(-\dfrac{\partial N_i^{\text{Reaction}}}{\partial z} + \displaystyle\sum_{j=1}^{N_{\text{Reaction}}} v_{i,j}r_j\right)$</td></tr>
<tr><td>Initial condition</td><td>$p_i^{\text{Reaction}}(z,r)\big|_{t=0} = p_i^{\text{Reaction, initial}}$</td></tr>
<tr><td>Boundary condition</td><td>$p_i^{\text{Reaction}}\big|_{z=0} = p_i^{\text{Feed}}$</td></tr>
<tr><td></td><td>$\dfrac{\partial p_i^{\text{Reaction}}}{\partial z}\big|_{z=L} = 0$</td></tr>
<tr><td rowspan="4">Permeation side</td><td>Equation</td><td>$\dfrac{\partial p_i^{\text{Permeation}}}{\partial t} = D_{i,z}\dfrac{\partial^2 p_i^{\text{Permeation}}}{\partial z^2} + D_{i,r}\dfrac{\partial^2 p_i^{\text{Permeation}}}{\partial r^2} - RT^{\text{Permeation}}\dfrac{\partial N_i^{\text{Permeation}}}{\partial z}$</td></tr>
<tr><td>Initial condition</td><td>$p_i^{\text{Permeation}}(z,r)\big|_{t=0} = p_i^{\text{Permeation, initial}}$</td></tr>
<tr><td>Boundary condition</td><td>$p_i^{\text{Permeation}}\big|_{z=0} = p_i^{\text{Sweep}}$ (Co-current flow)
 $p_i^{\text{Permeation}}\big|_{z=L} = p_i^{\text{Sweep}}$ (Counter-current flow)</td></tr>
<tr><td></td><td>$\dfrac{\partial p_i^{\text{Permeation}}}{\partial z}\big|_{z=L} = 0$</td></tr>
<tr><td rowspan="2">Tube side</td><td>Boundary condition</td><td>$\dfrac{\partial p_i}{\partial z}\bigg|_{R=\frac{ID_{\text{membrane}}}{2}} = 0 \qquad$ for $i \neq$ H$_2$</td></tr>
<tr><td></td><td>$\dfrac{\partial p_i}{\partial z}\big|_{R=0} = 0$</td></tr>
<tr><td rowspan="2">Annulus side</td><td>Boundary condition</td><td>$\dfrac{\partial p_i}{\partial z}\bigg|_{R=\frac{OD_{\text{membrane}}}{2}} = 0 \qquad$ for $i \neq$ H$_2$</td></tr>
<tr><td></td><td>$\dfrac{\partial p_i}{\partial z}\bigg|_{R=\frac{ID_{\text{Shell}}}{2}} = 0$</td></tr>
</table>

overcome by choosing smaller diameter of the membrane and bigger catalyst particles to avoid the pressure drop. In some other theoretical studies, utilization of specially structured catalyst bed that promotes the radial mixing properties was proposed to avoid concentration polarization [21].

8.2.1.3   Energy balance

The energy balances are used to model the thermal performance of non-isothermal reactors. The conditions in lab-scale reactors are near to isothermal conditions but in commercial size reactors this assumption is no more valid. Predicting the temperature variation in the MR is quite important because of the embrittlement of Pd

Table 8.4   Energy balance equations of a tubular Pd-alloy MR (1D)

<table>
<tr>
<td rowspan="3">Annulus</td>
<td>Equation</td>
<td>

$$\sum_{i=1}^{N_{\text{species}}} C_i C_{P_i} \frac{\partial T^{\text{Annulus}}}{\partial t} = -\sum_{i=1}^{N_{\text{species}}} N_i C_{P_i} \frac{\partial T^{\text{Annulus}}}{\partial z} + k_z \frac{\partial^2 T^{\text{Annulus}}}{\partial z^2}$$

$$+ \frac{U^{\text{Shell}} A^{\text{Shell}}}{V^{\text{Annulus}}} \left(T^{\text{Furnace}} - T^{\text{Annulus}} \right) - \frac{U^{\text{Membrane}} A^{\text{Membrane}}}{V^{\text{Annulus}}} \left(T^{\text{Annulus}} - T^{\text{Tube}} \right)$$

$$+ \Psi + \Phi \frac{A^{\text{Membrane}}}{V^{\text{Annulus}}}$$

</td>
</tr>
<tr>
<td>Initial condition</td>
<td>

$T^{\text{Annulus}}\big|_{t=0} = T^{\text{Annulus, initial}}$

</td>
</tr>
<tr>
<td>Boundary condition</td>
<td>

$T^{\text{Annulus}}\big|_{z=0} = T^{\text{Feed}}$ or T^{Sweep}

$\dfrac{\partial T^{\text{Annulus}}}{\partial z}\big|_{z=L} = 0$

</td>
</tr>
<tr>
<td rowspan="3">Tube</td>
<td>Equation</td>
<td>

$$\sum_{i=1}^{N_{\text{species}}} C_i C_{P_i} \frac{\partial T^{\text{Tube}}}{\partial t} = -\sum_{i=1}^{N_{\text{species}}} N_i C_{P_i} \frac{\partial T^{\text{Tube}}}{\partial z} + k_z \frac{\partial^2 T^{\text{Tube}}}{\partial z^2}$$

$$+ \frac{U^{\text{Membrane}} A^{\text{Membrane}}}{V^{\text{Tube}}} \left(T^{\text{Annulus}} - T^{\text{Tube}} \right) + \Psi + \Phi \frac{A^{\text{Membrane}}}{V^{\text{Tube}}}$$

</td>
</tr>
<tr>
<td>Initial condition</td>
<td>

$T^{\text{Tube}}\big|_{t=0} = T^{\text{Tube, initial}}$

</td>
</tr>
<tr>
<td>Boundary condition</td>
<td>

$T^{\text{Tube}}\big|_{z=0} = T^{\text{Feed}}$ or T^{Sweep}

$\dfrac{\partial T^{\text{Tube}}}{\partial z}\big|_{z=L} = 0$

</td>
</tr>
</table>

membranes that takes place at relatively low temperatures ($\approx$566 K) [22] and, on the other hand, when the temperature is too high and regarding composite Pd-based membranes, the interactions between the active layer and the support may affect negatively the whole structure of the membrane due to the different dilation. The maximum operating temperature for the activity of the catalysts is another factor that should be taken into account.

Energy balance should be solved simultaneously with the mass balance in order to predict the thermal behavior of the system. The terms of the energy balance equation will be: the input and output of enthalpy through the control volume, the heat generated by the reaction, the enthalpy variation due to hydrogen permeation, and the enthalpy accumulation in time.

Tables 8.4 and 8.5 present the energy balance equations of a tubular Pd-alloy MR respectively in 1D and 2D modeling approaches.

Table 8.6 expresses the terms involved in the energy balance equations.

In order to calculate the overall heat transfer coefficient, it is necessary to study the heat transfer mechanisms inside the reactor. Figure 8.7 shows the heat transfer resistances involved in the whole system from the furnace or the electric heater to the center of the reactor.

Table 8.5   Energy balance equations of a tubular Pd-alloy MR (2D)

<table>
<tr><td rowspan="6">Annulus side</td><td>Equation</td><td>

$$\sum_{i=1}^{N_{\text{species}}} C_i C_{P_i} \frac{\partial T^{\text{Annulus}}}{\partial t} = -\sum_{i=1}^{N_{\text{species}}} N_i C_{P_i} \frac{\partial T^{\text{Annulus}}}{\partial z}$$
$$+ k_z \frac{\partial^2 T^{\text{Annulus}}}{\partial z^2} + k_r \frac{\partial^2 T^{\text{Annulus}}}{\partial r^2} + \Psi$$
</td></tr>
<tr><td>Initial condition</td><td>$T^{\text{Annulus}}\big|_{t=0} = T^{\text{Annulus,initial}}$</td></tr>
<tr><td rowspan="4">Boundary condition</td><td>$T^{\text{Annulus}}\big|_{z=0} = T^{\text{Feed}}$ or T^{Sweep}</td></tr>
<tr><td>$\dfrac{\partial T^{\text{Annulus}}}{\partial z}\big|_{z=L} = 0$</td></tr>
<tr><td>$q\big|_{r=\frac{ID_{\text{Shell}}}{2}} = -U_{\text{shell}}\left(T^{\text{Annulus}} - T^{\text{Furnace}}\right)$</td></tr>
<tr><td>$q\big|_{r=\frac{OD_{\text{Membrane}}}{2}} = -U_{\text{Membrane}}\left(T^{\text{Tube}} - T^{\text{Annulus}}\right) + \Phi$</td></tr>
<tr><td rowspan="5">Tube side</td><td>Equation</td><td>

$$\sum_{i=1}^{N_{\text{species}}} C_i C_{P_i} \frac{\partial T^{\text{Tube}}}{\partial t} = -\sum_{i=1}^{N_{\text{species}}} N_i C_{P_i} \frac{\partial T^{\text{Tube}}}{\partial z}$$
$$+ k_z \frac{\partial^2 T^{\text{Tube}}}{\partial z^2} + k_r \frac{\partial^2 T^{\text{Tube}}}{\partial r^2} + \Psi$$
</td></tr>
<tr><td>Initial condition</td><td>$T^{\text{Tube}}\big|_{t=0} = T^{\text{Tube,initial}}$</td></tr>
<tr><td rowspan="3">Boundary condition</td><td>$T^{\text{Tube}}\big|_{z=0} = T^{\text{Feed}}$ or T^{Sweep}</td></tr>
<tr><td>$\dfrac{\partial T^{\text{Tube}}}{\partial z}\big|_{z=L} = 0$</td></tr>
<tr><td>$q\big|_{r=\frac{ID_{\text{Membrane}}}{2}} = -U_{\text{Membrane}}\left(T^{\text{Tube}} - T^{\text{Annulus}}\right) + \Phi$</td></tr>
<tr><td>$\dfrac{\partial T^{\text{Tube}}}{\partial r}\big|_{r=0} = 0$</td></tr>
</table>

8.2.1.4   Momentum balance

Due to the permeation through the membrane and non-ideal flow pattern (not plug), the flow rate changes along and in small diameters the bed porosity near the wall causes a non-uniform radial velocity profile, so it is necessary to include momentum balance in the equations. Marín *et al.* [23] used a 2D heterogeneous model to describe the behavior of a water gas shift (WGS) MR. It was observed that excluding momentum balance leads to misprediction in accurate results. However, including the momentum balances in literature is not very common despite its advantages. There are some works that have considered the momentum balance in their studies that are available in Tables 8.7 and 8.8.

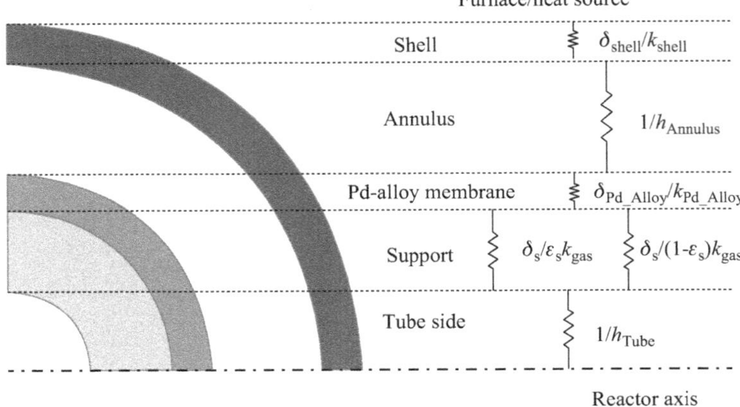

Figure 8.7   Scheme of the overall heat transfer resistances in an MR

Table 8.6   Energy balance equations of a tubular Pd-alloy MR (2D)

Term	**Expression**
$\Phi = \begin{cases} 0 & \text{reaction side} \\ J_{\mathrm{H_2}}^{\mathrm{Perm}}\left(H_{\mathrm{H_2}}^{T_{\mathrm{rxn}}} - H_{\mathrm{H_2}}^{T_{\mathrm{Perm}}}\right) & \text{permeation side} \end{cases}$	Temperature variation due to enthalpy flux associated with hydrogen permeation
$\Psi = \begin{cases} 0 & \text{permeation side} \\ \sum\limits_{j=1}^{N} r_j\left(-\Delta H_j\right) & \text{reaction side} \end{cases}$	Heat generated by the chemical reaction
$\sum\limits_{i=1}^{N_{\mathrm{species}}} C_i C_{P_i} \dfrac{\partial T^{\mathrm{Annulus}}}{\partial t}$	Accumulation of enthalpy, existing in transient state and zero at steady state
$-\dfrac{\partial N_i^{\mathrm{Reaction}}}{\partial z}$	Variation of the ith species along the reaction side by convective flux
$+\sum\limits_{j=1}^{N_{\mathrm{Reaction}}} \nu_{i,j} r_j$	Reaction term of ith species in all the reactions
$-\dfrac{A^{\mathrm{Membrane}}}{V^{\mathrm{Reaction}}} J_i^{\mathrm{Permeating}}$	Permeation term of the ith species through the membrane
$\dfrac{1}{RT^{\mathrm{Reaction}}} D_i \dfrac{\partial^2 p_i^{\mathrm{Reaction}}}{\partial z^2}$	Axial dispersion term, present in second-order model and is zero if Peclet number (Pe) is less than 1
U_{shell}	Overall heat transfer coefficient on the shell side
U_{membrane}	Overall heat transfer coefficient on the membrane side

Table 8.7   The summaries of some recent works on 1D MR modeling

Reaction	Reaction side	Flow mode	Mass	Momentum	Continuity	Convective flux	Rxn	With furnace	Through the membrane	Due to H_2 permeation	State	Reference
Methane steam reforming	Shell	Co-current	•			•	•	•			Steady state	[27]
Methane steam reforming	Shell	Co-current	•	•		•	•	•	•	•	Steady state	[35]
Water gas shift	Shell	Co-current	•	•		•	•	•	•	•	Steady state	[36]
Ethanol steam reforming	Tube	Counter-current	•								Steady state	[37]
Methane steam reforming	Tube	Co-current	•								Steady state	[24]
Water gas shift	Shell	Counter-current	•	•		•	•		•	•	Steady state	[32]
Water gas shift	Shell	Both	•			•	•		•	•	Steady state	[33]
Water gas shift	Tube	Co-current	•								Steady state	[38]
Water gas shift	Tube	Counter-current	•	•							Steady state	[39]
Tri-reforming of methane	Shell	Co-current	•			•	•	•	•	•	Steady state	[40]

(*Continues*)

Table 8.7   (Continued)

Reaction	Reaction side	Flow mode	Mass	Momentum	Continuity	Convective flux	Rxn	With furnace	Through the membrane	Due to H_2 permeation	State	Reference
Methane steam reforming	Tube	Co-current	•			•	•	•	•	•	Steady state	[41]
Methane steam reforming	Shell	Co-current	•			•	•	•	•	•	Steady state	[42]
Methanol steam reforming	Tube	Co-current	•								Steady state	[28]
Water gas shift	Shell	Counter-current	•								Steady state	[43]
Methanol steam reforming	Shell	Co-current	•								Steady state	[44]
Methane steam reforming	Shell	Co-current	•			•	•	•	•	•	Steady state	[34]
Methane steam reforming	Shell	Counter-current	•			•	•	•			Transient	[45]

8.3   Literature models

This section mainly discusses the recent studies on the MR modeling, particularly toward approaches including 1D and 2D models, in which some of them are summarized in Tables 8.7 and 8.8.

8.3.1   One-dimensional and two-dimensional models

A lot of theoretical studies in MR have considered 1D and 2D modeling due to their simplicity and easier solving methods. Iulianelli *et al.* [24] investigated methane steam reforming at relatively low pressures (1.0–3.0 bars) in a Pd–Ag MR. The permeation zone was described by a 1D and the reaction zone by a 2D pseudo-homogeneous model. The model was developed and successfully validated with the experimental results.

A 2D model for a WGS in an MR was proposed by Sanz *et al.* [25]. The model was validated with experimental data derived from a lab-scale setup. Some main parameters (such as pore tortuosity, internal particle porosity, or average pore size) were not easy to measure correctly, so they were determined by fitting the experimental data. The model was used for scaling up the reactor.

Several studies have compared the performance of different reactions in MRs and TRs [26]. Fernandes and Soares [27] developed a 1D, non-isothermal pseudo-homogeneous model for the steam reforming of methane to compare the yield enhancements of an MR and TR. The methane conversion achieved in the MR was considerably higher than the TR at the same operating conditions. The model also displayed that decreasing the membrane thickness (<0.1 μm) increases the methane conversion up to 90% at mild temperatures (≈ 850 K). Ghasemzadeh *et al.* [28] studied the methanol conversion and total yield of methanol steam reforming reaction in TR and MR via a 1D pseudo-homogeneous model in an isothermal condition. The experimental and modeling results showed better methanol conversions for the MR than the TR at the same conditions.

Few studies have investigated MRs on large scales. De Falco *et al.* [29,30] utilized a 2D pseudo-homogeneous model to simulate the axial and radial gradients in an MR and TR with methane steam reforming in a large-scale reactor. Both of them were considered to be a bundle of tubes while a tube represents others. The main advantages of MRs over the TRs were the higher conversion. But the disadvantage of MR is higher amount of steam required by the process. This drawback, the cost and separation efficiency of the membrane are the most essential parameters that should be taken to account while dealing with the industrial application.

The co-current and counter-current flow modes have been compared in some works. Basile *et al.* [31] developed a 2D pseudo-homogeneous model for WGS reaction in co-current and counter-current modes. CO conversion in both configurations was the same, but the hydrogen recovery from the end of the reactor in counter-current mode was more than in the concurrent mode. Piemonte *et al.* [32] examined different MR configurations using a 1D pseudo-homogeneous model for

Table 8.8   The summaries of some recent works on 2D MR modeling

Reaction	Reaction side	Flow mode	Balances								State	Reference
			Mass	Momentum	Continuity	Energy balance						
						Convective flux	Rxn	Heat exchange				
								With furnace	Through the membrane	Due to H$_2$ permeation		
Water gas shift	Tube	Both	•								Steady state	[31]
Auto-thermal reforming of methane	Tube	Co-current	•	•	•	•	•				Steady state	[20]
Methane steam reforming	Shell	Co-current	•	•		•	•	•	•	•	Steady state	[29,30]
Methane steam reforming	Tube	Co-current	•	•		•	•		•	•	Steady state	[46]
Water gas shift	Tube	Co-current	•	•	•	•	•		•		Steady state	[47]
Water gas shift	Shell	Counter-current	•	•		•	•	•	•	•	Steady state	[48]
Water gas shift	Tube	Co-current	•	•		•	•		•		Steady state	[25]
Methane steam reforming	Shell	Counter-current	•	•	•	•	•	•	•		Steady state	[49]

a WGS reaction. In counter-current mode, at the beginning of the reaction zone, the partial pressure of hydrogen is low while it is high in the permeation zone. This may cause reverse permeation of hydrogen and decrease the efficiency of the system. It was concluded that the most optimized use of the performance of the system is placing the membrane in the second part of the reactor where the partial pressure of hydrogen is high enough for permeation [32].

An analysis of the thermal behavior between the two flow modes in a multi-tubular MR for WGS reaction was done by Adrover *et al.* [33]. The simulation using a 1D pseudo-homogeneous model displayed that the temperature rise in the catalytic bed reduces in co-current mode, because of the heat exchanging with the sweep flow. On the other hand, in the counter-current mode, the preheating of the reaction stream facilitates reaction ignition. But it may cause higher temperature rise, possibly higher than adiabatic temperature.

Di Marcoberardino *et al.* [34] studied methane steam reforming in an MR via a 1D pseudo-homogeneous model. The validated model displayed that using hot gases instead of the electric oven to provide heat for the reaction increases the performance of the system. A counter-current flow caused higher methane conversion because at the bottom of the reactor the flow of hotter gases in the sweep increased the methane conversion beyond the equilibrium conversion.

8.3.2   Tri-dimensional (3D) models

In the MRs, the reactants flow is in the axial direction while the hydrogen permeation through the membrane is in the radial direction. So, the hydrogen flow through the membrane is a two-dimensional effect, and the real performance of the MRs are not fully understood via a 1D modeling task. In order to have a better understanding of such systems, it is necessary to apply 2D or 3D models [50]. At the moment, the number of studies on 3D modeling of MRs are not very much.

Byron Smith and Muruganandam [51] considered the temperature, pressure, time factor, sweep flow rate, and steam/carbon (S/C) ratio as the major operational parameters affecting the performance of the MR for WGS reaction. The optimum operating condition was introduced using a 3D model.

Chein *et al.* [52] investigated the syngas composition, sweep flow, membrane permeance, and S/C ratio effects on the WGS reaction that was carried out in a multi-tubular MR via a 3D model The results showed that higher permeance of the membrane, S/C ratio and the sweep flow specially in the counter-current mode would cause higher CO conversion and hydrogen recovery. On the other hand, increasing the CO_2 content of the feed would degrade the performance of the reactor. In another work, they studied the sweep flow effects as a primary parameter on the performance of a WGS reaction that was carried out in a 3D MR [53]. It was seen that the parallel flow (based on the inlet/outlet ports) is the optimum design for CO conversion and hydrogen recovery. It was also observed that when the membrane is only selective to hydrogen, the type of the sweep gas (N_2, steam, Argon) has no effects on the reactor performance. The high temperatures of the sweep gas would cause a reverse WGS reaction and the low temperatures would not be beneficial because of the reduction in the catalyst activity. Furthermore, it was concluded that by reducing the permeation side size, the flow rate is increased and causes more driving force for the permeation.

8.3.3   Effect of the operating conditions

Several studies have focused on determining the best operating conditions to maximize the MR performance. Pressure, temperature, space velocity, catalyst particle size, and S/C ratio in the reactor are some of the operating variables that influence the conversion and hydrogen recovery in the MRs.

In MRs, increasing the pressure in the reaction side increases the partial pressure of hydrogen and intensifies the driving force for the hydrogen permeation. Ghasemzadeh *et al.* [28] showed that in methanol steam reforming, although the reaction thermodynamically favored the lower number of moles in the product and increasing the pressure imposed a negative effect on methanol conversion, but the positive effect of the driving force for the permeation process improved the conversion.

Obviously, the usage of sweep gas enhances hydrogen recovery via decreasing the hydrogen partial pressure in the permeate side, but adding an additional component to the system is not beneficial for industrial applications since further separation of hydrogen from the sweep gas mixture is a significant issue [38].

Patrascu and Sheintuch [45] predicted the performance of an MR for steam reforming of methane by proposing a 1D transient pseudo-homogeneous model that took into account the axial dispersion for mass and energy balances. The pressure drop in reaction zone was neglected and no sweep was applied in the permeate. Working without sweep needed a higher reaction pressure to impose a positive driving force for hydrogen permeation. It was proved by the model and the experimental data that at any temperature a threshold pressure exists that this driving force is possible beyond this point.

Space velocity is another vital design factor because it is a representative of the residence time of the reactants in the reactor. Higher space velocities in the reaction side cause less residence time of the reactants in the reactor and less conversion. Piemonte *et al.* [48] studied the performance of a WGS reactor through a 2D non-isothermal model. It was displayed that at higher gas hourly space velocity (GHSV), CO conversion decreases as a result of less residence time of reactant in the rector. But in a fixed geometry, higher GHSV means higher reactants and consequently higher catalyst productivity. It was seen that a maximum CO conversion occurs at 600 K.

Cornaglia *et al.* [43] proposed a 1D heterogeneous model to describe the behavior of a laboratory scale Pd–Ag alloy MR for WGS reaction. The model was validated with the experimental data and the effect of space velocity and catalyst particle diameter on CO conversion and hydrogen recovery were investigated. Increasing the space velocity decreased the CO conversion and hydrogen recovery due to the decreasing in contact time of the process gas with the membrane and the catalyst. The model predicted that by increasing the particle size, the CO conversion drops. This is the result of the reduction in the external surface area per unit volume that made the mass transfer limitation important. So, this effect showed that in this case a pseudo-homogeneous model cannot correctly predict the results of the reactor.

In order to have a more quantitative investigation through the performance of MR, Damköhler's number (Da) (the ratio of the maximum reaction rate to the space velocity) or the product of Da and Pe (DaPe) as the dimensionless design parameters influencing the CO conversion and hydrogen recovery were chosen in some works [39,50].

Brunetti *et al.* [36] investigated the performance of water gas shift reaction using a Pd-alloy membrane by a 1D pseudo-homogeneous model. It was observed that at high Da (>10), a part of the MR would be remained unused for hydrogen permeation, on the other hand by low values of Da (e.g., $Da = 0.5$), lots of unconverted reactants would be seen in the outlet. The best operating conditions by $Da = 1$ was possible. Boutikos and Nikolakis [38] did a similar investigation of Da on the WGS.

Temperature is also an important operating condition. The wall temperature and therefore the operating temperature affect both the reaction kinetics and the membrane permeability, so the performance of the reactor is increased in terms of conversion and hydrogen recovery. In case of endothermic reactions, higher wall temperature means greater heat supply to the reactor that consequently rising the performance [54].

Gallucci *et al.* [37] studied ethanol steam reforming in a dense Pd–Ag MR by using a 1D pseudo-homogeneous model. The validated model demonstrated that higher ethanol conversion and pure hydrogen production is possible by increasing the temperature to have higher catalyst activity and hydrogen permeation. Rising the reaction pressure caused a higher hydrogen partial pressure resulting in more hydrogen permeation and ethanol conversion. It was displayed that by keeping constant the total molar feed rate, increasing the water/ethanol ratio of the feed, causes more time factor for the ethanol resulting in higher ethanol conversion, but the excess water reduced the hydrogen partial pressure in the reaction side and consequently the hydrogen recovery descended. These effects necessitated an optimization procedure to find the optimal feed rate. Even for exothermic reactions such as WGS, it was proved by Piemonte *et al.* [32] that CO conversion increases to the value of 95% for a wall temperature near to 680 K. They proposed that a convenient heating of the system rather than cooling has a positive effect on CO conversion.

Excess steam is always provided to the reformers to minimize the coke formation. Therefore, the ratio of the steam to carbon (S/C ratio) is another factor that controls the conversion in the reactor. Higher S/C ratios in reforming reactions enhance the WGS reaction involved in such reactors and consequently increases the methane conversion and hydrogen recovery [49]. But increased steam can cause lower conversion and hydrogen recovery due to the inhibition nature of the water on the permeation through the membrane [50]. Also worse performance results will be achieved from a separation point of view [29]. So, an optimized S/C ratio regarding all this situations should be chosen.

Hla *et al.* [55] developed a 2D axis-symmetric CFD model to investigate the effect of different operating parameters on the performance of WGS MR. Reducing the S/C ratio, reducing the permeate pressure, increasing the catalysts load, and inlet temperature increases the CO conversion and hydrogen recovery. But if these augmentations invade a particular limit, they cause undesirable temperature rise in the reactor that is a threat to the membrane and catalysts life.

8.3.4   Performance improvement

Improving the performance of the MRs has gained more attentions in recent years. Higher thermal efficiency of the system was achieved by an auto-thermal operation or coupling the endothermic reaction with an exothermic one to provide the required heat. Patel and Sunol [35] modeled an MR that was comprised of three channels. Methane combustion in the first channel provided the required heat for steam reforming of methane in the second channel, and the permeated hydrogen was collected from the third channel. The simulation results showed that, if a high enough fuel concentration is used, causing higher energy generation for the endothermic reaction and higher permeation of hydrogen, the required length of the reformer to get the highest conversion and hydrogen recovery is reduced. A similar approach was chosen by Chein *et al.* [44], who modeled an MR for methanol steam reforming. The reactor consisted of three coaxial tubes. In the outer tube, catalyzed combustion of methanol took place providing heat for methanol steam reforming in the middle tube. The produced hydrogen permeated via a membrane to the inner tube and was collected using a sweep gas. It was observed that a higher methanol/ air ratio in the combustor offers more heat for the reforming reaction causing greater conversion. Reformer pressure and the sweep gas flow rate also display positive effect on the performance of the reactor by increasing the driving force for the hydrogen permeation.

Abbasi *et al.* [42] proposed a new configuration for the improvement of the steam reforming of methane based on chemical looping combustion (CLC) and utilizing a Pd–Ag MR. In this thermally coupled configuration, CLC was carried out in the exothermic side to supply the required heat for the SRM process and Fe_2O_3 was used as an oxygen transfer to the fuel reactor. In this new configuration, methane conversion significantly raised due to the improved heat provided by the CLC and the shift effect of removing hydrogen.

Some other works proposed utilizing another membrane for introducing the feed gradually. Tiemersma *et al.* [20] investigated auto-thermal reforming of methane in a Pd-based MR via a 2D pseudo-homogeneous model. It was concluded that, by working in the auto-thermal mode, the energy efficiency is higher, but in this mode, larger temperature gradients inside the reactor are produced that are harmful to the membrane life. Using sweep gas decreased the temperature gradient along the membrane but the hydrogen recovery rate reduced. The best solution was to introduce oxygen in a staged feeding process to control the reactor temperature.

Adding an oxygen perm-selective membrane to provide the required oxygen for the reactor was proposed by Rahimpour *et al.* [40]. They developed a 1D heterogeneous model for the tri-reforming of methane as a combination of steam reforming, dry reforming, and partial oxidation of methane in a multi-tubular MR configuration. The air flowed in the inner tube providing oxygen for the partial oxidation. As downstream equipment cannot tolerate nitrogen, it is more beneficial to introduce the pure oxygen instead of the total air mixture. The oxygen permeated through a perovskite type ceramic membrane from the inner tube to the annulus (reaction side). The produced hydrogen permeated through a Pd-based membrane and was collected from the shell. Methane conversion and total hydrogen recovery showed higher values in this configuration with respect to a hydrogen perm-selective membrane and a TR.

8.4   Conclusions and future trends

The modeling task is a great tool for simulating the physical and chemical phenomena in an MR. A good prior knowledge to the system is acquired by a reliable model. This prior knowledge decreases the unnecessary consumption of time and energy in the experiments.

In this chapter, different modeling types of fixed bed MRs in recent studies have been investigated. Choosing the proper number of dimensions included in the model, the membrane permeability mechanism and hydrogen flux, reaction kinetics, and heat and mass transport inside the reactor and within the catalyst pellets are some of the most important parameters affecting the accuracy of the model. The validity of the model should be checked by a reliable set of experimental data.

Having the model enables the engineer to investigate the influence of different operating parameters on the performance of the reactor. Temperature, pressure, space velocity, and feed ratio are some of the factors that have been discussed in different studies recently.

Most of the modeling types of MR include 1 or 2D models due to their easier derivation. The future work on this area can be dedicated to more complex 3D modeling that enables the researchers to have a more accurate understanding toward the systems with asymmetry in angular directions that is from the geometry of the system for example, non-homogeneity of the catalyst bed or multi-tubular MRs.

Nomenclature

Symbol	Definition	Unit
A	reactor area	m^2
B	permeability	$\text{mol s}^{-1}\,\text{m}^{-1}\,\text{kPa}^{-0.5}$
C_i	molar concentration	mol m^{-3}

C_p	specific heat capacity of gas at constant pressure	J mol^{-1}
D	mass diffusion coefficient	m^2 s^{-1}
E_a	apparent activation factor	J mol^{-1}
h	heat transfer coefficient	W m^{-2} K^{-1}
H_i	molar enthalpy of component i	J mol^{-1}
J_{H2}	hydrogen permeation flux	mol m^{-2} s^{-1}
k	thermal conductivity	W m^{-1} K^{-1}
L	reactor length	m
N	molar flux	mol m^{-2} s^{-1}
P	pressure	kPa
p	partial pressure	kPa
R	gas universal constant	J mol^{-1} K^{-1}
r_i	reaction rate for component i	mol kg^{-1} s^{-1}
t	time	s
T	temperature	K
U	overall heat transfer coefficient	W m^{-2} K^{-1}
V	volume	m^3
Z	axial reactor coordinate	m

Subscripts

i	chemical species
r	radius direction
s	support
z	axial direction

Greek letters

ε	void fraction in the support	
δ	membrane thickness	m
ρ	density	kg m^{-3}
ν	stoichiometric coefficient	

Abbreviation

CLC	chemical looping combustion	
Da	Damköhler's number	
GHSV	gas hourly space velocity	h^{-1}
MR	membrane reactor	
Pe	Peclet number	
ODE	ordinary differential equation	
PDE	partial differential equation	
Rxn	reaction	
TR	traditional reactor	
WGS	water gas shift	

References

[1]   Nunes SP, Peinemann KV. Membrane technology in the chemical industry. 2nd. rev. and extended ed. Weinheim: Wiley-VCH; 2006. xiv, 340p.

[2]   Rahnama H, Farniaei M, Abbasi M, Rahimpour MR. Modeling of synthesis gas and hydrogen production in a thermally coupling of steam and tri-reforming of methane with membranes. Journal of Industrial and Engineering Chemistry. 2014;20(4):1779–92.

[3]   Farsi M, Jahanmiri A, Rahimpour MR. Simultaneous isobutane dehydrogenation and hydrogen production in a hydrogen-permselective membrane fixed bed reactor. Theoretical Foundations of Chemical Engineering. 2014;48(6):799–805.

[4]   Adris AM, Pruden BB, Lim CJ, Grace JR. On the reported attempts to radically improve the performance of the steam methane reforming reactor. The Canadian Journal of Chemical Engineering. 1996;74(2):177–86.

[5]   Rahimpour MR, Bayat M, Rahmani F. Dynamic simulation of a cascade fluidized-bed membrane reactor in the presence of long-term catalyst deactivation for methanol synthesis. Chemical Engineering Science. 2010;65 (14):4239–49.

[6]   Rahimpour MR, Elekaei H. Enhancement of methanol production in a novel fluidized-bed hydrogen-permselective membrane reactor in the presence of catalyst deactivation. International Journal of Hydrogen Energy. 2009;34 (5):2208–23.

[7]   Rahimpour MR, Rahmani F, Bayat M. Contribution to emission reduction of CO_2 by a fluidized-bed membrane dual-type reactor in methanol synthesis process. Chemical Engineering and Processing: Process Intensification. 2010;49(6):589–98.

[8]   Rahimpour MR, Mirvakili A, Paymooni K, Moghtaderi B. A comparative study between a fluidized-bed and a fixed-bed water perm-selective membrane reactor with in situ H_2O removal for Fischer–Tropsch synthesis of GTL technology. Journal of Natural Gas Science and Engineering. 2011;3(3):484–95.

[9]   Adris AM, Lim CJ, Grace JR. The fluidized bed membrane reactor system: a pilot scale experimental study. Chemical Engineering Science. 1994;49 (24):5833–43.

[10]   Adris AM, Elnashaie SSEH, Hughes R. A fluidized bed membrane reactor for the steam reforming of methane. The Canadian Journal of Chemical Engineering. 1991;69(5):1061–70.

[11]   Gallucci F, Paturzo L, Basile A. A simulation study of the steam reforming of methane in a dense tubular membrane reactor. International Journal of Hydrogen Energy. 2004;29(6):611–7.

[12]   Basile A, Curcio S, Bagnato G, Liguori S, Jokar SM, Iulianelli A. Water gas shift reaction in membrane reactors: theoretical investigation by artificial neural networks model and experimental validation. International Journal of Hydrogen Energy. 2015;40(17):5897–906.

[13]   Psichogios DC, Ungar LH. A hybrid neural network-first principles approach to process modeling. AIChE Journal. 1992;38(10):1499–511.

[14]   Agarwal M. Combining neural and conventional paradigms for modelling, prediction and control. International Journal of Systems Science. 1997;28(1):65–81.

[15]   Froment GF, Bischoff KB. Chemical reactor analysis and design. 2nd ed. New York: Wiley; 1990. xxxiv, 664p.

[16]   Koros WJ, Fleming GK. Membrane-based gas separation. Journal of Membrane Science. 1993;83(1):1–80.

[17]   Iulianelli A, Liguori S, Huang Y, Basile A. Model biogas steam reforming in a thin Pd-supported membrane reactor to generate clean hydrogen for fuel cells. Journal of Power Sources. 2015;273:25–32.

[18]   Mori N, Nakamura T, Noda K-i, *et al.* Reactor configuration and concentration polarization in methane steam reforming by a membrane reactor with a highly hydrogen-permeable membrane. Industrial & Engineering Chemistry Research. 2007;46(7):1952–8.

[19]   Dang NTY, Gallucci F, van Sint Annaland M. Micro-structured fluidized bed membrane reactors: solids circulation and densified zones distribution. Chemical Engineering Journal. 2014;239:42–52.

[20]   Tiemersma TP, Patil CS, Sint Annaland Mv, Kuipers JAM. Modelling of packed bed membrane reactors for autothermal production of ultrapure hydrogen. Chemical Engineering Science. 2006;61(5):1602–16.

[21]   Shigarov AB, Meshcheryakov VD, Kirillov VA. Use of Pd membranes in catalytic reactors for steam methane reforming for pure hydrogen production. Theoretical Foundations of Chemical Engineering. 2011;45(5):595–609.

[22]   Brodowsky H. On the non-ideal solution behavior of hydrogen in metals. Berichte der Bunsengesellschaft für Physikalische Chemie. 1972;76(8):740–6.

[23]   Marín P, Díez FV, Ordóñez S. Fixed bed membrane reactors for WGSR-based hydrogen production: optimisation of modelling approaches and reactor performance. International Journal of Hydrogen Energy. 2012;37(6):4997–5010.

[24]   Iulianelli A, Manzolini G, De Falco M, *et al.* H_2 production by low pressure methane steam reforming in a Pd–Ag membrane reactor over a Ni-based catalyst: experimental and modeling. International Journal of Hydrogen Energy. 2010;35(20):11514–24.

[25]   Sanz R, Calles JA, Alique D, Furones L, Ordóñez S, Marín P. Hydrogen production in a Pore-Plated Pd-membrane reactor: experimental analysis and model validation for the water gas shift reaction. International Journal of Hydrogen Energy. 2015;40(8):3472–84.

[26]   Gallucci F, Basile A. Pd–Ag membrane reactor for steam reforming reactions: a comparison between different fuels. International Journal of Hydrogen Energy. 2008;33(6):1671–87.

[27] Fernandes FAN, Soares Jr AB. Methane steam reforming modeling in a palladium membrane reactor. Fuel. 2006;85(4):569–73.

[28] Ghasemzadeh K, Liguori S, Morrone P, *et al.* H_2 production by low pressure methanol steam reforming in a dense Pd–Ag membrane reactor in co-current flow configuration: experimental and modeling analysis. International Journal of Hydrogen Energy. 2013;38(36):16685–97.

[29] De Falco M, Di Paola L, Marrelli L, Nardella P. Simulation of large-scale membrane reformers by a two-dimensional model. Chemical Engineering Journal. 2007;128(2–3):115–25.

[30] De Falco M, Nardella P, Marrelli L, Di Paola L, Basile A, Gallucci F. The effect of heat-flux profile and of other geometric and operating variables in designing industrial membrane methane steam reformers. Chemical Engineering Journal. 2008;138(1–3):442–51.

[31] Basile A, Paturzo L, Gallucci F. Co-current and counter-current modes for water gas shift membrane reactor. Catalysis Today. 2003;82(1–4):275–81.

[32] Piemonte V, De Falco M, Favetta B, Basile A. Counter-current membrane reactor for WGS process: membrane design. International Journal of Hydrogen Energy. 2010;35(22):12609–17.

[33] Adrover ME, Anzola A, Schbib S, Pedernera M, Borio D. Effect of flow configuration on the behavior of a membrane reactor operating without sweep gas. Catalysis Today. 2010;156(3–4):223–8.

[34] Di Marcoberardino G, Sosio F, Manzolini G, Campanari S. Fixed bed membrane reactor for hydrogen production from steam methane reforming: experimental and modeling approach. International Journal of Hydrogen Energy. 2015;40(24):7559–67.

[35] Patel KS, Sunol AK. Modeling and simulation of methane steam reforming in a thermally coupled membrane reactor. International Journal of Hydrogen Energy. 2007;32(13):2344–58.

[36] Brunetti A, Caravella A, Barbieri G, Drioli E. Simulation study of water gas shift reaction in a membrane reactor. Journal of Membrane Science. 2007;306(1–2):329–40.

[37] Gallucci F, De Falco M, Tosti S, Marrelli L, Basile A. Ethanol steam reforming in a dense Pd–Ag membrane reactor: a modelling work. Comparison with the traditional system. International Journal of Hydrogen Energy. 2008;33(2):644–51.

[38] Boutikos P, Nikolakis V. A simulation study of the effect of operating and design parameters on the performance of a water gas shift membrane reactor. Journal of Membrane Science. 2010;350(1–2):378–86.

[39] Mendes D, Sá S, Tosti S, Sousa JM, Madeira LM, Mendes A. Experimental and modeling studies on the low-temperature water–gas shift reaction in a dense Pd–Ag packed-bed membrane reactor. Chemical Engineering Science. 2011;66(11):2356–67.

[40] Rahimpour MR, Arab Aboosadi Z, Jahanmiri AH. Synthesis gas production in a novel hydrogen and oxygen perm-selective membranes tri-reformer for

methanol production. Journal of Natural Gas Science and Engineering. 2012;9:149–59.

[41]   Rahimpour MR, Aboosadi ZA, Jahanmiri AH. Differential evolution (DE) strategy for optimization of methane steam reforming and hydrogenation of nitrobenzene in a hydrogen perm-selective membrane thermally coupled reactor. International Journal of Energy Research. 2013;37(8):868–78.

[42]   Abbasi M, Farniaei M, Rahimpour MR, Shariati A. Enhancement of hydrogen production and carbon dioxide capturing in a novel methane steam reformer coupled with chemical looping combustion and assisted by hydrogen perm-selective membranes. Energy & Fuels. 2013;27(9):5359–72.

[43]   Cornaglia CA, Adrover ME, Múnera JF, Pedernera MN, Borio DO, Lombardo EA. Production of ultrapure hydrogen in a Pd–Ag membrane reactor using noble metals supported on La–Si oxides. Heterogeneous modeling for the water gas shift reaction. International Journal of Hydrogen Energy. 2013;38(25):10485–93.

[44]   Chein R-Y, Chen Y-C, Chung JN-C. Mathematical modeling of hydrogen production via methanol-steam reforming with heat-coupled and membrane-assisted reactors. Chemical Engineering & Technology. 2014;37(11):1907–18.

[45]   Patrascu M, Sheintuch M. On-site pure hydrogen production by methane steam reforming in high flux membrane reactor: experimental validation, model predictions and membrane inhibition. Chemical Engineering Journal. 2015;262:862–74.

[46]   Marín P, Patiño Y, Díez FV, Ordóñez S. Modelling of hydrogen perm-selective membrane reactors for catalytic methane steam reforming. International Journal of Hydrogen Energy. 2012;37(23):18433–45.

[47]   Chein R, Chen Y-C, Chung JN. Parametric study of membrane reactors for hydrogen production via high-temperature water gas shift reaction. International Journal of Hydrogen Energy. 2013;38(5):2292–305.

[48]   Piemonte V, De Falco M, Basile A. Performance assessment of water gas shift membrane reactors by a two-dimensional model. Energy Sources, Part A: Recovery, Utilization, and Environmental Effects. 2015;37(20):2174–82.

[49]   Said SAM, Simakov DSA, Mokheimer EMA, *et al.* Computational fluid dynamics study of hydrogen generation by low temperature methane reforming in a membrane reactor. International Journal of Hydrogen Energy. 2015;40(8):3158–69.

[50]   Byron Smith RJ, Muruganandam L, Murthy Shekhar S. CFD analysis of water gas shift membrane reactor. Chemical Engineering Research and Design. 2011;89(11):2448–56.

[51]   Byron Smith RJ, Muruganandam L. CFD based optimization of water gas shift membrane reactor. International Journal of ChemTech Research. 2011;3(3):1520–5.

[52]   Chein RY, Chen YC, Chyou YP, Chung JN. Three-dimensional numerical modeling on high pressure membrane reactors for high temperature water–gas shift reaction. International Journal of Hydrogen Energy. 2014;39 (28):15517–29.

[53]   Chein RY, Chen YC, Chung JN. Sweep gas flow effect on membrane reactor performance for hydrogen production from high-temperature water–gas shift reaction. Journal of Membrane Science. 2015;475:193–203.

[54]   De Falco M. Pd-based membrane steam reformers: a simulation study of reactor performance. International Journal of Hydrogen Energy. 2008;33 (12):3036–40.

[55]   Hla SS, Morpeth LD, Dolan MD. Modelling and experimental studies of a water–gas shift catalytic membrane reactor. Chemical Engineering Journal. 2015;276:289–302.

Chapter 9

Hydrogen production using micro-membrane reactors

Jianhua Tong[1]

Abstract

Hydrogen production using micro-membrane reactors or membrane microreactors (MMRs) is a significant component of membrane reactor research. The recent progress of three types of MMRs such as hollow-fiber, microchannel, and monolithic MMRs are reviewed using specific examples. The representative application of MMRs as fuel processors for proton exchange membrane fuel cells are summarized in detail. In addition, the modeling progress about MMRs is also briefly introduced.

9.1 Introduction

Pure hydrogen in the form of diatomic molecules (H_2) has found extensive applications in the fields of clean energy technologies [1–3], petroleum refining [4], petrochemical processing [4], Harber–Bosch ammonia synthesis [5], and semiconductor processing [4]. Because it is highly abundant and environmentally benign and has high specific energy and lightweight, H_2 has been commonly recognized as one of the best energy carriers [6]. However, although hydrogen element is a ubiquitous component of matter in the universe, very few of it exists in the form of H_2 on the earth. Most of the H_2 what we are using nowadays has to be produced from hydrogen containing sources, involving in the chemical reactions such as steam reforming of natural gas [7], gasification of coal [8], gasification of biomass [9], and electrolysis of water [10]. In order to satisfy the high H_2 purity requested by the specific application purposes, these H_2 production processes are often followed by additional H_2 separation or purification operation units.

In recent years, membrane reactors (MRs) are attracting increasing attention for pure H_2 production from chemical reactions because of the integration of

[1]Department of Materials Science and Engineering, Clemson University, 515 Calhoun Drive, Clemson, SC 29634, USA

reaction and separation into one single unit [11–14]. The MRs can realize the in situ removal of H_2 from reaction zone for overcoming the thermodynamic bottleneck and simultaneously achieve high H_2 yield and purity [15,16]. It has been accepted that the MRs are able to much more efficiently, cleanly, sustainably, and cost-effectively produce pure H_2 in a highly intensified single step. Different types of MRs have been developed for H_2 production in the literature [15,16]. Because of the easiness for fabrication, the planar MRs are frequently used in earlier laboratory research and development studies, whereas for medium scale and industrial scale, the tubular MRs are much preferred due to their higher surface area-to-volume ratio in comparison to the planar MRs. The major researches about MRs for pure H_2 production are focusing on the discovery of new membrane materials with high H_2 permeability and high H_2 selectivity [17], the fabrication of defect-free thin membranes with high H_2 permeance without marked deterioration of H_2 selectivity [18], and the optimization of operation conditions for enhanced H_2 production performance [19]. Simultaneously, the development of new MR configurations with high intensification is becoming one of the most promising research topics in the field of MRs, which allows one to provide much larger effective membrane area in the planar or tubular MRs with the same volume [16,18].

A novel concept of micro-membrane reactors or membrane microreactors (MMRs) has been developed and has attracted fast growing attention. MMRs can be defined as microreactors reinforced by membrane separation/purification or MRs miniaturized into characteristic dimensions of 1–1,000 μm, which combine the advantages of both MRs and microreactors, leading to a highly compacted operation unit. The advantages of MMRs include (1) improvement of mass/heat transfer owing to the reduction of the scale length; (2) removal of mass-transfer limitations owing to the mitigation of concentration polarization; (3) intensification of process owing to the integration of different steps in a small-scale device; (4) enhancement of surface area to volume ratio owing to extremely high intensification; (5) high reactant conversion or low reaction temperature owing to thermodynamic equilibrium shifting resulted from the removal of specified product from reacted mixture; and (6) high selectivity of aimed product owing to optimized reactant distribution resulted from the addition of specified reactant in a controllable way [20–26].

Among the versatile MMRs, the MMRs with H_2 separation function are under prevailing investigation and have found a number of applications such as H_2 production from water gas shift (WGS) reaction [27,28], H_2 production from methanol steam reforming (MSR) reaction [29,30] on-board fuel processing for portable proton exchange membrane fuel cells (PEMFCs) [31], production of moisture-free formaldehyde by the dehydrogenation of methanol [32], and dehydrogenation of cyclohexane to benzene [33]. In this chapter, in the following sections, the types of MMRs will be briefly introduced. Then the H_2 production and dehydrogenation reactions in different types of MMRs will be described using some specific examples. After that, the concept of MMRs using as the fuel processors for portable PEMFCs will be summarized. Finally, the modeling work of the MMRs for H_2 production will be introduced.

9.2   Types of MMRs

According to the basic search of keyword "membrane reactor" in topic on the premier research platform of the Web of Science, the total publications related to MRs in the recent 15 years (2001–2015) is ~30,000. Among them, about one-fourth is focused on the research of H_2, which clearly demonstrates the significance of MRs for H_2 production. In general, the MRs with modifiers of micro, hollow fiber, microchannel, nanotube, nanofiber, monolithic, etc. can be ascribed to be MMRs, which is close to 10% of the total publications of the MRs, representing a very significant component of MR research. Figure 9.1 shows the publication distribution of the different types of MMRs during the recent 15 years. About 900 publications have been found for the generic type of MMRs with micro as modifier before MR, which accounts for more than one-third the MMR publications. Hollow-fiber MMRs with hollow fiber as modifier, which is also thought to be MMRs based on their dimension, represents the largest portion of MMRs, accounting for almost half of the MMR publications. The publications of the more intensified microchannel and nanotube MRs both are ~7% of the total MMR publications, which are much lower than those for hollow-fiber MMRs because their fabrication process is complicated. The new concept of nanofiber MMRs and monolithic MMRs are very promising MMRs because of their extremely high intensification. However, the fabrication processes are so complicated that only 1%–2% of MMR publications are for nanofiber and monolithic MMRs. The publications of each type of MMRs related to H_2 are also provided in Figure 9.1. It is very easy to find out that H_2 MMRs are significant component for all the MMRs, which are 20%–45% of each type of MMRs. Some specific examples of the hollow-fiber, microchannel, and monolithic MMRs used for H_2 production will be briefly reviewed in the following subsections.

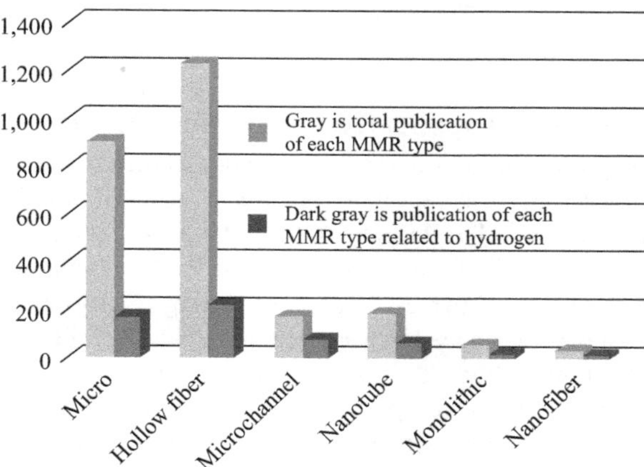

Figure 9.1   The publication distribution of different types of MMRs based on the basic search of web of science from 2001 to 2015

9.2.1   Hollow-fiber MMRs

In order to achieve the highly compact MRs, to decrease the diameter of the membrane tubes in tubular MRs is the most straightforward strategy and has been extensively studied [34,35]. When the diameter of a membrane tube is reduced to a certain level, that is ID < 1,000 μm, it becomes a hollow fiber and the fiber lumen may take effects of a microchannel. Catalysts can be coated on the inner surface of the hollow fibers or impregnated inside the porous wall, whereas the separation can be achieved by the porous hollow fibers themselves or by the membrane formed on the outer surface of the hollow fibers. This kind of hollow-fiber MRs can also be attributed to MMRs, then called hollow-fiber MMRs [18]. The hollow fibers are usually fabricated by a phase inversion process [36,37]. Solid powder is dispersed in a polymer solution to form a slurry, which is further spun through a spinneret. Therefore, the geometry of the resulting capillaries depends mainly on the geometry of the spinneret. By the variation of the slurry composition and precipitation conditions, the structure of the hollow fibers can be controlled from finger-like pores to foam-like structures. Therefore, the porosity can be varied from 50%–70% to 20%–30%. The finger-like structure is one of the main advantages of the ceramic support material chosen, since it gives the necessary stability for thin membrane layers, and at the same time the transport resistance over the support can be minimized so that high support permeability becomes possible. The fully dense metallic thin films (e.g., Pd and Pd alloys) and defect-free microporous thin films (e.g., zeolite) are commonly deposited on the outer surface of ceramic hollow fibers (e.g., Al_2O_3) for the fabrication of MMRs for H_2 production. The protonic ceramic hollow fibers with a thin dense outer surface layer are extensively studied to directly separate H_2 from mixture or reaction product. In recent years, the oxygen permeable mixed ionic and electronic conducting (MIEC) hollow fibers with configuration similar to the protonic ceramic hollow fibers have attracted increasing attention to produce pure H_2 by splitting water. In most recent, the cost-effective metal hollow fibers were also fabricated for H_2 separation. Therefore, in this section, we will give some detailed research examples for these five kinds of hollow-fiber MMRs.

9.2.1.1   Pd- and Pd–Ag-supported hollow-fiber MMRs

Pd and Pd–Ag supported hollow-fiber MMRs are one of the most popular MMRs for H_2 separation [38–40]. In the recent work of Li *et al.* [41–43], a hollow-fiber MMR (Figure 9.2) consisting of a thin Pd membrane and a 30% CuO/CeO_2 catalyst on an asymmetric Al_2O_3 hollow fiber was developed. The Pd membrane was deposited on outer surface of the hollow fiber by electroless plating (ELP) technique. The 30% CuO/CeO_2 catalyst was introduced by sol–gel impregnation of the finger pores in the inner surface of the hollow fiber. The catalytic activity of the hollow-fiber MMR was tested using WGS reaction and compared to that of a conventional fixed-bed reactor. In the conventional fixed-bed reaction, the maximum conversion of 23% was obtained at 450 °C, was much lower than the equilibrium value (~41%). In the hollow-fiber MMR, the highest conversion of 60% was obtained at 500 °C and a sweep gas flow of 75 mL min^{-1}.

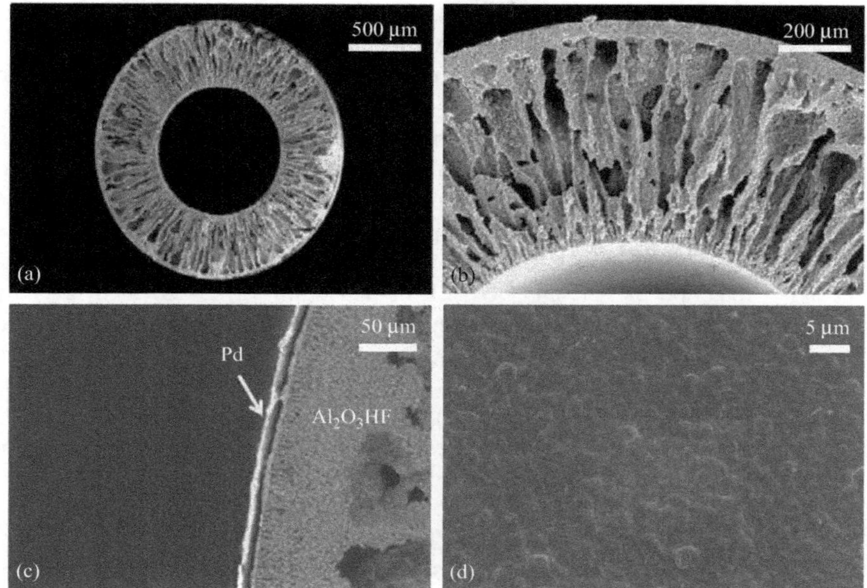

Figure 9.2   SEM pictures of Al$_2$O$_3$ hollow-fiber MMR; (a) and (b) cross section at different magnifications before Pd deposition, (c) cross section after Pd deposition, and (d) top of the Pd membrane surface. © 2010 Elsevier Ltd. Reprinted, with permission, from Reference 41

Another hollow-fiber MMR was constructed by Li *et al.* [44] using an yttria-stabilized zirconia hollow fiber as a substrate for the deposition of Pd/Ag membrane on the outer shell of the hollow fiber and for 10-wt.%NiO/MgO–CeO$_2$ catalyst inside the hollow fiber. The ethanol steam reforming (ESR) was carried out in this hollow-fiber MMR at temperatures ranging from 350 to 550 °C. Reference ESR reaction in a conventional fixed-bed reactor was performed at the same temperature range. At 510 °C, the flow rate of hydrogen produced in the hollow-fiber MMR was three times higher than that in the conventional fixed-bed reactor despite less amount of the catalyst was used in the hollow-fiber MMR. Moreover, during the operation of the hollow-fiber MMR, the high purity H$_2$ was obtained outside of the shell, and the yield was more than 53% of the total H$_2$ produced in the ESR.

They also constructed a two-zone fluidized bed reactor (TZFBR) using Pd hollow-fiber membrane for catalytic dehydrogenation of propane [45]. This design combined the in-situ catalyst regeneration provided by the regeneration zone with the conversion improvement achieved by a Pd membrane, which removes H$_2$ from the reactor, thus enhancing the reaction rate and in theory allowing even higher than equilibrium conversion. The experiments showed that the Pd hollow-fiber membrane acted effectively removing H$_2$ from the reaction media. However, with this catalyst the quicker coke formation caused by the removal of H$_2$ outweighed the beneficial effect of H_2 removal, and the yield achievable for a given feed was lower in the presence of the membrane. A conclusion from these results was that a catalyst with lower coke tendency was needed to apply this combination of Pd membrane and TZFBR.

Tanaka *et al.* [46] fabricated a H_2 permeable composite membranes by packing Pd nanoparticles into nanosized pores of alumina hollow-fiber support. The fabrication process involved the following steps: (1) coating a mesoporous γ-alumina layer (8–18 nm particle size) on a macroporous α-alumina tube; (2) activating the γ-alumina layer by seeding with palladium nuclei; (3) coating an additional γ-alumina layer as a protection barrier; and (4) ELP of palladium at the activated layer by applying a vacuum. Unlike metallic-colored palladium-film membranes, the obtained composite palladium membranes are black in color because very fine palladium particles are dispersed in the pores. The H_2 permeation rate and selectivity against N_2 was more than 10^{-6} mol m^{-2} s^{-1} Pa^{-1} and well over 1,000, respectively, at 300 °C using a 400-kPa pressure difference. These values were comparable to values for conventional Pd-film membranes with a similar thickness.

Nair and Harold [47] fabricated that "Pd encapsulated" and "Pd nanopore" membranes supported on α-Al$_2$O$_3$ hollow fibers using the similar method as Tanaka *et al.* [46]. The permeation characteristics (flux and permselectivity) of a series of unaged and aged encapsulated and nanopore membranes with different Pd loadings were compared to those of a conventional 1-μm Pd/4-μm γ-Al$_2$O$_3$/α-Al$_2$O$_3$ hollow-fiber membrane. The encapsulated membrane exhibited good performance with ideal H_2/N_2 separation factors of 3,000–8,000 and H_2 flux ∼0.4 mol m^{-2} s^{-1} at 370 °C and a transmembrane pressure gradient of 4×10^5 Pa. The Pd nanopore membranes had a lower initial flux and permselectivity but exhibited superior performance with extended use (200 h).

Sun *et al.* [48] systematically tuned the morphology of Al$_2$O$_3$ hollow fiber by varying both the internal and external coagulants as well as the calcination temperature in order to enhance the mechanical strength and improve the surface property of the Al$_2$O$_3$ hollow fiber. The bending test and scanning electron microscopy (SEM) results showed that the finger-like void-free Al$_2$O$_3$ hollow fiber calcined at 1,500 °C not only possessed very high bending strength but also uniform surface. The Al$_2$O$_3$ hollow fiber was successfully employed as a substrate to form an ultrathin Pd membrane by ELP without any modification process. The permeation test showed that this ultrathin Pd membrane supported on the Al$_2$O$_3$ hollow-fiber substrate exhibited high permeance and selectivity for separation of H_2 from a H_2/N_2 mixture.

Nair *et al.* [49] did a comprehensive study on ELP and permeation of a series of Pd and Pd/Ag membranes supported on α-Al$_2$O$_3$ hollow fibers. A multi-fiber ELP apparatus quantified Pd and Ag deposition rates over a range of conditions. The Pd deposition rate exhibited a maximum, whereas the Ag growth rate demonstrated autocatalytic behavior and was enhanced by Pd. The grain size and morphology were sensitive to the plating temperature. Three membrane types synthesized by ambient temperature ELP were of high quality determined by permeation measurements: Pd (13 μm)/α-Al$_2$O$_3$, Pd–(12 wt.%)Ag (11 μm)/α-Al$_2$O$_3$, and Pd (5 μm)/α-Al$_2$O$_3$. Initial binary separation factors of 1,000–10,000 were obtained with the low temperature ELP. The permselectivity for the Pd membrane decreased to 100 after 300 h of temperature and pressure cycling. The Pd/Ag alloy membrane exhibited high permselectivity for extended operation with the separation factor (α) decreasing from an initial value of 970 to 650 after 300 h with the H_2 flux sustained at 0.5 mol m^{-2} s^{-1} at 550 °C for a transmembrane pressure gradient (ΔpH_2) of $\sim 4 \times 10^5$ Pa. The thinner composite Pd/γ-Al$_2$O$_3$/α-Al$_2$O$_3$ exhibited high initial flux and permselectivity (0.5 mol m^{-2} s^{-1}, $\alpha \sim 1{,}000$ at 350 °C, $\Delta pH_2 \sim 4 \times 10^5$ Pa) with reduced nonselective transport compared to the Pd/α-Al$_2$O$_3$ fiber.

9.2.1.2 Protonic ceramic hollow-fiber MMRs

Although Pd- and Pd–Ag-supported hollow-fiber MMRs showed high H_2 permeability and high H_2 selectivity, they have to face with the challenges of high price and low operation temperature (<600 °C). Protonic ceramic membranes with mixed protonic and electronic conductivity allow for high temperature H_2 permeation and are thought to be one of the most promising high-temperature H_2 permeable membranes for MMRs [50–52].

The prototypical protonic ceramic material of SrCe$_{0.95}$Yb$_{0.05}$O$_{3-\delta}$ (SCYb) was prepared into hollow-fiber membranes by Liu *et al.* [50] using an immersion-induced phase inversion, in which polyethersulfone was used as a binder, *N*-methyl-2-pyrrolidone as a solvent, and poly(vinylpyrrolidone) (PVP (K90)) as an additive, respectively. After the membrane was properly characterized, a MR was subsequently constructed with the SCYb hollow fiber. The methane coupling reaction in the hollow-fiber MMR was investigated both experimentally and theoretically. The results showed that H_2 permeation through the membrane promoted the formations of ethylene and ethane. An increase of operating temperatures was favorable for the formation of ethylene over ethane. Due to the methane carbonization and the formation of higher hydrocarbons (C3–C5), the C2 selectivity decreased as the temperature was increased. When the temperature increased to 950 °C, the C2 yield was improved to a maximum of 13.4% and then decreased if the temperature was increased further. When increasing the methane feed flow rate, the ethylene formation rate was increased, but the ethane production was decreased.

In the work of Xue *et al.* [51], a U-shaped protonic ceramic hollow-fiber membrane with the composition La$_{5.5}$W$_{0.6}$Mo$_{0.4}$O$_{11.25-\delta}$ has been fabricated and used to construct hollow-fiber MMR. The nonoxidative methane dehydroaromatization (MDA) (6CH$_4$ $\rightleftarrows$ C$_6$H$_6$ + 9H$_2$) with 6 wt.% Mo/HZSM-5 as a catalyst was tested in this protonic ceramic hollow-fiber MMR. The H_2 was in situ removed to overcome thermodynamic constraints and shift the reaction equilibrium to the H_2 direction. The yield of aromatics (benzene, toluene, and naphthalene) in the MDA was increased in the beginning of the aromatization reaction by $\sim 50\%$–70%, in comparison with the conventional fixed-bed reactor, because 40%–60% of the produced H_2 has been extracted at 700 °C. These advantages of the MR operation decreased with time on stream, since the removal of H_2 boosted not only CH$_4$ conversion and yield of aromatics, but also catalyst deactivation by deposition of carbonaceous deposits.

In the work of Song *et al.* [52], BaCe$_{0.85}$Tb$_{0.05}$Co$_{0.10}$O$_{3-\delta}$ perovskite hollow-fiber membranes were also fabricated by a phase-inversion technique. The hollow-fiber surfaces were modified by coating Ni or Pd particles. H_2 permeation at 700–1,000 °C was improved because of the surface modification from the original 0.009–0.164 to 0.018–0.269 mL cm^{-2} min^{-1} for the Ni-coated membranes with maximum improvement by 64%, and to 0.1–0.42 mL cm^{-2} min^{-1} for the Pd-loaded membranes with maximum enhancement by 155%, respectively. Loading of the catalyst on the hollow-fiber outer surface was better than on the inner surface, but coating on both sides may enhance the H_2 permeation most effectively. The permeation enhancement relied on both the catalyst loading amount and its structure. The optimal Pd loading and coverage was around 0.667 mg cm^{-2} and 82%, respectively, for maximizing the permeation improvement.

9.2.1.3　Mixed ionic and electronic conducting hollow-fiber MMRs

MIEC membranes are commonly used for oxygen separation MRs for partial oxidation of methane (POM) and production of pure oxygen [53,54]. In the recent work of Jiang *et al.* [55], it was reported that the simultaneous production of hydrogen and synthesis gas in a hollow-fiber perovskite MIEC MR using methane to consume the permeated oxygen. At high temperatures, water dissociates into H_2 and O_2 on the membrane surface of the core side. Oxygen permeated from the core to the shell side of the hollow fiber, in which it was consumed by the POM to form synthesis gas according to $CH_4 + 1/2O_2 = CO + 2H_2$. Thus, when operating under equilibrium-controlled conditions, the water dissociation proceeded continuously as the oxygen was continuously consumed by the POM to produce synthesis gas. The advantage of this process is to produce pure H_2 as well as synthesis gas, which can be used to synthesize a wide variety of valuable hydrocarbons and oxygenates. A $BaCo_xFe_yZr_{1-x-y}O_{3-\delta}$ (BCFZ) perovskite hollow-fiber membrane was applied for the in-situ removal of the oxygen that was produced by high-temperature water splitting. The H_2 production rate increased as the temperature raised from 800 to 950 °C, and a H_2 flux as high as 3.1 mL min^{-1} cm^{-2} was obtained at 950 °C.

In the progressive work of Jiang *et al.* [56,57], a porous perovskite $BaCo_xFe_yZr_{0.9-x-y}Pd_{0.1}O_{3-\delta}$ (BCFZ-Pd) coating was deposited onto the outer surface of a BCFZ perovskite hollow-fiber membrane. The surface morphology of the modified BCFZ fiber was characterized by SEM, indicating the formation of a BCFZ-Pd porous layer on the outer surface of a dense BCFZ hollow-fiber membrane. The oxygen permeation flux of the BCFZ membrane with a BCFZ-Pd porous layer increased 3.5 times more than that of the blank BCFZ membrane when feeding reactive CH_4 onto the permeation side of the membrane. The blank BCFZ membrane and surface-modified BCFZ membrane were used as reactors to shift the equilibrium of thermal water dissociation for H_2 production, because they allow the selective removal of the produced oxygen from the water dissociation system. It was found that the H_2 production rate increased from 0.7 to 2.1 mL min^{-1} cm^{-2} at 950 °C after depositing a BCFZ-Pd porous layer onto the BCFZ membrane.

Reactor modules consisting of four gas-tight microtubular membranes made of the MIEC perovskite, $La_{0.6}Sr_{0.4}Co_{0.2}Fe_{0.8}O_{3-\delta}$, were fabricated by Franca *et al.* [58] and tested for oxygen permeation and H_2 production by membrane-based steam reforming, that is, simultaneous syngas production coupled with water splitting. The membranes were subjected to two known axial temperature profiles in the temperature of 900 °C. The microtubes showed good stability under reaction conditions, operating over a total operation period of ca. 400 h of oxygen permeation followed by ~400 h of steam reforming. An induction period of approximately 30 h was observed before steam reforming commenced. Both the methane and water side outlet gas compositions were measured, which allowed for accurate material balance, which indicated that the H_2 production occurred due to oxygen flux across the membrane and not just surface reaction. Post-operation analysis of the microtubes revealed the presence of a strontium-enriched dense layer on the water-exposed membrane surface and of crystallites enriched with cobalt and sulfur on the methane feed side surface.

9.2.1.4   Microporous MMRs

Microporous membranes are also commonly used for H_2 production and coupled into MMRs [59–61]. In the work of Hong *et al.* [59], Mordenite Framework Inverted (MFI) zeolite membranes were synthesized on porous α-alumina hollow fibers by in-situ hydrothermal synthesis. The membranes were further modified for H_2 separation by on stream catalytic cracking deposition of methyldiethoxysilane in the zeolitic pores. The separation performance of the modified membranes was characterized by separation of H_2/CO_2 gas mixture at 500 °C. Activation of MFI zeolite membranes by air at 500 °C was found to promote catalytic cracking deposition of silane in the zeolitic pores effectively, which resulted in significant improvement of H_2-separating performance. The H_2/CO_2 separation factor of 45.6 with H_2 permeance of 1.0×10^{-8} mol m^{-2} s^{-1} Pa^{-1} was obtained at 500 °C for a modified hollow-fiber MFI zeolite membrane. The as-made membranes showed good thermochemical stability for the separation of H_2/CO_2 gas mixture containing H_2O and H_2S, respectively.

Carbon membranes are less expensive than the completely selective Pd membrane and are significantly more selective than porous ceramics. Barbosa-Coutinho *et al.* [60] prepared carbon hollow-fiber membranes by the pyrolysis of polyetherimide hollow fibers. The influence of process variables on the final membrane morphology using a statistical experimental design was investigated. Sznejer *et al.* [61] fabricated molecular-sieve carbon membrane hollow-fiber module and the transport and separation of H_2 and light hydrocarbons were studied in a molecular-sieve carbon membrane hollow-fiber module, at the temperature range of 25–400 °C; nitrogen was used as a sweeping gas in the study of mixtures, and the fluxes of pure components were studied under a pressure gradient. The membrane selectivity, the ratio of H_2-to-hydrocarbon permeabilities, reached 100–1,000 in propane or butane with H_2, making the membrane an excellent candidate for a membrane dehydrogenation reactor.

9.2.1.5   Metallic hollow-fiber MMRs

In recent years, some cost-effective metals were directly made into hollow fibers for H_2 production [62,63]. In the work of Wang *et al.* [62], metallic nickel dense hollow-fiber membranes were fabricated by a combined spinning and high-temperature sintering technique. H_2 permeation through the nickel hollow-fiber membranes was measured at high temperatures up to 1,000 °C using H_2-containing gas mixtures fed on the shell side and N_2 as the sweep gas in the fiber lumen. The experimental results indicated that the sintering should be carried out at around 1,400 °C for 3 h, under a H_2-containing atmosphere so as to reach the required densification of the nickel hollow-fiber membranes. H_2 permeation through the dense nickel hollow-fiber membranes was controlled by H-atom diffusion through the membrane bulk and can be well described by the Sieverts' equation with the activation energy of 51.07 kJ mol^{-1}. For the hollow fiber with wall thickness of 256 μm, the H_2 permeation flux value reached up to 7.66×10^{-3} mol m^{-2} s^{-1} at 1,000 °C with 100% H_2 permselectivity. The Ni hollow membrane exhibited high stability in CO_2, CO or steam containing atmospheres.

Stainless steel (SS) is one of excellent materials for fabrication of hollow-fiber membranes because of its good mechanical property. In the work of Rui *et al.* [63], they prepared SS porous hollow-fiber membranes by combined dry-wet spinning and sintering technique. The influence of sintering atmosphere (air, CO_2, N_2, He, and H_2) on microstructure and properties of SS hollow fibers were investigated extensively. It was found that air and CO_2 could lead to metal oxidation, whereas inert atmospheres (He and N_2) caused carbon remaining in SS hollow fibers. By contrast, H_2 atmosphere was effective in removing organic additives without metal oxidation. High mechanical strength could be achieved for the hollow-fiber membranes sintered in H_2 atmosphere. At 1,100 °C for 1 h, the as-sintered SS hollow fiber showed high bending strength of 4,200 MPa and nitrogen permeance up to 9.5×10^{-5} mol s^{-1} m^{-2} Pa^{-1}.

9.2.2   Microchannel MMRs

Micro-electro-mechanical systems (MEMS) have been used in microfluidics to allow fluid handling in microscale channels and chambers etched in silicon substrates. This concept was also used to the fabrication of microchannel MMRs [64–69]. Figures 9.3 and 9.4 show the representative description of membrane fabrication procedure and the morphology of resulted membranes [65].

Kusakabe *et al.* [64] successfully demonstrated the feasibility of microchannel catalytic reactor system. In their work, a self-heating microreactor was constructed on a (1 0 0) silicon wafer by means of photolithography, wet etching, and sputtering techniques based on the MEMS concept. The reactor was used for catalytic reactions at elevated temperatures. A Pd membrane, approximately 3 μm in thickness, was also fabricated by a combination of electrolysis and photolithography. The separation factor of H_2 to N_2 was 120 at 600 °C. The results indicated that the construction of a microreactor combined with a small-sized membrane separator (microchannel MMR) was a distinct possibility.

Gielens *et al.* [70] fabricated high-flux Pd–Ag alloy membranes using MEMS technology. A dense Pd–Ag film (77-wt.%Pd/23-wt.%Ag) with a thickness of 750 nm was deposited on a nonporous 1-mm thick silicon nitride layer by cosputtering of a Pd and a Ag target. After sputtering, openings of 5 μm were made in the silicon nitride layer to create a clear passage to the Pd–Ag film. The as-fabricated Pd–Ag membranes were pinhole free and have a low resistance to mass transfer in the gas phase.

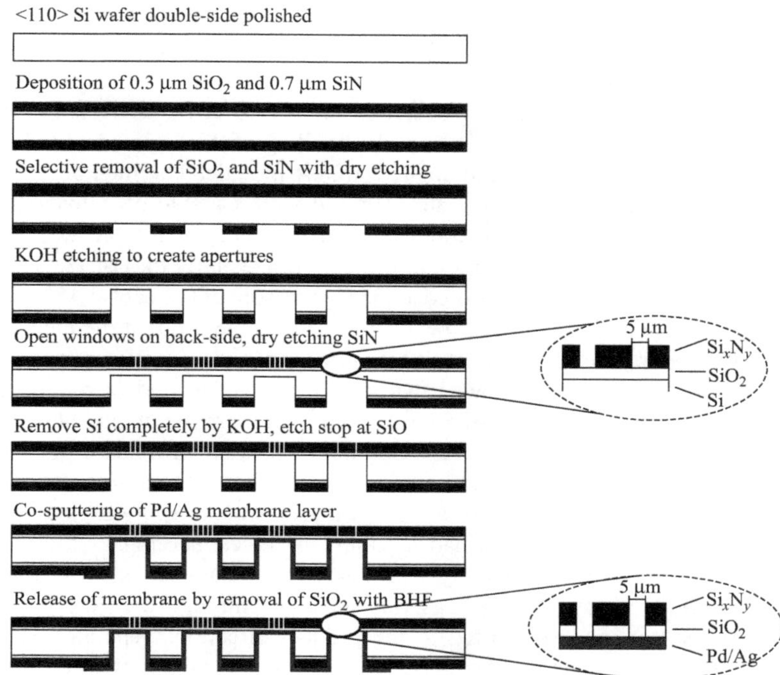

Figure 9.3   Steps followed to manufacture a microsieve-supported membrane by microsystem technology. © 2004 American Chemical Society. Reprinted, with permission, from Reference 65

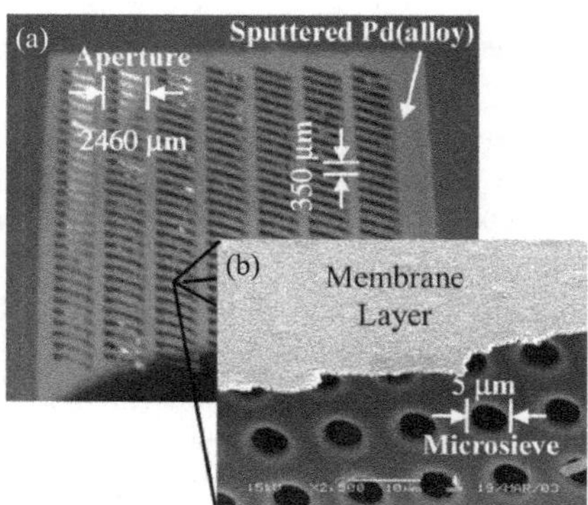

Figure 9.4   Morphology of a Pd microchannel membrane prepared by microsystem technology. (a) Top view of the aperture side and (b) SEM picture of the membrane layer deposited on top of the microsieve (membrane layer was partly removed to show the microsieve). © 2004 American Chemical Society. Reprinted, with permission, from Reference 65

The H_2 permeability of the Pd–Ag membranes ranged from 0.02 to 0.95 mol m^{-2} s^{-1} at varying temperatures of 350–450 °C. In addition, the H_2 selectivity over He was determined to be above 1,500.

Tong *et al.* [71] from the same research group with Gielens fabricated a similar wafer-scale Pd–Ag alloy membranes. Anodic bonding of thick glass to silicon was used to package the membrane and create a robust module. The membranes were found to have adequate mechanical strength and were capable of withstanding a pressure difference of 4–5 bars at room temperature. The microfabricated Pd–Ag membranes obtained a high permeation rate and high selectivity for H_2. Typical flow rates of 0.5 mol m^{-2} s^{-1} were measured at the H_2 pressure of 0.2 bars at 450 °C with a minimal selectivity of 550 for H_2 to He. The results indicated that the membranes were suitable for application in H_2 purification or in dehydrogenation reactors. The presented fabrication method allowed the development of a module for industrial applications consisting of a stack of a large number of glass/membrane plates.

Karnik *et al.* [27] fabricated a Pd-based microchannel MMR based on silicon substrate and integrated with Cu catalysts using a standard MEMS microfabrication method. The membrane was composed of four layers: Cu, Al, spin-on-glass (SOG), and Pd. Cu, Al, and SOG layers had a pattern of holes, serving as a structural support for the Pd membrane. Cu layer also acted as a catalyst for WGS reaction occurring in the channels. The H_2 gas was separated through the Pd membrane; thus, the WGS reaction was shifted to the product direction. Separated H_2 then flowed through corresponding channels. For example, an MMR with 66-nm Cu, 200-nm Al, 500-nm SOG, and 200-nm Pd was constructed. The H_2 permeation behavior was tested by using a mixture of 20-wt.% H_2 and 80-wt.% Ar. The H_2 flux was found to increase with temperature as well as the H_2 partial pressure gradient across the membrane. The H_2 flux at 100 °C under a H_2 partial pressure gradient of 8.3 psi was as high as 5.2 mol m^{-2} s^{-1}. The activation energy for the H_2 permeation was calculated to be 1.0 kJ mol^{-1}, which was much smaller than the reported values for Pd membranes and was attributed to the microstructure of the sputtering deposited film.

Mejdell *et al.* [67–69] constructed a microchannel MMR in a planar configuration from thin defect-free Pd–23-wt.%Ag membranes and tested the H_2 permeation behavior in detail. Pd–23-wt.%Ag dense thin films were prepared by DC magnetron sputtering on polished silicon single crystal substrates from a Pd–23-wt.%Ag target by using argon as sputtering gas. After sputtering, the Pd–Ag thin films were removed manually from the substrate and integrated to SS microchannels to produce a planar microchannel MMR. This microchannel MMR is consisted of a SS feed channel plate with six parallel channels with dimensions of 1 mm × 1 mm × 13 mm. The Pd–23-wt.%Ag membrane was placed between the channel housing and a SS plate with apertures corresponding to the channel geometry. The SS plate was employed for mechanical support. In total, the six channels provided for a 0.78 cm^2 active membrane area. Both the feed housing and the perforated steel plate were highly polished (down to 3 μm) before mounting. On the permeate side, an open housing was employed, and a copper gasket was placed between the perforated steel plate and the permeate housing for sealing. From measurements of pure H_2 in a microchannel MMR integrated with 1.4-μm Pd/23-wt%Ag membrane, a permeance of 1.7×10^{-2} mol m^{-2} s^{-1} Pa$^{-0.5}$ was obtained. This MMR was tested for $\sim$7 days at 300 °C, whereas the differential pressure >470 kPa was applied. The H$_2$/N$_2$ separation factor decreased from $\sim$5,700 to $\sim$390 at $\sim$300 kPa differential pressure. Furthermore, the effects of CO and CO$_2$ on hydrogen permeation behavior, through a $\sim$3-μm Pd/23-wt%Ag membranes employed in microchannel MMRs, were also investigated in detail. The membrane permeance was determined to 5.1×10^{-3} mol m^{-2} s^{-1} Pa$^{-0.5}$ at 300 °C under pure hydrogen. After the last experiment, a small leakage occurred, which reduced H$_2$/N$_2$ separation factor to $\sim$3,300 at a pressure gradient of 200 kPa. Both CO and CO$_2$ showed an inhibitive effect on hydrogen permeation. The CO effect was strongly dependent on both temperature (275–300 °C) and CO concentration (0–5 mol%). The CO inhibition occurred so rapidly that a sharp drop in the hydrogen permeation flux was found when CO concentration was between 0 and 0.25 mol%. The time required to restore the initial hydrogen permeation flux after CO exposure became longer when the exposure temperature was lowered. CO desorption was thought to be the main mechanism for hydrogen permeation flux restoration at the higher temperatures, whereas it was controlled by other, slower processes at the lower temperatures. The inhibitive effect of CO$_2$ was milder, and long-time exposure was necessary to reach apparently stable hydrogen permeation values. Only a weak effect was observed at 350 °C, whereas at 300 °C, a nearly linear decrease was observed over several days. The main inhibition mechanism was thought to be a slow formation and removal of some strongly adsorbed species rather than CO$_2$ competitive adsorption.

9.2.3 Monolithic MMRs

Honey-comb or straight-channel monoliths can provide an inexpensive and rapid means of constructing scalable two-dimensional arrays of identical square microchannels with diameters 500–5,000 μm and wall thickness 200–2,000 μm [23]. This kind of structures can be formed from a variety of porous ceramic materials such as cordierite, mullite, and alumina, which can realize large networks of MMRs. The monolithic MMRs can provide with much better mechanical stability than the hollow-fiber MMRs and much higher intensification than the planar microchannel MMRs [23].

In the research of Kim *et al.* [23], the dense thin Pd membranes were fabricated within the cordierite microchannel networks cut from honeycomb-monolith support, and the hydrogen permeation behavior was evaluated (Figure 9.5). Cordierite-extruded honeycomb monoliths (64 cpsi) were first washcoated with a micropowder γ-Al$_2$O$_3$ layer to form cylindrical surfaces for subsequent deposition of Pd films. After that, a nanopowder γ-Al$_2$O$_3$ layer was deposited in order to get a uniform surface for deposition of defect-free Pd films. Thin (8 μm) defect-free Pd films with crystallite sizes of $\sim$2 μm were deposited on nanopowder γ-Al$_2$O$_3$ layer by the conventional ELP under kinetic limited conditions. Analysis of resulting two-channel membrane systems for hydrogen separation at 350 °C demonstrated hydrogen fluxes of 1.0–5.5×10^{-3} mol m^{-2} s^{-1} and H$_2$/He separation factors of 40–360. The authors

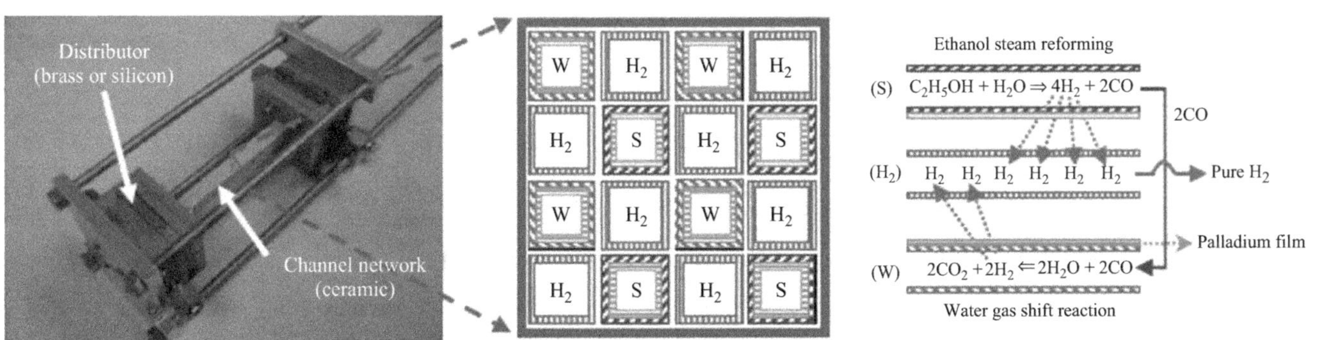

Figure 9.5   Picture of prototype of honeycomb-monolith MMRs, with inset showing schematic of integration schemes. © 2009 Elsevier Ltd. Reprinted, with permission, from Reference 23

also detailed a novel MMR strategy to employ resulting minichannel membrane networks as part of a thermally integrated portable reformer for high-purity hydrogen production.

9.3   MMRs as a fuel processor for portable PEMFCs

Current energy demand for portable electronic devices has encouraged researchers to investigate miniaturized fuel cell systems based on PEMFCs [72–76]. However, due to low-energy density of compressed H_2, it is not an ideal option. Liquid fuel, in particular methanol, has a much higher energy density and is easier to transport and handle, which makes it a very attractive for the on-board fuel for portable PEMPCs. Two types of portable PEMPC systems were proposed: direct methanol fuel cells have the advantage of room-temperature operation and dis-advantage of relatively low power density due to methanol crossover and low reaction rate of methanol oxidation. In contrast, the portable system of on-board reformer combined with PEMFCs can generate electricity in the fuel cell from concentrated hydrogen produced by steam reforming, for example, from methanol, which can achieve high power density. However, it is difficult to miniaturize the on-board reformer due to the complexity of the required structure, which includes not only a fuel reformer but also a H_2 purifier. In recent years, the H_2 permeable MMRs have been started to be used on-board reformer for portable PEMPCs.

Karnik *et al.* [72] explored the use of a microreactor to produce H_2 by the reaction of methanol with water. Their work concentrated on the design, fabrication, and performance evaluation of a Pd-based MMR for H_2 gas separation as well as WGS reaction in this miniature fuel processor. Integrated Pd-based micromembranes were microfabricated and tested for their mechanical strength. The device can be used for WGS reaction and H_2 gas separation in catalytic microreaction systems for methanol fuel reforming. The novelty of this structure is that we have integrated the WGS reactor as well as the H_2 gas separator in the same structure. This is because copper can act as a membrane support as well as a catalyst in the WGS reaction. For a particular combination of the thicknesses of its component layers, the micro-membrane rupture pressure was as high as 1 atm. The micromembrane separated H_2 from a 20% hydrogen balance nitrogen gas mixture at room temperature.

In the work of Kawamura *et al.* [25], a miniaturized methanol reformer was developed to provide H_2 for a small PEMFC. The microreformer was consisted of a catalyst-coated microchannel in a serpentine arrangement, with a length of 333 mm and cross-section of 0.6×0.4 mm^2. The microreformer was fabricated from silicon and glass substrates using a number of microfabrication techniques. Selective deposition of with the $Cu/ZnO/Al_2O_3$ catalyst in the microchannel was achieved by employing a photolithography technique. The overall size of the microreformer ($25 \times 17 \times 1.3$ mm^3) makes the small fuel cells suitable for application as a power source for portable electronic devices. MSR was tested in this microreformer, which demonstrated that the microreformer was capable of hydrogen production rates exceeding 0.05 mol h^{-1} at reactant feed rate of 1.6 mL h^{-1}. Based on the lower H_2 heating value of 241 kJ mol^{-1}, this H_2 production rate corresponded to 3.3 W$_{th}$ of H_2 power, in which electrical power greater than 1 W is expected assuming 45% fuel cell efficiency and 70% of hydrogen utilization, making it potentially applicable as a power source for a cell phone.

In the work of Ilinich *et al.* [29], a Pd-based catalyst was developed for the ceramic microreformer in a miniaturized PEMFC for the catalytic reaction MSR. In the microreformer, the catalyst was washcoated directly on the walls of the steam-reforming section, providing favorable conditions for efficient heat transfer between the heat-generating catalytic combustion section of the microreformer and its heat-consuming steam reforming section. The Pd-based catalyst showed activity and selectivity similar to those of Cu–Zn–Al catalysts but was more durable and stable under the duty-cycle conditions of a portable power source.

An integrated microchannel methanol processor consisting of fuel vaporizer, heat exchanger, catalytic combustor, and steam reformer was developed by Park *et al.* [77]. Unit reactor was fabricated by stacking and bonding several micro-channel plates. Commercially available $Cu/ZnO/Al_2O_3$ catalyst was coated inside microchannel of the unit reactor for steam reforming. Pt/Al_2O_3 pellets prepared by an incipient wetness method were filled in the cavity reactor for catalytic com-bustion. Other unit reactors were used as heat exchanger and fuel vaporizer. Those unit reactors were assembled, and two reactions occurred independently in the integrated reactor for H_2 production by MSR and for heat generation by catalytic combustion. The integrated reactor has the dimensions of 60 mm $\times$ 40 mm $\times$ 30 mm and produced 450-sccm-reformed gas including 73.3% H_2, 24.5% CO_2, and 2.2% CO at 230–260 °C which can produce a power output of 59 W.

Won *et al.* [78] fabricated that a microchannel reactor (MCR) with combustor for MSR was fabricated to produce H_2 for an on-board proton exchange membrane (PEM) fuel cell device. A commercial copper-containing catalyst ($Cu/ZnO/Al_2O_3$) and Pt/ZrO_2 were used as a catalyst for MSR and combustion reaction, respectively. It was found that the catalyst layer with zirconia sol solution in microchannel showed no crack on the surface of catalyst layer and an excellent adherence to SS microchannel even after reaction. The temperature of combustor could be controlled between 200 and 300 °C depending on the methanol feed rate. The H_2 flow of 3.9 L h^{-1} was obtained with the reforming feed flow rate of 3.65 mL h^{-1} at 270 °C.

Varady *et al.* [79] reported on the development of two MEMS H_2 generators with improved functionality achieved through an innovative process organization and system integration approach that exploited the advantages of transport and catalysis on the micro/nanoscale. One fuel processor design utilized transient, reverse-flow operation of an autothermal MEMS microreactor with an intimately integrated, micromachined ultrasonic fuel atomizer, and a Pd/Ag membrane for in situ H_2 separation from the product stream. The other design featured a simpler, more compact planar structure with the atomized fuel ejected directly onto the catalyst layer, which was coupled to an integrated hydrogen-selective membrane.

Yu *et al.* [80] constructed a methanol steam micro-reformer to produce H_2 for PEMFCs, in which a Cu/Zn/Al/Zr catalyst was supported on metal foams. To optimize the performance of the micro-reformer, the effect of metal foams on the catalytic property was investigated in detail, including catalytic activity, carbon monoxide (CO) selectivity, and mechanical properties. It was found that the strong interaction between the Ni or FeCrAl metal foams and catalyst resulted in the low activity for WGS reaction, therefore, increased the CO selectivity significantly. As a general principle, it was proposed that the metal materials poisoning or reducing the reforming activity should be excluded in the design of micro-reformers.

In the work of Pattekar and Kothare [81,82], the MEMS-based microreformers on silicon-chips were fabricated to supply pure hydrogen for small PEMFCs. The microreformers consisted of a network of Cu/ZnO catalyst-packed parallel micro-channels of depth 200–400 μm with a catalyst particle filter near the outlet. Issues related to microchannel and filter capping, on-chip heating and temperature sensing, introduction and trapping of catalyst particles in the microchannels, flow distribution, microfluidic interfacing and thermal insulation have been addressed. The catalytic reaction of MSR was carried in this microreformer. Experimental runs have demonstrated a methanol to hydrogen molar conversion of over 85% at flow rates enough to supply hydrogen to an 8–10-W micro-fuel cell. They also designed, fabricated, and tested a novel radial high throughput microreformer, which led to nearly an order of magnitude reduction in pressure drop with twice the throughput of methanol and conversions approaching 98%. The throughput from this single planar radial microreformer produced enough hydrogen for a 20-W small PEMFCs.

Ryi *et al.* [83] introduced a Pt–Zr catalyst–coated FeCrAlY mesh into the combustion outlet conduit of a MCR as an igniter of hydrogen combustion to decrease the start-up time (Figures 9.6 and 9.7). The catalyst was coated using a wash-coating

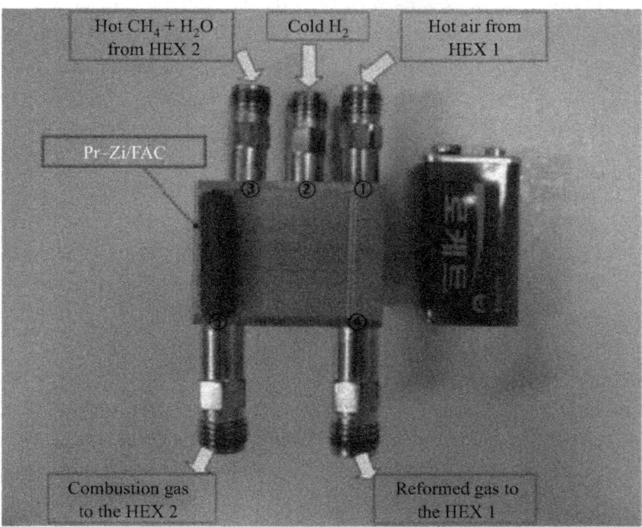

Figure 9.6   Igniter (Pt–Zr/FAC) equipped with a microchannel reactor for rapid start-up. Numbers indicate points at which temperature measured.
© 2006 Elsevier Ltd. Reprinted, with permission, from Reference 83

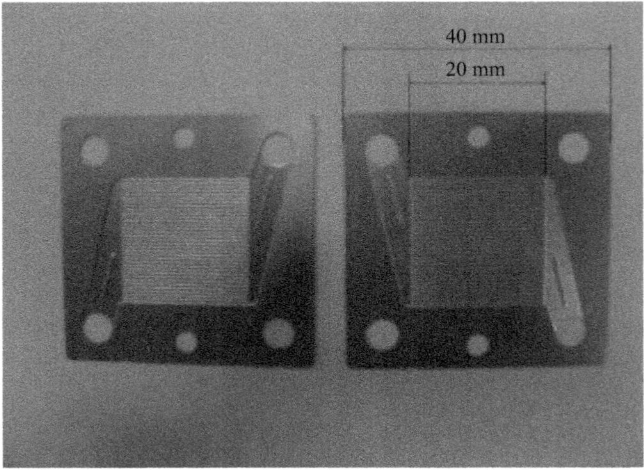

Figure 9.7   Two types of microchannel sheets used in heat-exchanger. Each sheet is a mirror image of the other. © 2006 Elsevier Ltd. Reprinted, with permission, from Reference 83

method. After installing the Pt–Zr/FeCrAlY mesh, the reactor was heated to its running temperature within 1 min with H_2 combustion. Two plate-type heat-exchangers were introduced at the combustion outlet and reforming outlet conduits of the MCR in order to recover the heat of the combustion gas and reformed gas, respectively. Using these heat-exchangers, methane steam reforming was carried out with H_2 combustion, and the reforming capacity and energy efficiency are enhanced by up to 3.4 and 1.7 times, respectively. A compact fuel processor and fuel-cell system using this reactor concept were expected to show considerable advancement.

Shah and Besser [84] studied a number of crucial challenges existing for the realization of practical portable fuel processors. Among these, the management of heat in a compact format was perhaps the most crucial challenge for portable fuel processors. In their study, a silicon microreactor-based catalytic MSR reactor was designed, fabricated, and demonstrated in the context of complete thermal integration to understand this critical issue and develop a knowledge base, which is required to rationally design and integrate the microchemical components of a fuel processor. Detailed thermal and reaction experiments were carried out to demonstrate the potential of microreactor-based on-demand H_2 generation. Based on thermal characterization experiments, the heat loss mechanisms and effective convective heat coefficients from the planar microreactor structure were determined, and suggestions were made for scale up and implementation of packaging schemes to reduce different modes of heat losses.

Belavic *et al.* [85] designed a ceramic chemical microreactor for the production of H_2 needed in portable PEMFCs. The microreactor was developed for the steam reforming of liquid fuels with water into hydrogen. The complex three-dimensional ceramic structure of the microreactor included evaporator(s), mixer(s), reformer, and combustor. Low-temperature co-fired ceramic (LTCC) technology was used to fabricate the ceramic structures with buried cavities and channels, and thick-film technology was used to make electrical heaters, temperature sensors, and pressure sensors. The final 3D ceramic structure consists of 45 LTCC tapes. The dimensions of the structure are $75 \times 41 \times 9$ mm^3 and the weight is about 73 g.

Kolb *et al.* [86] designed microstructured fuel processing reactors and tested them for the production of H$_2$ from fossil and renewable fuels for fuel cell applications. For airborne power generation, a prototype reactor for the partial dehydrogenation of kerosene was developed, which produced more than 100 L min^{-1} H$_2$. A system was under development for the integrated reforming of methanol in a high-temperature PEMFC, which will serve portable applications. To reduce the emissions of trucks during night time, a 5-kW auxiliary power unit was constructed for power generation during break intervals of the driver. A 250-W liquefied petroleum gas–based fuel processor/fuel cell system has been developed for recreational vehicles, which is now commercially available.

9.4   Mathematical modeling of the MMRs for H$_2$ production

Alfadhel *et al.* [87] studied the development of a mathematical model for the description of isothermal microfluidic steady flow in a membrane microreactor (MMR), i.e., a silicon microreactor that housed a permeable membrane in one wall. The model employed the Navier–Stokes equation with appropriate boundary conditions for fluid permeation through the membrane and velocity slip at the walls to account for high Knudsen number. The model equations were solved analytically using finite Fourier transforms. The model solution was used to evaluate the effect of fluid permeation through the membrane and the Knudsen number on the velocity profile and pressure drop. For the simplified cases of no permeation and/or no slip, the derived solution was in excellent agreement with published experimental and theoretical results available in the literature. The utility of the model was illustrated by applying the results to a membrane microseparator used to separate H$_2$ from the other effluents in a microreformer.

Alfadhel and Kothare [20] also studied the problem of modeling multi-component concentration profiles in a membrane-based microreactor. Using basic constitutive laws of mass balance, they derived a low order model of a generic MMR, incorporating chemical reaction and permeation through a selective micro-membrane and utilizing a pressure distribution formula for slipping flows from their previous work [87]. Without loss of generality, the model could address nonisothermal conditions and could be extended to allow flow compressibility. They studied the utility of their model in evaluating the optimal design and operation of a Pd-based MMR for conducting H$_2$ purification and WGS reforming in microfuel processing applications.

Wang *et al.* [88] prepared thin Pd/α-Al$_2$O$_3$ hollow-fiber membranes by an improved ELP technique. Pure gas permeation experiments showed that the membranes were defect free and had high H$_2$ permeance. Feeding an equimolar mixture of H_2/N_2, the total diffusion flow rate of hydrogen through the composite membranes was measured in dependence of the total feed flow rate at the inlet of the shell/retentate chamber, of the temperature, and of the total transmembrane pressure difference. Furthermore, a mathematical model was derived to describe the H_2 diffusion in pure and mixture gas experiments. The pure gas experiments were used to determine the membrane parameters. Applying the adapted model parameters, the fluxes of H_2 occurring in the mixture gas experiments were simulated. The comparison between experimental and calculated data showed a very good agreement in a broad parameter range.

Besides the experimental investigation, the mass transport processes were also analyzed for H_2 permeation through Pd and Pd/Ag hollow-fiber composite membranes (Figure 9.8) [89]. Experimental measurements of pure gas and two-component feeds quantify the extent of retentate-side transport limitations (concentration polarization) in reducing the H_2 flux. The effects of membrane thickness, feed composition and flow rate, temperature, and total pressure on the extent of concentration polarization are measured. The data reveal that concentration polarization increases with increasing temperature and total pressure and decreasing hydrogen feed concentration. A hierarchy of transport models of varying complexity is presented.

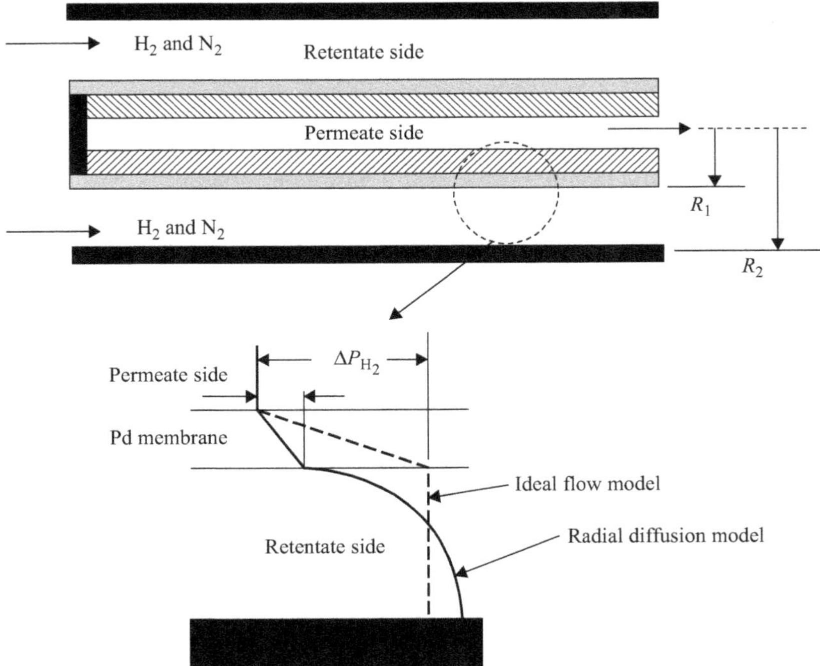

Figure 9.8   Simplified schematic of a membrane permeation process. The details of H_2 profiles are shown in the blown out view. © 2008 Elsevier Ltd. Reprinted, with permission, from Reference 89

The complex 2D models account for gas-phase axial and retentate-side radial concentration gradients, and both selective transport of hydrogen through the Pd membrane and non-selective transport through membrane defects. Satisfactory agreement between model predictions and experimental data is obtained using experimentally measured parameters. An analysis of the characteristic times of the pertinent transport processes identifies the rate-limiting regimes and helps to determine the conditions when gas-phase transport limits the overall productivity. The findings underscore the utility of small diameter fiber support in providing a high surface to volume ratio and reduced concentration polarization.

A microreactor consisting of two parallel channels, where MSR takes place in one channel, and the required heat is supplied by methanol oxidation in the other channel, was numerically simulated by Tadbir *et al.* [90]. Effects of different parameters on methanol conversion, H_2 yield, and CO concentration were examined. Results from the parametric study were then used to propose conditions for high methanol conversion and H_2 yield. A microreactor with enhanced output conditions is thus designed which was capable of producing a gas stream consisting of 74% hydrogen (dry). CO concentration in the generated synthesis gas stream was low enough to require only a PROX reactor for CO clean-up, eliminating the need for a bulky WGS reactor. The produced hydrogen from an assembly of such microreactors can feed a low-power PEM fuel cell. A cluster of these microreactors would take a volume of about 91 cm^3 to feed a typical 30-W PEM fuel cell.

Vigneault *et al.* [91] developed a steady-state 2D model for a multichannel membrane reactor (MCMR) to produce pure H_2. The model included one reforming channel coupled with a Pd–Ag membrane to produce H_2 and one combustion channel to generate the heat needed for the reforming. Both isothermal and non-isothermal simulations were applied in designing a laboratory-scale proof-of-concept reactor. Isothermal sensitivity analysis indicated parameter adjustments practically available to improve reactor performance. In nonisothermal simulations, catalyst layer thickness and kinetic pre-exponential factor were varied along the reactor length. Predictions indicated that the reforming methane conversion increased from 74% to 91%, while avoiding hot spots. Compared with other MRs, the MCMR had the potential for one to two orders of magnitude higher H_2 production per reactor volume and per mass of catalyst.

9.5   Conclusions

MMRs are very promising research area from both scientific and technological point of view, which represent one of the most promising research topics for MRs. The majority of the MMR research is focusing hollow-fiber MMRs, which includes Pd and Pd–Ag support hollow fiber, protonic ceramic, MIEC, microporous, and metallic hollow-fiber MMRs, and microchannel MMRs fabricated by MEMS technology. The new types of MMRs (monolithic, nanotube, and nanofiber MMRs) with higher intensification have not achieved any break-through progress yet because of the complicated fabrication process. The application of MMRs is centering on microchannel MMRs for fuel process for PEMFCs. The modeling work focusing on hollow-fiber and microchannel MMRs have achieved some promising progress. In future, the MMRs with higher intensification should still be pursued. The facile fabrication methods for those highly intensified MMRs should be developed. The applications should be expanded to other portable and distributable devices. The models with more accurate predicting function should be developed for MMRs.

List of abbreviations

APU	auxiliary power unit
BCFZ	$BaCo_xFe_yZr_{1-x-y}O_{3-\delta}$
BCFZ-Pd	$BaCo_xFeyZr_{0.9-x-y}Pd_{0.1}O_{3-\delta}$
BCTCO	$BaCe_{0.85}Tb_{0.05}Co_{0.10}O_{3-\delta}$
DMFCs	direct methanol fuel cells
ELP	electroless plating
ESR	ethanol steam reforming
ID	inner diameter
LPG	liquefied petroleum gas
LTCC	low-temperature co-fired ceramic
LWM0.4	$La_{5.5}W_{0.6}Mo_{0.4}O_{11.25-\delta}$
MCMR	microchannel membrane reactor
MDA	methane dehydroaromatization
MDES	methyldiethoxysilane
MIEC	mixed ionic and electronic conducting
MMRs	micro membrane reactors or membrane microreactors
MRs	membrane reactors
MSR	methanol steam reforming
NMP	*N*-methyl-2-pyrrolidone
PEMFCs	proton exchange membrane fuel cells
PESf	polyethersulfone
POM	partial oxidation of methane
PVP	poly (vinylpyrrolidone)
SCYb	$SrCe_{0.95}Yb_{0.05}O_{3-\delta}$
SEM	scanning electron microscopy
SOG	spin-on-glass
SS	stainless steel
TZFBR	two-zone fluidized bed reactor
WGS	was gas shift reaction

References

[1] B.C.H. Steele, A. Heinzel, "Materials for fuel-cell technologies," *Nature*, **2001**, 414, 345–352.

[2] L. Schlapbach, A. Züttel, "Hydrogen-storage materials for mobile applications," *Nature*, **2001**, 414, 353–358.

[3] J. Tong, Y. Matsumura, H. Suda, K. Haraya, "Thin and dense Pd/CeO_2/ MPSS composite membrane for hydrogen separation and steam reforming of methane," *Sep. Purif. Technol.*, **2005**, 46, 1–10.

[4] R. Ramachandran, R.K. Menon, "An overview of industrial uses of hydrogen," *Int. J. Hydrogen Energy*, **1998**, 23, 593–398.

[5] I. Rafiqul, C. Weber, B. Lehmann, A. Voss, "Energy efficiency improvements in ammonia production—perspectives and uncertainties," *Energy*, **2005**, 30, 2487–2504.

[6] K. Mazloomi, C. Gomes, "Hydrogen as an energy carrier: prospects and challenges," *Renewable Sustainable Energy Rev.*, **2012**, 16, 3024–3033.

[7] A.M. Amin, E. Croiset, W. Epling, "Review of methane catalytic cracking for hydrogen production," *Int. J. Hydrogen Energy*, **2011**, 36, 2904–2935.

[8] E. Mostafavi, N. Mahinpey, M. Rahman, M.H. Sedghkerdar, R. Gupta, "High-purity hydrogen production from ash-free coal by catalytic steam gasification integrated with dry-sorption CO_2 capture," *Fuel*, **2016**, 178, 272–282.

[9] M.A. Hamad, A.M. Radwan, D.A. Heggo, T. Moustaf, "Hydrogen rich gas production from catalytic gasification of biomass," *Renewable Energy*, **2016**, 85, 1290–1300.

[10] S. Nordmann, B. Berghoff, A. Hessel, *et al.*, "A monolithic all-silicon multi-junction solar device for direct water splitting," *Renewable Energy*, **2016**, 94, 90–95.

[11] J. Tong, Y. Matsumura, H. Suda, K. Haraya, "Experimental study of steam reforming of methane in a thin (6 μm) Pd-based membrane reactor," *Ind. Eng. Chem. Res.*, **2005**, 44, 1454–1465.

[12] J. Tong, Y. Matsumura, "Pure hydrogen production by methane steam reforming with hydrogen-permeable membrane reactor," *Catal. Today*, **2006**, 111, 147–152.

[13] Y. Liu, X. Tan, K. Li, "Mixed conducting ceramics for catalytic membrane processing," *Catal. Rev.*, **2006**, 48,145–198.

[14] S.M. Hashim, A.R. Mohamed, S. Bhatia, "Catalytic inorganic membrane reactors: present research and future prospects," *Rev. Chem. Eng.*, **2011**, 27, 157–178.

[15] F. Gallucci, E. Fernandez, P. Corengia, M. van Sint Annaland, "Recent advances on membranes and membrane reactors for hydrogen," *Chem. Eng. Sci.*, **2013**, 92, 40–66.

[16] N. Lu, D. Xie, "Novel membrane reactor concepts for hydrogen production from hydrocarbons: a review," *Int. J. Chem. React. Eng.*, **2016**, 14, 1–31.

[17] G. De Luca, "Assessment of the key properties of materials used in membrane reactors by quantum computational approaches," in A. Basile (Ed.), *Handbook of Membrane Reactors, vol. 1: Fundamental Materials Science, Design and Optimisation*. Woodhead Publishing Series in Energy, issue 55, Woodhead Publishing Limited, ISBN: 978-0-85709-414-8, **2013**, pp. 598–626.

[18] X. Tan, K. Li, "Membrane microreactors for catalytic reactions," *J. Chem. Technol. Biotechnol.*, **2013**, 88, 1771–1779.

[19] A. Qi, B. Peppley, K. Karan, "Integrated fuel processors for fuel cell application: a review," *Fuel Process. Technol.*, **2007**, 88, 3–22.

[20] K. Alfadhel, M.V. Kothare, "Modeling of multicomponent concentration profiles in membrane microreactors," *Ind. Eng. Chem. Res.*, **2005**, 44, 9794–9804.

[21] W.L. Allen, P.M. Irving, W.J. Thomson, "Microreactor systems for hydrogen generation and oxidative coupling of methane". In *IMRET 4: Fourth International Conference on Microreaction Technology*, AIChE Spring National Meeting, Atlanta, GA, March 5–9, **2000**, pp. 351–357.

[22] K.F. Jensen, "Microchemical systems: status, challenges and opportunities," *AIChE*, **1999**, 45, 2051–2054.

[23] D. Kim, A. Kellogg, E. Livaich, B.A. Wilhite, "Towards an integrated ceramic micro-membrane network: electroless plated palladium membranes in cordierite supports," *J. Membr. Sci.*, **2009**, 340, 109–116.

[24] G. Kolb, "Review: microstructured reactors for distributed and renewable production of fuels and electrical energy," *Chem. Eng. Process*, **2013**, 65, 1–44.

[25] Y. Kawamura, N. Ogura, T. Yamamoto, A. Igarashi, "A miniaturized methanol reformer with Si-based microreactor for small PEMFC," *Chem. Eng. Sci.*, **2006**, 61, 1092–1101.

[26] M.V. Kothare, "Dynamics and control of integrated microchemical systems with application to micro-scale fuel processing," *Comput. Chem. Eng.*, **2006**, 30, 1725–1734.

[27] S.V. Karnik, M.K. Hatalis, M.V. Kothare, "Towards a palladium micro-membrane for the water gas shift reaction: microfabrication approach and hydrogen purification results," *J. Microelectromech. Syst.*, **2003**, 12, 93–100.

[28] F.R. García-García, L. Torrente-Murcianob, D. Chadwick, K. Li, "Hollow fibre membrane reactors for high H_2 yields in the WGS reaction," *J. Membr. Sci.*, **2012**, 405–406, 30–37.

[29] O. Ilinich, Y. Liu, C. Castellano, G. Koermer, A. Moini, F. Farrauto, "A new palladium-based catalyst for methanol steam reforming in a miniature fuel cell power source," *Platinum Metals Rev.*, **2008**, 52, 134–143.

[30] D.R. Palo, R.A. Dagle, J.D. Holladay, "Methanol steam reforming for hydrogen production," *Chem. Rev.*, **2007**, 107, 3992–4021.

[31] A.V. Pattekar, M.V. Kothare, "Fuel processing microreactors for hydrogen production in micro fuel cell applications." In *Seventh International Conference on Microreaction Technology (IMRET-7)*, Lausanne, Switzerland, September, **2003**.

[32] R. Mauer, C. Claivaz, M. Fichtner, K. Schubet, R. Renken, "A micro-structured reactor system for the methanol dehydrogenation to water free formaldehyde". In *IMRET 4: Fourth International Conference on Micro-reaction Technology*, AIChE Spring National Meeting, Atlanta, GA, March 5–9, **2000**, pp. 100–105.

[33] A. Zheng, F. Jones, J. Fang, T. Cui, "Dehydrogenation of cyclohexane to benzene in a membrane microreactor". In *IMRET 4: Fourth International Conference on Microreaction Technology*, AIChE Spring National Meeting, Atlanta, GA, March 5–9, **2000**, pp. 284–292.

[34] J. Tong, L. Su, K. Haraya, H. Suda, "Thin and defect-free Pd-based composite membrane without any interlayer and substrate penetration by a combined organic and inorganic process," *Chem. Commun.*, **2006**, 1142–1144.

[35] J. Tong, L. Su, K. Haraya, H. Suda, "Thin Pd membrane on α-Al_2O_3 hollow fiber substrate without any interlayer by electroless plating combined with embedding Pd catalyst in polymer template," *J. Membr. Sci.*, **2008**, 310, 93–101.

[36] F.R. García-García, K. Li, "New catalytic reactors prepared from symmetric and asymmetric ceramic hollow fibers," *Appl. Catal. A: Gen.*, **2013**, 456, 1–10.

[37] V. Gepert, M. Kilgus, T. Schiestel, H. Brunner, G. Eigenberger, C. Merten, "Ceramic supported capillary Pd membranes for hydrogen separation: potential and present limitations," *Fuel Cells*, **2006**, 6, 472–481.

[38] S.H. Israni, B.K.R. Nair, M.P. Harold, "Hydrogen generation and purification in a composite Pd hollow fiber membrane reactor: experiments and modeling," *Catal. Today*, **2009**, 139, 299–311.

[39] E. Gbenedio, Z. Wu, I. Hatim, B.F.K. Kingsbury, K. Li, "A multifunctional Pd/alumina hollow fibre membrane reactor for propane dehydrogenation," *Catal. Today*, **2010**, 156, 93–99.

[40] G. Luo, I. Angelidaki, "Hollow fiber membrane based H_2 diffusion for efficient in situ biogas upgrading in an anaerobic reactor," *Appl. Microbiol. Biotechnol.*, **2013**, 97, 3739–3744.

[41] F.R. García-García, M.A. Rahman, B.F.K. Kingsbury, K. Li, "A novel catalytic membrane microreactor for CO_x free H_2 production," *Catal. Commun.*, **2010**, 12, 161–164.

[42] F.R. García-García, M.A. Rahman, I.D. González-Jiménez, K. Li, "Catalytic hollow fiber membrane microreactor: high purity H_2 production by WGS reaction," *Catal. Today*, **2011**, 171, 281–289.

[43] M.A. Rahman, F.R. García-García, M.D. Irfan Hatim, B.F.K. Kingsbury, K. Li, "Development of a catalytic hollow fiber membrane microreactor for high purity H_2 production," *J. Membr. Sci.*, **2011**, 368, 116–123.

[44] M.A. Rahman, F.R. García-García, K. Li, "Development of a catalytic hollow fiber membrane microreactor as a microreformer for automotive application," *J. Membr. Sci.*, **2012**, 390–391, 68–75.

[45] M.P. Gimeno, Z.T. Wu, J. Soler, J. Herguido, K. Li, M. Menéndez, "Combination of a two-zone fluidized bed reactor with a Pd hollow fibre

membrane for catalytic alkane dehydrogenation," *Chem. Eng. J.*, **2009**, 155, 298–303.

[46] D.A.P. Tanaka, M.A.L Tanco, T. Nagase, *et al.*, "Fabrication of hydrogen-permeable composite membranes packed with palladium nanoparticles," *Adv. Mater.*, **2006**, 18, 630–632.

[47] B.K.R. Nair, M.P. Harold, "Pd encapsulated and nanopore hollow fiber membranes: synthesis and permeation studies," *J. Membr. Sci.*, **2007**, 290, 182–195.

[48] G.B. Sun, K. Hidajat, S. Kawi, "Ultrathin Pd membrane on α-Al_2O_3 hollow fiber by electroless plating: high permeance and selectivity," *J. Membr. Sci.*, **2006**, 284, 110–119.

[49] B.K.R. Nair, J. Choi, M.P. Harold, "Electroless plating and permeation features of Pd and Pd/Ag hollow fiber composite membranes," *J. Membr. Sci.*, **2007**, 288, 67–84.

[50] Y. Liu, X. Tan, K. Li, "Nonoxidative methane coupling in a $SrCe_{0.95}Yb_{0.05}O_{3-\delta}$(SCYb) hollow fiber membrane reactor," *Ind. Eng. Chem. Res.*, **2006**, 45,3782–3790.

[51] J. Xue, Y. Chen, Y. Wei1, A. Feldhoff, H. Wang, J. Caro, "Gas to liquids: natural gas conversion to aromatic fuels and chemicals in a hydrogen-permeable ceramic hollow fiber membrane reactor," *ACS Catal.*, **2016**, 6, 2448–2451.

[52] J. Song, J. Kang, X. Tan, B. Meng, S. Liu, "Proton conducting perovskite hollow fibre membranes with surface catalytic modification for enhanced hydrogen separation," *J. Eur. Ceramic Soc.*, **2016**, 36, 1669–1677.

[53] J. Tong, W. Yang, B. Zhu, R. Cai, "Investigation of ideal zirconium-doped perovskite-type ceramic membrane materials for oxygen separation," *J. Membr. Sci.*, **2002**, 203, 175–189.

[54] J. Tong, W. Yang, R. Cai, B. Zhu, L. Lin, "Novel and ideal zirconium-based dense membrane reactors for partial oxidation of methane to syngas," *Catal. Lett.*, **2002**, 78, 129–137.

[55] H. Jiang, H. Wang, S. Werth, T. Schiestel, J. Caro, "Simultaneous production of hydrogen and synthesis gas by combining water splitting with partial oxidation of methane in a hollow-fiber membrane reactor," *Angew. Chem. Int. Ed.*, **2008**, 47, 9341–9344.

[56] H. Jiang, F. Liang, O. Czuprat, *et al.*, "Hydrogen production by water dissociation in surface-modified $BaCo_xFe_yZr_{1-x-y}O_{3-\delta}$ hollow-fiber membrane reactor with improved oxygen permeation," *Chem. Eur. J.*, **2010**, 16, 7898–7903.

[57] H. Jiang, Z. Cao, S. Schirrmeister, T. Schiestel, J. Caro, "A coupling strategy to produce hydrogen and ethylene in a membrane reactor," *Angew. Chem. Int. Ed.*, **2010**, 49, 5656–5660.

[58] R.V. Franca, A. Thursfield, I.S. Metcalfe, "$La_{0.6}Sr_{0.4}Co_{0.2}Fe_{0.8}O_{3-\delta}$ microtubular membranes for hydrogen production from water splitting," *J. Membr. Sci.*, **2012**, 389, 173–181.

[59] Z. Hong, F. Sun, D. Chen, C. Zhang, X. Gu, N. Xu, "Improvement of hydrogen-separating performance by on-stream catalytic cracking of silane

over hollow fiber MFI zeolite membrane," *Int. J. Hydrogen Energy*, **2013**, 38, 8409–8414.

[60] E. Barbosa-Coutinho, V.M.M. Salim, C.P. Borges, "Preparation of carbon hollow fiber membranes by pyrolysis of polyetherimide," *Carbon*, **2003**, 41, 1707–1714.

[61] G.A. Sznejer, I. Efremenko, M. Sheintuch, "Carbon membranes for high temperature gas separations: experiment and theory," *AIChE J.*, **2004**, 50, 596–610.

[62] M. Wang, J. Song, X. Wu, X. Tan, B. Meng, S. Liu, "Metallic nickel hollow fiber membranes for hydrogen separation at high temperatures," *J. Membr. Sci.*, **2016**, 509, 156–163.

[63] W. Rui, C. Zhang, C. Cai, X. Gu, "Effects of sintering atmospheres on properties of stainless steel porous hollow fiber membranes," *J. Membr. Sci.*, **2015**, 489, 90–97.

[64] K. Kusakabe, S. Morooka, H. Maeda, "Development of a microchannel catalytic reactor system," *Korean J. Chem. Eng.*, **2001**, 18, 271–276.

[65] J.T.F. Keurentjes, F.C. Gielens, H.D. Tong, C.J.M. van Rijn, M.A.G. Vorstman, "High-flux palladium membranes based on microsystem technology," *Ind. Eng. Chem. Res.*, **2004**, 43, 4768–4772.

[66] C.B. Lee, S.W. Lee, D.W. Lee, S.K. Ryi, J.S. Park, S.H. Kim, "Hydrogen production from methane steam reforming in combustion heat assisted novel microchannel reactor with catalytic stacking," *Ind. Eng. Chem. Res.*, **2013**, 52, 14049–14054.

[67] A.L. Mejdell, M. Jøndahl, T.A. Peters, R. Bredesen, H.J. Venvik, "Experimental investigational of microchannel membrane configuration with a 1.4 μm Pd/Ag23 wt.% membrane-effects of flow and pressure," *J. Membr. Sci.*, **2009**, 327, 6–10.

[68] A.L. Mejdell, M. Jøndahl, T.A. Peters, R. Bredesen, H.J. Venvik, "Effects of CO and CO_2 on hydrogen permeation through a ~3 μm Pd/Ag23 wt.% membrane employed in a microchannel membrane configuration," *Sep. Purif. Technol.*, **2009**, 68, 178–184.

[69] A.L. Mejdell, T.A. Peters, M. Stange, H.J. Venvik, R. Bredesen, "Performance and application of thin Pd-alloy hydrogen separation membranes in different configurations," *J. Taiwan Inst. Chem. E*, **2009**, 40, 253–259.

[70] F.C. Gielens, H.D. Tong, C.J.M. van Rijn, M.A.G. Vorstman, J.T.F. Keurentjes, "High-flux palladium–silver alloy membranes fabricated by microsystem technology," *Desalination*, **2002**, 147, 417–423.

[71] H.D. Tong, J.W. Ervin Berenschot, M.J. De Boer, *et al.*, "Microfabrication of palladium–silver alloy membranes for hydrogen separation," *J. Microelectromech. Syst.*, **2003**, 12, 622–629.

[72] S.V. Karnik, M.K. Hatalis, M.V. Kothare, "Palladium based micro-membrane hydrogen gas separator-reactor in a miniature fuel processor for micro fuel cells," *Mater. Res. Soc. Symp. Proc.*, **2002**, 687, B7.2.1–B7.2.6.

[73] A. Kundu, J.H. Jang, J.H. Gil, *et al.*, "Micro-fuel cells—current development and applications," *J. Power Sources*, **2007**, 170, 67–78.

[74] G. Kolb, K.P. Schelhaas, M. Wichert, J. Burfeind, C. Heßke, G. Bandlamudi, "Development of a micro-structured methanol fuel processor coupled to a high-temperature proton exchange membrane fuel cell," *Chem. Eng. Technol.*, **2009**, 32, 1739–1747.

[75] M. Karakaya, A.K. Avci, "Comparison of compact reformer configurations for on-board fuel processing," *Int. J. Hydrogen Energy*, **2010**, 35, 2305–2316.

[76] J. Xuan, D.Y.C. Leung, M.K.H. Leung, M. Ni, H. Wang, "Chemical and transport behaviors in a microfluidic reformer with catalytic-support membrane for efficient hydrogen production and purification," *Int. J. Hydrogen Energy*, **2012**, 37, 2614–2622.

[77] G.G. Park, S.D. Yima, Y.G. Yoon, *et al.*, "Hydrogen production with integrated microchannel fuel processor for portable fuel cell systems," *J. Power Sources*, **2005**, 145, 702–706.

[78] J.Y. Won, H.K. Jun, M.K. Jeon, S.I. Woo, "Performance of microchannel reactor combined with combustor for methanol steam reforming," *Catal. Today*, **2006**, 111, 158–163.

[79] M.J. Varady, L. McLeod, J.M. Meacham, F.L. Degertekin, A.G. Fedorov, "An integrated MEMS infrastructure forfuel processing: hydrogen generation and separation for portable power generation," *J. Micromech. Microeng.*, **2007**, 17, S257–S264.

[80] H. Yu, H. Chen, M. Pan, *et al.*, "Effect of the metal foam materials on the performance of methanol steam micro-reformer for fuel cells," *Appl. Catal. A: Gen.*, **2007**, 327, 106–113.

[81] A.V. Pattekar, M.V. Kothare, "A microreactor for hydrogen production in micro-fuel cell applications," *J. Microelectromech. Syst.*, **2004**, 13, 7–18.

[82] A.V. Pattekar, M.V. Kothare, "A radial microfluidic fuel processor," *J. Power Sources*, **2005**, 147, 116–127.

[83] S.K. Ryi, J.S. Park, S.H. Cho, S.H. Kim, "Fast start-up of microchannel fuel processor integrated with an igniter for hydrogen combustion," *J. Power Sources*, **2006**, 161, 1234–1240.

[84] K. Shah, R.S. Besser, "Key issues in the microchemical systems-based methanol fuel processor: energy density, thermal integration, and heat loss mechanisms," *J. Power Sources*, **2007**, 166, 177–193.

[85] D. Belavic, M. Hrovat, G. Dolanc, M. Santo Zarnik, J. Holc, K. Makarovic, "Design of LTCC-based ceramic structure for chemical microreactor," *Radioengineering*, **2012**, 21, 195–200.

[86] G. Kolb, S. Keller, M. O'Connell, *et al.*, "Microchannel fuel processors as a hydrogen source for fuel cells in distributed energy supply systems," *Energy Fuels*, **2013**, 27, 4395–4402.

[87] K.A. Alfadhel, M.V. Kothare, "Microfluidic modeling and simulation of flow in membrane microreactors," *Chem. Eng. Sci.*, **2005**, 60, 2911–2926.

[88] W.P. Wang, S. Thomas, X.L. Zhang, X.L. Pan, W.S. Yang, G.X. Xiong, "H_2/N_2 gaseous mixture separation in dense Pd/α-Al_2O_3 hollow fiber membranes: experimental and simulation studies," *Sep. Purif. Technol.*, **2006**, 52, 177–185.

[89]   B.K.R. Nair, M.P. Harold, "Experiments and modeling of transport in composite Pd and Pd/Ag coated alumina hollow fibers," *J. Membr. Sci.*, **2008**, 311, 53–67.

[90]   M.A. Tadbir, M.H. Akbari, "Integrated methanol reforming and oxidation in wash-coated microreactors: a three-dimensional simulation," *Int. J. Hydrogen Energy*, **2012**, 37, 2287–2297.

[91]   A. Vigneault, S.S.E.H. Elnashaie, J.R. Grace, "Simulation of a compact multichannel membrane reactor for the production of pure hydrogen via steam methane reforming," *Chem. Eng. Technol.*, **2012**, 35, 1520–1533.

Chapter 10

Perovskite membrane reactors

*Kamran Ghasemzadeh[1], M. Nasiri Nezhad[1]
and Angelo Basile[2]*

Abstract

As one of the most promising strategies in chemical process intensification, membrane reactor (MR) technology has attracted considerable worldwide researches in the last three decades, and this subject is still currently undergoing rapid development and innovation. Nevertheless, inorganic MRs such as perovskite MRs have not achieved any large-scale commercial applications, which implies that there are still a lot challenges to their practical applications. In contrast, several novel perovskite membranes and MRs have been developed in recent years. Therefore, this chapter addresses research and development of perovskite MR applications, in which can permeate oxygen and hydrogen at high temperatures. Indeed, in this chapter, is introduced the structure, transport mechanisms, and performance of various perovskite membranes, followed by evaluation of employing perovskite membranes for both oxidative and non-oxidative reactions. In this viewpoint, the perovskite membrane role of either removing a reactant to shift the equilibrium or adding a reactant to control the reaction mechanism and associated side reactions is significant. Furthermore, the advantages and disadvantages of perovskite MRs are mentioned as a developed technology compared to the traditional reactors and the main challenges that must be overcome for industrial startup of MR technology.

10.1   Introduction

In general, dense ceramic membranes are fabricated from composite oxides having a perovskite crystalline structure. Attention to prepared membranes from perovskite materials for oxygen enrichment arose after the 1980s when

[1]Chemical engineering department, Urmia University of Technology, Urmia, Iran
[2]CNR-ITM, Via P. Bucci c/o University of Calabria Cubo 17/C, Rende (CS) 87046, Italy

many perovskite materials were developed [1]. Indeed, perovskites are crystalline ceramic materials that can permeate oxygen ions and hydrogen protons at high temperatures. However, depending on the electronic flux style of perovskite materials, the perovskite membranes can be categorized into three types [1]:

- The mixed ionic electron conducting (MIEC) perovskite membrane; a material displaying mixed ionic and electronic conductivity.
- Dual-phase perovskite membrane; a dual-phase composite consisting of percolating phases of an ionic conductor and an electronic conductor.
- Electrolyte perovskite membrane; a pure ionic conductor with suitable electrodes connected to an external circuit for the electronic current.

In general, for cases the MIEC and dual-phase perovskite membranes, oxygen ions or hydrogen protons can permeate through the membranes under a partial pressure gradient at high temperatures without any external electrical circuits. This aspect makes the separation performance of perovskite membrane much simpler and consequently the operating cost can be remarkably reduced. In fact, the MIEC type of perovskite membrane presents several significant advantages over the industrial-scale conventional process such as pressure swing adsorption (PSA) and cryogenic distillation, including high energy-efficiency, simplified operation, and cost-effective process. For instance, the cost of oxygen production via the perovskite membrane process can be theoretically reduced by around 35% compared to the PSA or cryogenic distillation technologies [2,3]. On the other hand, many industry plants during various processes such as metallurgical, chemical and petrochemical and glass/concrete consume high pure oxygen. Moreover, the processing of hydrogen, as a clean fuel, also has many potential applications, particularly in transport systems using fuel cell technology.

Meanwhile, regarding to the special characteristic of perovskite membranes/membrane reactors (MRs), they have potential applications in both gas separation and reaction processes. The perovskite membranes have received considerable attention in the last two decades owing to their potential applications in the production of pure oxygen and hydrogen streams and also for partial oxidation or autothermal reforming of hydrocarbons or alcohols to produces syngas [1–3]. However, all hydrogenation or dehydrogenation processes can be carried out by using perovskite MRs. Indeed, the interest towards this kind of MRs is testified by the growing number of scientific publications, as reported in Figure 10.1. In this figure, reported scientific publications are only related to perovskite MR applications.

Therefore, this chapter will describe extensively the perovskite MRs from their principles, fabrication, and design to their applications. The prospects and critical issues of the perovskite MRs in commercial applications will also be presented and discussed at the end of the chapter.

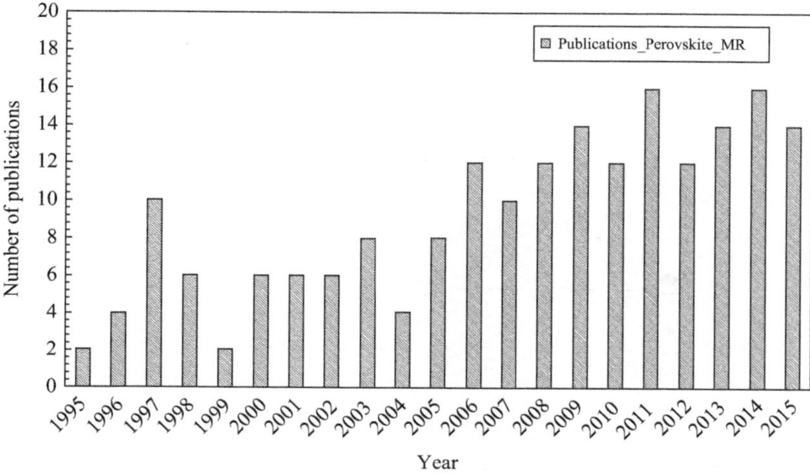

Figure 10.1   Number of scientific publications about the perovskite MRs versus year, Scopus database: www.scopus.com

10.2   Structure and material of perovskite membranes

It is well known that most of perovskite types were originally recognized on the basis of a mineral oxide, namely $CaTiO_3$. The basic structure of this mineral is found that orthorhombic and the name perovskite has been retained for this type of structure. It should be noted that this kind of structure makes the material very stable at high temperatures. A simple cubic structure highlighting the coordination situation about the A site cation is indicated in Figure 10.2. It can be observed that the A site cation is corresponding with twelve oxygen ions forming a cuboctahedral geometry, whereas the B site cation is corresponding with six oxygen ions with an octahedral coordination. This structure provides a sign of chemical composition and the structural unit formed when preparing the materials [4]. Indeed, an ideal perovskite contains of ABO_3 units, but the chemical composition can differ depending on the valencies of the A and B site cations. Components, such as $A^{1+}B^{5+}O^3$, $A^{2+}B^{4+}O^3$, and $A^{3+}B^{3+}O^3$, are normally presented. In most cases, the A site can be generally occupied by 2+ large alkali earth metals such as barium, lanthanum or strontium, and the B site by 4+ smaller first row transition elements such as cerium, iron, or cobalt. Although the same structure is retained, the properties can be very changed. For instance, the $SrCoO_3$ and $BaCoO_3$ can be used in oxygen separation, whereas the $SrCeO_3$ and $BaCeO_3$ are useful in hydrogen separation.

Therefore, in a perovskite structure, the A site cation is normally larger than the B site cation due to the different coordination. It was investigated that there is a correlation between the electrical conductivity and related parameters such as the

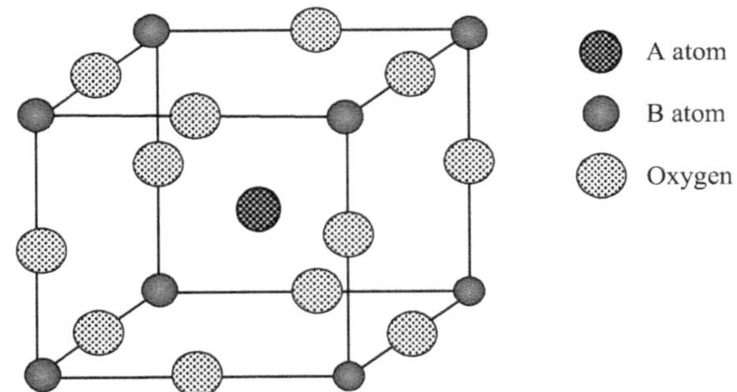

Figure 10.2   Ideal crystal structure of a perovskite compound

tolerance factor and the specific free volume of the lattice [4]. The tolerance factor is defined for indicating the deformation from the cubic lattice as

$$\text{Tolerance factor} = (R_A/R_O)/(2^{0.5}) \times (R_B + R_O) \tag{10.1}$$

where R_A, R_B, and R_O are the ionic radius of the A site cation, B site cation and oxygen. High conductivity is achieved by a compositions with large specific free volumes and with tolerance factors of around 0.96. Hence, the following strategies were suggested [4]:

- To obtain the large free volume, a site cation should have a large ionic radius.
- To adjust the tolerance factor to around 0.96, B site cation should be selected.

However, a perovskite with the ideal structure as shown in Figure 10.2 does not present the capability to conduct oxide ions; there must be a considerable amount of imperfections or defects produced owing to the nonstoichiometry for diffusion to take place. Indeed, the basic concepts of defect chemistry have been adapted to describe the permeation properties of perovskite materials. On the other hand, based on the Goldschmidt number, a large group of perovskite compositions have been reported for the preparation of perovskite membranes, which, as a summarized list of perovskite types, is given in Table 10.1.

To author's best knowledge, perovskites of barium strontium cobalt iron (BSCF) and lanthanum strontium cobalt iron (LSCF) oxides have attracted main interest from researchers and engineers of industry. Regarding various research results, the major attraction of these perovskites containing Sr or Co is that their mixed conductivity properties increase as a function of Sr and Co content. In addition, the perovskite membrane structure can be stabilized by partial substituting Co with Fe and is further improved by the partial replacing of Sr with Ba, thus founding the BSCF perovskite [5]. This composition has been fully investigated for its potential to present high oxygen fluxes, although BSCF suffers from thermal and chemical stability problems during prolonged disclosure to lower temperatures (<900 °C) [4]. Even though presenting lower oxygen fluxes than BSCF membranes, LSCF membranes have been confirmed to be stable for operating at 800 °C after 3,000 h [16].

Table 10.1   Representative oxygen fluxes for perovskite membranes

Compound	Temperature (K)	Flux (mL min^{-1} cm^{-2})	Reference
$BaBi_{0.4}Co_{0.2}Fe_{0.4}O_{3-s}$	1,200	0.8044	[5]
$BaCe_{0.15}Fe_{0.85}O_{3-s}$	1,225	0.5230	[6]
$BaCo_{0.4}Fe_{0.5}Zr_{0.1}O_{3-s}$	1,225	0.9157	[7]
$BaTi_{0.2}Co_{0.4}Fe_{0.4}O_{3-s}$	1,225	8.9994	[8]
$SrCe_{0.8}Fe_{0.2}O_{3-s}$	1,140	3.3398	[9]
$SrCe_{0.89}Fe_{0.1}Cr_{0.01}O_{3-s}$	1,150	0.6317	[10]
$Ba_{0.5}Sr_{0.5}Co_{0.8}Fe_{0.2}O_{3-s}$	1,225	4.3895	[11]
$Ba_{0.5}Sr_{0.5}Zn_{0.2}Fe_{0.8}O_{3-s}$	1,250	3.4998	[12]
$Gd_{0.6}Sr_{0.4}CoO_{3-s}$	1,090	1.5846	[13]
$La_{0.6}Sr_{0.4}CoO_{3-s}$	1,120	2.9998	[13]
$La_{0.6}Sr_{0.4}Co_{0.8}Cu_{0.2}O_{3-s}$	1,130	1.9044	[13]
$La_{0.6}Sr_{0.4}Co_{0.8}Ni_{0.2}O_{3-s}$	1,130	1.4461	[14]
$La_{0.6}Ba_{0.4}Co_{0.8}Fe_{0.2}O_{3-s}$	1,130	2.0644	[13]
$La_{0.6}Ca_{0.4}Co_{0.8}Fe_{0.2}O_{3-s}$	1,130	1.8332	[13]
$Nd_{0.6}Sr_{0.4}CoO_{3-s}$	1,090	1.0290	[15]

Compared with a variety of perovskite materials for oxygen conducting, the choice of proton-conducting perovskites is limited so far. Several perovskite types have been identified as having good abilities in protonic (hydrogen) conduction including doped $BaCeO_3$ [17], $BaZrO_3$ [18], $BPrO_3$ [19], $SrZrO_3$, $CaZrO_3$ [20], and $SrTiO_3$ [21], since the first protonic conductor, $SrCe_{0.95}Yb_{0.05}O_{3-\delta}$, was reported. Hence, the general formula is written as $AB_{1-x}M_xO_{3-\delta}$, in which the A element is usually selected from the group consisting of calcium, strontium, or barium; the B element is selected from the group consisting of cerium, terbium, zirconium or thallium; the M element is selected from the group consisting of titanium, thulium, chromium, manganese, cobalt, nickel, copper, aluminum, gallium, yttrium, ytterbium, indium or tin, and normally x is less than 0.2, and δ is the oxygen deficiency per perovskite unit [22]. In general, their protonic conductivities in a hydrogen atmosphere are of the order of 10^{-3}–10^{-2} S cm^{-1} at 600–1,000 °C. Hence, the $BaCeO_3$-based oxides indicate the highest conductivity, but oxygen ions contribute to the conduction as the temperature is increased, and result in the transport number of protons decreasing by enhancement of the temperature [23]. Even if, the conductivities of $SrCeO_3$-based oxides are lower, the transport number of protons is higher than that of $BaCeO_3$-based ones. Moreover, the zirconate-based oxides with lower conductivity such as $SrZrO_3$ or $CaZrO_3$ indicate good chemical stability and

Table 10.2   Properties of typical perovskite membranes

Perovskite types	T (°C)	Hydrogen proton flux (mol cm^{-2} s^{-1})	Reference
$BaCe_{0.95}Y_{0.05}O_{3-\delta}$	800	8.4×10^{-9}	[26]
$BaCe_{0.95}Nd_{0.05}O_{3-\delta}$	900	2.2×10^{-8}	[27]
$SrCe_{0.95}Yb_{0.05}O_{3-\delta}$	900	1.7×10^{-8}	[26]
$SrCe_{0.95}Tm_{0.05}O_{3-\delta}$	900	2.9×10^{-8}	[28]

mechanical strength, and also they are more stable against carbon dioxide gas that reacts with cerate materials below 800 °C [24,25]. The reported flux of some proton conducting perovskite membranes and their condition are summarized in Table 10.2.

10.3   Transport mechanism of perovskite membranes

10.3.1   Oxygen transport mechanism

Regarding results of various studies, oxygen permeation through perovskite membranes occurs by the transmission of oxygen ions through oxygen vacancies in the crystal structure generated at high temperatures (>600 °C). Indeed, the perovskite materials catalyze dissociation of the O_2 molecule into oxygen ions (O^{2-}), allowing the latter to diffuse ions through the perovskite structure defects. On the other hand, the permeation of species across a membrane can only occur under the influence of a driving force which, in the case of perovskite membranes, takes the form of an O_2 partial pressure gradient along the membrane. As per the author knowledge, the O_2 permeation through perovskite membranes can consist of five main steps as illustrated in Figure 10.3 and is described in the following text:

First step: Diffusion to membrane surface; in this step, the oxygen molecules diffuse to the surface of the perovskite membrane.

Second step: Dissociation (surface reaction); in this stage, the oxygen molecule adsorbs the membrane surface and then disassociates due to catalytic activity of the perovskite material. Afterwards, the oxide ion incorporates into a lattice oxygen vacancy.

Third step: Bulk diffusion; the oxygen ions diffuse through the membrane, driven by a partial pressure gradient of oxygen across the membrane and simultaneously, electrons are transported in the opposite direction to retain electrical neutrality.

Fourth step: Combination (surface reaction); the oxygen ions recombine into oxygen molecules and desorb from the membrane surface to gas bulk.

Fifth step: Diffusion to gas balk; in this step, the oxygen molecules diffuse to the permeate gas phase from membrane surface.

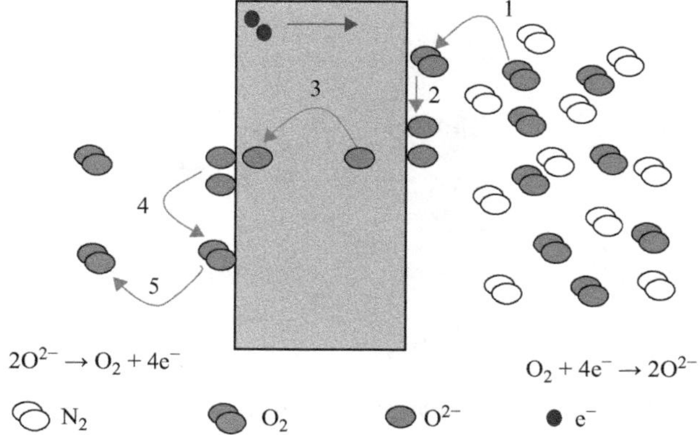

Figure 10.3   Schematic of O_2 permeation mechanism through a perovskite membrane

It should be noted that the oxygen flux in steps of 2 and 4 is controlled by the kinetics of the catalytic surface reaction, whereas the flux in step 3 is controlled by bulk diffusion and membrane thickness. Hence, the oxygen flux can be increased by decreasing the perovskite membrane thickness or by integration of catalyst layer on the membrane surface to improve the kinetics of the surface reaction. In fact, one of other most significant concepts of perovskite membranes is that the critical length, which is defined as the thickness of a membrane at which the transport resistance, owing to the surface kinetics, is equal to the transport resistance related with bulk diffusion. In particular, for BSCF membranes, the critical length was reported around 0.7–1.1 mm between 800 and 900 °C [29].

10.3.2   Hydrogen transport mechanism

Aliovalent-doped perovskite membranes are classic high-temperature proton conductors and mostly operate on the principle that oxygen vacancies can adsorb water molecules, resulting in protonic defects in the crystal lattice OH^+. In the crystal structure, the hydrogen is sited in between two oxygen atoms, forming a hydrogen-bond-like arrangement with one of the oxygens. Hydrogen permeation through the perovskite membrane proceeds in a similar style with oxygen transport. As illustrated in Figure 10.4, the gas species diffuse to the perovskite surface, followed by the adsorption and catalytic dissociation of hydrogen from water.

Indeed, protonic transport through the perovskite membrane can be carried out via two different mechanisms. In the first approach, the protonic defect or lattice OH^+ can diffuse through the membrane in a manner, similar to O^{2-} diffusion, in which the proton remains connected to the oxygen ion. In the second approach, the proton can diffuse through the membrane via the dissociating and reforming of the OH^+ defect at adjacent oxygen sites through the lattice [4]. In fact, this second type of diffusion can be considered as the rate-limiting component of the transport

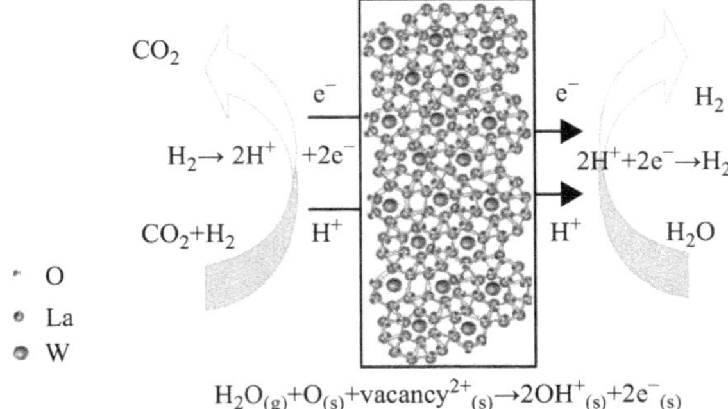

Figure 10.4   Schematic of hydrogen proton transport mechanism through the perovskite membranes

mechanism. In contrast, to oxygen transport, the electrons must simultaneously accompany the protonic defect as it diffuses through the membrane to balance the electrical charge.

The protonic defects then recombine into an adsorbed water molecule, which decomposes to release hydrogen on the permeate stream. As an oxygen ion, the oxygen atom from the water molecule remains in the perovskite membrane structure. In general, in perovskites material, the highest conductivities are observed with cubic crystal structures or reduced symmetry. On the other hand, the nature of the oxidation and reduction stages in hydrogen transport, in combination with the intrinsically reducing environment of the permeate stream, means that these membrane structure must be stable in both highly reducing and oxidizing conditions. In addition, the chemical expansion related with the protonic defects should be small to retain the mechanical integrity of the membranes [4].

10.4   Performance of perovskite membranes

In this section, performance of the perovskite membranes in oxygen and hydrogen separations is investigated. To this evaluation, various types of suitable membrane performances are studied. Hence, as shown in Figures 10.5 and 10.6, trade-off diagrams can indicate the high performance of perovskite membrane with respect to other membranes in oxygen separation, whereas its performances in hydrogen separation are not acceptable, compared with the silica and palladium membranes.

10.5   Perovskite MRs

During the past 20 years, considerable attention has been sited on the integration of the membrane process into the chemical reactors as MR systems, which combine reaction and separation or combine distribution and reaction in one unit operation. These kinds of reactors have been used to enhance the reaction conversion and

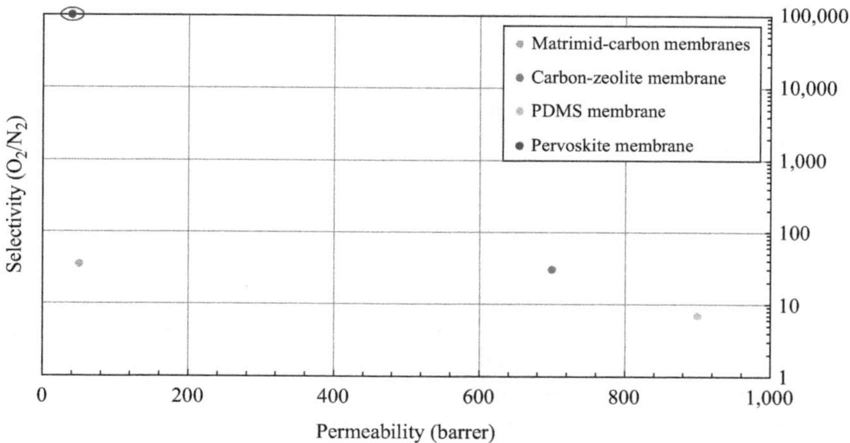

Figure 10.5   Evaluation of various membranes for O$_2$ separation

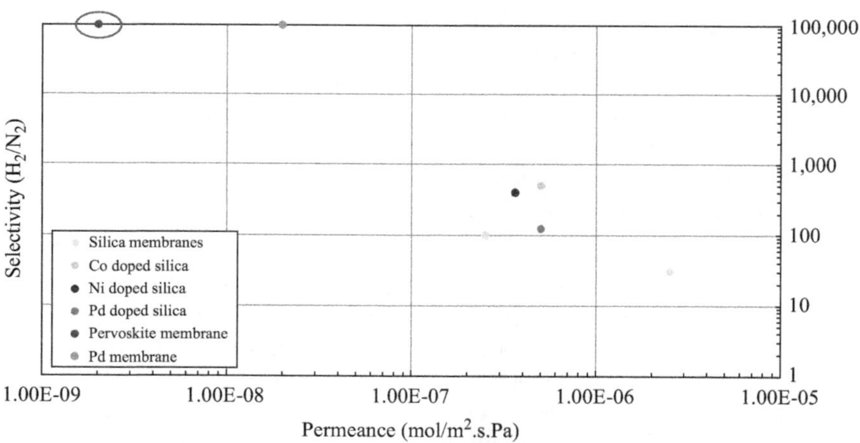

Figure 10.6   Evaluation of various membranes over to per for H$_2$ separation

reaction yield or to control the reaction mechanism by tracing a reactant into the reaction zone as a controlled manner [30].

Hence, perovskite MRs can be operated in either packed bed MR (PBMR) or catalytic MR (CMR) configurations, in which the PBMRs are mostly applied in practical use of perovskite MRs. Indeed, the reactions carried out in the catalyst bed, whereas the membrane mainly operates as an oxygen and hydrogen extractor or distributor. As the catalyst is physically separated from the membrane, the separation role of the membrane and the catalytic properties of catalysts can be modulated, in order that the perovskite MR performance will be optimized. In CMRs, it is well known that perovskites are fundamentally catalytic to oxidation reactions. Therefore, perovskite membranes may serve as both catalyst and separator, and no other catalysts are used in the MR module. However, as chemical reactions occur on the membrane surface, it is required to have a much more porous membrane surface in order to provide a sufficient quantity of active sites. This can be obtained in the membrane preparation process, or by coating a porous membrane surface after preparation. The key potential problems for this kind of configuration are that the membrane may not have adequate catalytic activity, and the catalytic selectivity cannot be adapted over to the considered reactions [4].

10.5.1   Applications of perovskite MRs

Oxygen permeable perovskite MRs have been extensively studied for potential applications, such as partial oxidation of hydrocarbons and autothermal reforming, in which the membrane acts as an oxygen distributor. From viewpoint of a membrane performance, the motivations consist of the following:

- To supply oxygen for the reaction system in a more controllable procedure and to maintain oxygen concentration at a low value, and leading to higher selectivity.
- To use air directly as the oxygen source without contaminating the products with nitrogen and nitrogen oxides, resulting in remarkably reduced capital investment and operation costs.
- To avoid premixing of hydrocarbon feed with oxygen and consequently to reduce the formation of hot spots as encountered in a co-feed reactor, and cause to safer operation.
- To decrease strongly the atmosphere created by the reaction products and provide a large oxygen potential gradient to facilitate oxygen transport through the membrane.

On the other hand, so far, hydrogen permeable perovskite membranes have been applied as product separators in methane steam reforming (MSR) and the coupling of methane to produce valuable C_2 products (ethane or ethylene). The methane coupling reaction is a difficult one, only taking place at high temperatures (>650 °C). Compared with Pd-based membranes, hydrogen permeable perovskite membranes have good thermochemical stability at high temperatures and also show some catalytic activities in methane coupling, and therefore, are more suitable for this reaction, as dense metal membranes are only appropriate for application at a temperature range of 300–600 °C.

10.5.1.1   **Partial oxidation of methane**

Regarding reaction enthalpy, the direct catalytic partial oxidation of methane (POM) is a slightly exothermic:

$$CH_4 + \frac{1}{2}O_2 \leftrightarrow CO + 2H_2, \qquad \Delta H^0_{298} = -36 \text{ kJ mol}^{-1} \tag{10.2}$$

It yields a suitable feedstock ($H_2/CO = 2:1$) for the Fischer–Tropsch reaction to produce linear hydrocarbons and also for methanol synthesis. The application of perovskite MRs makes it possible to integrate oxygen separation from air and POM

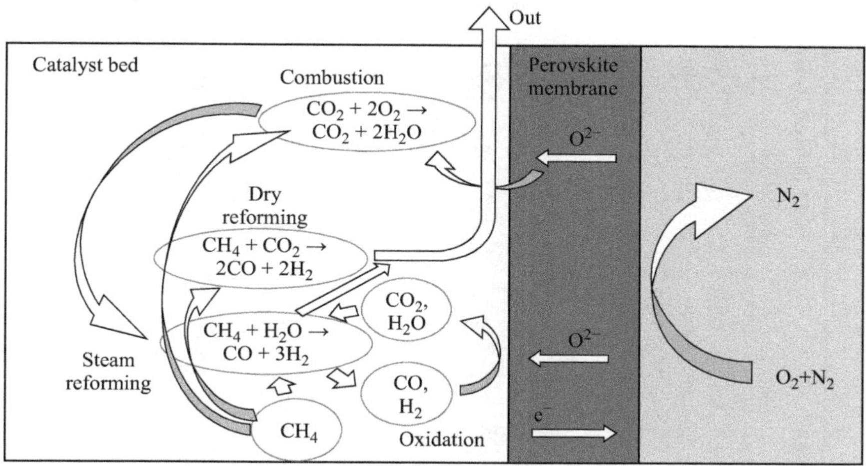

Figure 10.7   Schematic diagram for possible reaction pathways of the POM reaction in a perovskite MR

reaction in a single step, thus enabling major reductions of capital investment in the gas-to-liquid (GTL) industry [31]. Although perovskite membranes may display catalytic activity in the POM reaction, a POM catalyst is usually applied to improve the CO selectivity and methane conversion. Figure 10.7 indicates the process of the POM reaction in a perovskite MR packed with oxygen separation.

As illustrated in Figure 10.7, methane and air are respectively fed to opposite sides of the perovskite MR and under an electrochemical potential gradient, the oxygen in air is permeated through the membrane at a high temperature to the reaction side and consequently reacts with methane to produce syngas. Indeed, in the POM process, the perovskite membrane provides only molecular oxygen to react at the catalyst because the lattice oxygen of the membrane is not active in breaking the C–H bond but specially oxidizes hydrogen to water [32]. Therefore, some of the CO and H_2 formed by the reforming reactions are oxidized at the membrane surface into CO_2 and H_2O:

$$CO + O^{2-} \leftrightarrow CO_2 + 2e$$
$$H_2 + O^{2-} \leftrightarrow H_2O + 2e \tag{10.3}$$

As a result of Gong and Hong [33] works, the oxygen permeation flux under POM reaction conditions is much higher than that when helium is used as sweep gas at the same temperature. Furthermore, Ishihara *et al.* [34] indicated that the POM catalyst also has a significant influence on the oxygen permeation of the perovskite membrane, but the mechanism is not very clear.

Of all the potential applications of perovskite MRs, the POM reaction to syngas is thought to be the most commercially important one. Excessive efforts have been carried out in this field during the last two decades, as summarized in Table 10.3.

Table 10.3 Perovskite MRs for POM reactions

Membrane composition	Reactor or configuration	catalyst	T (°C)	Main results	Stability (h)	Reference
$La_{0.4}Ba_{0.6}Fe_{0.8}Zn_{0.2}O_{3-\delta}$	Disk; PBMR	Ni–Ca	900	$X \to 100\%$; $S_{CO} > 95\%$	500	[33]
$BaCo_{0.4}Fe_{0.4}Zr_{0.2}O_{3-\delta}$	Disk; PBMR	–	850	$X = 98\%$; $S_{CO} = 100\%$; $J_{O_2} = 5.6$ mL cm^{-2} min^{-1}	2,200	[34]
$La_{0.8}Sr_{0.2}Fe_{0.7}Ga_{0.3}O_{3-\delta}$	Tube; CMR	$La_{0.8}Sr_{0.2}Fe_{0.7}Ni_{0.3}O_{3-\delta}$	900	$X = 74\%$; $S_{H_2} > 50\%$	142	[35]
$SrCo_{0.4}Fe_{0.5}Zr_{0.1}O_{3-\delta}$	Disk; PBMR	NiO/Al_2O_3	950	$S_{CO} > 90\%$	–	[36]
$SrFeCo_{0.5}O_x$; $SrFe_{0.2}Co_{0.8}O_x$	Tube; PBMR	Ru-based	850	$X > 99\%$; $S_{CO} > 98\%$	>1,000	[37]
$Ce_{0.8}Sm_{0.2}O_{2-\delta}$; $La_{0.8}Sr_{0.2}CrO_{3-\delta}$	Tube; PBMR	$Ca_{0.8}Sr_{0.2}TiO_3$	950	$X = 17\%$; $S_{H_2} > 75\%$	–	[38]
$La_{0.6}Sr_{0.4}Co_{0.4}Fe_{0.8}O_{3-\delta}$	Tube; PBMR	$Ni/\gamma\text{-}Al_2O_3$	825–885	$X > 96\%$; $S_{CO} > 97\%$	3–7	[39]
La_2NiO_4	Tube	–	900	$X = 89\%$; $S_{CO} = 96\%$; $H_2/CO = 1.5$	–	[40]
$Ba_{0.5}Sr_{0.5}Co_{0.8}Fe_{0.2}O_{3-\delta}$	Tube; PBMR	$LiLaNiO/\gamma\text{-}Al_2O_3$	875	$X = 94\%$; $S_{CO} > 95\%$; $J_{O_2} = 8.0$ mL cm^{-2} min^{-1}	500	[41]
$YSZ\text{–}SrCo_{0.4}Fe_{0.6}O_{3-\delta}$	Disk; PBMR	NiO/Al_2O_3	750–850	$X = 64\%$; $S_{CO} \sim 100\%$	220	[42]
$Ca_{0.8}Sr_{0.2}Ti_{0.7}Fe_{0.3}O_{3-\alpha}$	Disk; CMR	$Ni\text{–}Ca_{0.8}Sr_{0.2}Ti_{0.9}Fe_{0.1}O_{3-\delta}$	900	$X = 13.7\%$; $S_{CO} = 98\%$	–	[43]
$YBa_2Cu_3O_{7-\alpha}$	Disk; PBMR and CMR	Ni/ZrO_2	875	$X = 100\%$; $S_{CO} = 95\%$	5	[44]
$SrFe_{0.7}Al_{0.3}O_{3-\delta}$	Disk; PBMR	$SrFe_{0.7}Al_{0.3}O_{3-\delta}$	950	$X = 65\%$; $S_{CO} = 48\%$	–	[45]

Membrane material	Configuration	Catalyst	Temperature (°C)	Performance		Reference
Ba(Co, Fe, Zr)O$_{3-\delta}$	Hollow fiber; PVMR	Ni-catalyst	925	$X = 96\%$; $S_{CO} = 97\%$; $H_2/CO \sim 2$	–	[46]
Sm$_{0.15}$Ce$_{0.85}$O$_{1.925}$/ Sm$_{0.6}$Sr$_{0.4}$Fe$_{0.7}$Al$_{0.3}$O$_{3-\delta}$	Disk; PBMR	LiLaNiO/γ-Al$_2$O$_3$	950	$X > 98\%$; $S_{CO} > 98\%$; $H_2/CO = 2$ $J_{O_2} = 4.3$ mL cm^{-2} min^{-1}	1,100	[47]
BaCo$_{0.7}$Fe$_{0.2}$Nb$_{0.1}$O$_{3-\delta}$	Disk; PBMR	Ni-catalyst	975	$X = 92\%$; $S_{H_2} > 90\%$; $J_{O_2} = 15$ mL cm^{-2} min^{-1}	550	[48]
Ce$_{0.85}$Sm$_{0.15}$O$_{1.925}$/ Sm$_{0.6}$Sr$_{0.4}$FeO$_{3-\delta}$	Disk; PBMR	LiLaNiO/γ-Al$_2$O$_3$	950	$X > 98\%$; $S_{CO} > 98\%$	500	[49]
BaCo$_{0.7}$Fe$_{0.2}$Ta$_{0.1}$O$_{3-\delta}$	Disk; PBMR	Ni-catalyst	900	$X = 900\%$; $S_{H_2} > 94\%$; $J_{O_2} = 16.2$ mL cm^{-2} min^{-1}	400	[50]
BaCe$_{0.1}$Co$_{0.4}$Fe$_{0.5}$O$_{3-\delta}$	Disk; PBMR	NiLaNiO/γ-Al$_2$O$_3$	950	$X = 99\%$; $S_{CO} > 93\%$; $J_{O_2} = 9.5$ mL cm^{-2} min^{-1}	1,000	[51]
3%Al$_2$O$_3$-doped SrCo$_{0.8}$Fe$_{0.2}$O$_3$	Tube; PBMR	Ni-catalyst	900	$X = 99\%$; $S_{CO} > 93\%$	–	[52]

In general, the performance of perovskite MRs in terms of methane conversion and CO selectivity is strongly dependent on the MR design and operating conditions [35]. In most cases, the membrane mainly functions as an oxygen supplier and distributor, whereas its catalytic properties are less important due to the high activity of the reforming catalyst. Moreover, the simulation results indicate that the reactor with smaller diameter (D) and greater length-to-diameter ratio (L/D) may give better performance in terms of high hydrogen recovery and high methane conversion [36].

As well, the amounts of catalyst packed and the feed contact time have significant effects on the methane conversion, CO selectivity and oxygen permeation rate [52,53]. To achieve better performance, the amount of catalyst must match well with the available membrane surface area. Nevertheless, it is generally considered that the methane conversion is mainly controlled by the oxygen permeation rate rather than the reaction rate at the catalyst surface.

10.5.1.2   Oxidative coupling of methane

According to literature [54], oxidative coupling of methane (OCM) to C_2 products such as C_2H_4 and C_2H_6 represents one of the most effective approaches to convert natural gas to more useful products:

$$2CH_4 + \frac{1}{2}O_2 \rightarrow C_2H_6 + H_2O, \qquad \Delta H^0_{298} = -177 \text{ kJ mol}^{-1} \tag{10.4}$$

$$2CH_4 + O_2 \rightarrow C_2H_4 + 2H_2O, \qquad \Delta H^0_{298} = -282 \text{ kJ mol}^{-1} \tag{10.5}$$

It is considered that the OCM process may be commercially utilized if a single-pass conversion of 35%–37%, selectivity of 88%–85% and C_2 yield of 30% are achieved. Most of the previous studies [54–62] have focused on finding suitable catalysts for the selective methane conversion. However, it is difficult to obtain C_2 yields higher than 25% in a conventional reactor (CR). This may be related to the competition between the coupling and the combustion reactions.

In general, it is accepted that the first step in the catalytic OCM reaction includes the hemolytic dissociation of a C–H bond on the catalyst surface to form •CH_3 radicals, which may undergo coupling to consume ethane in the gaseous phase. In the presence of molecular oxygen, the intermediate radicals and their products may undergo strong oxidation to carbon oxides. It should be noted that to improve the OCM reaction selectivity, in the gas phase, the oxygen concentration should be as low as possible, whereas the amount of methane provided should also be enough for high methane conversion. Therefore, applications of perovskite MRs to control oxygen concentration along the reactors offer the possibility of attaining much higher C_2 hydrocarbon selectivity and yield for OCM [3]. Moreover, compared with other membranes, the ionic conduction of the perovskite membranes delivers the oxygen into the reaction zone in the form of dissociated and ionized oxygen. This ionized oxygen reacts with methane on the membrane surface, following a different reaction mechanism with the purpose of the formation of CO_x from by-reactions owing to the presence of gas-phase oxygen is inhibited. Figure 10.8 demonstrates carrying out the OCM reaction in the perovskite MRs.

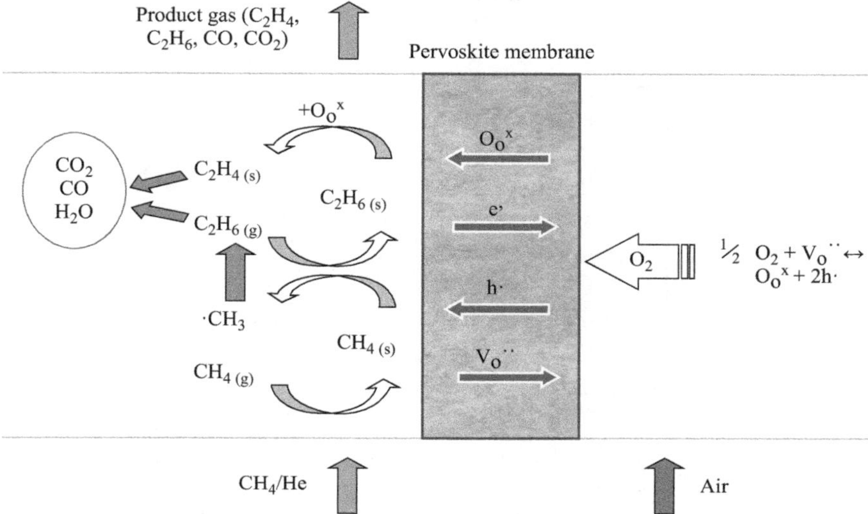

Figure 10.8   The mechanism of OCM in the perovskite MR

As indicated in this figure, methane is adsorbed, and it reacts with the lattice oxygen ($O_o^{\times}$) and electron holes (h•) to form methyl radicals, which are then coupled in gas phase to form C_2 products, or further react with gaseous oxygen to form carbon oxides. On the other hand, Table 10.4 summarizes the literature results of OCM in perovskite MRs. It is found that the perovskite membrane may intrinsically exhibit catalytic activity toward the OCM reaction, since no OCM catalyst is applied. Furthermore, the C_2 yield is very sensitive to the membrane characteristics, reaction conditions, and reactor design [54].

Regarding to the complete oxidation reactions occurring in the gas phase, and partially on the catalyst surface, may present lower C_2 selectivity, especially under conditions of high pressure and temperature, to obtain high C_2 yields; the oxygen permeation flux, methane flow rate, and intrinsic reaction rate must match well with each other. Indeed, insufficient oxygen supply results in poor conversion, but a high oxygen flux may result in low selectivity due to the complete oxidation reactions; especially at high pressure and temperature [55].

This refers that the oxygen permeance of the perovskite membrane has to match with the catalytic activation of the membrane surface [56]. For a prepared composite membrane, the oxygen permeance can be readily improved by improving the surface exchange kinetics or decreasing the membrane thickness. Therefore, the selection of a membrane material with good catalytic properties or the modification of these high-oxygen-permeable perovskite membrane surfaces with a suitable OCM catalyst has become the most critical step in the development of perovskite MRs for OCM reaction. Among the various membranes developed, $Bi_{1.5}Y_{0.3}Sm_{0.2}O_{3-\delta}$ exhibits not only high oxygen permeance and catalytic activity but also high chemical and mechanical stability under OCM reaction conditions.

Table 10.4 Perovskite MRs for OCM reactions

Membrane	Reactor configuration	Temp. (°C)	Main results	Reference
$Ba_{0.5}Sr_{0.5}Co_{0.8}Fe_{0.2}O_{3-\delta}$	Disk coated with La–Sr/CaO catalyst	950	$Y_{C_2} = 18\%$; $S_{C_2} > 65\%$	[56]
$Ba_{0.5}Sr_{0.5}Co_{0.8}Fe_{0.2}O_{3-\delta}$	Tube, no catalyst, or packed with La–Sr/CaO catalyst	800–900	$S_{C_2} = 62\%$; or $Y_{C_2} = 13\% - 15\%$; $S_{C_2} = 54\%$	[58]
$BaCe_{0.8}Gd_{0.2}O_{3-\delta}$	Tube	778	$Y_{C_2} = 16.5\%$; $S_{C_2} = 62.5\%$	[59]
$La_{0.6}Sr_{0.4}Co_{0.8}Fe_{0.2}O_{3-\delta}$	Disk, no catalyst	800–900	$Y_{C_2} = 1\% - 3\%$; $S_{C_2} \rightarrow 70\%$	[60]
Y-doped Bi_2O_3	Disk	750–950	$Y_{C_2} = 16\% - 14\%$; $S_{C_2} = 20\% - 90\%$	[61]
$La_{0.8}Sr_{0.2}Co_{0.6}Fe_{0.4}O_{3-\delta}$	Disk	850	$Y_{C_2} = 10\% - 18\%$; $S_{C_2} = 70\% - 90\%$	[62]
$La_{0.8}Sr_{0.2}CoO_3$	Disk	800–850	$Y_{C_2} = 12 - 14$; $S_{C_2} = 40\% - 56\%$	[63]
$Bi_{1.5}Y_{0.3}Sm_{0.2}O_{3-\delta}$	Tube, no catalyst	900	$Y_{C_2} = 35\%$; $S_{C_2} = 54\%$	[64]
$La_{0.6}Sr_{0.4}Co_{0.2}Fe_{0.8}O_3$	Hollow fiber packed with $SrTi_{0.9}Li_{0.1}O_3$ catalyst	780–980	$Y_{C_2} \rightarrow 21\%$; $S_{C_2} \rightarrow 71.9\%$	[65]

Regarding this membrane type, the C_2 yield in the $Bi_{1.5}Y_{0.3}Sm_{0.2}O_{3-\delta}$ MR was achieved 35% [57].

In the SOFC-type MR prepared from pure ionic conducting membranes such as Yttria-stabilized zirconia (YSZ), electrical power can be co-generated accompanied by the OCM reaction. In this case, an YSZ tube with one dead-end was applied as the electrolyte, and $La_{0.85}Sr_{0.15}MnO_3$ powder is milled and mixed with glycerol, pasted into thin film on the outside of the YSZ tube, and heated at high temperature to form the cathode. In fact, $La_{1.8}Al_{0.2}O_3$ prepared on the inside of the YSZ tube by a mist pyrolysis method is used as dual roles—the OCM catalyst and anode roles. Moreover, Pt wire is connected to platinum mesh placed on both electrodes to serve as the current collector. Oxygen ions are transferred from the cathode through the perovskite membrane to the anode side and react with CH_4 to achieve C_2 products [66–68]. In this case, the theoretical electromotive force can be calculated by:

$$E = -\frac{\Delta G}{nF} \tag{10.6}$$

where ΔG is the Gibbs free energy, and n is the number of electrons. Furthermore, F, the oxygen permeation rate is determined by the electrical current. Indeed, the anode catalyst plays a key role in the C_2 selectivity. For instance, when silver was applied as electrode and 1 wt% of Sr/La_2O_3–Bi_2O_3 as catalyst, an electric current of 20–40 mA with C_2 selectivity of 90%–94% and C_2 yield of 0.2%–1% was achieved at 1,000 K. Although, by using $La_{1.8}Al_{0.2}O_3$ as anode and catalyst, the C_2 yield and electric current could reach 4% and 180 mA, respectively. However, all the perovskite MRs tested in practice have not presented very high C_2 yields, so far. This was probably related to the low oxygen permeance that did not match the methane catalytic activation on the membrane surface. Indeed, if an external power source is used to create a pervoskite MR, the catalytic activity and C_2 selectivity of the metal and metal oxide catalysts can be dramatically altered and reversibly owing to supplying more active oxygen species, resulting in much higher C_2 yields. In general, the SOFC-type MR requires an operating temperature approximately 200 K higher than the others; the electricity simultaneously produced as a by-product still makes it attractive [67–70].

10.5.1.3 Oxidative dehydrogenation of alkanes

Oxidation dehydrogenation of alkanes such as ethane and propane to corresponding olefins is a significant catalytic process:

$$C_2H_6 + \frac{1}{2}O_2 \leftrightarrow C_2H_4 + H_2O, \qquad \Delta H^0_{298} = -105 \text{ kJ mol}^{-1} \tag{10.7}$$

$$C_3H_8 + \frac{1}{2}O_2 \leftrightarrow C_3H_6 + H_2O, \qquad \Delta H^0_{298} = -136 \text{ kJ mol}^{-1} \tag{10.8}$$

The principle of the perovskite MRs for selective oxidation of ethane and propane is similar to the OCM process, but without the presence of methane coupling reactions.

Table 10.5   Oxidative dehydrogenation of ethane/propane in dense ceramic MRs

Reaction	Membrane	Reactor configuration	Catalyst	Temp (°C)	Main results	Reference
$C_2H_6 \rightarrow C_2H_4$	BSCF	Tube	–	650	$S = 90\%$	[67]
	BYS	Tube with a dead-end	–	875	$Y = 56\%$; $S = 80\%$	[71]
	BCFZ	Hollow fiber	–	800	$S = 64\%$; $X = 63\%$	[72]
	BSCF	Disk; coated catalyst	V/MgO	770	$Y = 75\%$; $S > 92\%$	[73]
$C_3H_8 \rightarrow C_3H_6$	BSCF	Tube with a dead-end	–	750	$S = 23.8\%$–40.2\%; $X = 71.8\%$–29.0\%	[74]

BSCF, $Ba_{0.5}Sr_{0.5}Co_{0.8}Fe_{0.2}O_{3-\delta}$; BCFZ, $BaCo_xFe_yZr_zO_{3-\delta}$ $(x + y + z = 1)$; BYS, $Bi_{1.5}Y_{0.3}Sm_{0.2}O_3$.

On the oxygen-rich side, molecular oxygen is firstly adsorbed on the membrane surface, reduced to O^{2-}, and then permeate through the membrane to the reaction side surface. On the reaction side, alkane is oxidized by the surface O^{2-}. By depleting the surface oxygen from permeate side, the oxygen molecules diffuse from the oxygen-rich side to fill in the oxygen vacancies.

Therefore, this kind of operation permits complete control over the contact approach of reactants with each other, and with the catalytically active surface, the selectivity of the oxidation reaction can be controlled at a very high level [67]. A summary for the results of oxidative dehydrogenation of ethane/propane in perovskite MRs is reported in Table 10.5.

It should be noted that the performance of the perovskite MR can be changed with application of surface catalyst. For instance, by using BSCF membranes with V/MgO micrometer grain or Pd nanocluster-modified surfaces, 75% ethylene yield can be achieved at 1,040 or 1,050 K, respectively. However, Ni cluster deposition results in a reduction in ethane conversion compared with the simple membrane without changing the ethylene selectivity. In addition, the contact time between the reactant and the membrane plays a key role in reaction selectivity. Therefore, hollow-fiber configuration in MRs presents lower selectivity than disk-shaped ones owing to their longer contact time [73].

10.5.1.4   Decomposition of H_2O, NO_x, and CO_2

Oxygen-permeable perovskite membranes can also be applied as extractors to selectively remove the produced oxygen during reactions to overcome the thermodynamic limitation or kinetic limitation and improve the product yields. Hence, Table 10.6 reports the application of perovskite MRs as an oxygen extractor during various reactions.

Table 10.6   Applications of perovskite MRs for decomposition of H_2O, NO_x, and CO_2

Reaction	Membrane	Configuration	Catalyst	T (°C)	Main results	Reference
$H_2O \rightarrow H_2 + 1/2O_2$	Gd-doped CeO_2–40%Ni	Disk (0.13 mm)	–	900	$r_{H_2} = 4.46$ µmol cm^{-2} s^{-1}	[75]
$H_2O \rightarrow H_2 + 1/2O_2$	$SrFeCo_{0.5}O_x$	Disk (0.09 mm)	–	900	$r_{H_2} = 7.44$ µmol cm^{-2} s^{-1}	[76]
$H_2O \rightarrow H_2 + 1/2O_2$	GDC-GSTA	Disk (25 µm coated support)	–	900	$r_{H_2} = 7$ µmol cm^{-2} s^{-1}	[77]
$H_2O \rightarrow H_2 + 1/2O_2$	BCFZ	Hollow fiber (0.17 mm)	–	950	$r_{H_2} = 2.31$ µmol cm^{-2} s^{-1}	[78]
$N_2O \rightarrow N_2$ $C_2H_6 \rightarrow C_2H_4$	BCFZ	Hollow fiber (0.17 mm)	Ni/Al_2O_3	875	$X_{N_2O} = 100\%$; $X_{N_2O} = 91\%$; $S_{C_2H_4} = 80\%$	[78]
$CO_2 \rightarrow CO$ $CH_4 \rightarrow$ syngas	SCFA	Tube	Ni/Al_2O_3	900	$X_{CO_2} = 12.4\%$; $X_{CH_4} = 86\%$; $S_{CO} = 93\%$; $H_2/CO = 1.8$	[51]

GSC, $Gd_{0.2}Ce_{0.8}O_{1.9-\delta}$; GSTA, $Gd_{0.08}Sr_{0.88}Ti_{0.99}Al_{0.05}O_{3-\delta}$; BCFZ, $BaCo_xFe_yZr_{1-x-y}O_{1-\delta}$; SCFA, 3% Al_2O_3 doped $SrCo_{0.8}Fe_{0.2}O_{3-\delta}$.

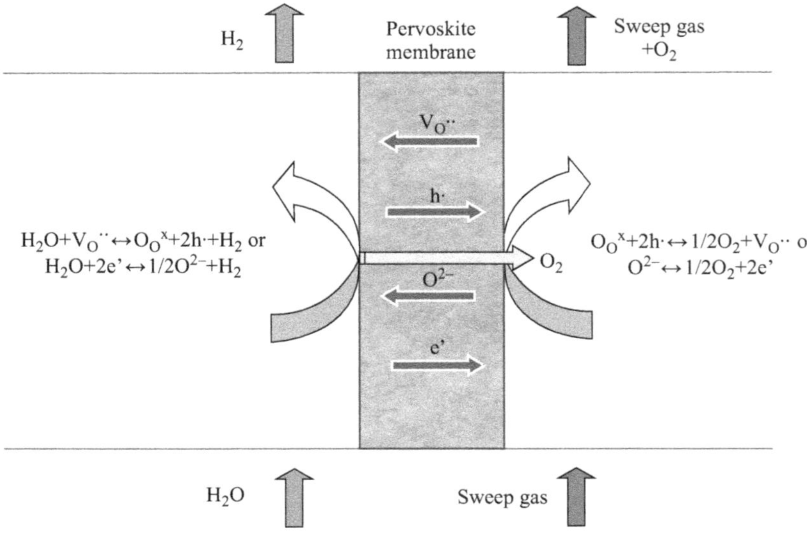

Figure 10.9   *Hydrogen production during water decomposition by perovskite MR*

At high temperatures, water dissociates into oxygen and hydrogen. Generally, very low concentrations of oxygen and hydrogen are generated even at high temperatures (e.g., 0.1% and 0.042% for hydrogen and oxygen, respectively, at 1,900 K), owing to the small equilibrium constant. If the equilibrium is shifted toward dissociation by removing either oxygen or hydrogen, significant volumes of hydrogen or oxygen can be produced at lower temperatures. This can be obtained using a perovskite membrane without the need for electrical power or circuitry, as indicated in Figure 10.9. For oxygen permeation, driving force may be achieved by using an inert sweep gas or a reducing gas such as methane in permeate side.

Indeed, the hydrogen production rate mainly depends on the rate at which oxygen is removed from the water dissociation zone. Hence, to reach a high hydrogen production rate, the perovskite membrane should possess high electron and oxygen-ion conductivities and good surface exchange properties. In addition, the hydrogen production rate can also be improved by reducing the membrane thickness, increasing the active surface area of the membrane, or applying a water-dissociation catalyst to the surface of the membrane.

In literatures [75–78], several perovskite membranes such as Ga-doped CeO_2–Ni and $Gd_{0.2}Ce_{0.8}O_{1.9-\delta}$–$Gd_{0.08}Sr_{0.88}Ti_{0.95}Al_{0.05}O_{3-\delta}$ were investigated for hydrogen production during water decomposition, and the results are summarized in Table 10.6. The maximum hydrogen production rate achieved 7.44 μmol cm^{-2} s^{-1}. On the other hand, nitrogen oxides (i.e., NO, NO_2, and N_2O) are considered as major air pollutants responsible for photochemical smog, acid rain, ozone depletion, as well as climate change. The conventional approach to remove NO_x pollution is to reduce NO_x catalytically into N_2 by using ammonia, hydrogen, carbon monoxide, and

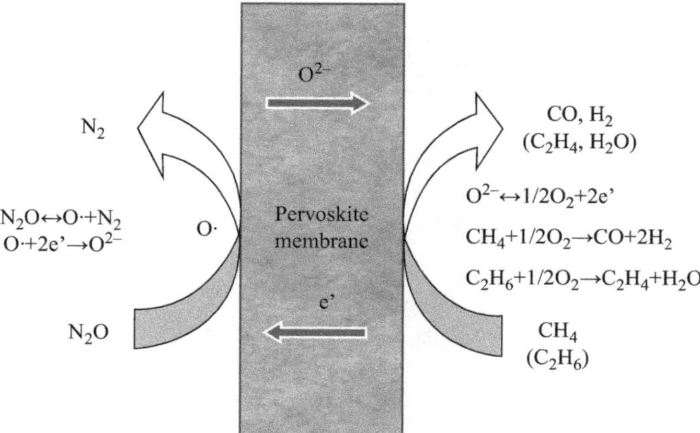

Figure 10.10 N_2O decomposition in perovskite MR enhanced by coupling with the partial oxidation of hydrocarbons

hydrocarbons as reducing agents. The N_2O decomposition is a kinetically limited reaction and inhibited by the oxygen molecule produced:

$$N_2O(g) \leftrightarrow N_2 + \frac{1}{2}O_2 \tag{10.9}$$

However, most perovskite catalysts cannot endure the coexistence of oxygen because the adsorbed oxygen blocks the catalytically active sites for N_2O decomposition. Accordingly, the total decomposition of N_2O can be obtained in the perovskite MR [53]. To enhance the driving force for oxygen transport through the perovskite membrane, methane or ethane can be fed to the permeate side of the membrane to consume the permeated oxygen. Hence, Figure 10.10 illustrates the principle of the perovskite MR for N_2O decomposition accompanied by the partial oxidation of hydrocarbons.

It should be mentioned that the water decomposition for hydrogen production can also be enhanced by coupling with the partial oxidation of hydrocarbons. In recent years, more attention has been focused on CO_2 capture and sequestration. One potential strategy for the consumption of CO_2 is the thermal decomposition of CO_2 to CO and O_2, because CO can be utilized as a raw material in the production of significant basic chemical products:

$$2CO_2 \leftrightarrow 2CO + O_2, \qquad \Delta H^0_{298} = 552 \, \text{kJ mol}^{-1} \tag{10.10}$$

However, this reaction is highly endothermic, taking place only at high temperature, and is not easy to achieve in CRs. In a perovskite MR, the CO_2 decomposition reaction can be coupled with POM to syngas, as illustrated in Figure 10.11. In this process, the decomposition reaction takes place on one side of the membrane.

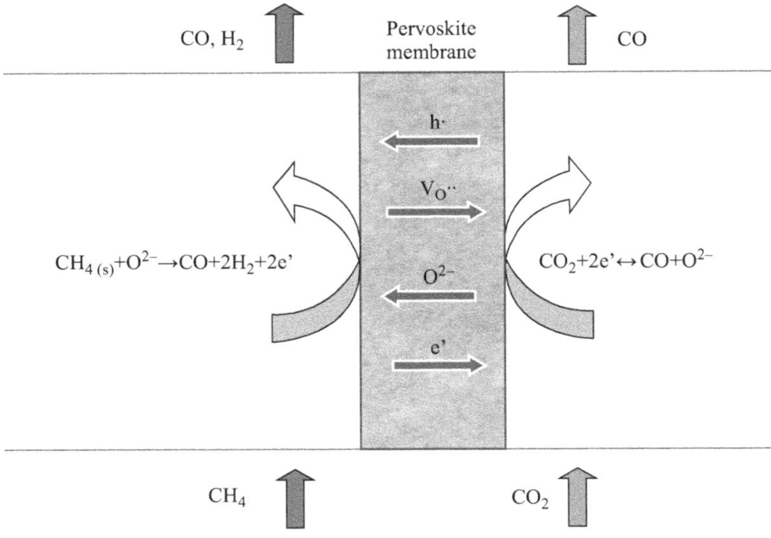

Figure 10.11   Decomposition of CO$_2$ coupled with POM in perovskite MR

The produced oxygen from the decomposition can permeate through the perovskite membrane to the other side of the membrane and reacts with CH$_4$ to produce syngas over catalyst. Therefore, CO$_2$ actually acts as the oxygen source for the POM reaction. The oxygen permeation rate in the perovskite membrane plays a significant role in the CO$_2$ conversion [51].

10.5.1.5   Methane coupling

In general, during methane coupling reaction, methane and air are fed into two sections separated by the perovskite membrane at high temperatures as indicated in Figure 10.12.

As methane stream flows over a proton–hole conducting perovskite membrane surface at high temperature, it is catalytically dissociated into methane radicals by combining with a proton–hole transported from the air side [30]. Indeed, it should be noted that dissociation reaction is actually the combination of the following two steps:

$$CH_4 \rightarrow \bullet CH_3 + H^{\bullet} + e \tag{10.11}$$

$$H^{\bullet} + e \leftrightarrow \text{nil} \tag{10.12}$$

The proton is permeated to the oxygen side concerning the electrochemical potential gradient and is consequently oxidized by oxygen to form water. Meanwhile, the methane radicals are released into the gas phase, in which the coupling reaction to form ethane is occurred:

$$2 \bullet CH_3 \leftrightarrow C_2H_6 \tag{10.13}$$

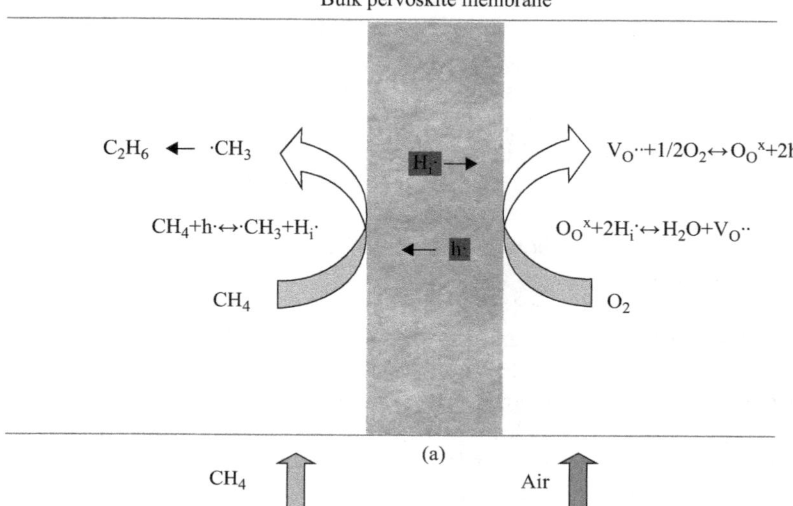

Figure 10.12   Methane coupling reaction in the perovskite MR

Following the similar mechanism, ethylene may be formed from ethane by the formation of ethane radicals. It should be mentioned that the higher carbon products ($>C_2$) such as propane and butane can be further produced following a same procedure. Moreover, if some particular catalyst is applied to the membrane surface, even benzene or other valuable aromatic hydrocarbons can be produced [30].

However, some other applications of perovskite MRs during other reactions such as reforming reactions were presented in recent years [79,80]. In one case, the feasibility of MSR at low temperatures (450–650 °C) was studied by using a Ni-BZCY72/BZCY72/Cu perovskite MR, which allowed for the simultaneous separation of hydrogen by Kyriakou *et al.* [79]. Their results showed that the methane conversion and hydrogen yield were improved by up to 50% in the temperature range of 550–650 °C with respect to the CR. Moreover, Jin *et al.* [80] investigated a BSCF Perovskite-type during autothermal ethanol reforming for hydrogen production. Their results indicated the BSCF membrane holding the high oxygen permeation flux, the excellent phase reversibility and the good stability under the highly reducing atmosphere, showed great application potential for hydrogen production via the autothermal reforming reaction.

10.6   Conclusion and future trends

There has been considerable research interest in recent decades to integrate perovskite membranes into MR unit operations for equilibrium limited or otherwise complex reaction systems involving oxygen and hydrogen species. Indeed, perovskite MRs have presented significant improvement with respect to CRs for both oxidative and non-oxidative reactions including lower weight hydrocarbons, especially methane. By either selective removing products or selective controlling reactants during related reactions, the production of desired component can be improved and competing reactions suppressed, respectively. However, these evaluations have only taken place at the laboratory scale, primarily owing to engineering and material limitations. In particular, the high temperature, operating conditions imposes the novel engineering design, and consequently scaled-up solutions for industrial testing of perovskite MRs at the pilot scale.

In fact, the advantages presented by MRs for integrating separation and reaction systems in a single-unit operation are driving research and development (R&D) in this field, particularly when high yields, conversions, and selectivities make the MR concept economically attractive. Currently, in terms of the perovskite membrane materials, the stability of perovskites against thermo-mechanical and chemical degradation at high temperatures is crucial for successful utilization of this technology. On the other hand, in terms of perovskite MR operation, the performance of the membrane must be balanced versus the performance of the catalysts. For membrane transport, the relative importance of the convective transport of the feed stream against permeation is defined by Peclet number (*Pe*).

A second important aspect of perovskite MR operation is related to fluid dynamics and heat controlling. Because of the membrane operation, gas velocities can be varied along the membrane length and may be complicated the fluid dynamics and convective transport within the MR. Furthermore, managing heat through the perovskite MR becomes dominant, particularly when exothermic and endothermic reactions occur on either side or even on the same side of the membrane. Finally, one of the most important viewpoints is related to the process start-up and shutdown, heating cycles, catalyst regeneration cycles, and related transient conditions that may degrade and foul the perovskite membranes. Therefore, the membranes themselves will have to exhibit robustness to withstand long-term operation and process variations for successful industrial applications of perovskite MRs.

Acronyms

BSCF	barium strontium cobalt iron
LSCF	lanthanum strontium cobalt iron
CMR	catalytic membrane reactor
CR	conventional reactor
CO_x	carbon oxides
ESR	ethanol steam reforming
GTL	gas to liquid
MIEC	mixed ionic electron conducting
MR	membrane reactor
MSR	methane steam reforming
NO_x	nitrate oxides

OCM   oxidative coupling of methane
PBMR   packed bed membrane reactor
POM   partial oxidation of methane
PSA   pressure swing adsorption
YSZ   Yttria-stabilized zirconia

References

[1] Jiang, Q., Faraji, S., Slade, D.A. and Stagg-Williams, S.M. (2011). A review of mixed ionic and electronic conducting ceramic membranes as oxygen sources for high-temperature reactors. In Membrane Science and Technology, Elsevier B.V., Amsterdam, pp. 235–273.

[2] Mundschau, M.V., Xie, X., Evenson, C.R. and Sammells, A.F. (2006). Dense inorganic membranes for production of hydrogen from methane and coal with carbon dioxide sequestration. Catalysis Today, 118, 12–23.

[3] Tan, X. and Li. K. (2015). Dense Ceramic Oxygen-Permeable MRs, Inorganic MRs: Fundamental and Applications, Wiley, New York, 143–179.

[4] Smart, S., Liu, S., Basile, A. and diniz da Costa, J.C. (2014). Perovskite MRs: fundamentals and applications for oxygen production, syngas production and hydrogen processing. In Membranes for Clean and Renewable Power Applications, Woodhead Publishing Limited, Sawston, 182–217.

[5] Shao, Z., Xiong, G., Cong, Y. and Yang, W. (2000). Synthesis and oxygen permeation study of novel perovskite-type $BaBi_xCo_{0.2}Fe_{0.8-x}O_{3-\delta}$. Journal of Membrane Science, 164, 167–176.

[6] Tong, J., Yang, W., Zhu, B. and Cai, R. (2002). Investigation of ideal zirconium-doped perovskite-type ceramic membrane materials for oxygen separation. Journal of Membrane Science, 203, 175–189.

[7] Zhu, X., Wang, H. and Yang, W. (2004). Novel cobalt-free oxygen permeable membrane. Chemical Communications, 9, 1130–1131.

[8] Tong, J., Yang, W., Cai, R., Zhu, B., Xiong, G. and Lin, L. (2003). Investigation on the structure stability and oxygen permeability of titanium-doped perovskite type oxides of $BaTi_{0.2}Co_xFe_{0.8-x}O_{3-s}$ ($x = 0.2$–0.6). Separation and Purification Technology, 32, 289–299.

[9] Teraoka, Y., Zhang, H.M. and Yamazoe, N. (1985). Oxygen-sorptive properties of defect perovskite-type $La_{1-x}Sr_xCo_{1-y}Fe_yO_{3-\delta}$. Chemistry Letters, 14, 1367–1370.

[10] Kharton, V.V., Yaremchenko, A.A., Kovalevsky, A.V., Viskup, A.P., Naumovich, E.N. and Kerko, P.F. (1999). Perovskite-type oxides for high-temperature oxygen separation membranes. Journal of Membrane Science, 163, 307–317.

[11] Liu, S. and Gavalas, G.R. (2005). Oxygen selective ceramic hollow fiber membranes. Journal of Membrane Science, 246, 103–108.

[12] Wang, H., Cong, Y. and Yang, W. (2005). Oxidative coupling of methane in $Ba_{0.5}Sr_{0.5}Co_{0.8}Fe_{0.2}O_{3-\delta}$ tubular MRs. Catalysis Today, 104, 160–167.

[13]   Teraoka, Y., Nobunaga, T. and Yamazoe, N. (1988). Effect of cation substitution on the oxygen semi-permeability of perovskite-type oxides. Chemistry Letters, 17, 503–506.

[14]   Miura, N., Okamoto, Y., Tamaki, J., Morinaga, K. and Yamazoe, N. (1995). Oxygen semi-permeability of mixed-conductive oxide thick-film prepared by slip casting. Solid State Ionics, 79, 195–200.

[15]   Ishihara, T., Yamada, T., Arikawa, H., Nishiguchi, H. and Takita, Y. (2000). Mixed electronic-oxide ionic conductivity and oxygen permeating property of Fe-, Coor Ni-doped $LaGaO_3$ perovskite oxide. Solid State Ionics, 135, 631–636.

[16]   Schlehuber, D., Wessel, E., Singheiser, L. and Markus, T. (2010). Long-term operation of a $La_{0.58}Sr_{0.4}Co_{0.2}Fe_{0.8}O_{3-s}$-membrane for oxygen separation. Journal of Membrane Science, 351, 16–20.

[17]   Slade, R.C.T. (1993). The perovskite-type proton-conducting solid electrolyte $BaCe_{0.90}Y_{0.10}O_{3-\alpha}$ in high temperature electrochemical cells. Solid State Ionics, 61, 111–114.

[18]   Schober, T. and Bohn, H.G. (2000). Water vapor solubility and electrochemical characterization of the high temperature proton conductor $BaZr_{0.9}Y_{0.1}O_{2.95}$. Solid State Ionics, 127, 351–360.

[19]   Fukui, T., Ohara, S. and Kawatsu, S. (1999). Ionic conductivity of gadolinium-doped barium praseodymium oxide. Solid State Ionics, 116, 331–337.

[20]   Yajima, T., Kazeoka, H., Yogo, T. and Iwahara, H., (1991). Proton conduction in sintered oxides based on $CaZrO_3$. Solid State Ionics, 47, 271–275.

[21]   Sata, N., Hiramoto, K., Ishigamw, M., Hosoya, S., Nijmuma, N. and Shin, S., (1996). Site identification of protons in $SrTiO_3$: mechanism for large protonic conduction. Physics Reviews B, 54 (22), 15795–15799.

[22]   Norwick, A.S. and Du, Y. (1995). High-temperature protonic conductors with perovskite-related structures. Solid State Ionics, 77, 137–146.

[23]   Yajima, T., Iwahara, H. and Uchida, H. (1991). Protonic and oxide ionic conduction in $BaCeO_3$-based ceramics – effect of partial substitution for Ba in $BaCe_{0.9}O_{3-s}$ with Ca. Solid State Ionics, 47, 117–124.

[24]   Yajima, T., Suzuki, H., Yogo, T. and Iwahara, H. (1992). Protonic conduction in $SrZrO_3$-based oxides. Solid State ionics, 51, 101–107.

[25]   Matzke, T. and Cappadonia, M. (1996). Proton conductive perovskite solid solutions with enhanced mechanical stability. Solid State Ionics, 86, 659–663.

[26]   Guan, J., Dorris, S.E., Balachandran, U. and Liu, M. (1997). Transport properties of $BaCe_{0.95}Y_{0.05}O_{3-s}$ mixed conductors for hydrogen separation. Solid State Ionics, 100, 45–52.

[27]   Liu, J.F. and Norwick, A.S. (1992).The incorporation and migration of protons in Nd-doped $BaCeO_3$. Solid State Ionics, 50, 131–138.

[28]   Qi, X and Lin, Y.S. (2000). Electrical conduction and hydrogen permeation through mixed proton-electron conducting strontium cerate membranes. Solid State Ionics, 130 (1–2), 149–156.

[29] Hong, W.K. and Choi, G.M. (2010). Oxygen permeation of BSCF membrane with varying thickness and surface coating. Journal of Membrane Science, 346, 353–360.

[30] Li, K. (2007). Ceramic Membranes for Separation and Reaction, John Wiley & Sons Ltd., New York, 169–239.

[31] Ishihara, T. and Takita, Y. (2000). Partial oxidation of methane into syngas with oxygen permeating ceramic MRs. Catalysis Surveys from Japan, 4, 125–133.

[32] Czuprat, O., Caro, J., Kondratenko, V.A. and Kondratenko, E.V. (2010). Dehydrogenation of propane with selective hydrogen combustion: a mechanistic study by transient analysis of products. Catalysis Communications, 11, 1211–1214.

[33] Gong, Z. and Hong, L. (2011). Integration of air separation and partial oxidation of methane in the $La_{0.4}Ba_{0.6}Fe_{0.8}Zn_{0.2}O_{3-\delta}$ MR. Journal of Membrane Science, 380, 81–86.

[34] Ishihara, T., Tsuruta, Y., Todaka, T., Nishiguchi, H. and Takita, Y. (2002). Fe doped $LaGaO_3$ perovskite oxide as an oxygen separating membrane for CH_4 partial oxidation. Solid State Ionics, 152, 709–714.

[35] Delbos, C., Lebain, G., Richet, N. and Bertail, C. (2010). Performances of tubular $La_{0.8}Sr_{0.2}Fe_{0.7}Ga_{0.3}O_{3-\delta}$ mixed conducting MR for under pressure methane conversion to syngas. Catalysis Today, 156, 146–152.

[36] Hoang, D. and Chan, S. (2006). Effect of reactor dimensions on the performance of an O_2 pump integrated partial oxidation reformer – a modelling approach. International Journal of Hydrogen Energy, 31, 1–12.

[37] Tian, T., Wang, W., Zhan, M. and Chen, C. (2010). Catalytic partial oxidation of methane over $SrTiO_3$ with oxygen-permeable MR. Catalysis Communications, 11, 624–628.

[38] Ikeguchi, M., Mimura, T., Sekine, Y., Kikuchi, E. and Matsukata, M. (2005). Reaction and oxygen permeation studies in $Sm_{0.4}Ba_{0.6}Fe_{0.8}Co_{0.2}O_{3-\delta}$ MR for partial oxidation of methane to syngas. Applied Catalysis A: General, 290, 212–220.

[39] Zhu, D.C., Xu, X.Y., Feng, S.J., Liu, W. and Chen, C.S. (2003). La_2NiO_4 tubular MR for conversion of methane to syngas. Catalysis Today, 82, 151–156.

[40] Wang, H., Cong, Y. and Yang, W. (2003). Investigation on the partial oxidation of methane to syngas in a tubular $Ba_{0.5}Sr_{0.5}Co_{0.8}Fe_{0.2}O_{3-\delta}$ MR. Catalysis Today, 82, 157–166.

[41] Gu, X.H., Jin, W.Q., Chen, C.L., Xu, N.P., Shi, J. and Ma, Y.H. (2002). YS Z–$SrCo_{0.4}Fe_{0.6}O_{3-\delta}$ membranes for the partial oxidation of methane to syngas. AIChE Journal, 48, 2051–2060.

[42] Ritchie, J.T., Richardson, J.T. and Luss, D. (2001). Ceramic MR for synthesis gas production. AIChE Journal, 47, 2092–2101.

[43] Hu, J., Xing, T., Jia, Q., *et al.* (2006). Methane partial oxidation to syngas in $YBa_2Cu_3O_{7-x}$ MR. Applied Catalysis A: General, 306, 29–33.

[44] Kharton, V.V., Yaremchenko, A.A., Valente, A.A., *et al.* (2005). Methane oxidation over Fe-, Co-, Ni- and V-containing mixed conductors. Solid State Ionics, 176, 781–791.

[45] Wang, H., Tablet, C., Schiestel, T., Werth, S. and Caro, J. (2006). Partial oxidation of methane to syngas in a perovskite hollow fiber MR. Catalysis Communications, 7, 907–912.

[46] Zhu, X., Li, Q., Cong, Y. and Yang, W. (2008). Syngas generation in a MR with a highly stable ceramic composite membrane. Catalysis Communications, 10, 309–312.

[47] Zhang, Y., Liu, J., Ding, W. and Lu, X. (2011). Performance of an oxygen-permeable MR for partial oxidation of methane in coke oven gas to syngas. Fuel, 90, 324–330.

[48] Zhu, X., Li, Q., He, Y., Cong, Y. and Yang, W. (2010). Oxygen permeation and partial oxidation of methane in dual-phase MRs. Journal of Membrane Science, 360, 454–460.

[49] Luo, H., Wei, Y., Jiang, H., *et al.* (2010). Performance of a ceramic MR with high oxygen flux Ta-containing perovskite for the partial oxidation of methane to syngas. Journal of Membrane Science, 350, 154–160.

[50] Li, Q., Zhu, X., He, Y. and Yang, W. (2010). Partial oxidation of methane in $BaCe_{0.1}Co_{0.4}Fe_{0.5}O_{3-\delta}$ MR. Catalysis Today, 149, 185–190.

[51] Zhang, C., Jin, W., Yang, C. and Xu, N. (2009). Decomposition of CO_2 coupled with POM in a thin tubular oxygen-permeable MR. Catalysis Today, 148, 298–302.

[52] Tan, X. and Li, K. (2009). Design of mixed conducting ceramic membranes/reactors for the partial oxidation of methane (POM) to syngas. AIChE Journal, 55, 2675–2685.

[53] Zhang, P., Chang, X.F., Wu, Z.T., Jin, W.Q. and Xu, N.P. (2005). Effect of the packing amount of catalysts on the partial oxidation of methane reaction in a dense oxygen permeable MR. Industrial and Engineering Chemistry Research, 44, 1954–1959.

[54] Wang, W. and Lin, Y.S. (1995). Analysis of oxidative coupling of methane in dense oxide MRs. Journal of Membrane Science, 103, 219–233.

[55] Haag, S., van Veen, A.C. and Mirodatos, C. (2007). Influence of oxygen supply rates on performances of catalytic MRs. Application to the oxidative coupling of methane. Catalysis Today, 127, 157–164.

[56] Olivier, L., Haag, S., Mirodatos, C. and van Veen, A.C. (2009). Oxidative coupling of methane using catalyst modified dense perovskite MRs. Catalysis Today, 142, 34–41.

[57] Akin, F.T. and Lin, Y.S. (2002). Oxidative coupling of methane in dense ceramic MR with high yields. AIChE Journal, 48, 2298–2306.

[58] Wang, H., Cong, Y. and Yang, W. (2005). Oxidative coupling of methane in $Ba_{0.5}Sr_{0.5}Co_{0.8}Fe_{0.2}O_{3-\delta}$ tubular MRs. Catalysis Today, 104, 160–167.

[59] Lu, Y.P., Dixon, A.G., Moser, W.R., Ma, Y.H. and Balachandran, U. (2000). Oxidative coupling of methane using oxygen-permeable dense MRs. Catalysis Today, 56, 297–305.

[60] Ten Elshof, J.E., Bouwmeester, H.J.M. and Verweij, H. (1995). Oxidative coupling of methane in a mixed-conducting perovskite MR. Applied Catalysis A: General, 130, 195–212.

[61] Zeng, Y. (2000). Oxygen permeation and oxidative coupling of methane in yttria doped bismuth oxide MR. Journal of Catalysis, 193, 58–64.

[62] Zeng, Y., Lin, Y.S. and Swartz, S.L. (1998). Perovskite-type ceramic membrane: synthesis, oxygen permeation and MR performance for oxidative coupling of methane. Journal of Membrane Science, 150, 87–98.

[63] Lin, Y.S. and Zeng, Y. (1996). Catalytic properties of oxygen semipermeable perovskite-type ceramic membrane materials for oxidative coupling of methane. Journal of Catalysis, 164, 220–231.

[64] Tan, X., Pang, Z., Gu, Z. and Liu, S. (2007). Catalytic perovskite hollow fibre MRs for methane oxidative coupling. Journal of Membrane Science, 302, 109–114.

[65] Tagawa, T., Moe, K.K., Ito, M. and Goto, S. (1999). Fuel cell type reactor for chemicals – energy co-generation. Chemical Engineering Science, 54, 1553–1557.

[66] Bhatia, S., Thien, C.Y. and Mohamed, A.R. (2009). Oxidative coupling of methane (OCM) in a catalytic MR and comparison of its performance with other catalytic reactors. Chemical Engineering Journal, 148, 525–532.

[67] Xui-Mei, G., Hidajat, K. and Ching, C.-B. (1999). Simulation of a solid oxide fuel cell for oxidative coupling of methane. Catalysis Today, 50, 109–116.

[68] Eng, D. and Stoukides, M. (1991). Catalytic and electrochemical oxidation of methane on platinum. Journal of Catalysis, 130, 306–309.

[69] Kiatkittipong, W., Tagawa, T., Goto, S., Assabumrungrat, S., Silpasup, K. and Praserthdam, P. (2005). Comparative study of oxidative coupling of methane modeling in various types of reactor. Chemical Engineering Journal, 115, 63–71.

[70] Akin, F.T. and Lin, Y.S. (2002). Selective oxidation of ethane to ethylene in a dense tubular MR. Journal of Membrane Science, 209, 457–467.

[71] Wang, H., Cong, Y. and Yang, W. (2002). High selectivity of oxidative dehydrogenation of ethane to ethylene in an oxygen permeable MR. Chemical Communications, 14, 1468–1469.

[72] Akin, F.T. and Lin, Y.S. (2002). Selective oxidation of ethane to ethylene in a dense tubular MR. Journal of Membrane Science, 209, 457–467.

[73] Rebeilleau-Dassonneville, M., Rosini, S., Veen, A.C.v., Farrusseng, D. and Mirodatos, C. (2005). Oxidative activation of ethane on catalytic modified dense ionic oxygen conducting membranes. Catalysis Today, 104, 131–137.

[74] Wang, H., Tablet, C., Schiestel, T. and Caro, J. (2006). Hollow fiber MRs for the oxidative activation of ethane. Catalysis Today, 118, 98–103.

[75] Balachandran, U. (2004). Use of mixed conducting membranes to produce hydrogen by water dissociation. International Journal of Hydrogen Energy, 29, 291–296.

[76]   Balachandran, U., Lee, T. and Dorris, S. (2007). Hydrogen production by water dissociation using mixed conducting dense ceramic membranes. International Journal of Hydrogen Energy, 32, 451–456.

[77]   Wang, H., Gopalan, S. and Pal, U.B. (2011). Hydrogen generation and separation using $Gd_{0.2}Ce_{0.8}O_{1.9-\delta}$–$Gd_{0.08}Sr_{0.88}Ti_{0.95}Al_{0.05}O_{3-s}$ mixed ionic and electronic conducting membranes. Electrochimica Acta, 56, 6989–6996.

[78]   Jiang, H., Wang, H., Liang, F., *et al.* (2010). Improved water dissociation and nitrous oxide decomposition by in situ oxygen removal in perovskite catalytic MR. Catalysis Today, 156, 187–190.

[79]   Kyriakou, V., Garagounis, I., Vourros, A., *et al.* (2016). Methane steam reforming at low temperatures in a $BaZr_{0.7}Ce_{0.2}Y_{0.1}O_{2.9}$ proton conducting membrane reactor. Applied Catalysis B: Environmental, 186, 1–9.

[80]   Jin, Y., Rui, Z., Tian, Y., Lin, Y.S. and Li, Y. (2015). Autothermal reforming of ethanol in dense oxygen permeation membrane reactor. Catalysis Today, 264, 214–220.

Chapter 11

Polymeric membrane materials for hydrogen separation

Yuri Yampolskii[1] and Victoria Ryzhikh[1]

Abstract

Contemporary membrane processes for separation of hydrogen from different industrial streams are based on use of polymeric membrane materials. The subject of this chapter is consideration of properties of various membrane materials in respect of hydrogen and other light gases. The effects of properties of gases, polymers, and conditions of separation on the gas permeation parameters are considered. Possible options for improvement of these parameters include crosslinking and introduction of nanoparticles into polymer matrices. The problem of separation of hydrogen isotopes is briefly discussed. The main message of this review is that many existing and widely applied hydrogen-separating membranes can be replaced by novel ones based on polymers with enhanced permeability and/or selectivity.

11.1   Introduction

Membrane gas separation as a large-scale industrial process started to exist in the early 1980s. The first demonstrations of the success of this novel separation technology were in the field of separation of hydrogen, its recovery, and return into technological streams. Due to many advantages of the membrane processes, this technology was successfully applied very rapidly in numerous technological processes of chemistry, refinery, and petroleum chemistry: hydrocracking, hydrodesulfurization, and hydrotreating of gas oils and oil residues, hydrogenation, pyrolysis, ammonia synthesis, methanol synthesis, and others.

Membrane gas separation of the streams containing hydrogen has many advantages as compared to other separation technologies (e.g., cryogenic rectification and pressure swing adsorption (PSA)) [1]. Among these advantages are lower energy expenses, continuous separation process, mild conditions of separation, and

[1]A.V. Topchiev Institute of Petrochemical Synthesis, Russian Academy of Sciences, Moscow, Russia

capability of scaling. Technical and economic assessment made by Bernardo and Drioli [2] showed that mass intensity (required steam and cooling water) is lowest for membrane separation. In addition, productivity/footprint ratio is 10-fold lower for membrane processes than that of PSA. It means that a membrane system occupies a much smaller area than PSA or cryogenic plants.

When the first membrane gas separation units with hollow-fiber permeators were created by Monsanto Co in the early 1980s [3], the chemical and refinery markets were ready to accept these novelties. In many industrial processes, there were technological streams or purge gases of high pressure that include relatively high partial pressure of hydrogen, the driving force of potential membrane separation. So there was no need to spend energy for compression of feed streams to be directed to membranes. Let us consider, as an example, ammonia synthesis plant. In this large industrial process, a mixture of H_2 (from steam reforming of methane) and N_2 (from air) is circulated around the apparatus of catalytic synthesis. Conversion to ammonia at every step is incomplete, so unreacted gas is recycled back into the reactor. In the process of this circulation, "inert" impurities (CH_4, Ar) are accumulating in the mixture, thus reducing the partial pressure of hydrogen and yield of ammonia. To overcome this unpleasant phenomenon, continuous gas purge is maintained to keep the level of hydrogen sufficiently great. This purge gas contains hydrogen, which is usually lost or used as plant fuel, because conventional separation technologies do not allow its economic extraction and subsequent use as a by-product or valuable recycle directed back into the process. The situation changed dramatically with appearance of membranes.

Figure 11.1 shows a typical configuration of the process of ammonia synthesis with added a "loop" with membrane module.

Introduction of membrane module provides a possibility to return into the process about 95% of hydrogen [4]. Composite hollow-fiber membrane developed by Monsanto Co (or its daughter company Permea) was based on polysulfone asymmetric hollow fiber with a thin layer of polydimethylsiloxane coated over it to diminish or prevent the detrimental role of defects [3,5]. In glassy polymers, the permeability coefficients of hydrogen $P(H_2)$ as well as the separation factors $\alpha(H_2/N_2)$, $\alpha(H_2/CH_4)$, $\alpha(H_2/CO)$ are large, so it favored fast development of membrane separation plants for improvements of diverse processes and by several companies in the United States, Europe, and Japan (MEDAL (DuPont + Air liquid), MTR Inc., Air Products, GKSS, Ube, and others).

It should be noted that several years earlier than the first developments of Monsanto's technology, the first gas separation membrane produced in industrial scale was created in USSR. That was a flat-sheet asymmetric membrane based on poly(vinyltrimethyl silane) which had relatively high permeance in respect of hydrogen (20 m^3(STP)/m^2 h MPa). That was successfully used for obtaining technical nitrogen and oxygen enriched streams from air and separation of hydrogen from its mixtures with methane and nitrogen [6,7]. But of course the scale of practical application of these membrane units was much smaller than that of Prism permeators of Monsanto.

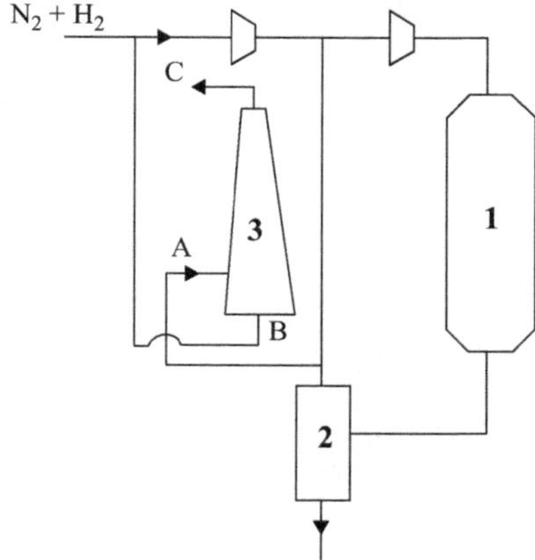

*Figure 11.1   Scheme of a plant of ammonia synthesis with added membrane
module: 1 column of ammonia synthesis, 2 separator of liquid
ammonia, 3 membrane separator; A line of feedstock (purge) gas,
total pressure 13 MPa, [H₂] 61%, B line of permeate, total pressure
7 MPa, [H₂] 92%, C line of retentate, total pressure 13 MPa, [H₂]
7% (adapted from Reference 4)*

This was a very brief description of the first and glorious days of industrial
membrane technology. Since that time, many things happened. According to the
prediction of Baker [8], the total market of membrane gas separation in 2010 would
be 350 million USD, whereas the actual rate of growth was much faster. The current state of this technology has grown into an industry with system sales close to
1 billion USD [9]. The number of membrane units has strongly increased. Thus,
Permea, now a part of Air Products, has produced more than 500 Prism membrane
systems. History of H_2 commercial plants created by MEDAL started in 1990 with
several relatively small units used in petrochemistry. Now the total number of
membrane systems made by this company is more than 120, which were applied in
such industries as ammonia synthesis, refining, and syngas [10].

It is worth noting, especially for this chapter, that virtually all practical applications of membrane gas separation were accomplished by polymeric membranes.
Meanwhile, other types of membranes were extensively studied. In many cases,
some of them demonstrate much better performance for hydrogen separation and
recovery as compared to nonporous polymeric membranes. This is true for metallic
membranes, carbon molecular sieves, and other types of membranes actively discussed in the literature [11–13]. All these subjects will be beyond this chapter,
because they are discussed in other chapters of this volume. We plan to focus on
membrane materials for separation of the mixtures H_2/CH_4, H_2/N_2 and H_2/CO.
A very important process of separation of H_2/CO_2 mixtures has been considered
recently in much detail by Li *et al.* [13], so we can refrain from discussing this
subject here.

11.2 Short background

Since the nineteenth century, it was known that the flux through membranes J is determined by formula:

$$J = P(\Delta p/l) \tag{11.1}$$

that is, the flux is proportional to the pressure drop $\Delta p = p_2 - p_1$, where p_2 and p_1 characterize pressure before and after membrane and inversely proportional to the thickness of the membrane. The proportionality constant P was named the permeability coefficient. Obviously, it is desired that a membrane would be not only highly permeable (large P values) but also selective. The ideal selectivity is defined as:

$$\alpha = P_i/P_j \tag{11.2}$$

that is, the ratio of the permeability coefficients of ith and jth gases. About in the same time, it became clear that the permeability coefficient includes the thermodynamic component, solubility coefficient S, which determines the driving force of the process or the concentration gradient in the membrane and kinetic (mobility) component, the diffusion coefficient that determines the rate of transport caused by this gradient. In other words, the following equation holds for the permeability coefficient:

$$P = DS \tag{11.3}$$

Therefore, it is possible to speak of selectivity of diffusion and sorption:

$$\alpha = (D_i/D_j)(S_i/S_j) \tag{11.4}$$

In the SI system, permeability coefficients are expressed in the following units: $[P] = \text{mol/m s Pa}$. However, a more widely used and accepted unit for P is Barrer:

$$1 \text{ Barrer} = 10^{-10} \text{ cm}^3(\text{STP})\text{cm}/(\text{cm}^2 \text{ s cmHg})$$

Permeability coefficients of common gases in polymers span a range of about seven orders of magnitude, from 10^{-3} to 10^4 Barrer or more.

In the case of membranes, the thickness of the selective layer is unknown, so the pressure-normalized steady-state flux is characterized by permeance Q or P/l. The accepted units for Q are $\text{mol/(m}^2 \text{ s Pa)}$ (SI) or $\text{cm}^3(\text{STP})/(\text{cm}^2 \text{ s (cmHg))}$. Permeance is also expressed in so-called Gas Permeation Units (GPU), where $1 \text{ GPU} = 10^{-6} \text{ cm}^3(\text{STP})/(\text{cm}^2 \text{ s (cmHg))}$. A permeance of 1 GPU corresponds to a membrane exhibiting an intrinsic permeability of 1 Barrer and having a selective layer thickness of 1 μm.

The relation (11.2) is very useful for light gases (e.g., hydrogen, nitrogen) that have low solubility in membrane materials; hence, they only weakly affect the property and behavior of polymers and do not influence the mutual diffusion and sorption parameters in the process of simultaneous transport of gases in mixture separation. Hence, the ideal selectivity can be used in description of the processes of separation of mixtures of light gases. However, for heavy vapors or gases with great solubility, the applicability of this parameter is less predictable. For mixture separation, the following equation is often used:

$$\alpha^* = (y_i/y_j)/(x_i/x_j) \tag{11.5}$$

Here y_i and y_j are the mole fractions in the permeate (the stream penetrated through the membrane). As for x_i and x_j, two different definitions are considered. According to Koros *et al.* [14], this ratio is called *separation coefficient* if x_i and x_j characterize the composition in the feed stream and *separation factor* if x_i and x_j characterize the composition in the retentate (the stream that did not pass through the membrane).

11.3   Gas permeation properties of polymers in respect of hydrogen

Gas permeation parameters depend on the following:

- conditions of the transport process;
- properties of gases;
- properties of polymers.

Let us consider all these issues with respect to hydrogen and main components of the mixtures with H_2 to be separated using membranes.

11.3.1   Effects of conditions of the transport process

Permeability and diffusion coefficients depend on temperature according to Arrhenius equations:

$$P = P_o\exp(-E_P/RT) \tag{11.6}$$

$$D = D_o\exp(-E_D/RT) \tag{11.7}$$

where E_P and E_D are the activation energies of permeation and diffusion. By combining (11.3), (11.6), and (11.7), one obtains:

$$S = S_o\exp(-\Delta H_s/RT) \tag{11.8}$$

where ΔH_s is the enthalpy of sorption. The E_D values increase for larger penetrants [15,16], whereas dependence of E_P on penetrants' size is much less noticeable [17]. It can be added that E_D values are always positive and are in the range of 1.4–34 kJ/mol for glassy polymers, whereas the sign of E_P depends on the relative magnitude of E_D and ΔH_s [17]. As H_2 molecule is one of the smallest ones, its diffusion coefficients increase rather weakly when temperature increases. For strongly size-sieving gas separation polymers, such as those used for H_2 removal from gas mixtures, $E_D > |\Delta H_s|$ and E_P values are positive, so hydrogen permeability increases when temperature increases.

Permeability and diffusion coefficients can depend on gas pressure. However, this does not hold for hydrogen and other light gases, components of hydrogen-containing mixtures to be separated (N_2, CH_4) [15]. Pressure dependence is, however, characteristic for carbon dioxide, mixture of which with hydrogen must be separated using membranes [13]. Pressure dependence is also typical for permeability coefficients of hydrocarbons, usual impurities in industrial streams H_2/CH_4.

11.3.2   Effects of properties of gases

The influences of different physical properties of gases on the permeation parameters (P, D, S) are well known. The diffusion coefficients of gases in polymers strongly depend on the size of molecules of penetrants. The following dependence has been proposed [18] and extensively used:

$$\ln D = a - bd^2 \tag{11.9}$$

where d is the penetrant gas kinetic diameter and $b > 0$. As the d value of hydrogen is very small (2.14 Å), the diffusion coefficients of hydrogen are among the largest for every polymer.

Opposite is true for the solubility coefficients of hydrogen in polymers. General correlations of the type:

$$\ln S = M + NT_c \tag{11.10}$$

where T_c is the critical temperature of gas, and $N > 0$ are known for the solubility of gases in polymers and liquids. Similar correlations with boiling points T_b and parameters of the potential of Lennard–Jones ε/k hold. As hydrogen has very low critical temperature 33.24 K, its solubility coefficients in polymers are very small (much less than 1 cm^3(STP)/cm^3 atm) [15].

Combination of these two factors determines the variations of the permeability coefficients of hydrogen in polymers. It should be emphasized that the role of the diffusion coefficients is prevailing so the $P(H_2)$ values in many polymers is in the range of 10^3–10^4 Barrer.

11.3.3   Effects of properties of polymers

It is known that there exists a "trade-off" between permeability and selectivity of separation of light gases in various polymers [19,20]: the most permeable materials are characterized by lower selectivity and vice versa. An example of such diagram "permeability–selectivity" is shown in Figure 11.2 for the pair H_2/N_2.

An analysis of this figure indicates that glassy polymers have preferences as membrane materials for separation of hydrogen in comparison with rubbers: they are more permselective and in many cases more permeable than rubbers. A wide range of variation of both parameters $P(H_2)$ and $P(H_2)/P(N_2)$ in glassy polymers can be noted. Depending on the chemical structure of polymers the values of $P(H_2)$ are in the range from s0.01 to 10,000 Barrer. For most polymers, the diffusion coefficients $D(H_2)$ are limited in the range of 10^{-7}–10^{-4} cm^2/s.

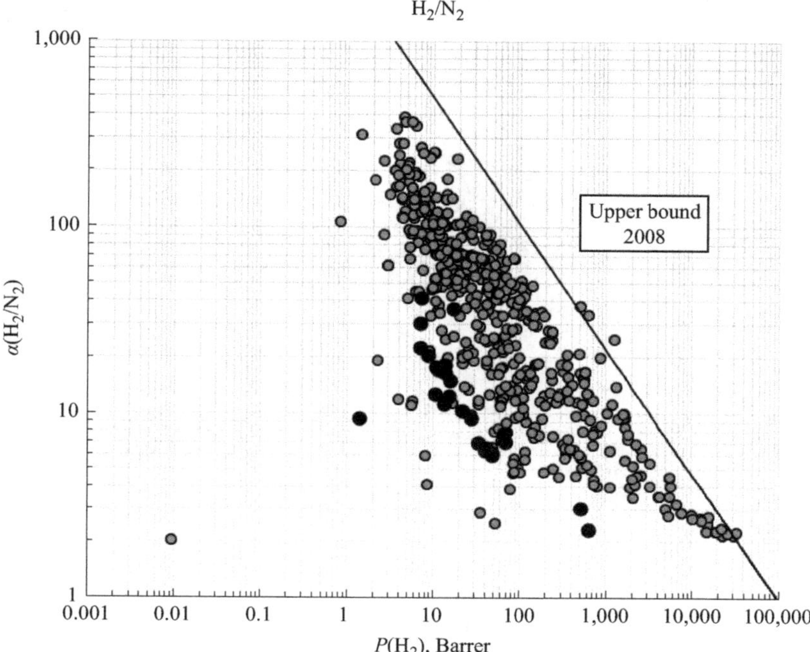

Figure 11.2   Diagram "permeability–selectivity" for gas pair H$_2$/N$_2$, Source is the Database [17]; gray points glassy polymers, black points rubbers. Line is Upper Bound of 2008 [20]

Mentioned in Section 11.1, industrial companies use a quite limited set of polymers as membrane materials (polysulfone, polyimides (PIs), cellulose acetate, and some others) for separation and recovery of hydrogen from streams of chemical and refinery plants. Meanwhile, today a number of polymers can be considered as suitable replacement for existing membrane materials for separation of hydrogen. Figure 11.3 can serve as an illustration of this statement. It shows the results of separation of equimolar mixture H$_2$/CH$_4$ after a single pass through membrane based on glassy polymers with different permeability coefficients P(H$_2$) and selectivity P(H$_2$)/P(CH$_4$). The indicated purity of hydrogen in permeate corresponds to a single pass through the membrane at recovery close to zero. Obviously, in industrial gas separation, recovery must be higher, so these values characterize the upper boundary of achieved purity of produced hydrogen. On the other hand, in many instances, the feed streams contain hydrogen in greater concentration, so this estimate can be further improved. The fact that membrane industry still uses membranes developed decades ago is an indication of its great conservatism and, maybe, in some cases economic factors. However, as Baker and Low reminded recently [9], even very expensive membrane materials costing as much as $50,000/kg can be used in industrial plants.

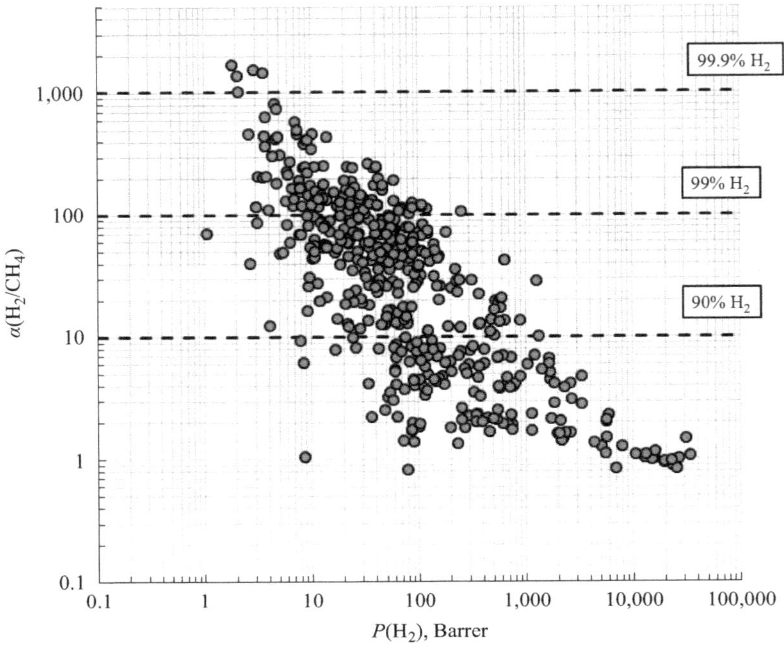

Figure 11.3   Possibilities of efficient separation of mixture H₂:CH₄ = 1:1 using membrane based on glassy polymers. Source is the Database [17]. The dashed line and concentrations of H₂ indicate the composition of permeate obtained after passage through membranes with certain selectivity at zero recovery

There are several technical problems in which membrane separation of hydrogen is important. Extremely relevant and still difficult is the task of separation of H_2/CO_2 mixtures that are formed in the process of hydrogen production by steam reforming of natural gas. One of the difficulties of this task is caused by relatively close permeability coefficients of hydrogen and carbon dioxide in many polymers: both gases can be considered as "fast" components of mixture, H_2 due to its great diffusivity and CO_2 due to its large solubility coefficients. This issue has been discussed in a recent review [13], so we shall refrain from discussing it in more detail.

Another kindred problem is the separation of the mixtures H_2/CO. This separation was one of the first realized in industry (see, e.g., Reference 21). The problems solved included recovery of hydrogen from purge gas of methanol synthesis (the task similar to the treatment of purge gases in ammonia synthesis) and adjustment of the composition of synthesis gas. The process of production of hydrogen via steam reforming of methane results in obtaining the synthesis gas with the ratio $H_2:CO = 3:1$. Meanwhile, in the process of Oxo Synthesis (hydroformylation of olefins with production of aldehydes), another ratio is needed: $H_2:CO = 1:1$.

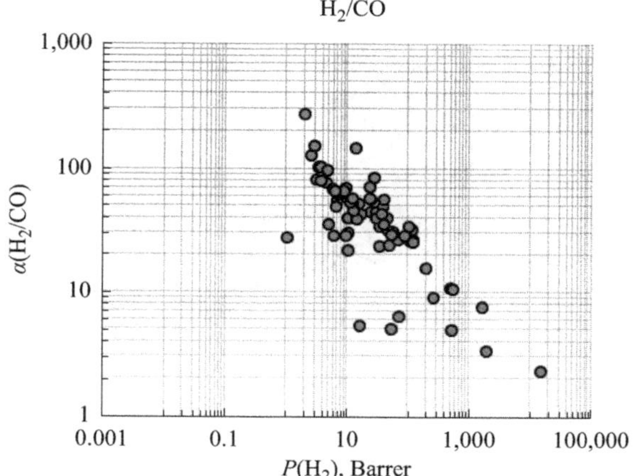

Figure 11.4   Diagram "permeability–selectivity" for the gas pair H$_2$/CO in glassy polymers. Source is the Database [17]

Use of membranes is an economic alternative to cryogenic separation of this mixture. The practical implementations of membrane processes were realized by Monsanto and later by MEDAL.

Figure 11.4 shows the diagram "permeability–selectivity" for the gas pair H$_2$/CO. Because physical properties of CO and N$_2$ are rather similar, this diagram resembles that shown in Figure 11.2 though the number of the data points here is smaller.

Another interesting example of mixtures to be separated using membranes is the gas pair He/H$_2$, as has been discussed by Merkel *et al.* [22]. In combination with other gases, hydrogen was also the "fast" component. Here as it is seen from Figure 11.5, the selectivity α(He/H$_2$) can be either larger than 1 or smaller than 1 depending on the nature of membrane material.

Thermodynamics of sorption of hydrocarbons in perfluorinated polymers revealed that these systems do not obey classical regular solution theory [23]. This was manifested by reduced solubility coefficients of hydrocarbon gases and liquids in perfluorinated polymers. Consequences of this behavior were special positions of fluorocarbons in various correlations, decreased permeability of hydrocarbons in perfluorinated polymers which made them as excellent barrier materials. A compelling elucidation of these phenomena has not been obtained in spite of extensive efforts [22,24]. Significantly, hydrogen behaves like aliphatic hydrocarbons; hence, selectivity of perfluorinated polymers is systematically higher than that of glassy and rubbery materials as is shown in Figure 11.5 and has been discussed by Merkel *et al.* [22]. Perfluorinated materials form a separate line located between Upper Bound of 1991 and 2008. Although He/H$_2$ separation is not a large industrial application, there is need to separate these gases in, for example, the space industry [25].

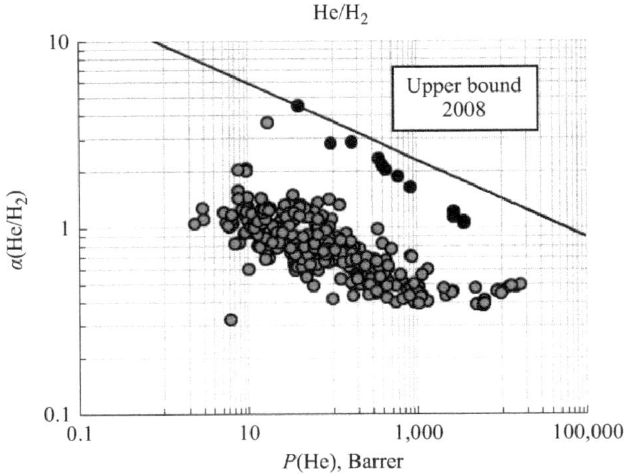

Figure 11.5   Diagram "permeability–selectivity" for the gas pair He/H₂.
Gray points glassy polymers, black points perfluorinated polymers,
and copolymers

It is of interest to compare the achieved level of selectivity of studied polymeric membrane materials for separation of different gas pairs. The Database [17] allows such possibility. It contains about 500 records for permeability coefficients of hydrogen in various glassy homopolymers. In this list, the most abundant class is PIs (185 structures). Relatively much data are available for polyacetylenes (83) and polynorbornenes (44). Table 11.1 presents a comparison of groups of the most selective polymers for three gas pairs from the Databases. It can be seen that among these, most selective polymers, PIs appear more frequently than polymers of other classes. The absolute values of $\alpha(H_2/CO)$ are somewhat lower than for other gas pairs, but maybe this can be explained by a smaller number of reported $P(CO)$ values.

11.4   Routes for improvement of permeation parameters of hydrogen in polymers

Analysis of Figures 11.2–11.5 leaves no doubt that existing polymers reveal a very large range of permeability and selectivity and some of them have excellent transport parameters. Nevertheless, other types of membrane materials (carbon molecular sieves, zeolites, and metallic membranes) show often better performance [13], so researchers active in the field of polymer membranes keep trying to improve the properties of polymeric membrane materials. There are several directions of such endeavor. Synthetic and physical chemists attempt to improve selectivity, permeability, or both of polymeric materials. These efforts will be considered in the following sections.

Table 11.1   Examples of glassy polymers that exhibit the greatest selectivity in separation of different gases

Structure	α_{ij}	Ref.
H_2/CH_4		
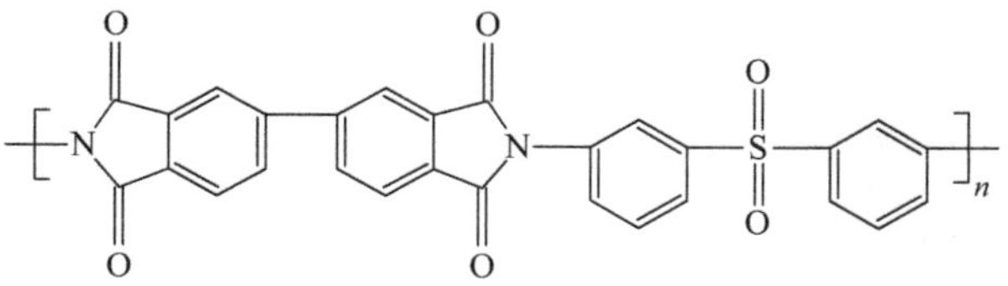	1,673	[26]
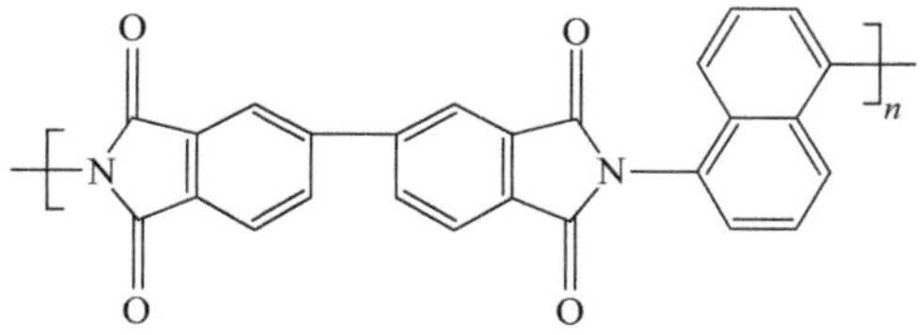	1,543	[27]
H_2/N_2		
$-[CH_2-C(CH_3)(COOCH_3)]_n-$	385	[28]
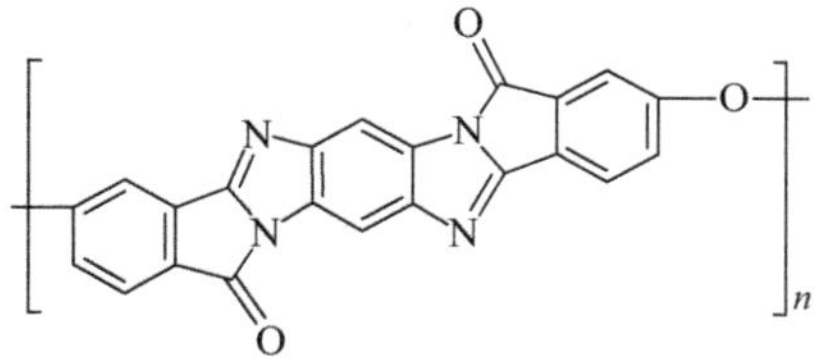	362	[29]
H_2/CO		
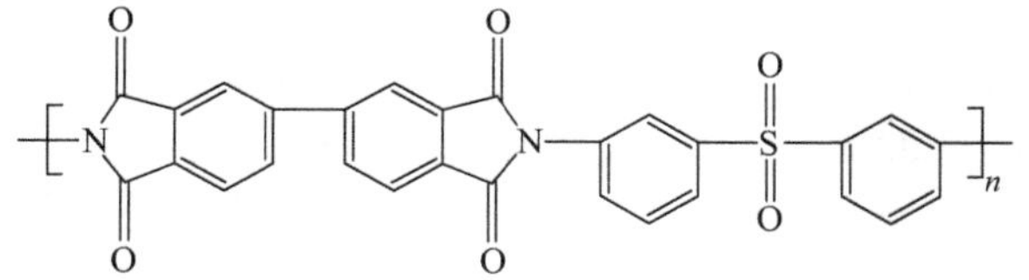	275	[27]
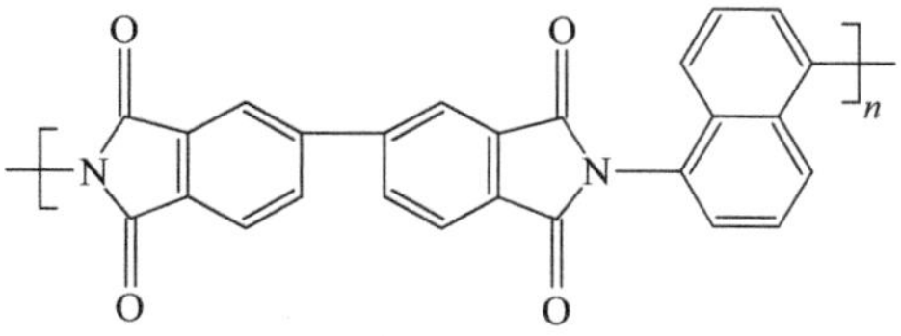	153	[27]

11.4.1   Cross-linking

Cross-linking of polymers is a standard approach for improvement of their properties. Virtually all the fields of application of rubbers employ cross-linked materials. Cross-linking makes polymers less inclined to plasticization, a basic reason for decreasing selectivity in separation of gas mixtures [30]. Cross-linked polymers are characterized by more rigid structure, and it is manifested in increases in selectivity of gas separation. The most abundant results were obtained for PIs. This class of polymers is located in the left "flank" and in the middle of the "clouds" of the data points in the diagrams "permeability–selectivity." Many points of PIs are very close to or on the Upper Bound 2008 [20], so even relatively small changes of selectivity can move the corresponding points above these Upper Bound.

Several approaches have been used to realize cross-linking of PIs: photochemical reactions induced by ultra violet (UV) radiation, use of diamines as reagents especially in the cases of functionalized PIs, application of dendrimers, and hyperbranched PIs.

Photoinduced cross-linking resulted in obtaining materials with extremely high selectivity. Thus, copolymers of polyimide based on benzophenone tetracarboxylic dianhydride and trimethyl-*p*-phenylene diamine and polyimide based on pyromellitic dianhydride and trimethyl-*p*-phenylene diamine were subjected to UV irradiation for different times [31]. The obtained materials showed very high selectivity, for example, $\alpha(H_2/N_2) = 410$ and 731. In spite of the fact that simultaneously permeability coefficients $P(H_2)$ decreased by factors 4–8, the corresponding data points for the cross-linked materials were above Upper Bound 2008. Cross-linking can also be conducted in the presence of photosensitizers (e.g., benzophenone) [32]. It was shown that the observed effect depends on the permeability of the original PI. As the PI studied in this work had rather high $P(H2) = 570$ Barrer, the observed increases in $\alpha(H_2/N_2)$ were rather modest (only 5-fold). On the other hand, relatively low permeability of original PI cannot guarantee moving of the data point above Upper Bound after cross-linking [33]. Probably, the highest selectivity in cross-linked PIs was observed by Kita *et al.* [34]. Thus, photoinduced cross-linking of 3,3′,4,4′-benzophenonetetracarboxylic dianhydride-tetramethyl-p-phenylenediamine (BTDA–TMPD) led to the following results: $\alpha(H_2/CH_4) = 3,000$ and $P(H_2) = 20$ Barrer. The corresponding data point is above the Upper Bound.

The work by Shao *et al.* [35] can serve as an example of cross-linking of a PI (in particular highly permeable polyimides based on hexafluoroisopropane dianhydride (6FDA)-durene PI) by additives of dendrimers. Again, highly permeable polymer shows relatively weak tendency to increase selectivity of gas separation. The same 6FDA-durene PI can be cross-linked also by using different diamines as cross-linking agents (ethylene diamine, 1,3-propane diamine, etc.) [36]. The observed selectivity and permeability depend to some extent on the nature of diamine, though the effects again are rather modest.

11.4.2   High free volume polymers

During the last decade, very interesting results were obtained on preparation and investigation of highly permeable, high free volume polymers. Some of them are known as polymers of intrinsic microporosity. This group of polymers forms the right "flank" of the cloud of the data points (highly permeable polymers with low selectivity). Nonetheless, for some of them the data points overcome Upper Bounds, that is, these polymers combine great permeability with somewhat improved selectivity. Table 11.2 shows the structures of some of high free volume polymers.

Table 11.2   Formulas of some highly permeable polymers

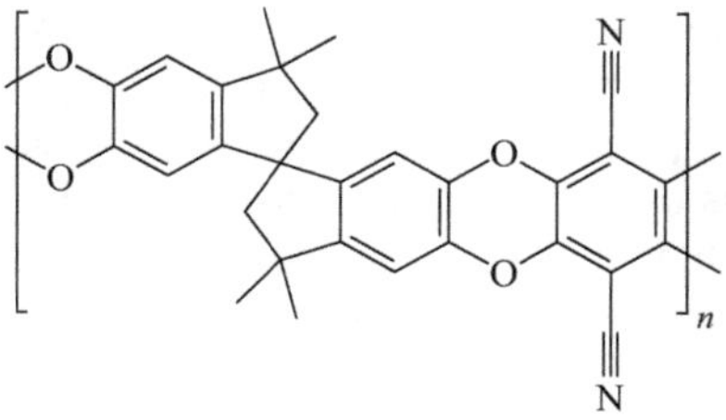

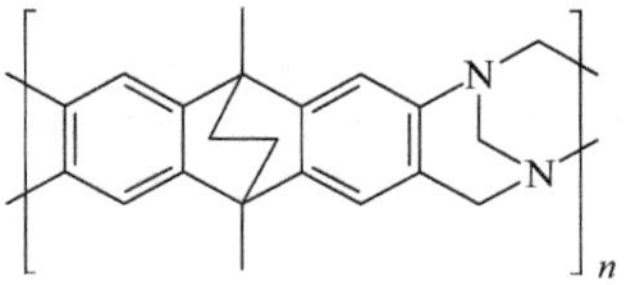

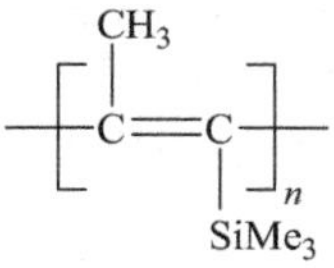

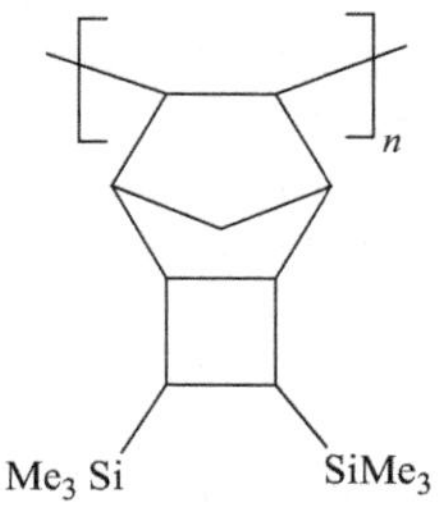

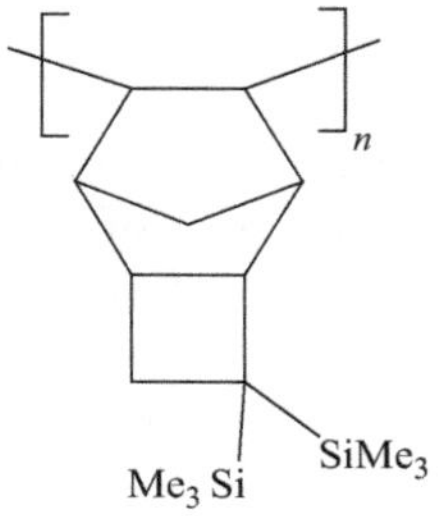

All these polymers have very rigid main chains: their glass transitions are not observed until the onset of thermal decomposition (usually $T_g > 300$ °C). They have large free volume as estimated via positron annihilation or Bondi's methods. Their great permeability is the result of large solubility coefficients, diffusion coefficients, or both. Although as has been mentioned, some of them overcome Upper Bound, the observed selectivity is not really great. Addressing Figure 11.3 shows that the content of hydrogen in permeate at such selectivity is not great (about 90% or somewhat more). Hence, recovery hydrogen from purge gases and its return in the processes are possible if some technical problems are solved. One of them is a tendency to aging: this is a well-known feature of high free volume polymers such as poly(trimethylsilyl propyne). Aging is revealed in fast decrease in the permeability coefficients and is especially characteristic for thin films or thin selective layers of composite membranes [44]. Some of these polymers are rather expensive materials, which also hampers their practical application.

11.4.3   Mixed matrix membranes

Preparation and investigation of mixed matrix membranes (MMMs), composites containing nanoparticles of different nature dispersed in continuous polymer matrices attracted much attention as one of the methods for affecting gas permeation properties of membrane materials. Several reviews have been published, for examples [45,46]. In numerous studied MMM, different nanoparticles (oxides like SiO_2 or TiO_2, zeolites, carbon molecular tubes, metal organic frameworks, or MOFs) were introduced into various mainly glassy polymers. The observed influence on P_i and α_{ij} was different depending on the nature of polymers and nanoparticles: there are increases in P_i and decreases in α_{ij} [47–49] and opposite trends [50]. On the contrary, sometimes a sad situation takes place: joint reduction of both parameters [51]. Of course, the case of joint increases of P_i and α_{ij} (e.g., Reference 52) is the most attractive one. On the contrary, the scale of changes of the transport parameters even in such cases is relatively small: thus, in Reference 52 the permeability coefficient $P(H_2)$ increases by 62% and the factor $\alpha(H_2/N_2)$ by 26.5%.

A specific and very promising type of MMM is the one, in which MOFs are used as nanoadditives. Several thousands of different MOFs have been prepared, which have different structural peculiarities (size of cages, of windows, etc.). So they can be selected for deliberate sorption of penetrants of certain size. Obviously, the number of polymers that could be combined with these MOFs is also very big, so the number of possible combination is huge. Experimental studies and modeling (see, e.g., Reference 53) showed that there are cases when MMMs containing MOF in different concentrations overcame Upper Bounds due to the increases in P_i with constant α_{ij} or even joint increases in both parameters. An example of such trend was observed in investigation of the MMM based on polymer of intrinsic microporosity and MOF zeolitic imidazolate framework [43]. In this case, $P(H_2)$ increased from 1,630 Barrer for pure polymer to 6,680 Barrer for the composition with 43 vol.% of the additive. The corresponding increase in selectivity $\alpha(H_2/N_2)$ was from 9.1 to 19.1. Because of this the data points for this MMM moved

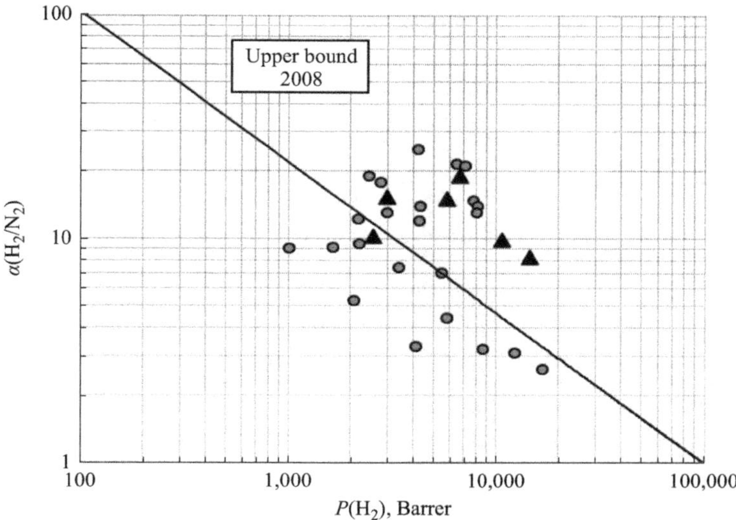

Figure 11.6   A part of "permeability–selectivity" diagram for highly permeable glassy polymers (circles) [37–42] and mixed matrix membranes (triangles) [43]

markedly above the Upper Bound for this gas pair (Figure 11.6). Further works with various MOFs are very promising.

However, although preparation of MMM is an interesting novel approach for modifying properties of membrane materials, it is hardly possible to change significantly (say by an order) the parameters of obtaining membrane materials and make them much more attractive for separation of hydrogen-containing mixtures. Also mechanical properties of many MMMs are not as good as desirable.

11.5   Separation of isotopes of hydrogen

An interesting and seldom considered problem is related to gas permeation parameters of hydrogen isotopes. Beyond common hydrogen (H_2), the data are available for deuterium (D_2) and to much less extent for tritium containing molecules (T_2 and HT). Although small isotope effects can be anticipated (and as it is seen from the following text this is true), significantly smaller isotope effects have not prevented solving a membrane process of an extremely important practical task: separation of ^{235}U and ^{238}U isotopes.

Sufficiently detailed information is available for permeability of H_2 and D_2 gases in rubbers. The results are summarized in Table 11.3 at temperatures in the range 23–35 °C.

Table 11.3   Permeability and diffusion coefficients of stable hydrogen isotopes in rubbers

Polymer	P (Barrer)		D 10^7 (cm²/s)		Reference
	H_2	D_2	H_2	D_2	
Butyl rubber	13.3	11.9	28.5	25.7	[54]
Hypalon	11.0	10.2	26.5	23.2	[54]
Viton[a] (F-containing rubber)	10.6	10.0	18.7	17.9	[54]
EPDM[b]	49.4	45.8	50.2	47.1	[54]
Neoprene	27.6	26.7	93.3	107	[54]
Polyethylene	10.2	9.3	58/5	46.6	[55]
Acrylonitrile–butadiene rubber	–	–	34.0	30.4	[56]

[a]Copolymer of hexafluoropropylene and ethylidene fluoride.
[b]Copolymer of ethylene and propylene.

Table 11.4   Permeability and diffusion coefficients of stable hydrogen isotopes in glassy polymers

Polymer	P (Barrer)		D 10^7 (cm²/s)		Reference
	H_2	D_2	H_2	D_2	
PETP[a]	0.813	0.71	5.9	4.9	[55]
PTFE[b]	16.6	13.9	1.5	1.16	[55,57]
CA[c]	8.3	7.7	38.0	34.0	[55]

[a]Poly(ethylene terephthalate).
[b]Poly(tetrafluoroethylene).
[c]Cellulose acetate.

Less abundant data have been reported for glassy polymers (Table 11.4). They show a wider variation of the permeability and diffusion coefficients.

It is seen that both in rubbery and glassy materials permeability and diffusion coefficients of hydrogen are greater than those for deuterium. The values of isotope effects in permeation and diffusion are summarized in Table 11.5.

Analysis of numerous data for P and D in various polymers led to the conclusion [54,55] that isotope effects of the solubility coefficients $S(H_2)/S(D_2)$ are less than unity. This result is consistent with lower critical point T_c of hydrogen (33.24 K) as compared to T_c of deuterium (38.35 K).

Much less results were obtained for permeability and diffusivity of tritium-containing molecules: T_2 and HT. Meanwhile, they are necessary for radiochemical industry for assessment of tritium leaking. Nakagawa *et al.* [56] found the following values for the permeability coefficient $P(T_2) = 0.66$ Barrer and isotope effect $P(H_2)/P(T_2) = 1.06$ in a blend of acrylonitrile—butadiene rubber and poly

Table 11.5   Isotope effects in permeability and diffusion coefficients

Polymer	$P(H_2)/P(D_2)$	$D(H_2)/D(D_2)$	Reference
Rubber			
Butyl rubber	1.12	1.11	[54]
Hypalon	1.08	1.14	[54]
Viton[a] (F-containing rubber)	1.06	1.04	[54]
EPDM[b]	1.08	1.07	[54]
Neoprene	1.03	0.87	[54]
Polyethylene	1.11	1.25	[55]
Acrylonitrile–butadiene rubber	–	1.12	[56]
Glassy polymers			
PETP	1.14	1.20	[55]
PTFE	1.19	1.29	[55,57]
CA	1.08	1.12	[55]

[a]Copolymer of hexafluoropropylene and ethylidene fluoride.
[b]Copolymer of ethylene and propylene.

(vinyl chloride) modified by *N,N*-dimethyldithiocarbamate (radiation reducing additive). Identical values of the diffusion coefficient $D(T_2) = 0.52 \times 10^{-7}$ cm^2/s were reported in polytetrafluoroethylene by Watanabe *et al.* [57] and Miyake *et al.* [58]. Stodilka *et al.* [59] measured the diffusion coefficient of HT in this polymer, and the value 0.9×10^{-7} cm^2/s was obtained. Much higher value was obtained by the same authors for diffusion coefficient of HT in ethylene-propylene rubber with additive of diene: 65×10^{-7} cm^2/s.

An innovated paper [60] published recently showed that much greater selectivity in separation of hydrogen and deuterium isotopes can be achieved using a composition membrane consisting of a layer of graphene superimposed over Nafion membrane with Pt catalyst. This part of the membrane, as in common fuel cells, produces H$^+$ and D$^+$ ions that can penetrate through graphene film in contrast to larger particles (e.g., H$_2$ and D$_2$ and other gases). It was shown that selectivity of the fluxes $Q(H^+)/Q(D^+)$ is equal to 10 ± 0.8 and is explained by the difference of energy barriers of the process of activation diffusion $\Delta E = E_D - E_H = 60$ meV.

It can be assumed that the data considered in this section would be useful in the processes of the treatment of the stable and radioactive isotopes of hydrogen.

11.6   Concluding remarks

It seems obvious that the most promising polymeric membrane materials for separation of hydrogen containing mixtures are PIs. In spite of the fact that they are the most studied, their potential is not exhausted. As has been discussed by Alentiev *et al.* [61], most linear PIs are based on seven dianhydride and 70 diamine. Not all the combinations are tested, applications of copolyimides, hyperbranched, and cross-linked PIs also provide some unexplored possibilities [62]. And even many polymers described in the literature show much better transport parameters than those membrane materials currently used in industry.

Acknowledgment

The authors acknowledge useful discussion with Prof. A. Alentiev. This article was partly supported by RSF project of TIPS headed by Yu. Yampolskii.

List of symbols

Latin symbols

a	parameter in (11.9)
b	parameter in (11.9)
d	parameter in (11.9)
D	diffusion coefficient (cm^2/s)
D_o	prefactor in Arrhenius equation for D (cm^2/s)
E_D	activation energy of diffusion (kJ/mol)
E_P	activation energy of permeation (kJ/mol)
J	flux through the membrane
l	thickness of a film (μm)
M	parameter of (11.10)
N	parameter of (11.10)
P	permeability coefficient (Barrer)
P_o	prefactor in Arrhenius equation for permeability coefficient (Barrer)
R	universal gas constant (J/mol K)
S	solubility coefficient (cm^3(STP)/cm^3 atm)
S_o	prefactor in van't Hoff equation (cm^3(STP)/cm^3 atm)
T	absolute temperature (K)
T_c	critical temperature of penetrants (K)
x	mole fraction in retentate or feed
y	mole fraction in permeate

Greek symbols

α	separation factor or selectivity
Δp	pressure gradient (atm)
ΔH_s	enthalpy of sorption (kJ/mol)

List of acronyms/abbreviations

BTDA-3MPDA	polyimide based on benzophenone tetracarboxylic dianhydride and trimethyl-*p*-phenylene diamine
CA	cellulose acetate
6FDA	polyimides based on hexafluoroisopropane dianhydride

EPDM	ethylene-propylene rubber with additive of diene
Hypalon	rubber of Du Pont with structure $-CHCl-(CH_2)_3-CH(SO_2Cl)-$
MMM	mixed matrix membrane
MOF	metal organic framework
PETP	polyethylene terephthalate
PI	polyimide
PIM-1	polymer of intrinsic microporosity
PSA	pressure swing adsorption
PMDA-3MPDA	polyimide based on pyromellitic dianhydride and trimethyl-*p*-phenylene diamine
PTFE	polytetrafluoroethylene
UV	ultra violet
Viton	copolymer of vinylidene fluoride and hexafluoropropylene
ZIF-8	zeolitic imidazolate framework

References

[1] Spillman R. W. Economics of gas separation membranes, *Chem. Eng. Prog.*, 85 (1989) 41–62.

[2] Bernardo P., Drioli E. Membrane gas separation progresses for process intensification strategy in the petrochemical industry, *Pet. Chem.*, 50 (2010) 271–282.

[3] Henis J. M. S. Commercial and practical aspects of gas separation membranes, in D. R. Paul, Yu. P. Yampolskii (eds) *Polymeric Gas Separation Membranes*, Boca Raton: CRC Press; 1994, pp. 441–512.

[4] MacLean D. L., Prince C. E., Chae Y. C. Energy saving modification in ammonia plants, *Chem. Eng. Prog.*, 76 (1980) 98.

[5] Henis J. M. S., Tripody M. K. The development technology of gas separation membranes, *Science*, 220 (1983) 11–17.

[6] Yampolskii Yu. P. Advances in membrane gas separation in the USSR, *Filtrieren und Separieren*, 4 (1990) 327–331.

[7] Yampolskii Yu. P., Volkov V. V. Studies in gas permeability and membrane gas separation in the Soviet Union, *J. Membr. Sci.*, 64 (1991) 191–228.

[8] Baker R. W. Future directions of membrane gas separation technology, *Ind. Eng. Chem. Res.*, 41 (2002) 1393–1411.

[9] Baker R. W., Low B. T. Gas separation membrane materials: a perspective, *Macromolecules*, 47 (2014) 6999–7013.

[10] Fleming G. Membrane technology for hydrogen recovery, 2006, available from: http://www.topsoe.com/sites/default/files/fleming.pdf.

[11] Adhikari S., Fernando S. Hydrogen membrane separation techniques, *Ind. Eng. Chem. Res.*, 45 (2006) 875–881.

[12] Ockwig N. W., Nenoff T. M. Membrane for hydrogen separation, *Chem Rev.*, 107 (2007) 4078–4110.

[13] Li P., Wang Z., Qiao Z., *et al.* Recent developments in membranes for efficient hydrogen purification, *J. Membr. Sci.*, 495 (2015) 130–168.

[14] Koros W. J., Ma Y. H., Shimidzu T. Terminology for membranes and membrane processes, *Pure Appl. Chem.*, 68 (1996) 1479–1489.

[15] Matteucci S., Yampolskii Yu., Freemen B. D., Pinnau I. Transport of gases and vapors in glassy and rubbery polymers, in Yu. Yampolskii, B. D. Freeman, I. Pinnau, (eds) *Materials Science of Membranes for Gas and Vapor Separation*, Chichester: Wiley; 2006, pp. 1–47.

[16] Mears P. The diffusion of gases through polyvinyl acetate, *J. Am. Chem. Soc.*, 76 (1954) 3415–3422.

[17] Gas permeation parameters of glassy polymers, *Database # 3585*, Informregistr RF, Moscow, 1998.

[18] Teplyakov V. V., Mears P. Correlation aspects of the selective gas permeabilities of polymeric materials and membranes, *Gas Sep. Purif.*, 4 (1990) 66–73.

[19] Robeson L. M. Correlation of separation factor versus permeability for polymeric membranes, *J. Membr. Sci.*, 62 (1991) 165–185.

[20] Robeson L. M. The upper bound revisited, *J. Membr. Sci.*, 320 (2008) 390–400.

[21] Kurz G., Teuner S. Calcor process for CO production, *Erdöl und Kohle*, 43 (1990) 171–172.

[22] Merkel T. C., Pinnau I., Prabhakar R., Freeman B. D. Gas and vapor transport properties of perfluoropolymers, in Yu. Yampolskii, B. D. Freeman, I. Pinnau (eds) *Materials Science of Membranes for Gas and Vapor Separation*, Chichester: Wiley; 2006, pp. 251–270.

[23] Scott R. L. The anomalous behavior of fluorocarbon solutions, *J. Phys. Chem.*, 62 (1958) 136–145.

[24] Song W., Rossky P. J., Maroncelly M. Modeling alkane + perfluoroalkane interactions using all-atom potentials: failure of the usual combination rules, *J. Chem. Phys.*, 119 (2003) 9145–9162.

[25] Slattery D., Hampton M. *Metal Hydrides for Hydrogen Separation, Recovery and Purification*, University of Central Florida, FL, USA (2005).

[26] Tanaka K., Kita H., Okamoto K., Nakamura A., Kusuki Y. Gas permeability and permselectivity in homo- and copolyimides from 3,3′,4,4′-biphenyltetracarboxylic dianhydride and 3,3′- and 4,4′-diaminodiphenylsulfones, *Polym. J.*, 22 (1990) 381–385.

[27] Yosihiro K., Kaniji M., Toshimune Y. Preparation of asymmetric polyimide hollow fiber membrane. *Proceeding of the 1990 International Congress on Membranes and Membrane Processes ICOM'90*, Chicago, USA, 20–24 August 1990. pp. 1025–1026.

[28] Min K. E., Paul D. R. Effect of tacticity on permeation properties of poly (methyl methacrylate), *J. Polym. Sci. Part B: Polym. Phys.*, 26 (1988) 1021–1033.

[29] Gao X., Tan Z., Lu F. Gas permeation properties of some polypyrrolones, *J. Membr. Sci.*, 88 (1994) 37–45.

[30] Wind J. D., Staudt-Bickel C., Paul D. R., Koros W. J. Solid-state covalent cross-linking of polyimide membranes for carbon dioxide plasticization reduction, *Macromolecules*, 36 (2003) 1882–1888.

[31] Liu Y., Pan C., Ding M., Xu J. Gas permeability and permselectivity of photochemically crosslinked copolyimides, *J. Appl. Polym. Sci.*, 73 (1999) 521–526.

[32] Matsui S., Sato H., Nakagawa T. Effects of low molecular weight photosensitizer and UV irradiation on gas permeability and selectivity of polyimide membrane, *J. Membr. Sci.*, 141 (1998) 31–43.

[33] Matsui S., Nakagawa T. Effect of ultraviolet light irradiation on gas permeability in polyimide membranes. II. Irradiation of membranes with high-pressure mercury lamp, *J. Appl. Polym. Sci.*, 67 (1998) 49–60.

[34] Kita H., Inada T., Tanaka K., Okamoto K. Effect of photocrosslinking on permeability and permselectivity of gases through benzophenone containing polyimide, *J. Membr. Sci.*, 87 (1994) 139–147.

[35] Shao L., Chung T.-S., Goh S. H., Pramoda K. P. Transport properties of cross-linked polyimide membranes induced by different generations of diaminobutane (DAB) dendrimers, *J. Membr. Sci.*, 238 (2004) 153–163.

[36] Shao L., Liua L., Cheng S.-X., Huang Y.-D., Ma L. Comparison of diamino cross-linking in different polyimide solutions and membranes by precipitation observation and gas transport, *J. Membr. Sci.*, 312 (2008) 174–185.

[37] Carta M., Malpass-Evans R., Croad M., *et al.* An efficient polymer molecular sieve for membrane gas separation, *Science*, 339 (2013) 303–307.

[38] Swaodan R., Ghanem B., Liweller E., Pinnau I. Physical aging, plasticization and their effects on gas permeation in "rigid" polymers of intrinsic microporosity, *Macromolecules*, 48 (2015) 6553–6561.

[39] Tocci E., De Lorenzo L., Bernardo P., *et al.* Molecular modeling and gas permeation properties of a polymer of intrinsic microporosity composed of ethaneanthracene and Tröger base units, *Macromolecules*, 47 (2014) 7900–7916.

[40] Finkeshtein E., Makovetskii K., Gringolts M., *et al.* Addition-type polynorbornenes with $Si(CH_3)_3$ side groups: synthesis, gas permeability and free volume, *Macromolecules*, 39 (2006) 7022–7029.

[41] Gringolts M., Bermeshev M., Yampolksii Yu., Starannikova L., Shantarovich V., Finkelshtein E. New highly permeable addition poly(tricyclononenes) with $Si(CH_3)_3$ side groups. Synthesis, gas permeation parameters and free volume, *Macromolecules*, 43 (2010) 7165–7172.

[42] Chapala P., Bermeshev M., Starannikova L., *et al.* A novel, highly gas-permeable polymer representing a new class of silicon-containing polynorbornenes as efficient membrane materials, *Macromolecules*, 48 (2015) 8055–8061.

[43] Bushell A. F., Attfeld M. P., Mason C. R., *et al.* Gas permeation parameters of mixed matrix membranes based on the polymer of intrinsic microporosity PIM-1 and the zeoliticimidazolate framework ZIF-8, *J. Membr. Sci.*, 427 (2013) 48–62.

[44] Pfromm P. H. The impact of physical aging of amorphous glassy polymers on gas separation membranes, in Yu. Yampolskii, B. D. Freeman, I. Pinnau (eds) *Materials Science of Membranes for Gas and Vapor Separation*, Chichester: Wiley; 2006. pp. 293–306.

[45] Vinh-Thang H., Kaliaguine S. Predictive models for mixed matrix membrane performance: a review, *Chem. Rev.*, 113 (2013) 4980–5028.

[46] Yampolskii Yu., Starannikova L., Belov N. Hybrid gas separation polymeric membranes containing nanoparticles, *Pet. Chem.*, 54 (2014) 637–651.

[47] Ahn J., Chung W.-J., Pinnau I., Guiver M. D. Polysulfone/silica nanoparticle mixed matrix membranes for gas separation, *J. Membr. Sci.*, 314 (2008) 123–133.

[48] Zornoza B., Irusta S., Tellez C., Coronas J. Mesoporous silica sphere – polysulfone mixed matrix membranes for gas separation, *Langmuir*, 25 (2009) 5903–5909.

[49] Ahn J., Chung W.-J., Pinnau I., *et al.* Gas transport behavior of mixed matrix membranes composed of silica nanoparticles in a polymer of intrinsic microporosity, *J. Membr. Sci.*, 346 (2010) 280–287.

[50] Sen D., Kalipcilar H., Yilmaz L. Development of polycarbonate based zeolite 4A filled mixed matrix gas separation membranes, *J. Membr. Sci.*, 303 (2007) 194–203.

[51] Ahmad J., Hägg M. B. Preparation and characterization of polyvinylacetate/ zeolite 4A mixed matrix membrane for gas separation, *J. Membr. Sci.*, 427 (2013) 73–84.

[52] Ahmad J., Hägg M. B. Polyvinylacetate/titanium dioxide nanocomposite membranes for gas separation, *J. Membr. Sci.*, 445 (2013) 200–210.

[53] Erucar I., Keskin S. Computational screening of metal organic frameworks for mixed matrix membrane applications, *J. Membr. Sci.*, 407 (2012) 221–230.

[54] Fitch M. W., Koros W. J., Nolen R. L., Carnes J. R. Permeation of several gases through elastomers, with emphasis on deuterium/hydrogen pair, *J. Appl. Polym. Sci.*, 47 (1993) 1033–1046.

[55] Mercea P. Permeation of H_2 and D_2 through polymers, *Isotopenpraxis*, 19 (1983) 153–155.

[56] Nakagawa T., Yoshida M., Kidokoro K. Development of rubbery materials with excellent barrier properties to H_2, D_2 and T_2, *J. Membr. Sci.*, 52 (1990) 263–274.

[57] Watanabe K., Matsuyama M., Ashida K., Takeuchi T. Diffusion of hydrogen, deuterium and tritium in poly(tetrafluoroethylene), *J. Nucl. Mater.*, 99 (1981) 320–323.

[58] Miyake H., Matsuyama M., Ashida K., Watanabe K. Permeation, diffusion, and solution of hydrogen isotopes, methane, and inert gases in/through polytetrafluoroethylene and polyethylene, *J. Vac. Sci. Technol.*, 1 (1983) 1447–1451.

[59] Stodilka D. O., Kherani N. P., Shmayda W. T., Thorpe S. J. A tritium tracer technique for the measurement of hydrogen permeation in polymeric materials, *Int. J. Hydrogen Energy*, 25 (2000) 1129–1136.

[60] Lozanda-Hidalgo M., Hu S., Marshall O., *et al.* Sieving hydrogen isotopes through two-dimensional crystals, *Science*, 351 (2016) 68–70.

[61] Alentiev A., Loza K., Yampolskii Yu. Development of the methods for prediction of gas permeation parameters of glassy polymers: polyimides as alternative co-polymers, *J. Membr. Sci.*, 167 (2000) 91–106.

[62] Kanehashi S., Sato S., Nagai K. Synthesis and gas permeability of hyperbranched and cross-linked polyimide membranes, in Yu. Yampolskii, B. Freeman (eds) *Membrane Gas Separation*, Chichester: Wiley; 2010, pp. 3–27.

Chapter 12

Industrial membranes for hydrogen separation

*Hamid Reza Rahimpour[1], Mahshid Nategh[1] and
Mohammad Reza Rahimpour[1]*

Abstract

Production of hydrogen, as an environmentally benign alternative for fossil fuels that mainly contribute to the growing pollutant emissions, has been considered specifically in the last decades. As a result of being associated with other gases, such as CO_2, CO and other impurities, the produced hydrogen must be separated and purified before being utilized by various processes. For this purpose, adsorption-based and cryogenic processes are the most conventional methods which encounter some restrictions, related to the required energy and time that make these processes not economically lucrative in some circumstances. As a result, recently, the membrane technology as well as membrane reactors has emerged to deal with these limitations. Among the common types of membranes including organic and inorganic membranes and their subgroups, which offer high selectivity and permeability to hydrogen, the stable, energy-efficient and cost-effective ceramic membranes, which are unaffected by the existing poisonous gases in the gas mixtures, are the most promising candidates for hydrogen separation in an effective manner in near future.

12.1   Introduction

Finding a sustainable solution to this universal concern is of great importance nowadays due to the increasing consumption of fossil-based fuels and the resultant pollution (Song, 2003; Adhikari and Fernando, 2006). There are numerous alternative energy resources for fossil fuels such as solar, wind, hydro, marine, geo-thermal and bio-energy (Ellabban *et al.*, 2014). Recently, hydrogen as the most abundant element, in its monatomic form, in the universe has received a lot of attention worldwide (Palmer, 1997). Hydrogen only produces water which is a useful and non-pollutant by-product. Therefore, it can solve the major issues related to the climate change and air pollution (Adhikari and Fernando, 2006; Song, 2010).

[1]Department of Chemical Engineering, Shiraz University, Shiraz 71345, Iran

The most important reasons for increasing the popularity of hydrogen is its availability, renewability, being environmentally friendly, high efficiency, high energy content and having no harmful by-products. Of course, like any other choices for energy, hydrogen has its own disadvantages such as high flammability, high cost of production, storage and substitution and in some cases dependency on fossil fuels; all of which are being investigated for further improvement by numerous researchers (Chang *et al.*, 2008).

Hydrogen was first discovered in 1766 by Henry Cavendish as 'inflammable air'; but it was named in 1783 by Antoine Lavoisier. It has a Greek origin and is composed of 'hydro' (i.e. 'water') and 'genes' (i.e. 'generator'). The nomination is based on the production of water in hydrogen burning process which is associated with energy production named as hydrogen energy (Song, 2003).

Owing to the fact that pure hydrogen does not exist naturally on Earth in large quantities, it can be considered as an energy carrier, not an energy resource, and H_2 must be produced using other energy resources which itself is an energy-consuming process (Song, 2003). Hydrogen can be produced from the following resources and processes (Song, 2003; Bolland and Undrum, 2003; Robert, 2004; Penner, 2006; Ritter and Ebner, 2007):

1. Gasification of coal
2. Steam reforming, partial oxidation, autothermal reforming or plasma reforming of natural and propane gas
3. Dehydrocyclization, aromatization, oxidative steam reforming or pyrolytic decomposition of petroleum fractions
4. Gasification, steam reforming or biological conversion of biomass
5. Electrolysis, photocatalytic conversion, chemical and catalytic conversion of water.

Because of immaturity and high cost of hydrogen production from renewable energy resources, production of H_2 is mainly done using the fossil fuel resources such as natural gas, coal and petroleum fractions. The hydrogen production reaction from fossil fuels is associated with by-products such as CO_2, CO, impurities (O_2, H_2O, N_2, SO_x, NO_x and VOCs) and sulphur-containing substances (Tao *et al.*, 2015). Therefore, the produced hydrogen from the above processes must be separated from the gaseous products.

As a result of the difference in the nature of the coexisting gaseous products, different methods of hydrogen separation should be used (Song, 2003). These separation methods are categorized into conventional methods such as pressure swing adsorption (PSA), temperature swing adsorption (TSA), electrical swing adsorption (ESA), vacuum swing adsorption (VSA), cryogenic separation and modern separation methods such as membrane separation (Faraji *et al.*, 2005; Adhikari and Fernando, 2006; Uehara, 2006a, 2006b; Holmes and Erickson, 2010; Finamore *et al.*, 2011).

The purpose of this chapter is to introduce the conventional methods for hydrogen separation and their challenges, with our focus on the novel technology used in hydrogen separation, that is membrane separation. Membranes have the ability to efficiently and economically separate the hydrogen with a high selectivity. This chapter also reports the performance of selected membranes in terms of hydrogen selectivity, permeability, cost and being affected by poisonous gases. Furthermore, membrane reactor (MR) for highly efficient hydrogen production and separation is described.

12.2   Conventional methods for hydrogen separation

H_2 separation is a major issue in industrial processes as the various coexisting gaseous by-products lower the efficiency of hydrogen (Song, 2003). The difference between 'separation' and 'purification' processes is that the first is used for first-stage hydrogen concentration, and the last is used for second-stage upgrading of the hydrogen, respectively (Uehara, 2006a).

Hydrogen separation processes are classified on the basis of different criteria. The conventional methods for hydrogen separation are cryogenic method and adsorption methods such as PSA, VSA, TSA and ESA, which are further described in the next sections (Faraji *et al.*, 2005; Adhikari and Fernando, 2006; Uehara, 2006a, 2006b; Holmes and Erickson, 2010; Finamore *et al.*, 2011).

12.2.1   Cryogenic process

The cryogenic or partial condensation process for hydrogen separation is based on the difference in volatility of the components present in the gaseous mixture at low temperatures, which results in the condensation of other components while separating the hydrogen. The volatility for hydrogen separation, although depends on the operating pressure, is very high (Tomlinson and Finn, 1990).

In the other words, successive temperature reduction in the cryogenic process leads to condensation of components such as water vapour, hydrocarbons, CO and N_2 due to their lower volatility compared to hydrogen. Because of higher volatility of hydrogen, it remains in the gaseous form at the end of the process and separates from the other gas impurities (Faraji *et al.*, 2005; Uehara, 2006b). The efficiency of the cryogenic process depends on various factors such as operating temperature and pressure of the process, and feed composition (Uehara, 2006b).

Like any other methods, cryogenic process has some limitations. One of these limitations is the low pressure of gas mixture which must be increased using compressors (Engineering Data Book, 2004). Also, supplying the very low operating temperatures and pressurizing the inlet gas mixture is an expensive and energy-consuming task in the cryogenic process (Engineering Data Book, 2004; Adhikari and Fernando, 2006). In addition, the flammable and toxic nature of the cryogenic fluids, the risk of blockage of process equipment due to freezing of the possible existing water or CO_2 in the inlet mixture are other challenges in this process (Hands, 1986; Ebenezer and Gudmunsson, 2006).

12.2.2   Pressure swing adsorption

One of the most famous methods for hydrogen separation is the PSA (Ratan and Wentink, 2001; Sircar, 2002). This method, first patented in the 1930s (Hasche and Dargan, 1931; Finlayson and Sharp, 1932; Perley, 1933) but applied in 1960 (Skarstrom, 1960; Montgareuil and Domine, 1964), is carried out by pressure change during the process at isothermal condition (Grande, 2012) and is based on the difference in physical adsorption affinities of different components for the adsorbent; that is the impurities like N_2, CO_2, CO, hydrocarbons and water vapour adsorb on the surface of the adsorbent whereas hydrogen with high volatility and low polarity does not adsorb and thus can be separated from the initial gaseous mixture. The driving force for the adsorption depends on the gas component, type of adsorbent material, partial pressure of the gas component and operating temperature. A schematic of PSA process flow diagram is illustrated in Figure 12.1.

The main stages of a PSA process are adsorption of gas components, pressure equilibrium, desorption of gases (i.e. regeneration) and repressurization (Yang and Lee, 1997; Uehara, 2006a; Grande, 2012). In the first stage, the gas impurities are adsorbed on the adsorbent and after a while the surface of the adsorbents is saturated, and no more gas molecules can be adsorbed. In this condition, the adsorbent must be regenerated by reducing the pressure slightly above the atmospheric pressure, which results in desorption of gases. Finally, the operating pressure is increased again to the adsorption level, and the cycle is repeated (Ruthven *et al.*, 1993; Stocker and Miller, 1998).

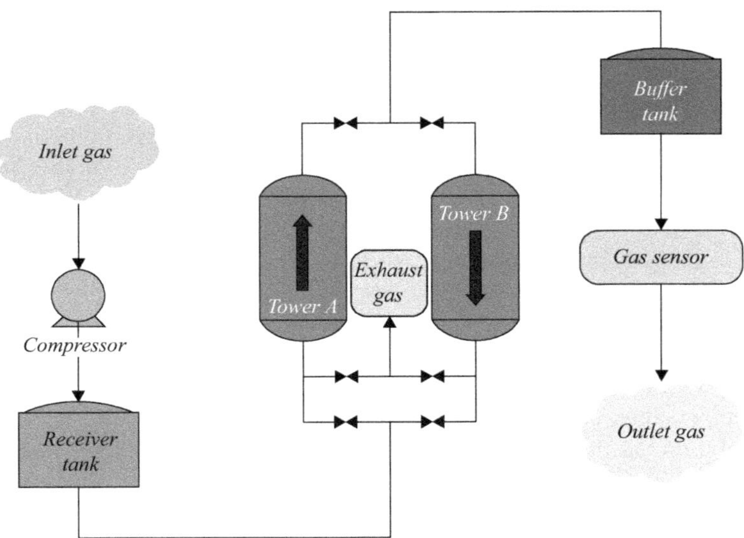

Figure 12.1   A schematic of PSA flow diagram (Gas Separation by Pressure Swing Adsorption)

Choosing the best adsorbent for PSA units is a complex process and depends on different parameters such as cost, capacity, selectivity, compatibility, regenerability and kinetics. However, the selected adsorbent may not have the best yield for all of the mentioned parameters. The most common adsorbent used in PSA units are inorganic materials such as zeolite, silica, alumina and organic materials like polymers and activated carbon (Siriwardane *et al.*, 2001; Knaebel, 2004).

Despite the advantages, PSA also has some major drawbacks which are as follows (Ruthven, 1984, 1993; Jain *et al.*, 2001; Yang, 2003; Shimekit and Mukhtar, 2012):

1. High production cost because of low recovery rates
2. High production of CO_2 as a pollutant due to its greenhouse effect
3. High levels of noise as a result of successive compressions and expansions in cycles
4. Problems due to the leakage during the opening and closing of the valves
5. Reduction in the adsorbent capacity and a difficult desorption process caused by the temperature change with time and the position in exothermic adsorption stage
6. Lack of general easy-to-use design rules for PSA unit attributable to the complicated nature of the process and large number of decision parameters.

However, by considering the heat effects in designing a PSA unit, the heat wave generated in the process can be used for a faster desorption (Yang and Cen, 1986). Also in order to overcome some related issues, a hybrid process combining PSA and membrane separation technologies with lower operating cost and higher overall H_2 recovery and purity with a CO_2 outlet stream ready for carbon capture and sequestration purpose has been developed (Sircar *et al.*, 1999; Esteves and Mota, 2002; Shimekit and Mukhtar, 2012).

12.2.3 *Vacuum swing adsorption*

VSA may actually be best described as a subset of the larger category of PSA. The main principles of VSA process are the same as PSA, that is hydrogen is separated from a gas mixture due to the different amounts of adsorption on the adsorbent. But the difference between these two processes is that VSA unit operates at near-ambient temperatures and pressures to below the atmospheric pressure (Chou and Chen, 2004; Tlili *et al.*, 2009; Grande, 2012; Ribeiro *et al.*, 2014). The other difference is that PSA typically vents to atmospheric pressures and uses a high pressure gas mixture as a feed, but VSA uses a vacuum to feed the gas through the separation process. The common adsorbents used in this method are the same as PSA.

12.2.4 *Temperature swing adsorption*

TSA process is also another cyclic process with similar approach as the PSA and VSA processes. However, the changing factor for regeneration of adsorbent in this method is the operating temperature instead of pressure, that is the temperature is low in the adsorption stage, whereas it is high in the regeneration stage. This is based on the fact that higher temperature acts in favour of increasing gas desorption. Choosing the best adsorbent for TSA process is similar to PSA, and the selected adsorbent should have high surface area to give high loadings. The isotherm data for common adsorbents such as activated carbon, silica gel, alumina and zeolite are given in the literature (Breck, 1974; Ruthven, 1984; Sinnott, 2009).

TSA has lower operating pressure and operating cost but higher initial cost than PSA. However, TSA is not widely used because of high-energy consumption and large adsorbent inventories (Bonjour *et al.*, 2002). In addition, it has very long cycles as a result of the time-consuming heating and cooling processes which lead to the larger amount of adsorbent and higher investment needed for the process. However, this method is chosen when high product purities are not achievable with PSA.

The reduced adsorption capacity or thermal ageing of adsorbent due to the repeated PSA cycles results in the same issues in TSA (Cavalcante, 2000). In order to increase the adsorbent temperature and achieve a more effective heating, direct electrical heating (Fabuss and Dubois, 1971; Petkovska, 1991), induced electrical heating (Moskal and Nastaj, 2007) and microwave heating (Reuss *et al.*, 2002) methods have been proposed, although they do not solve all the related problems completely and have some issues themselves.

12.2.5   Electrical swing adsorption

The idea of using ESA was first proposed by Fabuss and Dubois (1971). ESA is a specific case of TSA process in which the need for transporting or heating the adsorbent, large temperature and pressure changes, or concentration of the system is eliminated due to in-situ heat generation as a result of applied electric current and the conductivity of adsorbent based on Joule effect (Finamore *et al.*, 2011; Tlili *et al.*, 2012; Ribeiro *et al.*, 2014). ESA uses a highly conductive new material named 'carbon-bonded activated carbon fibre' as an adsorbent. This adsorbent allows rapid desorption of adsorbed gases by low-voltage electrical current, whereas the system pressure is constant, and the temperature is modified to some extent (Pinto *et al.*, 2011).

The ESA processes have advantages over traditional TSA processes, some of which are higher heating efficiency caused by direct energy delivery to the adsorbent, higher rate of heating which leads to smaller systems, independence of heating rate to heat transfer rate between source/adsorbent and heat capacity of the heat source, identical direction of heat and mass fluxes in favour of better desorption due to the effects of thermal and gas diffusion (Fabuss and Dubois, 1971; Petkovska *et al.*, 1991; Sullivan *et al.*, 2004).

All of the mentioned swing adsorption methods have alternatives, with different supplying heat power based on product specification, energy of separation, types of adsorbent and the sequences of process operations (Lee, 2003). Some of these methods are direct electrical heating (Fabuss and Dubois, 1971; Petkovska *et al.*, 1991), induced electrical heating (Moskal and Nastaj, 2007) and microwave heating (Reuss *et al.*, 2002).

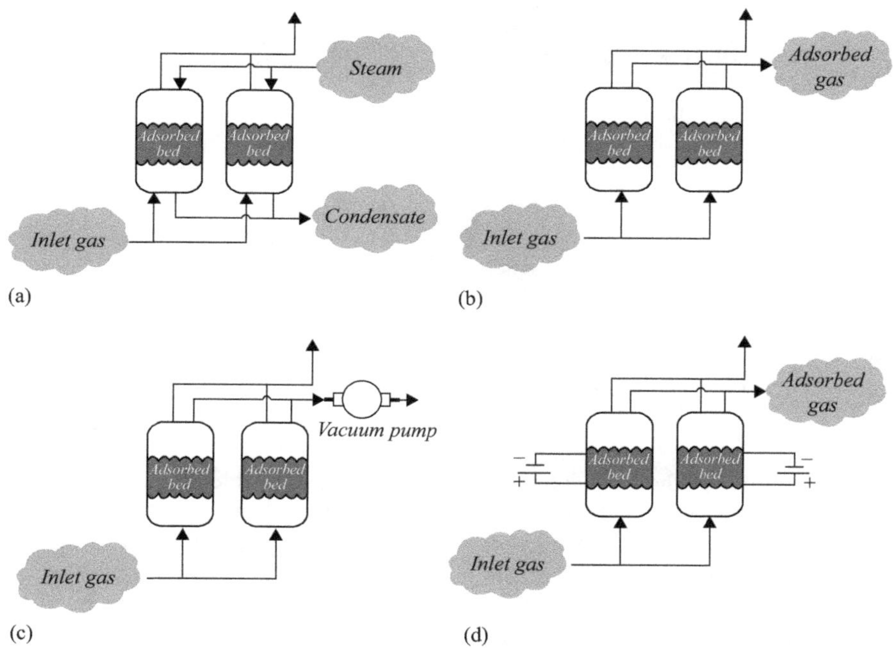

Figure 12.2   Schematic diagrams of adsorption processes for hydrogen separation: (a) PSA, (b) TSA, (c) VSA, (d) ESA (Songolzadeh et al., 2014)

The schematic diagrams of adsorption processes for hydrogen separation are revealed in Figure 12.2.

12.3   Membrane technology: separation of hydrogen

As mentioned in the previous sections, all of the conventional methods have major drawbacks which may cause some issues in their implementation in industrial scale. The drawbacks are mainly related to the time and cost of these processes, but there are also challenges with security of these processes. Recently, membrane technology has emerged as a promising candidate for a new generation of hydrogen separation methods.

A membrane is defined as a thin film of material that allows certain substances to pass through and acts like a barrier to the transport of other substances (see Figure 12.3). Therefore, it is a useful tool for selective separation of the desired components from other components in a gas or liquid mixture (Tao *et al.*, 2015). The driving force for separation process can be partial pressure, concentration, temperature or electrical potential gradient, whereas the first one is the most common in gas separation processes.

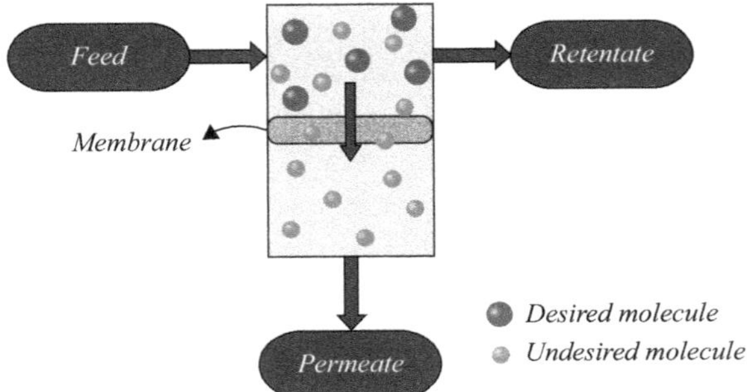

Figure 12.3   A scheme of membrane for separation of the desired component (Tan and Li, 2015)

Monsanto was the first company that used polymeric membranes for large-scale gas separation in 1980 (Coker *et al.*, 1998; Nunes and Peinemann, 2006). Nowadays, membrane technology is widely used for separation of hydrogen because of its advantages such as simplicity, high-energy efficiency, low capital and operating costs, low maintenance, low weight and space, high process flexibility and easy intensification (Stern, 1994; Koros and Mahajan, 2000; Baker, 2002; Penner, 2006; Fawas *et al.*, 2007; Viano *et al.*, 2015). For example, hydrogen separation using membranes requires less than half the energy required for the PSA process (Noble, 1995).

Two important characteristics of a membrane are permeability and selectivity which determine the overall yield in a separation process. The first one is an indicator of the component flux through the membrane, and the last one indicates the tendency of a membrane for separation of one component from another component in a mixture (Tan, 2015). The rate of permeation across the membrane is proportional to the pressure gradient, solubility of gas in the membrane or diffusivity of gas through the membrane and is inversely proportional to membrane thickness. Also the permeation flux of the component is affected by the membrane material, the microstructure and the operating temperature (Hung *et al.*, 2014).

The separation mechanisms in gas separation processes using membranes are molecular sieving, solution-diffusion, surface diffusion, viscous flow and Knudsen diffusion (Koros and Fleming, 1993). The operating principle of hydrogen separation membranes is presented in Figure 12.4.

Membranes can be classified on the basis of various criteria such as thickness, phase, symmetry and polarity (Tan, 2015). However, the most common way to classify the membranes is based on the type of materials they are made up of. The three groups of membranes are organic (polymeric), inorganic (metal, metal alloy, zeolite, carbon molecular sieve and ceramic) and composite or hybrid membranes which are described in detail in the next sections (Sanchez Marcano and Tsotsis, 2002; Richter and Hoyer, 2013; Wijenayake *et al.*, 2014; Al-Mufachi *et al.*, 2015).

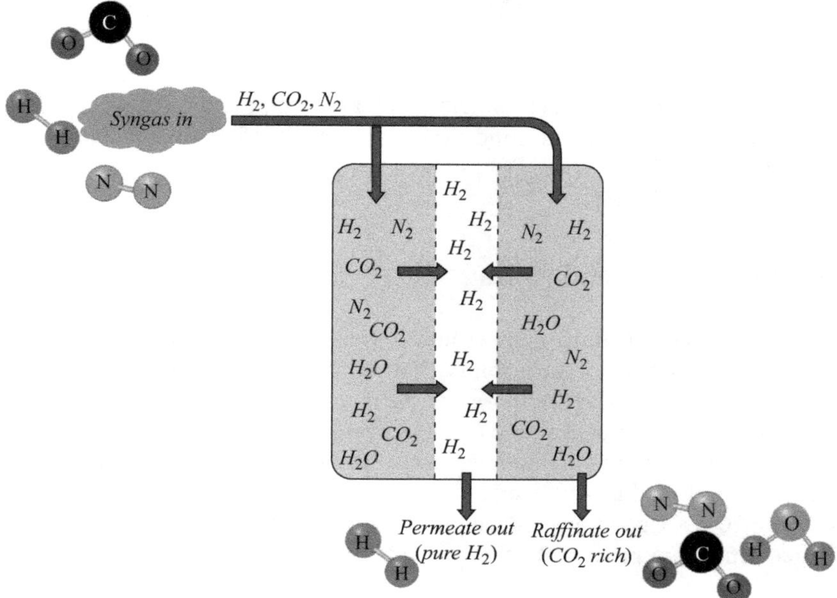

Figure 12.4   The operating principle of hydrogen separation membranes (Holmes and Erickson, 2010; Stanislowski and Laumb 2009)

12.4   Organic membranes

Polymeric membranes were the first and most used membranes for gas separation purposes (Lin *et al.*, 2006; dos Santos, 2009; Yampolskii, 2012; Sanders *et al.*, 2013) due to the reasonable gas selectivity, good mechanical properties and low operating temperature (below 110 °C) (Tao *et al.*, 2015). Two main types of polymeric membranes are porous polymeric membranes with high flux and low selectivity and glassy polymeric membranes with low permeability and high selectivity (Rautenbach *et al.*, 1998; Lababidi, 2000). Hydrogen separation through polymeric membranes is carried out by a solution-diffusion mechanism; that is the hydrogen in the inlet gas mixture dissolves in the polymer, diffuses to the other side and desorbs to the exiting flow (Lin and Freeman, 2011).

Choosing the most suitable polymer for the purpose of membrane fabrication is a determining task in hydrogen separation and is done on the basis of compatibility, binding affinity towards the separated molecules, withstanding the cleaning conditions, pH, operating temperature, chain interactions, chain rigidity, polarity of functional group and stereoisomerism (Zeaman and Zydney, 1996). Some of the typical synthetic and natural polymers used in organic membrane synthesis are cellulose acetate, nitrocellulose and cellulose esters, polysulfone, polyether sulfone, polyacrylonitrile, polyamide, polyimide, polyethylene and polypropylene, polytetrafluoroethylene, polyvinylidene fluoride and polyvinylchloride (Naylor, 1996; Ulbricht, 2006).

The major drawbacks of polymeric membranes are the limited lifetime, susceptibility to the syngas components such as sulphur and mercury, as well as inability to operate in harsh conditions as a result of chemical and thermal instability (Rautenbach *et al.*, 1998; Lababidi, 2000; Robeson, 2008; Koonaphapdeelert *et al.*, 2008).

12.5   Inorganic membranes

Seeking for higher selectivity and permeation rate of membranes, combined with a desired thermal and chemical stability, has led to the development of inorganic membranes (dos Santos, 2009). Inorganic membranes, including metal, metal alloy, zeolite, molecular sieve carbon and ceramic membranes, first were used over half a century ago for separating uranium isotopes and then in the 1980s for separation of solids from fluids and mixtures of fluid (de Lange *et al.*, 1995; Fain, 2000; Ismail and David, 2001). Due to the wide application of inorganic membrane in industrial hydrogen separation process, all the types are explained in the next sections.

12.5.1   Metal and metal-alloy membranes

One of the common methods for purification of hydrogen is to pass the gas through dense metal membranes (Yun and Oyama, 2011; Gao and Wang, 2014; Al-Mufachi *et al.*, 2015; Viano *et al.*, 2015). An important characteristic of metals which makes them desirable for fabrication of membranes is their ability to react with hydrogen at moderately high temperatures to form metal hydrides (Pan *et al.*, 2005). The main steps in hydrogen gas separation using metal membranes are as follows (Ward and Dao, 1999; Gepert *et al.*, 2006):

1. External diffusion of H_2 from gas phase to the vicinity of the membrane surface which occurs in the high pressure conditions
2. Reversible dissociative adsorption of hydrogen on the membrane surface
3. Reversible dissolution of hydrogen atoms from surface to the bulk metal
4. Diffusion of hydrogen atoms through the membrane layer
5. Mass transfer of the hydrogen atoms from the metal layer to the low pressure surface
6. Recombinative desorption of hydrogen from the low pressure side of the membrane.

The easier movement of atomic hydrogen around the metal lattice leads to its separation from other components in the mixture (Ward and Dao, 1999).

There are requirements which must be investigated before selection of a proper metal for membrane fabrication. High solubility and diffusivity of hydrogen, catalytic activity of the surface for hydrogen gas dissociation and mechanical strength in order to withstand the harsh atmosphere of hydrogen are some of these requirements. Based on the mentioned criteria, the most common metals used in the fabrication of membranes are palladium, platinum, tantalum, niobium, vanadium and nickel which have high permeability, diffusivity and solubility. Pd and its alloys with excellent catalytic surface property and relatively high hydrogen permeability and selectivity can withstand the hydrogen atmosphere more than the other metals (Sholl and Ma, 2006).

Metal-alloy membranes are also used in hydrogen separation processes. In this type of membranes, the connected porous layer to metal leads to excellent sealing of the membrane. Metal membranes are fabricated by electroless plating, chemical vapour deposition (CVD), physical vapour deposition, electroplating deposition etc. (Yun and Oyama, 2011).

Table 12.1 presents the permeation data of different Pd-based membranes produced by different techniques.

The advantages of metal and metal-alloy membranes over the other types of membranes are cost effectiveness, high flux, producing high-purity H_2, strong surface which hinders the fouling, ease of connection to a module, etc. (Sholl and Ma, 2006; Ockwig and Nenoff, 2007b; Tao *et al.*, 2015). However, there are some drawbacks for Pd membranes. Some of them are membrane degradation due the irreversible change in Pd lattice structure during thermal cycling (Tao *et al.*, 2015), the phase change at low temperatures (below 300 °C) which can be eliminated by alloying palladium with other metals such as Cu, Ag and Au, and poisoning which can be overcome by gas cleaning procedure before membrane separation (Atsonios *et al.*, 2015).

12.5.2   *Zeolite membranes*

Zeolite, first synthesized in 1940, is one of the main materials used in the fabrication of the industrial membranes (An *et al.*, 2011; Wang *et al.*, 2013; Dong *et al.*, 2015; Kosinov *et al.*, 2015). Zeolite is a microporous crystalline aluminosilicate made up of a 3D framework that forms uniform and molecular size pores (usually 0.3–1.3 nm). The general formula of the zeolite is $M_{x/n}[(AlO_2)_x(SiO_2)_y] \cdot zH_2O$; where M is the compensating cation, and n indicates the valence (Daramola *et al.*, 2012). Transport of molecules within zeolite crystals is controlled by an adsorption–diffusion mechanism. Two main controlling factors that determine the properties of zeolites such as adsorbing, catalytic and ion-exchange properties and hydrophobicity, acidity, chemical and structural stability are the Si/Al ratio (from 1 to infinity) and the amount of the cations (Dong *et al.*, 2008; Yu *et al.*, 2011; Daramola *et al.*, 2012).

The main methods for synthesis of zeolite that have been reported so far are Liquid-phase hydrothermal technique, vapour phase transport technique, secondary seeded growth technique and pore-plugging hydrothermal technique (Daramola *et al.*, 2012).

Zeolites have characteristics which have made them attractive in gas separation processes. Some of these properties are uniform, small pores, excellent thermal, mechanical and chemical stability (Yu *et al.*, 2011). Despite the growing number of researches on zeolite membranes, reproducibility of this type of membranes still remains as a major problem to be solved in near future. Limited pore

Table 12.1 Permeation data of different palladium-based membranes (Basile et al., 2011b)

Membrane type	T (°C)	ΔP (bar)	δ (µm)	J_{H_2} (mol/m^2 s)	Pe_{H_2} (mol m/m^2 s Pa)	α_{H_2/N_2}	Preparation method
Pd/PSS	520	1.5	10	1.8×10^{-1}	1.2×10^{-11}	–	ELP
Ti–Ni–Pd	450	3	45	$\sim3.3 \times 10^{-3}$	1.7×10^{-10}	∞	Cold rolling
Pd/PSS–YSZ	400	–	7–10	2.5×10^{-2}	4.7×10^{-9}	800–900	ELP
Pd/Al$_2$O$_3$	200	0.1	15	2.2×10^{-1}	3.3×10^{-10}	7	ELP
Pd/glass	350–500	4	2	–	3.4×10^{-12}	1140–12,900	ELP
Pd/Al$_2$O$_3$	450	–	4.8	–	1.4×10^{-11}	60	ELP
Pd/Al$_2$O$_3$	300	0.3	2–4	$1–2 \times 10^{-1}$	$1.3–2.7 \times 10^{-11}$	5000	CVD
Pd/Al$_2$O$_3$	528	–	2–3	–	3.5×10^{-12}	<18	ELP
Pd/Al$_2$O$_3$	400	1	5	1.6×10^{-1}	7.8×10^{-12}	100–200	ELP
Pd/BaZrO$_3$	600	–	41	–	–	5.7	CVD
Pd/MPSS	500	1	6	3×10^{-1}	1.8×10^{-11}	–	ELP
Pd/PNS	500	3.6	–	8.3×10^{-2}	–	3.7	MS
Pd/ZrO$_2$/PSS	500	1	10	8.3×10^{-2}	8.3×10^{-11}	–	ELP
Pd/αAl$_2$O$_3$	370	2.9	1	4×10^{-1}	–	3000–8000	ELP
Pd$_{84}$–Cu$_{16}$/ZrO$_2$–PSS	480	2.5	5	6×10^{-1}	2.6×10^{-9}	∞	ELP
Pd$_{90}$–Ag$_{10}$/αAl$_2$O$_3$	200–343	0.8–2.5	20	1.4×10^{-1}	2.5×10^{-11}	30–178	ELP
Pd–Ag/Al$_2$O$_3$	–	1.4	10	1×10^{-1}	1×10^{-11}	1500	ELP
Pd–Ag/PSS	400–500	1	2–3	3×10^{-1}	6×10^{-12}	–	ELP
Pd/αAl$_2$O$_3$	550	4	11	7×10^{-2}	–	~1000	ELP
Pd–Cu/αAl$_2$O$_3$	450	3.5	11	8×10^{-1}	2.6×10^{-11}	1150	ELP

size and high production costs are the other major drawbacks of zeolite membranes (Caro and Noack, 2010). In addition, ceramic support materials for zeolite-based membranes are very costly, which has contributed to unwillingness in applying this membrane in industrial scale (Maloncy *et al.*, 2005; Ockwig and Nenoff, 2007a; Gascon *et al.*, 2012).

12.5.3   Carbon molecular sieve membranes

Carbon molecular sieve membranes (CMSMs), with pore sizes smaller than 2 nm, are used for gas separation with separation mechanisms of molecular sieving, surface diffusion and Knudsen diffusion (Tao *et al.*, 2015). This type of membrane is categorized in two major groups including supported membranes with high mechanical stability and un-supported membranes. Supported membranes involve flat and tube configurations, whereas un-supported ones include flat film, hollow fibre and capillary tubes (Meinema *et al.*, 2005; Scholz *et al.*, 2011).

CMSMs have received much attention for application in gas separation processes due to high flux and selectivity (Fuertes *et al.*, 1999; Ismail and David, 2001; Barsema *et al.*, 2002; Kishore *et al.*, 2003), rigidity to retain their stability in aggressive and harsh conditions that are results of presence of solvent, acid and base and operating temperature and pressure (Itoh and Haraya, 2000; Koros and Mahajan, 2000; Kishore *et al.*, 2003; Saufi and Ismail, 2004). CMSMs can lead to a good separation process as a result of their porous structure and molecular sieving morphology which provide high permeability and selectivity of the desired component, respectively. The properties of CMSMs can be manipulated to achieve the intended characteristics (Singh and Koros, 1996). The common method for fabrication of CMSMs is carbonization of polymeric precursors (Tin *et al.*, 2004).

Besides the advantages of CMSMs, there are some limitations including the very high production cost which is between 1 and 3 orders of magnitude greater than polymeric membranes, membrane brittleness, decreased performance in the presence of strongly adsorbing vapours, such as H_2S, NH_3 or chlorofluorocarbons (Koros and Mahajan, 2000; Adhikari and Fernando, 2006). Also carbon membranes, although more stable compared to zeolites membranes, have a limited thermal stability to 300 °C (Jüttke *et al.*, 2013).

12.5.4   Ceramic membranes

Another candidate for application in high temperature gas separation processes is ceramic membrane which has combined the advantages of other types of membranes that have led to the higher strength and permeability (Li *et al.*, 1996; Jiang and Chan, 2004; Riedel *et al.*, 2006; Fontaine *et al.*, 2008; Prasad *et al.*, 2010). Ceramic membranes are classified into porous (mainly composed of silica) and dense categories. Silica in the porous ceramic category exhibits appropriate thermal and chemical stability under severe situations (Diniz da Costa *et al.*, 2002; Lee and Oyama, 2002); however, dense ceramic membranes, with a proton exchange mechanism, show a high hydrogen purity, while considering the fact that these membranes are required to operate under high temperatures (around 900 °C) (Gallucci *et al.*, 2013).

The driving force for permeation of hydrogen in mixed protonic and electronic conducting ceramics is the chemical potential gradient across the membrane (Polfus *et al.*, 2015). The separation mechanism is similar to separation using other types of membranes, that is dissociative adsorption of hydrogen molecule on the membrane surface which forms protons and electrons, diffusion of electrons and protons to the other side of the membrane surface and the recombinative desorption (Tao *et al.*, 2015) (see Figure 12.5).

Ceramic membranes are superior than other membranes due to lower required cost and energy, higher selectivity, durability and stability in the presence of carbon monoxide and hydrogen sulphide which affect the metallic membranes. Therefore, ceramic membranes, inert to poisonous gases, are desirable for industrial hydrogen separation purposes (Zhang *et al.*, 2003; Zuo *et al.*, 2006a, 2006b; Liu *et al.*, 2009; Fang *et al.*, 2010; Zhu *et al.*, 2015). Despite the excellent features of ceramic membranes for hydrogen separation, some modifications need to be considered in future studies in order to increase the performance of separation processes. Some of them are modification of the chemical stability of the membrane in acidic gas atmospheres, modification of the thermal and mechanical stability for long-term operation (Tao *et al.*, 2015).

Among the ceramic membranes, silica-based ones reveal a great potential for hydrogen separation (Suda *et al.*, 2006). Various researchers have investigated the efficiency of silica membranes (de Vos and Verweij, 1998a; Yoshida *et al.*, 2001; Kanezashi and Asaeda, 2006; Barboiu *et al.*, 2009; Battersby *et al.*, 2009). Two common techniques for fabrication of silica membranes are sol–gel (Battersby *et al.*, 2009; Barboiu *et al.*, 2009) and CVD technique (Moore *et al.*, 2004). Silica membranes exhibit a good stability towards higher temperatures, harsh and corrosive atmospheres, high H_2 permeability and high selectivity over larger gas

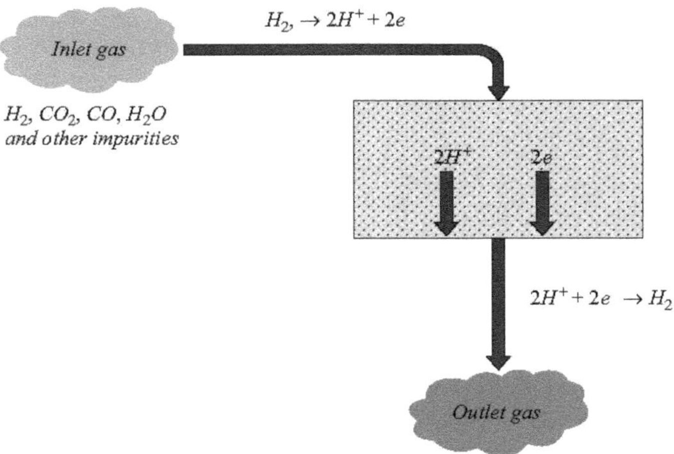

Figure 12.5   A proton-electron conducting ceramic membrane (Tao et al., 2015)

molecules present in the mixture due to their porosity (Burggraaf, 1996; de Vos and Verweij, 1998b). Besides the aforementioned advantages, there are some limitations for silica membranes such as being hydrothermally unstable (Dong *et al.*, 2008; Bighane, 2012).

12.6   Composite (hybrid) membranes

Due to the limitations of some membranes, for example polymeric ones which do not pass the upper bound trade-off of Robeson's curve, a reasonable choice is using composite membranes to modify the characteristics of these membranes. For example, a combination of two polymer substances (co-polymer), a polymer and an inorganic substance or a polymer with a nano-structured inorganic porous material within its matrix (mixed matrix membrane (MMM)) have been proposed as a substitute for the existing membranes (see Figure 12.6). So MMMs have the higher permeability, chemical and thermal stability of the inorganic substances in addition to better mechanical properties of the polymers (Chung *et al.*, 2007).

Another type of hybrid membrane are metal–organic framework membranes, built of inorganic metal or metal oxide centres and organic linkers by coordinate bonds, with the uniform pore structures, large surface area and very high adsorption affinities (Li *et al.*, 2015). There are various studies that have reported modification of the composite membranes in comparison with the common membranes (Mahajan and Koros, 2002; Wang *et al.*, 2002; Li *et al.*, 2005; Husain and Koros, 2007; Adams *et al.*, 2010; Khan *et al.*, 2010; Bastani *et al.*, 2013; Lin *et al.*, 2014).

A comparison between properties of four most common membranes in hydrogen separation process is demonstrated in Table 12.2.

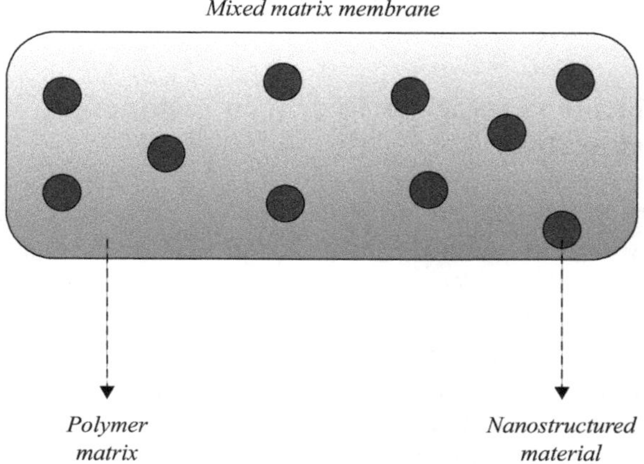

Figure 12.6   Schematic of a mixed matrix membrane (Bastani et al., 2013)

Table 12.2   Properties of different hydrogen separation membranes (McLeod, 2008)

	Membrane type			
	Dense polymer	**Porous ceramic**	**Metallic**	**Porous carbon**
Temperature	<100 °C	200–600 °C	200–600 °C	500–900 °C
H$_2$ selectivity	5–500	10–5,000	>1,000	10–1,000
H$_2$ flux[a]	0.1–1	60–300	100–1,000	10–200
Stability issues	Swelling, compaction	Water vapour	Embrittlement	Brittle, oxidizing environments
Poisoning issues	HCl, SO$_x$	None	H$_2$S, HCl, CO	Strongly adsorbing vapours
Materials	Polymers	Silica, alumina, zirconia or titania in a metal oxide form	Pd alloys	Carbon
Transport mechanism	Solution-diffusion	Molecular sieving	Solution-diffusion	Molecular sieving

[a]Units are 10^{-3} mol/m^2/s at a pressure difference of 101 kPa.

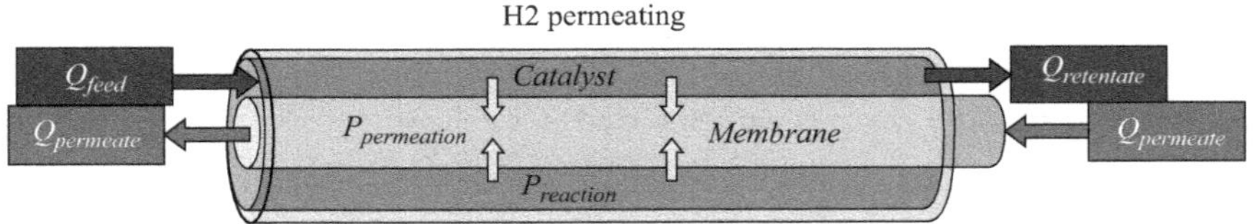

Figure 12.7   Schematic of a membrane reactor (MR) (Drioli et al., 2012)

12.7   Membrane reactor for hydrogen production and separation

Research efforts for increasing the efficiency of reaction systems as well as decreasing the volume required for equipment such as reactors and separators has led to development of the MRs which allows the simultaneous chemical reaction and membrane-based separation in one integrated unit (closed architecture) or two units in a close vicinity (open architecture) (Koros *et al.*, 1996; Cao *et al.*, 2007). The major difference between these two configurations is that in the open architecture arrangement, the reaction and separation processes take place independently in two units (i.e. reactor and membrane separator) which necessitate providing a larger space in comparison with the closed architecture system. As a result, the closed architecture form is preferable because it is more cost-effective and needs lower space. A schematic of a MR is presented in Figure 12.7.

Table 12.3   Comparison of hydrogen separation techniques (McLeod, 2008)

Process	H_2 production rate (N m³/h)	Recovery (%)	Purity (%)	Capital investment	Operating cost
PSA	>100	75–90	>99.99	Medium	High
Cryogenic distillation	>100	90–98	95–99	High	High
Membrane separation	Variable	70–95	70–100	Low	Low

The advantages of MRs are listed as follows (Coronas and Santamaria, 1999; Julbe *et al.*, 2001; Sanchez Marcano and Tsotsis, 2002; Mendes *et al.*, 2010; Basile *et al.*, 2011a, 2011b):

1. Shift in thermodynamic equilibrium towards the product side due to hydrogen separation according to Le Chatelier's principle
2. Higher conversion factor
3. More compact arrangement
4. Simpler operation
5. Lower operating temperatures
6. Higher energy efficiency and capital cost.

The most important factors that must be considered in a MR are membrane pore size and thickness, membrane fouling and operating temperature; each of them has a considerable effect on hydrogen production. Specially, the first parameter determines the highest hydrogen permeability that can be achieved in a MR (Rahimpour and Nategh, 2016).

There are many investigations devoted to the hydrogen production by different reactions such as water–gas-shift and reforming, utilizing the MR concept (Abu El Hawa *et al.*, 2015; Bakhtyari *et al.*, 2015; Ghasemzadeh *et al.*, 2015; Marcoberardino *et al.*, 2015; Patrascu and Sheintuch, 2015). In these studies, reactors with different configurations such as packed-bed MRs, membrane micro-reactors as well as fluidized-bed MRs have been used; each of them has its own advantages and limitations (Gallucci *et al.*, 2013). In micro-reactors, difficulty arises with matching the permeation rates and the required catalyst volume/activity (Marra *et al.*, 2014). Packed-bed reactors involve with significant heat and mass-transfer limitations (Vigneault *et al.*, 2012), whereas a good heat and mass-transfer characteristics are offered by fluidized-bed reactors. In addition, the required catalyst volume and membrane area can be easily adjusted in fluidized-bed MRs (Gallucci *et al.*, 2010). However, in this configuration the membrane may undergo damages due to the fact that collisions of moving particles with the membrane surface make holes in the surface. This limitation can be overcome by using a membrane with higher thickness or implementing an additional porous layer with smaller pore size compared to the particles in order to avoid blockage of the pores. Nevertheless, the first mentioned solution may face some economic issues due to the lower flux (Arratibel *et al.*, 2015).

After describing the conventional and membrane hydrogen separation methods in detail, a general comparison of these techniques is tabulated in Table 12.3 in order to summarize the differences between these methods.

12.8   Conclusion

Hydrogen gas produced from common production processes is associated with coexisting gases which lower the yield. The conventional hydrogen separation methods (adsorption-based, cryogenic, etc.) have some limitations such as energy and economic issues. Therefore, proposing a novel solution for elimination or minimization of such problems is very important. An alternative is membrane technology which offers higher simplicity, flexibility, energy efficiency and lower costs, maintenance, weight and space as well as easier intensification in comparison with the conventional separation methods.

Among the common organic and inorganic membranes, metallic membranes and especially palladium and its alloys provide high purity of hydrogen, resistance to corrosion, excellent catalytic surface properties and high selectivity to hydrogen. However, carbon monoxide and hydrogen sulphide present in the inlet gas mixtures affect the metallic membranes. Also the application of polymeric membranes is limited due to their limited lifetime, poor thermal and chemical stability, poor durability, decomposition or failure above 100–300 °C, catalytic deactivation and susceptibility to the present impurities in the gas mixtures. On the other hand, CMSMs have high costs, limited hydrothermal stability and decreasing yield in presence of H_2S, NH_3 and CO. Therefore, ceramic membranes, with lower required cost and energy, higher selectivity, durability and stability, and specifically being inert to poisonous gases, which is a very useful characteristic in hydrogen separation processes, are desirable for industrial applications. Finally, using the best membrane for hydrogen production in a combined hydrogen generation and separation process (process intensification) can reduce the capital and operating cost of hydrogen production to a large extent.

List of abbreviations

PSA	pressure swing adsorption
TSA	temperature swing adsorption
VSA	vacuum swing adsorption
ESA	electrical swing adsorption
CCS	CO_2 capture and sequestration
CA	cellulose acetate
CN	nitrocellulose
CE	cellulose esters
PS	polysulfone
PES	polyether sulfone
PAN	polyacrylonitrile
PE	polyethylene
PP	polypropylene

PTFE	polytetrafluoroethylene
PVDF	polyvinylidene fluoride
PVC	polyvinylchloride
ELP	electroless plating
CVD	chemical vapour deposition
PVD	physical vapour deposition
EPD	electroplating deposition
CMSMs	carbon molecular sieve membranes
LH	liquid-phase hydrothermal
VPT	vapour phase transport
PH	pore-plugging hydrothermal
MMM	mixed matrix membrane
MOF	metal–organic framework
MR	membrane reactor
WGS	water-gas-shift

References

Abu El Hawa, H.W., Paglieri, S.N., Morris, C.C., Harale, A. and Douglas Way, J. (2015) 'Application of a Pd–Ru Composite Membrane to Hydrogen Production in a High Temperature Membrane Reactor', *Separation and Purification Technology*, 47, 388–397.

Adams, R., Carson, C., Ward, J., Tannenbaum, R. and Koros, W. (2010) 'Metal Organic Framework Mixed Matrix Membranes for Gas Separations', *Microporous and Mesoporous Materials*, 131(1–3), 13–20.

Adhikari, S. and Fernando, S. (2006) 'Hydrogen Membrane Separation Techniques', *Industrial and Engineering Chemistry Research*, 45, 875–881.

Al-Mufachi, N.A., Rees, N.V. and Steinberger-Wilkens, R. (2015) 'Hydrogen Selective Membranes: A Review of Palladium-Based Dense Metal Membranes', *Renewable and Sustainable Energy Reviews*, 47, 540–551.

An, W., Swenson, P., Wu, L., Waller, T., Ku, A. and Kuznicki, S.M. (2011) 'Selective Separation of Hydrogen from C_1/C_2 Hydrocarbons and CO_2 through Dense Natural Zeolite Membranes', *Journal of Membrane Science*, 369, 414–419.

Arratibel, A., Astobieta, U., Tanaka, D.A.P., van Sint Annaland, M. and Gallucci, F. (2015) 'N_2, He and CO_2 Diffusion Mechanism through Nanoporous YSZ/ γ-Al_2O_3 Layers and Their Use in a Pore-filled Membrane for Hydrogen Membrane Reactors', *International Journal of Hydrogen Energy*, 41(20), 1–13.

Atsonios, K., Panopoulos, K.D., Doukelis, A., *et al.* (2015) *Chapter 1: Introduction to Palladium Membrane Technology*, in *Palladium Membrane Technology for Hydrogen Production, Carbon Capture and Other Applications. Principles, Energy Production and Other Applications*, Doukelis, A., Panopoulos, K.,

Koumanakos, A. and Kakaras, E. (Editors), A Volume in Woodhead Publishing Series in Energy, Oxford, UK, 1–21.

Baker, R.W. (2002) 'Future Directions of Membrane Gas Separation Technology', *Industrial and Engineering Chemistry Research*, 41, 1393–1411.

Bakhtyari, A., Mohammadi, M. and Rahimpour, M.R. (2015) 'Simultaneous Production of Dimethyl Ether (DME), Methyl Formate (MF) and Hydrogen from Methanol in an Integrated Thermally Coupled Membrane Reactor', *Journal of Natural Gas Science and Engineering*, 26, 595–607.

Barboiu, C., Sala, B., Bec, S., *et al.* (2009) 'Structural and Mechanical Characterizations of Microporous Silica–Boron Membranes for Gas Separation', *Journal of Membrane Science*, 326(2), 514–525.

Barsema, J.N., Van der Vegt, N.F.A., Koops, G.H. and Wessling, M. (2002) 'Carbon Molecular Sieve Membranes Prepared from Porous Fiber Precursor', *Journal of Membrane Science*, 205, 239–246.

Basile, A., Gallucci, F., Téllez, C. and Menéndez, M. (2011a) *Membranes for Membrane Reactors: Preparation, Optimization and Selection*, Basile, A. and Gallucci, F. (Editors), John Wiley & Sons, Ltd, Chichester, UK.

Basile, A., Iulianelli, A., Longo, T., Liguori, S. and De Falco, M. (2011b) *Chapter 2: Pd-based Selective Membrane State-of-the-Art* in *Membrane Reactors for Hydrogen Production Processes*, De Falco, M., Marrelli, M. and Iaquaniello, G. (Editors), Springer, London, UK.

Bastani, D., Esmaeili, N. and Asadollahi, M. (2013) 'Polymeric Mixed Matrix Membranes Containing Zeolites as A Filler for Gas Separation Applications: A Review', *Journal of Industrial and Engineering Chemistry*, 19, 375–393.

Battersby, S., Tasakia, T., Smarta, S., *et al.* (2009) 'Performance of Cobalt Silica Membranes in Gas Mixture Separation', *Journal of Membrane Science*, 329(1–2), 91–98.

Bighane, N. (2012) *Novel Silica Membranes for High Temperature Gas Separations*, A Master thesis in the School of Chemical & Biomolecular Engineering, Georgia Institute of Technology.

Bolland, O. and Undrum, H. (2003) 'A Novel Methodology for Comparing CO_2 Capture Options for Natural Gas-Fired Combined Cycle Plants', *Advances in Environmental Research*, 7, 901–911.

Bonjour, J., Chalfen, J.-B. and Meunier, F. (2002) 'Temperature Swing Adsorption Process with Indirect Cooling and Heating', *Industrial & Engineering Chemistry Research*, 41, 5802–5811.

Breck, D.W. (1974) *Zeolite Molecular Sieves: Structure, Chemistry, and Use*, New York, John Wiley & Sons.

Burggraaf, A.J. (1996) *Important Characteristics of Inorganic Membranes*, in *Fundamentals of Inorganic Membrane Science and Technology, Membrane Science and Technology Series*, Volume 4, Burggraaf, A.J. and Cot, L. (Editors), Elsevier Science, Amsterdam, 21–34.

Cao, P., Tremblay, A.Y., Dubé, M.A. and Morse, K. (2007) 'Effect of Membrane Pore Size on the Performance of a Membrane Reactor for Biodiesel Production', *Industrial & Engineering Chemistry Research*, 46, 52–58.

Caro, J. and Noack, M. (2010) *Zeolite Membranes-Status and Prospective*, in *Advances in Nanoporous Materials*, Ernst, S. (Editor) Volume 1, Elsevier, Oxford, UK, 1–96.

Cavalcante, C.L. (2000) 'Industrial Adsorption Separation Processes: Fundamentals, Modeling and Applications', *Latin American Applied Research*, 30(4), 357–364.

Chang, K., Li, Q. and Li, Q. (2008) 'Refrigeration Cycle for Cryogenic Separation of Hydrogen from Coke Oven Gas', *Frontiers of Energy and Power Engineering in China*, 2(4), 484–488.

Chou, C.T. and Chen, C.Y. (2004) 'Carbon Dioxide Recovery by Vacuum Swing Adsorption', *Separation & Purification Technology*, 39, 51–65.

Chung, T.-S., Jiang, L.Y., Li, Y. and Kulprathipanja, S. (2007) 'Mixed Matrix Membranes (MMMs) Comprising Organic Polymers with Dispersed Inorganic Fillers for Gas Separation', *Progress in Polymer Science*, 32, 483–507.

Coker, D.T., Freeman, B.D. and Fleming, G.K. (1998) 'Modeling Multicomponent Gas Separation Using Hollow-Fiber Membrane Contactors', *AIChE Journal*, 44, 1289–1302.

Coronas, J. and Santamaria, J. (1999) 'Catalytic Reactors Based on Porous Ceramic Membranes', *Catalysis Today*, 51, 377–389.

Daramola, M.O., Aransiola, E.F. and Ojumu, T.V. (2012) 'Potential Applications of Zeolite Membranes in Reaction Coupling Separation Processes', *Materials*, 5(11), 2101–2136.

De Lange, R.S.A., Keizer, K. and Burggraaf, A.J. (1995) 'Analysis and Theory of Gas Transport in Microporous Sol–Gel Derived Ceramic Membranes', *Journal of Membrane Science*, 104(1–2), 81–100.

De Vos, R.M. and Verweji, H. (1998a) 'High-Selectivity, High-Flux Silica Membranes for Gas Separation', *Science*, 279, 1710–1711.

De Vos, R.M. and Verweij, H. (1998b) 'Improved Performance of Silica Membranes for Gas Separation', *Journal of Membrane Science*, 143, 37–51.

Diniz da Costa, J.C., Lu, G.Q., Rudolph, V. and Lin, Y.S. (2002) 'Novel Molecular Sieve Silica (MSS) Membranes: Characterization and Permeation of Single-Step and Two-Step Sol–Gel Membranes', *Journal of Membrane Science*, 198, 9–21.

Domine, D. and Montgareuil, P.G.D. (1964) 'Process for Separating a Binary Gaseous Mixture by Adsorption', *Patent US3155468 A*.

Dong, J., Lin, Y.S., Kanezashi, M. and Tang, Z. (2008) 'Microporous Inorganic Membranes for High Temperature Hydrogen Purification', *Journal of Applied Physics*, 104, 121301: 1–17.

Dong, X., Wang, H., Rui, Z. and Lin, Y.S. (2015) 'Tubular Dual-Layer MFI Zeolite Membrane Reactor for Hydrogen Production Via the WGS Reaction: Experimental and Modeling Studies', *Chemical Engineering Journal*, 268, 219–229.

Dos Santos, M.M.C.C. (2009) *Carbon Molecular Sieve Membranes for Gas Separation: Study, Preparation and Characterization*, LEPAE-Laboratory of

Engineering Processes, Environment and Engineering, Chemical Engineering Department, Faculty of Engineering, University of Porto.

Drioli, E., Brunetti, A., Di Profio, G. and Barbieri, G. (2012) 'Process Intensification Strategies and Membrane Engineering', *Green Chemistry*, 14, 1561–1572.

Ebenezer, S.A. and Gudmunsson, J.S. (2006) 'Removal of Carbon Dioxide from Natural Gas for LPG Production', Semester project work, Institute of Petroleum Technology, Norwegian University of Science and Technology, Trondheim, Norway, 74 pages.

Ellabban, O., Abu-Rub, H. and Blaabjerg, F. (2014) 'Renewable Energy Resources: Current Status, Future Prospects and Their Enabling Technology', *Renewable and Sustainable Energy Reviews*, 39, 748–764.

Engineering Data Book published as a service to The Gas Processing & Related Process Industries, Gas Processors Suppliers Association, (2004), 12th Edition, Volumes I and II, Tulsa, Oklahoma.

Esteves, I.A.A.C. and Mota, J.P.B. (2002) 'Simulation of a New Hybrid Membrane/Pressure Swing Adsorption Process for Gas Separation', *Desalination*, 148, 275–280.

Fabuss, B.M. and Bois, W.C.D. (1971) 'Apparatus and Process for Desorption of Filter Beds by Electric Current', *Patent US3608273 A*.

Fain, D.E. (2000) 'Mixed Gas Separation Technology Using Inorganic Membranes', *Membrane Technology*, 120, 9–13.

Fang, S., Bi, L., Yan, L., Sun, W., Chen, C. and Liu, W. (2010) 'CO_2-Resistant Hydrogen Permeation Membranes Based on Doped Ceria and Nickel', *The Journal of Physical Chemistry C*, 114, 10986–10991.

Faraji, S., Sotoudeh-Gharebagh, R. and Mostoufi, N. (2005) 'Hydrogen Recovery from Refinery off-gases', *Journal of Applied Sciences*, 5(3), 459–464.

Fawas, E.P., Kapantaidakis, G.C., Nolan, J.W., Mitropoulos, A.C. and Kanellopoulos, N.K. (2007) 'Preparation, Characterization and Gas Permeation Properties of Carbon Hollow Fiber Membranes Based on Matrimid (R) 5218 Precursor', *Journal of Materials Processing Technology*, 186, 102–110.

Finamore, N.K., Liu, C., Mohanty, P., Moore, D.T. and Landskron, K. (2011) *Electric Field Swing Adsorption for Carbon Capture Applications*, Bethlehem: Department of Chemistry, Lehigh University.

Fontaine, M., Norby, T., Larring, Y., Grande, T. and Bredesen, R. (2008) 'Oxygen and Hydrogen Separation Membranes Based on Dense Ceramic Conductors', *Membrane Science and Technology*, 13, 401–458.

Fuertes, A.B., Nevskaia, D.M. and Centeno, T.A. (1999) 'Carbon Composite Membranes from Matrimid and Kapton Polyimides for Gas Separation', *Microporous & Mesoporous Materials*, 33, 115–125.

Gallucci, F., Fernandez, E., Corengia, P. and Annaland, M.V.S. (2013) 'Recent Advances on Membranes and Membrane Reactors for Hydrogen Production', *Chemical Engineering Science*, 92, 40–66.

Gallucci, F., Van Sint Annaland, M. and Kuipers, J.A.M. (2010) 'Theoretical Comparison of Packed Bed and fluidized Bed Membrane Reactors for Methane Reforming', *International Journal of Hydrogen Energy*, 35, 7142–7150.

Gao, H. and Wang, L. (2014) 'Analysis of H_2S Tolerance of Pd–Cu Alloy Hydrogen Separation Membranes', *Chinese Journal of Chemical Engineering*, 22(5), 503–508.

Gascon, J., Kapteijn, F., Zornoza, B., Sebastian, V., Casado, C. and Coronas, J. (2012) 'Practical Approach to Zeolitic Membranes and Coatings: State of the Art, Opportunities, Barriers and Future Perspectives', *Chemistry of Materials*, 24, 2829–2844.

Gas Separation by Pressure Swing Adsorption, Japan Envirochemicals, Activated Carbon Business Division, https://www.jechem.co.jp/shirasagi_e/tech/psa.html.

Gepert, V., Kilgus, M., Schiestel, T., *et al.* (2006) 'Ceramics Supported Capillary Pd Membranes for Hydrogen Separation: Potentials and Present Limitations', *Fuel Cells*, 6(6), 472–481.

Ghasemzadeh, K., Morrone, P., Babalou, A.A. and Basile, A. (2015) 'A Simulation Study on Methanol Steam Reforming in the Silica Membrane Reactor for Hydrogen Production', *International Journal of Hydrogen Energy*, 40(10), 3909–3918.

Grande, C.A. (2012) 'Advances in Pressure Swing Adsorption for Gas Separation', International Scholarly Research Network, Article ID 982934, 13 pages, doi:10.5402/2012/982934.

Hands, B.A. (1986) *Cryogenic Engineering*, Academic Press, London, UK.

Hasche, R.L. and Dargan, W.H. (1931) 'Separation of Gases', *Patent US1794377 A*.

Holmes, M. and Erickson, T. (2010) *Hydrogen Separation Membranes*, Energy & Environmental Research Center, Grand Forks, North Dakota, USA.

Hung, I.-M., Chiang, Y.-J., Jang, S.-C.J., *et al.* (2014) 'The Proton Conduction and Hydrogen Permeation Characteristic of Sr(Ce0.6Zr0.4)0.85Y0.15O3−δ Ceramic Separation Membrane', *Journal of the European Ceramic Society*, 35(1), 163–170.

Husain, S. and Koros, W.J. (2007) 'Mixed Matrix Hollow Fiber Membranes Made with Modified HSSZ-13 Zeolite in Polyetherimide Polymer Matrix for Gas Separation', *Journal of Membrane Science*, 288(1–2), 195–207.

Ismail, A.F. and David, L.I.B. (2001) 'A Review on the Latest Development of Carbon Membranes for Gas Separations', *Journal of Membrane Science*, 193, 1–18.

Itoh, N. and Haraya, K. (2000) 'A Carbon Membrane Reactor', *Catalysis Today*, 56(1–3), 103–111.

Jae-Yun, H., Chang-Hyun, K., Sang-Ho, K. and Dong-Won, K. (2014) 'Development of Pd Alloy Hydrogen Separation Membranes with Dense/Porous Hybrid Structure for High Hydrogen Perm-Selectivity', *Advances in Materials Science and Engineering*, Article ID 438216, 10 pages, doi:10.1155/2014/438216.

Jain, S., Moharir, A.S., Li, P. and Wozny, G. (2003) 'Heuristic Design of Pressure Swing Adsorption: A Preliminary Study', *Separation and Purification Technology*, 33(1), 25–43.

Jiang, S.P. and Chan, S.H. (2004) 'A Review of Anode Materials Development in Solid Oxide Fuel Cells', *Journal of Materials Science*, 39(14), 4405–4439.

Julbe, A., Farrusseng, D. and Guizard, C. (2001) 'Porous Ceramic Membranes for Catalytic Reactors e Overview and New Ideas', *Journal of Membrane Science*, 181, 3–20.

Jüttke, Y., Richter, H., Voigt, I., *et al.* (2013) 'Polymer Derived Ceramic Membranes for Gas Separation', *Chemical Engineering Transactions*, 32, 1891–1896.

Kanezashi, M. and Asaeda, M. (2006) 'Hydrogen Permeation Characteristics and Stability of Ni-Doped Silica Membranes in Steam at High Temperature', *Journal of Membrane Science*, 271(1–2), 86–93.

Khan, A.L., Cano-Odena, A., Gutiérrez, B., Minguillón, C. and Vankelecom, I.F.J. (2010) 'Hydrogen Separation and Purification using Polysulfone Acrylate–Zeolite Mixed Matrix Membranes', *Journal of Membrane Science*, 350, 340–346.

Kishore, N., Sachan, S., Rai, K.N. and Kumar, A. (2003) 'Synthesis and Characterization of a Nanofiltration Carbon Membrane Derived from Phenol-Formadehyde Resin', *Carbon*, 41, 2961–2972.

Knaebel, K.S. (2003) *Adsorbent Selection*, Adsorption Research, Inc., Dublin, Ohio, 43016, 1–23.

Koonaphapdeelert, S., Tan, X., Wu, Z. and Li, K. (2008) 'Solvent Distillation by Ceramic Hollow Fibre Membrane Contactor', *Journal of Membrane Science*, 314(1–2), 58–66.

Koros, W.J. and Fleming, G.K. (1993) 'Membrane-based Gas Separation', *Journal of Membrane Science*, 83, 1–80.

Koros, W.J. and Mahajan, R. (1994) 'Pushing the Limits on Possibilities for Large Scale Gas Separation: Which Strategies?', *Journal of Membrane Science*, 175, 181–196.

Koros, W.J., Ma, Y.H. and Shimidzu, T. (1996) 'Terminology for Membranes and Membrane Processes', *Pure and Applied Chemistry*, 68, 1479–1489.

Kosinov, N., Auffret, C., Borghuis, G.J., Sripathi, V.G.P. and Hensen, E.J.M. (2015) 'Influence of the Si/Al Ratio on the Separation Properties of SSZ-13 Zeolite Membranes', *Journal of Membrane Science*, 484, 140–145.

Lababidi, H.M.S. (2000) 'Air Separation by Polysulfone Hollow Fibre Membrane Permeators in Series-Experimental and Simulation Results', *Chemical Engineering Research and Design*, 78, 1066–1076.

Lee, C.-H. (Editor) (2003) *Proceedings of the Third Pacific Basin Conference on Adsorption Science & Technology*, Kyongju, Korea, 25–29 May.

Lee, D. and Oyama, S.T. (2002) 'Gas Permeation Characteristics of a Hydrogen Selective Supported Silica Membrane', *Journal of Membrane Science*, 210, 291–306.

Li, K. (2007) *Ceramic Membranes for Separation and Reaction*, England, John Wiley & Sons Ltd.

Li, W., Zhang, Y., Li, Q. and Zhang, G. (2015) 'Metal–Organic Framework Composite Membranes: Synthesis and Separation Applications', *Chemical Engineering Science*, Berlin Heidelberg, 135, 232–257.

Li, Y., Chung, T.-S., Cao, C. and Kulprathipanja, S. (2005) 'The Effects of Polymer Chain Rigidification, Zeolite Pore Size and Pore Blockage on Polyethersulfone (PES)-Zeolite: A Mixed Matrix Membranes', *Journal of Membrane Science*, 260, 45–55.

Li, Z.Y., Kusakabe, K. and Morooka, S. (1996) 'Preparation of Thermostable Amorphous Si–C–O Membrane and Its Application to Gas Separation at Elevated Temperature', *Journal of Membrane Science*, 118(2), 159–168.

Lin, H.Q. and Freeman, B.D. (2011) *Permeation and Diffusion*, in *Handbook for Materials Measurement Methods*, Czichos, H., Saito, T. and Smith, L. (Editors), Springer, Berlin Heidelberg, 426–444.

Lin, H., Van Wagner, E., Freeman, B.D., Toy, L.G. and Gupta, R.P. (2006) 'Plasticization-enhanced Hydrogen Purification Using Polymeric Membranes', *Science*, 311, 639–642.

Lin, R., Ge, L., Hou, H., Strounina, E., Rudolph, V. and Zhu, Z. (2014) 'Mixed Matrix Membranes with Strengthened MOFs/Polymer Interfacial Interaction and Improved Membrane Performance', *ACS Applied Materials & Interfaces*, 6, 5609–5618.

Liu, K., Song, C. and Subramani, V. (2010) *Hydrogen and Syngas Production and Purification Technologies*, Hoboken, NJ, John Wiley & Sons Inc.

Mahajan, R. and Koros, W.J. (2002) 'Mixed Matrix Membrane Materials with Glassy Polymers. Part 2', *Polymer Engineering & Science*, 42(7), 1432–1441.

Maloncy, M.L., Maschmeyer, T. and Jansen, J.C. (2005) 'Technical and Economical Evaluation of a Zeolite Membrane Based Heptane Hydroisomerization Process', *Chemical Engineering Journal*, 106, 187–195.

Marcoberardino, G.D., Sosio, F., Manzolini, G. and Campanari, S. (2015) 'Fixed Bed Membrane Reactor for Hydrogen Production from Steam Methane Reforming: Experimental and Modeling Approach', *International Journal of Hydrogen Energy*, 40(24), 7559–7567.

McLeod, L.S. (2008) *Hydrogen Permeation through Microfabricated Palladium–Silver Alloy Membranes*, A PhD dissertation in George W. Woodruff School of Mechanical Engineering, Georgia Institute of Technology.

Meinema, H.A., Dirrix, R.W.J., Brinkman, H.W., Terpstra, R.A., Jekerle, J. and Kösters, P.H. (2005) 'Ceramic Membranes for Gas Separation – Recent Developments and State of the Art', *InterCeram: International Ceramic Review*, 54(2), 86–91.

Mendes, D., Mendes, A., Madeira, L.M., Iulianelli, A., Sousa, J.M. and Basile, A. (2010) 'The Water–Gas Shift Reaction: From Conventional Catalytic Systems to Pd-Based Membrane Reactors: A Review', *Asia-Pacific Journal of Chemical Engineering*, 5, 111–137.

Moore, T., Damlea, S., Williamsa, P.J. and Koros, W.J. (2004) 'Characterization of Low Permeability Gas Separation Membranes and Barrier Materials: Design and Operations', *Journal of Membrane Science*, 245, 227–231.

Moskal, F. and Nastaj, J.F. (2007) 'Internal Heat Source Capacity at Inductive Heating in Desorption Step of ETSA Process', *International Communications in Heat and Mass Transfer*, 34(5), 579–586.

Naylor, T.D.V. (1995) *Polymer Membranes: Materials, Structures and Separation Performance*, Rapra Technology Ltd, Shawbury, Shrewsbury, Shropshire, UK.

Noble, R.D. and Stern, S.A. (1995) *Membrane Separations Technology: Principles and Applications*, Elsevier Science, Amsterdam, the Netherlands.

Nunes, S.P. and Peinemann, K.V. (2006) *Gas Separation with Membranes,* in *Membrane Technology in the Chemical Industry*, Nunes, S.P. and Peinemann, K.V. (Editors), Wiley-VCH, Weinheim, Germany, 53–75.

Ockwig, N.W. and Nenoff, T.M. (2007a) 'Chemistry of Hydrogen Separation Membranes', *Chemical Reviews*, 107, 4078–4110.

Ockwig, N.W. and Nenoff, T.M. (2007b) 'Membranes for Hydrogen Separation', *Chemical Reviews*, 107, 4078–4110.

Palmer, D. (1997) *Hydrogen in the Universe*, 13 September, NASA.

Pan, X., Kilgus, M. and Goldbach, A. (2005) 'Low-temperature H_2 and N_2 Transport through thin $Pd_{66}Cu_{34}H_x$ Layers', *Catalysis Today*, 104, 225–230.

Patrascu, M. and Sheintuch, M. (2015) 'On-Site Pure Hydrogen Production by Methane Steam Reforming in High Flux Membrane Reactor: Experimental Validation, Model Predictions and Membrane Inhibition', *Chemical Engineering Journal*, 262, 862–874.

Penner, S.S. (2006) 'Steps toward the Hydrogen Economy', *Energy*, 31, 33–43.

Perley, A. (1933) 'Method of Making Commercial Hydrogen', *Patent US1896916 A*.

Petkovska, M., Tondeur, D., Grevillot, G., Granger, J. and Mitrovic, M. (1991) 'Temperature-swing Gas Separation with Electrothermal Desorption Step', *Separation Science and Technology*, 26(3), 425–444.

Pinto, F., André, R., Costa, P., Carolino, C., Lopes, H. and Gulyurtlu, I. (2011) *Chapter 7: Gasification Technology and Its Contribution to Deal with Global Warming*, in *Solid Biofuels for Energy: A Lower Greenhouse Gas Alternative*, Grammelis, P. (Editor). Springer-Verlag, London, UK, 151–175.

Polfus, J.M., Xing, W., Fontaine, M.-L., Denonville, C., Henriksen, P.P. and Bredesen, R. (2015) 'Hydrogen Separation Membranes Based on Dense Ceramic Composites in the $La_{27}W_5O_{55.5}$–$LaCrO_3$ System', *Journal of Membrane Science*, 479, 39–45.

Prasad, R.M., Iwamoto, Y., Riedel, R. and Gurlo, A. (2010) 'Multilayer Amorphous Si–B–C–N/Gamma-Al_2O_3/Alpha-Al_2O_3 Membranes for Hydrogen Purification', *Advanced Engineering Materials*, 12, 522–528.

Rahimpour, M.R. and Nategh, M. (2016) *Chapter 18: Hydrogen Production from Pyrolysis-derived Bio-Oil Using Membrane Reactors,* in *Membrane Technologies for Biorefining*, 1st Edition, Figoli, A., Cassano, A. and Basile, A. (Editors), Woodhead Publishing Series, Oxford, UK.

Ratan, S. and Wentink, P. (2001) 'Cost Effective Hydrogen from Refinery Off-gases' in *Proceedings of Fourth National Rubber Conference*, February, 20–21, Iran, 131–137.

Rautenbach, R., Struck, A. and Roks, M.F.M. (1998) 'A Variation in Fiber Properties Affects the Performance of Defect-free Hollow Fiber Membrane Modules for Air Separation', *Journal of Membrane Science*, 150, 31–41.

Reuss, J., Bathen, D. and Schmidt-Traub, H. (2002) 'Desorption by Microwaves: Mechanisms of Multicomponent Mixtures', *Chemical Engineering & Technology*, 25(4), 381–384.

Ribeiro, R.P.P.L., Grande, C.A. and Rodrigues, A.E. (2014) 'Electric Swing Adsorption for Gas Separation and Purification: A Review', *Separation Science and Technology*, 49, 1985–2002.

Richter, H. and Hoyer, T. (2013) *Mixed Matrix Membranes*, Fraunhofer IKTS Annual Report.

Riedel, R., Mera, G., Hauser, R. and Klonczynski, A. (2006) 'Silicon-Based Polymer-Derived Ceramics: Synthesis Properties and Application – A Review', *Journal of the Ceramic Society of Japan*, 114(6), 425–444.

Ritter, J.A. and Ebner, A.D. (2007) 'State-of-the-Art Adsorption and Membrane Separation Processes for Hydrogen Production in the Chemical and Petrochemical Industries', *Separation Science and Technology*, 42(6), 1123–1193.

Robert, F. (2004) The Hydrogen Backlash, in *Science*. 13 August.

Robeson, L.M. (2008) 'The Upper Bound Revisited', *Journal of Membrane Science*, 320(1–2), 390–400.

Ruthven, D.M. (1984) *Principles of Adsorption and Adsorption Processes*, New York, John Wiley & Sons.

Ruthven, D.M., Farooq, S. and Knaebel, K.S. (1994) '*Pressure Swing Adsorption*', New York, Wiley-VCH.

Sanchez Marcano, J.G. and Tsotsis, T.T. (2002) *Chapter 1: The Coupling of the Membrane Separation Process with a Catalytic Reaction, in Catalytic Membranes and Membrane Reactor*, Weinheim, Wiley-VCH, 5–14.

Sanders, D.F., Smith, Z.P., Guo, R., *et al.* (2013) 'Energy-Efficient Polymeric Gas Separation Membranes for A Sustainable Future: A Review', *Polymer*, 54, 4729–4761.

Saufi, S.M. and Ismail, A.F. (2004) 'Fabrication of Carbon Membranes for Gas Separation-1 Review', *Carbon*, 42(2), 241–259.

Scholz, M., Wessling, M. and Balster, J. (2011) *Chapter 5: Design of Membrane Modules for Gas Separations, in Membrane Engineering for the Treatment of Gases, Volume 1: Gas-separation Problems with Membranes*, Drioli, E. and Barbieri, G. (Editors), The Royal Society of Chemistry, Cambridge, UK.

Sharp, A.J., Finlayson, D. and Celanese, B. (1932) 'Improvements in or Relating to the Treatment of Gaseous Mixtures for the Purpose of Separating Them into Their Components or Enriching Them with Respect to One or More of Their Components', *United Kingdom Patent 365092-A*.

Shimekit, B. and Mukhtar, H. (2012) *Chapter 9: Natural Gas Purification Technologies – Major Advances for CO_2 Separation and Future Directions*, in *Advances in Natural Gas Technology*, 1st Edition, Al-Megren, H.A. (Editor), Intech.

Sholl, D.S. and Ma, Y.H. (2006) 'Dense Metal Membranes for the Production of High-Purity Hydrogen', *Materials Research Society Bulletin*, 31(10), 770–773.

Singh, A. and Koros, W.J. (1996) 'Significance of Entropic Selectivity for Advanced Gas Separation Membranes', *Industrial & Engineering Chemistry Research*, 35(4), 1231–1234.

Sinnott, R. (2009) *Chemical Engineering Design*, Butterworth-Heinemann, Oxford, Uk, 1280 pages.

Sircar, S. (2002) 'Pressure Swing Adsorption', *Industrial & Engineering Chemistry Research*, 41, 1389–1392.

Sircar, S., Waldron, W.E., Rao, M.B. and Anand, M. (1999) 'Hydrogen Production by Hybrid SMR-PSA-SSF Membrane System', *Separation and Purification Technology*, 17, 11–20.

Siriwardane, R.V., Shen, M.-S., Fisher, E.P. and Poston, J.A. (2001) 'Adsorption of CO_2 on Molecular Sieves and Activated Carbon', *Energy & Fuels*, 15, 279–284.

Skarstrom, C.W. (1960) 'Method and Apparatus for Fractionating Gas Mixtures by Adsorption', *Patent US2944627 A*.

Song, C. (2003) 'Overview of Hydrogen Production Options for Hydrogen Energy Development, Fuel-Cell Fuel Processing and Mitigation of CO Emissions', in *Proceedings of the 20th International Pittsburgh Coal Conference*, Pittsburgh, PA, USA.

Songolzadeh, M., Soleimani, M., Takht Ravanchi, M. and Songolzadeh, R. (2014) 'Carbon Dioxide Separation from Flue Gases: A Technological Review Emphasizing Reduction in Greenhouse Gas Emissions', Hindawi Publishing Corporation, *Scientific World Journal*, 2014 (Article ID 828131), 34 Pages.

Stanislowski, J.J. and Laumb, J.D. (2009) 'Gasification of Lignites to Produce Liquid Fuels, Hydrogen, and Power', *Presented at the Pittsburgh Coal Conference*, September.

Stern, S.A. (1994) 'Polymers for Gas Separations: The Next Decade', *Journal of Membrane Science*, 94, 1–65.

Stocker, J., Whyshall, M. and Miller, G. (1998) '30 Years of PSA Technology for Hydrogen Purification', UOP LLC.

Suda, H., Yamauchi, H., Uchimaru, Y., Fujiwara, I. and Haraya, K. (2006) 'Preparation and Gas Permeation Properties of Silicon Carbide-based Inorganic Membranes for Hydrogen Separation', *Desalination*, 193, 252–255.

Sullivan, P.D., Rood, M.J., Grévillot, G., Wander, J.D. and Hay, K.J. (2004) 'Activated Carbon Fiber Cloth Electrothermal Swing Adsorption System', *Environmental Science & Technology*, 38(18), 4865–4877.

Tan, X. and Li, K. (2015) *Inorganic Membrane Reactors: Fundamentals and Applications*, London, John Wiley & Sons, Ltd.

Tao, Z., Yan, L., Qiao, J., Wang, B., Zhang, L. and Zhang, J. (2015) 'A Review of Advanced Proton-Conducting Materials for Hydrogen Separation', *Progress in Material Science*, 74, 1–50.

Tin, P.S., Chung, T.S. and Hil, A.J. (2004) 'Advanced Fabrication of Carbon Molecular Sieve Membranes by Nonsolvent Pretreatment of Precursor Polymers', *Industrial & Engineering Chemistry Research*, 43, 6476–6483.

Tlili, N., Grevillot, G. and Vallieres, C. (2009) 'Carbon Dioxide Capture and Recovery by Means of TSA and/or VSA', *International Journal of Greenhouse Gas Control*, 3(5), 519.

Tlili, N., Grevillot, G., Latifi, A. and Vallieres, C. (2012) 'Electrical Swing Adsorption Using New Mixed Matrix Adsorbents for CO_2 Capture and Recovery: Experiments and Modeling', *Industrial Engineering Chemistry Research*, 51, 15729–15737.

Tomlinson, T.R. and Finn, A.J. (1990) 'H_2 Recovery Processes Compared', *Oil & Gas Journal*, 88(3), 35–39.

Uehara, I. (2006a) *Hydrogen Separation and Handling, in Energy Carriers and Conversion Systems, in Encyclopedia of Life Support Systems (EOLSS)*. Ohta, T. (Editor), Developed under Auspices of UNESCO, Oxford, UK, Eolss Publishers.

Uehara, I. (2006b) *Separation and Purification of Hydrogen, in Energy Carriers and Conversion Systems, in Encyclopedia of Life Support Systems (EOLSS)*. Ohta, T. (Editor), Developed under Auspices of UNESCO, Oxford, UK, Eolss Publishers.

Ulbricht, M. (2006) 'Advanced Functional Polymer Membranes. *Polymer*, 47(7), 2217–2262.

Viano, D.M., Dolan, M.D., Weiss, F. and Adibhatla, A. (2015) 'Asymmetric Layered Vanadium Membranes for Hydrogen Separation', *Journal of Membrane Science*, 487, 83–89.

Vigneault, A., Elnashaie, S.S.E.H. and Grace, J.R. (2012) 'Simulation of a Compact Multichannel Membrane Reactor for the Production of Pure Hydrogen Via Steam Methane Reforming', *Chemical Engineering Technology*, 35, 1520–1533.

Wang, H., Dong, X. and Lin, Y.S. (2014) 'Highly Stable Bilayer MFI Zeolite Membranes for High Temperature Hydrogen Separation', *Journal of Membrane Science*, 450, 425–432.

Wang, H.T., Holmberg, B.A. and Yan, Y.S. (2002) 'Homogeneous Polymer-Zeolite Nanocomposite Membranes by Incorporating Dispersible Template-Removed Zeolite Nanocrystals', *Journal of Materials Chemistry*, 212(12), 3640–3643.

Ward, T.L. and Dao, T. (1999) 'Model of Hydrogen Permeation Behavior in Palladium Membranes', *Journal of Membrane Science*, 153, 211–231.

Wijenayake, S.N., Panapitiya, N.P., Nguyen, C.N., *et al.* (2014) 'Composite Membranes with a Highly Selective Polymer Skin for Hydrogen Separation', *Separation and Purification Technology*, 135, 190–198.

Yampolskii, Y. (2012) 'Polymeric Gas Separation Membranes', *Macromolecules*, 45(8), 3298–3311.

Yang, J. and Lee, C.-H. (1997) 'Separation of Hydrogen Mixtures by a Two-Bed Pressure Swing Adsorption Process Using Zeolite 5A', *Industrial & Engineering Chemistry Research*, 36, 2789–2798.

Yang, L., Wang, S.Z., Blinn, K., *et al.* (2009) 'Enhanced Sulfur and Coking Tolerance of a Mixed Ion Conductor for SOFCs: BaZr0.1Ce0.7Y0.2–xYbxO3-delta', *Science*, 326(5949), 126–129.

Yang, R.T. (2003) *Adsorbents. Fundamentals and Applications*, New Jersey, USA, John Wiley & Sons.

Yang, R.T. and Cen, P.L. (1986) 'Improved Pressure Swing Adsorption Processes for Gas Separation: By Heat Exchange between Adsorbers and by High-Heat-Capacity Inert Additives', *Industrial & Engineering Chemistry Process Design and Development*, 25(1), 54–59.

Yoshida, K., Hirano, Y., Fujii, H., Tsuro, T. and Asaeda, M. (2001) 'Hydrothermal Stability and Performance of Silica–Zirconia Membranes for Hydrogen Separation in Hydrothermal Condition', *Journal of Chemical Engineering of Japan*, 34(4), 523–530.

Yu, M., Noble, R.D. and Falconer, J.L. (2011) 'Zeolite Membranes: Microstructure Characterization and Permeation Mechanisms', *Accounts of Chemical Research*, 44(11), 1196–1206.

Yun, S. and Oyama, S.T. (2011) 'Correlations In Palladium Membranes for Hydrogen Separation: A Review', *Journal of Membrane Science*, 375, 28–45.

Zeman, L.J. and Zydney, A.L. (1996) *Microfiltration and Ultrafiltration*, New York, NY, Marcel-Dekker.

Zhang, G., Dorris, S., Balachandran, U. and Liu, M. (2003) 'Interfacial Resistances of Ni–BCY Mixed-Conducting Membranes for Hydrogen Separation', *Solid State Ionics*, 159, 121–134.

Zhu, Z., Sun, W., Wang, Z., Cao, J., Dong, Y. and Liu, W. (2015) 'A High Stability NieLa0.5Ce0.5O2D Asymmetrical Metal-Ceramic Membrane for Hydrogen Separation and Generation', *Journal of Power Sources*, 281, 417–424.

Zuo, C.D., Dorris, S.E., Balachandran, U. and Liu, M.L. (2006) 'Effect of Zr-Doping on the Chemical Stability and Hydrogen Permeation of the Ni–BaCe0.8Y0.2O3−δ Mixed Protonic-Electronic Conductor', *Chemistry of Materials*, 18, 4647–4650.

Zuo, C.D., Lee, T.H., Dorris, S.E., Balachandran, U. and Liu, M.L. (2006) 'Composite Ni–Ba (Zr0.1Ce0.7Y0.2)O3 Membrane for Hydrogen Separation', *Journal of Power Sources*, 159, 1291–1295.

Chapter 13

Multifunctional hybrid sorption-enhanced membrane reactor

P. Ribeirinha[1], M. Boaventura[1], José M. Sousa[1,2] and A. Mendes[1]

Abstract

The growth of the global hydrogen market demands more efficient industrial processes for its production. Hydrogen can be produced from renewable or nuclear sources, using electricity as an intermediate energy carrier. However, industrially is produced mainly by steam reforming of methane or other hydrocarbons and also by gasification of coal and oil refining residues. Methane steam reforming (MSR) is being used for decades, despite the severe operating conditions (high temperatures and pressures) and low-energy efficiency, which challenges the development of more efficient and reliable processes.

The present chapter provides an overview of hydrogen production via MSR, purification processes and procedures for enhancing the hydrogen production. Sorption-enhanced and membrane-enhanced reactors, considering selective CO_2 sorption removal from the reaction bulk and selective hydrogen membrane permeation are, respectively, addressed. Particular attention was paid to the recently proposed hybrid sorption-enhanced membrane reactor (HSEMR), in which sorption and permeation processes occur inside the reforming reactor. This technology allows lower operating temperatures, produces hydrogen with higher purity and exhibits higher reaction conversions than sorption or membrane reactors. The major contributions in this field are reviewed and the advantages and drawback of each approach are discussed in detail.

13.1 Introduction

Since the industrial revolution, the world energy consumption relied mostly on fossil fuels. The worldwide implementation of more restricted environmental

[1]Laboratório de Engenharia de Processos, Ambiente, Biotecnologia e Energia (LEPABE), Faculdade de Engenharia do Porto, Rua Roberto Frias, 4200-465 Porto, Portugal
[2]Departamento de Química, Escola de Ciências da Vida e do Ambiente, Universidade de Trás-os-Montes e Alto Douro, Apartado 1013, 5001-801 Vila Real Codex, Portugal

regulations during the last decades led to a growing effort to develop technologies that take advantage of renewable energy sources. New technologies, such as fuel cells, made hydrogen an attractive energy carrier, for both mobile and stationary applications.

The use of hydrogen in transport applications requires the construction of infrastructures for hydrogen production and delivery, analogous to the ones that exist nowadays for fossil fuels. Hydrogen production in situ by steam reforming of hydrocarbons is also being considered as a feasible alternative [1,2].

Presently, most of the hydrogen is produced by steam reforming of methane over supported nickel catalysts in packed-bed reactors. This hydrogen-rich stream is further upgraded through several steps: water gas shift (WGS) reaction, performed in two reactors at different temperatures – the lower temperature reactor allows to attain higher conversions since the WGS is an exothermal equilibrium limited reaction; the preferential oxidation (PROX, a process for decreasing the amount of CO by chemical oxidation); and the pressure swing adsorption unit (PSA, a cyclic adsorption purification process, which preferentially adsorbs all other components but hydrogen).

The overall methane steam reforming (MSR) process comprehends then several individual process units running at different temperatures and pressures, which influence negatively the overall efficiency [3,4]. As a result, several approaches have been attempted to develop a more efficient reforming process for hydrogen production. These include the use of membrane reactors [5,6], adsorptive reactors [7–9], hybrid adsorptive membrane reactor [10–12] and fluidized bed adsorptive membrane reactors [13,14]. These technologies present several benefits, namely: (i) increase of the reaction conversion by shifting the equilibrium towards the reaction products and increase of the reactants residence time due to selective removal of one or more reaction products; (ii) reduction of the downstream purification requirements by, for example, separation of the desired product in situ; (iii) reduction of the operating temperature.

Membrane reactors are already a well-developed technology, showing very good results for hydrogen production. Prove of that is the semi-industrial plant developed by KT – Kinetics Technology in Chieti Scalo (Italy), based on a membrane reforming reactor for hydrogen production with a capacity of 20 m^3 h^{-1}. Significant advances in membrane reactors were achieved as a result of using Pd and Pd alloys (Ag, Cu and Au) membranes. They present high selectivity towards hydrogen and high permeation at temperatures above 250 °C. But Pd-based membranes have a high cost, suffer from hydrogen embrittlement at low temperatures (lower if appropriate alloys are used) and deactivate in the presence of coke and sulphur [15]. Microporous membranes, although cheaper than dense Pd membranes, are difficult to prepare without defects and present low selectivity towards hydrogen. Adsorptive reactors, on the other hand, have been pointed as the best approach for producing a purified reformate rather than membrane reactors [16]. Typically the most used absorbents are based on calcium oxide [17], hydrotalcite [18], lithium zirconates [19] and lithium silicates [19]. Nevertheless, these sorbents have poor capacity to CO_2 at high temperatures and show progressive loss on their sorption capacity under consecutive adsorption/desorption cycles.

The merge of membrane and sorption reactors in a single unit has potentially significant advantages, namely:

1. one single step process for producing high purity hydrogen;
2. less active membrane area is required;
3. higher hydrogen permeation;
4. higher conversions;
5. lower operating temperatures;
6. environment friendlier due to CO_2 capture.

This chapter addresses the study of the new field of multifunctional reactors, which combines membrane and sorption reactors in a single unit named hybrid sorption-enhanced membrane reactor (HSEMR), applied to the hydrogen production.

13.2   Hydrogen production via methane steam reforming and purification

The process for converting hydrocarbons into hydrogen in the presence of steam was first described in 1868 by Tessie du Motay and Marechal [20]. The catalytic process was later developed in the first quarter of the twentieth century by BASF and was implemented for the first time in 1931 in Baton Rouge by Standard Oil of New Jersey (Exxon). The feedstock for this process included methane, naphtha and fuel oil [21]. Depending on the feedstock, the fraction of hydrogen provided the steam changes, being 50% and 89% when methane and coal were used as feed, respectively [22]. Methane, from natural gas, is the most extensively used fossil fuel to produce hydrogen, due to its availability and low price.

13.2.1   Methane steam reforming process

The current MSR is a multiple step process with severe operating conditions, which includes steam reforming, WGS and H_2 post-process purification (Figure 13.1). In addition, and not represented in Figure 13.1, before being fed to the reforming reactor, the natural gas is desulfurized and the steam pre-heated.

The reforming reaction occurs typically over supported nickel catalysts in a packed-bed reactor and can follow two different reactions, depending on the steam availability:

$$CH_4 + H_2O \leftrightarrow CO + 3H_2 \qquad \Delta H^{\circ}_{298} = 206.2 \text{ kJ mol}^{-1} \tag{13.1}$$

$$CH_4 + 2H_2O \leftrightarrow CO_2 + 4H_2 \qquad \Delta H^{\circ}_{298} = 165 \text{ kJ mol}^{-1} \tag{13.2}$$

The MSR is a highly endothermic reaction that normally occurs between 800 and 1,000 °C. The high temperature is achieved by burning additional natural gas.

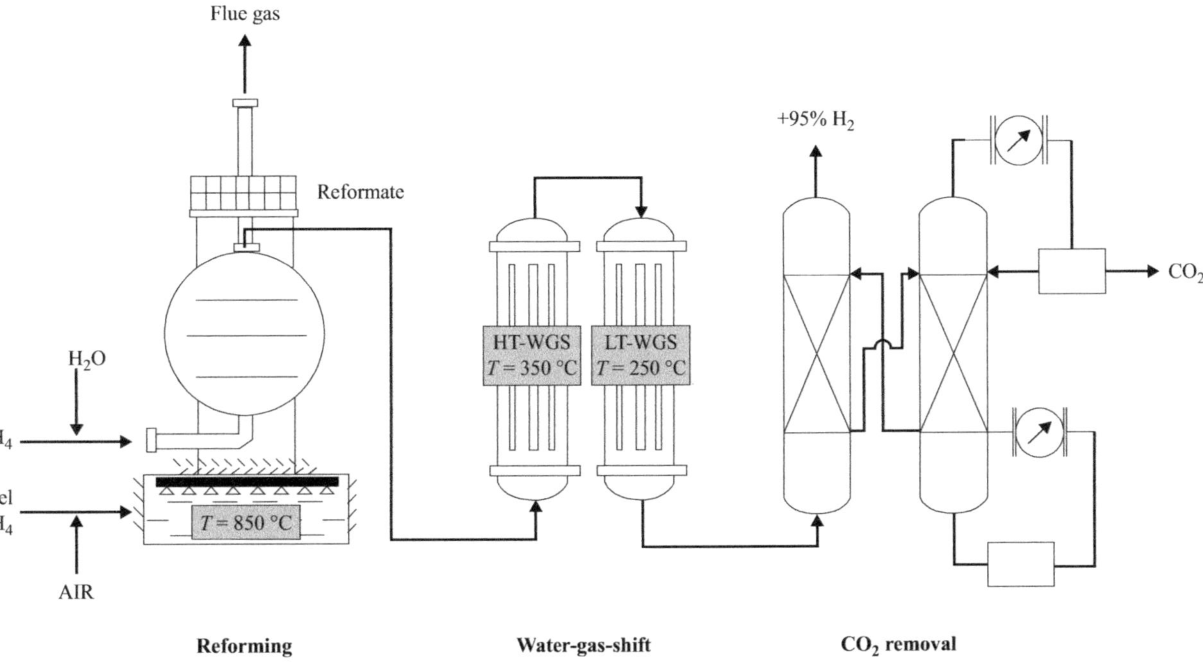

Figure 13.1   Diagram of a conventional methane steam reforming process

The natural gas and steam are fed at pressures between 14 and 20 bar with steam to carbon ratios (S/C) of 1.2–2 [23]. The molar composition of the dry reformate is 76% in H_2, 1.3% in CH_4, 12% in CO and 10% in CO_2 [23]. The second step is the exothermic reaction of carbon monoxide oxidation in the presence of steam, known as WGS, (13.3).

$$CO + H_2O \leftrightarrow CO_2 + H_2 \quad \Delta H^\circ_{298} = -41.2\,\mathrm{kJ\,mol^{-1}} \tag{13.3}$$

The WGS reaction is favoured at lower temperatures and normally is carried out in two steps, using fixed-bed adiabatic reactors (Figure 13.1). The first step runs at higher temperatures, *ca.* 350 °C, over chromium promoted iron oxide catalyst; the second step is conducted at *ca.* 250 °C over a $CuO/ZnO/Al_2O_3$ catalyst. The product molar composition after the WGS reaction is 86% in H_2, 1.3% in CH_4, 0.25% in CO and 12% in CO_2 [23].

The last step of the hydrogen production process is the H_2 purification, in which several techniques can be implemented, depending on the final application or/and needed purity. Polymer electrolyte membrane fuel cell for automotive applications requires a hydrogen stream with CO concentration lower than 0.2 ppm, according to the International Organization for Standardization (ISO 14687) [24]. To obtain such hydrogen purity, methanation or PROX reactions are required, followed by a PSA or amine scrubbing process to remove CO_2.

13.2.2   Methane steam reforming kinetic model

Extensive studies have been made concerning the reaction mechanism and reaction rate of MSR [3,25–28]. The first kinetic model proposed for the MSR was developed by Akers *et al.* [25] in 1955, but the most complete and accepted model was proposed by Xu and Froment [3,28]. This model comprehends two endothermic reforming reactions (see (13.1) and (13.2)), and the exothermic WGS reaction (see (13.3)); all kinetic equations are equilibrium limited.

$$r_1 = \left(\frac{k_1}{p_{H_2}^{2.5}}\right) \frac{(p_{CH_4}p_{H_2} - (p_{H_2}^3 p_{CO}/K_{eq1}))}{DEN^2}, \qquad K_{eq1} = e^{30.114-(26,830/T)} \tag{13.4}$$

$$r_2 = \left(\frac{k_2}{p_{H_2}^{3.5}}\right) \frac{(p_{CH_4}p_{H_2O}^2 - (p_{H_2}^4 p_{CO_2}/K_{eq2}))}{DEN^2}, \qquad K_{eq2} = e^{-4.036+(4,400/T)} \tag{13.5}$$

$$r_3 = \left(\frac{k_3}{p_{H_2}}\right) \frac{(p_{CO}p_{H_2O} - (p_{H_2}p_{CO_2}/K_{eq3}))}{DEN^2}, \qquad K_{eq3} = K_{eq1} \times K_{eq2} \tag{13.6}$$

$$DEN = 1 + K_{CO}p_{CO} + K_{H_2}p_{H_2} + K_{CH_4}p_{CH_4} + \frac{K_{H_2O}p_{H_2O}}{p_{H_2}} \tag{13.7}$$

where r_i ($i = 1, 2, 3$) are the reaction rates of reactions (13.1), (13.2) and (13.3), respectively; p_j ($j = CH_4$, H_2O, H_2, CO, CO_2) are the partial pressures of species j; k_i ($i = 1, 2, 3$) are the reaction rate constants at a defined temperature; $K_{eq,i}$ ($i = 1, 2, 3$) are the equilibrium constants for the reactions described by (13.1)–(13.3); K_j ($j = CH_4$, H_2O, H_2, CO, CO_2) are the adsorption constants of species j at a defined temperature.

The increase of the hydrogen partial pressure has a negative and significant impact on the reaction rates r_1 and r_2 due to the high exponent of the hydrogen partial pressure. The selective removal of hydrogen increases significantly the reaction rate and consequently the methane conversion. The CO_2 removal has also a positive effect in the reaction rates r_1 and r_2 but lower than hydrogen removal due the lower exponent of the carbon dioxide. Formation rates of H_2, CO and CO_2 products and disappearance rates of CH_4 and H_2O are given by the following equations:

$$r_{H_2} = 3r_1 + 4r_2 + r_3 \tag{13.8}$$

$$r_{CO} = r_1 - r_3 \tag{13.9}$$

$$r_{CO_2} = r_2 + r_3 \tag{13.10}$$

$$r_{CH_4} = -r_1 - r_2 \tag{13.11}$$

$$r_{H_2O} = -r_1 - 2r_2 - r_3 \tag{13.12}$$

13.2.3   PSA principles (hydrogen purification)

PSA is nowadays the most used process for hydrogen purification from the reformate stream of the methane reformation process [29,30], taking advantage of hydrogen being less adsorbed when compared to other species. The first PSA for hydrogen purification was built in the late 1960s with three- or four-bed units with relative modest performance ($\sim$70%) [30]. The main applications of PSA technology are the recovery and purification of hydrogen, carbon dioxide removal and purification, methane purification as well as nitrogen and oxygen production. Moreover, the PSA principles can also be applied in a sorption-enhanced reactor or a sorption-enhanced membrane reactor as it will be discussed later.

In all adsorption-based separation processes, the essential requirement of a PSA is an adsorbent that preferentially adsorbs one component (or one family of related components) from a mixture [29]. This selectivity may depend on a difference in adsorption equilibrium or in sorption kinetics. In the case of hydrogen purification, the most used adsorbents are activated carbons and zeolites that perform an equilibrium-based separation [31]. The PSA unit has two or more adsorption beds, as depicted in Figure 13.2, commonly loaded with a layer of zeolites, activated carbon, carbon molecular sieve (CMS), alumina or silica [31]. As all impurities adsorb more than hydrogen, during the production step it emerges purified from the adsorption column.

The operating principles of an equilibrium-based PSA unit are rather simple. Inside of one PSA column bed, at the high operating pressure occurs a selective adsorption of one or more components of the gas feed mixture (adsorption step), so a current enriched in the desired product – the less adsorbed one – leaves the

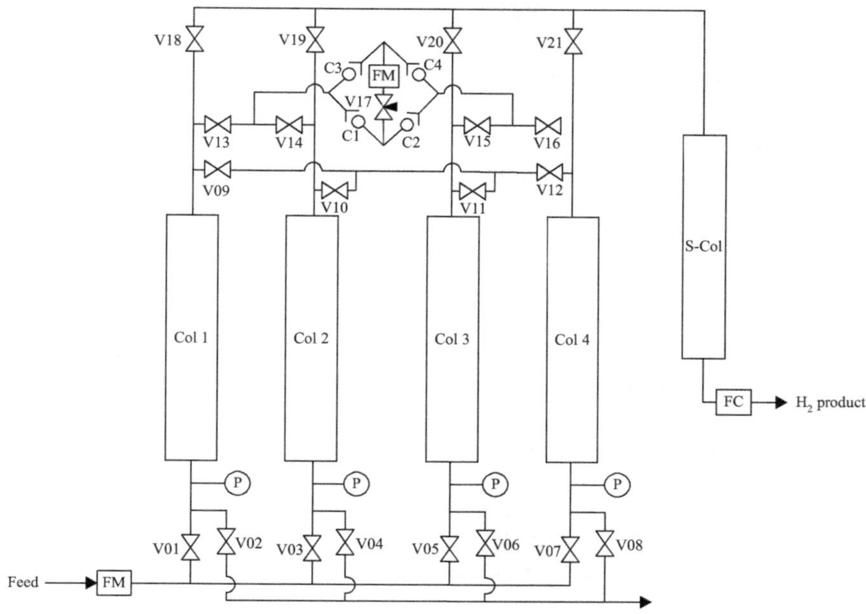

Figure 13.2   Diagram of an equilibrium-based PSA unit

column. The adsorption step is interrupted before the mass-transfer zone reaches the end of the column to avoid saturation of adsorbent and consequent contamination of the product. Forthwith, the column is depressurized to a lower operating pressure (desorption step) and the species adsorbed desorb, thus partially regenerating the adsorbent for use in the next cycle. Following, a fraction of the raffinate product is used to counter-currently purge the depressurized column (purge step), which further regenerates. It is possible to obtain useful products from either the adsorption and desorption steps or from both. Finally, the column is pressurized again, for the adsorption step. It can be pressurized firstly, with a stream from another column (equalization step), when the high and low pressure beds are either connected through their product ends or feed ends, or both. It can also be pressurized at the product end with part of the product flowing counter-currently (backfill step). Instead, or after the equalization and backfill steps, the column is further pressurized with feed stream through the column end until the high adsorption pressure is reached. This cycle is repeated in a time interval pre-set, as a PSA unit goes under cyclic operation, repeating the steps for which is set up. The continuous regime is achieved using multiple adsorption beds operating in lagged mode. The definition of the optimum cycle must be substantiated by previous studies. The processes differ from one to another in the sequence of the elementary steps and in the way these steps are carried out, as well as by the adsorbent (or adsorbents) used in the process.

In short, the most common PSA steps are follows:

1. pressurization with equalization;
2. feed pressurization;
3. adsorption (production);
4. depressurization with equalization;
5. depressurization;
6. counter-current purge.

All beds undergo these steps and the sequence is phased in such a way that a quasi-continuous flow of product in maintained.

A PSA unit can have two, four or more beds, so-called polybed, which includes seven to ten beds and three pressure equalization steps. The operating principle is mainly the same but more beds allow higher recoveries and higher purities. However, the operational and implementation costs increase considerably, so a balance between performance and cost must be carried out. For optimized conditions, a 4-adsorption column PSA unit, the hydrogen purity can reach 99.99% with a recovery of 70%–80%, whereas for a polybed process the hydrogen purities can achieve 99.9999% with a recovery of 85%–90% [29].

13.3   Enhanced methane steam reforming processes

As mentioned before, MSR has been carried out for decades in packed-bed reactors, though the process raised several concerns since the implementation at the industrial scale. However, the high operating temperature (800–1,000 °C) and pressure (14–20 bar) are severe operating conditions for the materials involved in the process and energetically inefficient [32]. The heat transfer is typically a limitation in packed bed reactors due to their low thermal conductivities [33]. Therefore, the heat transfer from the metallic wall of the reactor to the gas phase and catalyst pellets is a slow process. This condition gains extra-importance for highly endothermic reactions, such as the MSR reaction. To avoid thermal profiles inside the reactor, long narrow tubes made of a super alloy are used. Coke formation is one of many concerns related to this reaction. It may cause the blockage of the flow passage increasing the pressure drop; the coke deposit on the catalyst active surface deactivates it reducing the methane conversion and originating hotspots that overheat the metal tube eventually leading the plant to shut down. To avoid coke formation, alkali promoters such as KOH are added to the catalyst to increase its resistance to coking formation [34].

New and reliable solutions with reduced overall costs are then desirable. Within the conventional process, several approaches have been attempted, namely improving the catalyst performance, enhancing the properties of the reactor tubes to withstand higher stresses at high temperatures and thermal flux and finally improving the reformer configuration. Improving the catalyst properties is a constant seek for all catalytic reactions, although for MSR small advancements in the catalyst development that fits the industrial requirements is observed. The improvement of the catalyst performance has been focused in increasing the activity and consequently lowering the operating MSR temperatures, increasing the mechanical strength, improving the resistance to carbon formation and sulphur poisoning and improving the catalyst effectiveness by optimizing the pellet configuration [32]. The most important modifications suggested for the reforming reactor concern the use of a fluidized-bed reactor instead of the conventional fixed-bed reactor, use of membrane and adsorptive reactors and changing from external firing to internal heat supply. In the following section, the implementation of adsorptive, membrane and hybrid sorption membrane reactor are presented.

13.3.1   Sorption-enhanced reactor

The use of sorbents in hydrocarbon steam reforming was firstly suggested by Tessie du Motay and Marechal in 1868 [20], but the use of sorbents was only described in a patent by Gluud *et al.* [35] for MSR and later, in 1963, used by Gorin and Retallick [36] in a patent involving a fluidized-bed reactor containing both catalyst and a carbon dioxide acceptor. Brun-Tsekhovoi *et al.* [37] reported energy savings of about 20% with catalytic steam-reforming of hydrocarbons in the presence of a carbon dioxide acceptor compared with the conventional processes. The process involves two reactors, a primary for the reaction and a secondary for sorbent regeneration. In the primary reactor, loaded with Ni-based reforming catalyst and Ca-based CO_2 sorbent, the MSR, WGS and CO_2 removal occur simultaneously. The sorbent particles are regenerated in the secondary reactor and continuously fed to primary reactor counter-currently to the reactant-gas flow. The size and density differences between catalyst and sorbent allows their separation at the reactor inlet, and then the sorbent particles are carried pneumatically to the secondary reactor.

In the last few years, several new reports concerning the use of sorbents for hydrocarbons steam reforming have been published [37–40]. For this type of application the sorbent must have the following characteristics:

- stability during the production/regeneration cycles;
- low-temperature interval between sorption and desorption steps;
- low cost;
- high sorption capacity;
- CO_2 selective sorption.

Sorbents such as calcium carbonate and dolomite have high sorption capacity (more accurately chemisorption capacity) and are relatively cheap. The regeneration process of these sorbents named 'calcination' (described by (13.13)) occurs at very high temperature causing sintering and the consequent decay in the capacity upon multiple re-calcinations. Moreover, kinetics of adsorption process after several cycles of carbonation/calcination becomes extremely slow.

$$CaO(s) + CO_2(g) \leftrightarrow CaCO_3(s) \quad \Delta H^{\circ}_{298} = -178\,kJ\,mol^{-1} \tag{13.13}$$

Other sorbents such as hydrotalcite are also considered for this application. Hydrotalcite is an anionic clay with positively charged layers of metal oxide (or metal hydroxide) with anionic inter-layers typically carbonates. Hydrotalcite has fast sorption/desorption kinetics that can be carried out at temperatures around 400 °C, it is quite stable for long-term operation but shows a low adsorption capacity, *ca.* 0.3–0.45 mmol g^{-1}, restricting its potential at industrial scale [40].

Besides MSR, chemical acceptors or sorbents can also be employed for enhancing other reactions. Han and Harrison [41] studied the hydrogen production via WGS reaction using CaO as CO_2 sorbent in a tubular reactor. The results showed CO conversions above the thermodynamic equilibrium, based on the feed conditions. Goto *et al.* [42] reported the use of CaNi$_5$ alloy as hydrogen acceptor in dehydrogenation of cyclohexane over a Pt-alumina catalyst. Carvil *et al.* [43] were able to reduce to operating temperature of the CO production via reverse WGS by using NaX zeolite as water adsorbent.

Sircar *et al.* [40,43,44] proposed the sorption-enhanced reaction process (SERP) for MSR, which involves pressure and concentration swing adsorption for reaction enhancement; these authors used a hydrotalcite-based CO_2 adsorbent and a commercial Ni-based catalyst. For a CH_4 conversion of 82%, the operating temperature was decreased from 650 °C of a conventional SMR reactor to 450 °C using the SERP. Johnsen *et al.* [45] also assessed the SERP in an atmospheric-pressure bubbling fluidized bed reactor obtaining a product stream with a hydrogen concentration >98% on a dry basis at 600 °C and 100 kPa with dolomite as the CO_2 acceptor, Figure 13.3.

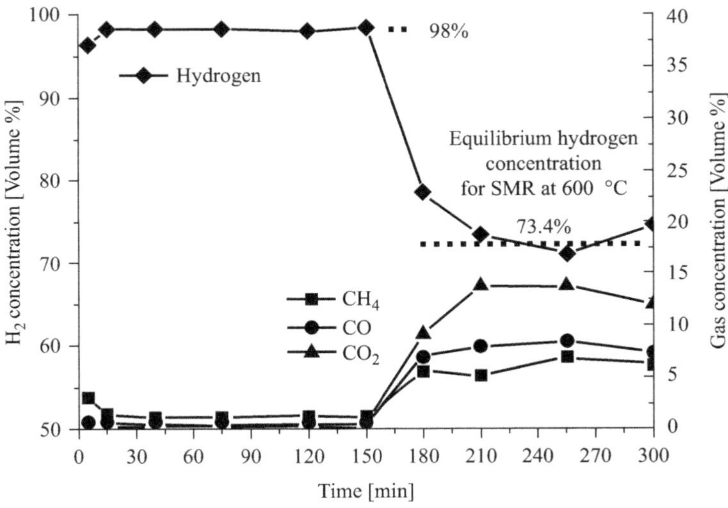

Figure 13.3   Outlet composition (dry basis) as a function of time at 600 °C. Reprinted, with permission, from Reference 45

The sorption-enhanced reforming process for H_2 production has several potential advantages but also a few drawbacks when compared to the conventional reforming process. The five steps required in a conventional steam reforming process (reforming, two shift steps, H_2 purification and steam stripping) are reduced to two steps, one reaction/adsorption step and one sorbent regeneration step. Sorption enhancement allows lower operating temperatures, which reduces catalyst coking and sintering, saves energy and uses less expensive materials to manufacture the reactor. The heat released by the exothermic carbonation reaction and WGS nearly compensates the heat required for the endothermic reforming reaction. Though the sorbent being regenerated by calcination to its oxide form at high temperatures, the sorption-enhanced reforming process uses less *ca.* 20%–25% of energy than the conventional process. In a sorption-enhanced reforming process based on the PSA [40,43,44] the regeneration of the adsorbent (counter-current purge) is performed with a mixture of 5%–10% H_2 balanced with steam to desorb the CO_2. The vented gas mixture (mostly CO_2 but also CH_4, H_2 and H_2O) can be used as fuel after removing the steam. Because the reaction temperature in a sorption enhanced reforming process is lower than in the conventional process, in the range of 450–650 °C, CO has nearly complete conversion. Sorption enhanced reforming can produce hydrogen with high purity, *ca.* 90%–95% [40], but most of the hydrogen applications require hydrogen purities of at least 99.99%. Thus, a PSA unit is required for extra purification, which implies hydrogen losses of *ca.* 15% and energy consumption in pumping [40]. Other concerns relates to the adsorption capacity and the adsorbent chemical stability during carbonation/calcination cycles.

13.3.2　Membrane-enhanced reactor

The membrane reactors technology has been successfully applied to a wide range of reactions [15,46] such as: dehydrogenation of ethane, cyclohexane, ethylbenzene and acetylene, CO production via the SMR and WGS reactions and H_2 production via hydrocarbon steam reforming and WGS reactions.

Membranes for hydrogen production via steam reforming are typically based on palladium or palladium alloys. What makes these membranes interesting is their extremely high selectivity to hydrogen and very high permeability. Microporous membranes are also used in steam reforming application for hydrogen removal from the reaction bulk; they are cheaper than Pd membranes, but difficult to prepare without defects and present low selectivity towards hydrogen [47].

Porous membranes filled with ionic liquids have also been suggested for hydrogen production via low temperature methanol steam reforming [15]. These membranes have the particularity of being selective to CO_2, allowing air to be used as sweep gas and then avoiding the usage of vacuum pumps. Although ionic liquid membranes are still in an early stage for steam reforming applications, they represent a very promising technology.

Pd-based membrane reactors applied to MSR can produce pure hydrogen and enhance the methane conversion. This enhancement is attributed to the partially suppression of the MSR backward reaction and the increase of the residence time due to the partial removal of hydrogen from the reaction bulk. The membrane enhanced steam reforming has the advantage of incorporating in a single step the five steps required in a conventional steam reforming. The new technology developed by Praxair and Argonne National Laboratory [48] is a good example of the membrane reactor capabilities for industrial applications. For example, in a process disclosed by Praxair, syngas is produced from the partial oxidation of natural gas with oxygen supplied using an ionic conducting solid oxide membrane and a suitable catalyst. A second membrane, palladium-based, is used to selectively remove hydrogen.

The main disadvantages of using a membrane reactor based on Pd membranes are the irreversible damage caused by impurities to the membrane, mainly CO. Thus, in many cases, Pd membranes are considered only for H_2 purification after reforming and CO removal. Several challenges such as membrane fouling, thermal and mechanical stability, hydrogen embrittlement, and the energy for driving the hydrogen permeation (either using steam at the permeate side or high vacuum) are limiting the implementation of membrane reactors at the industrial scale.

13.3.3   *Hybrid sorption-enhanced membrane reactor*

The HSEMR incorporates in the reforming reactor a sorption and permeation processes. The concept is quite recent and was firstly suggested by Park and Tsotsis [10] (2004) for the esterification of acetic acid with ethanol. In this process, acetic acid and ethanol are fed to the PBMR, where the esterification reaction is catalysed by sulphuric acid producing ethyl acetate and water. A polyetherimide membrane supported on a porous alumina permeates the ethyl acetate and water; anhydrous $CaSO_4$ placed in retentate side adsorbs the water, while the ethyl acetate was extracted using vacuum pump. Integrating reaction, pervaporation, and adsorption processes in one single step significantly improved the performance and reduced the operating temperature. In the same year, Park [49] suggested the use of HSEMR concept for the hydrogen production (Figure 13.4). The hydrogen production from MSR was modelled using a HSEMR considering the adsorbent inserted in permeate side or mixed with the catalyst in retentate side. The author considered a continuous operation with a single HSEMR introducing a continuous pulse-type or a sinusoidal type sweep gas in permeate-side of the HSEMR.

These authors claim that the HSEMR technology can be more advantageous than membrane reactors or adsorptive reactors in terms of reagents conversion, selectivity and operating temperature. In addition, this technology is more flexible than other processes; for example, the adsorbent can be placed in the retentate side mixed with the catalyst or in the permeate side using the membrane as a barrier to separate the catalyst from the adsorbent, allowing regeneration in situ of the adsorbent which can be achieved by pressure or thermal (calcination) swing adsorption (PSA or TSA) processes. Since the reaction and purification are carried out in a single step, it is expected an efficiency increase and a cost decrease.

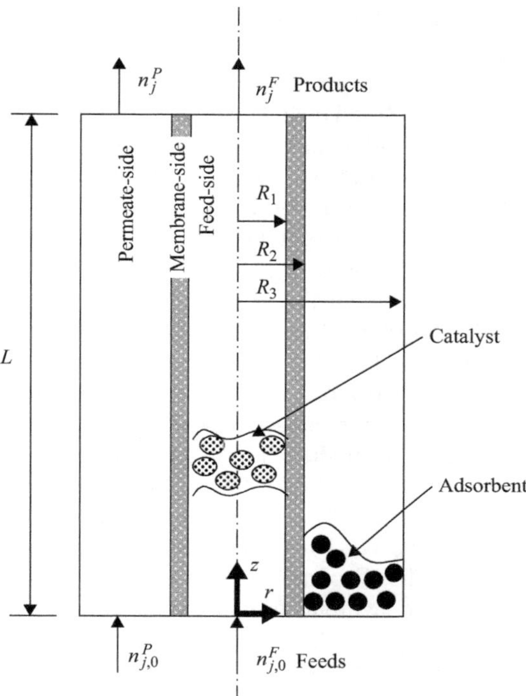

Figure 13.4   Hybrid sorbent-membrane reactor. Reprinted, with permission, from Reference 49

The HSEMR technology is still in its early stage, but it has been already considered for several applications such as, hydrogen production [11,12,16,50–52], CO_2 capturing [53] and water treatment [54], showing always very promise results. In the field of hydrogen production, the main advances have been made by Tsotsis *et al.* [11,12,52] and research efforts on a fluidized bed sorption-enhanced membrane reactor is undergoing at the University of British Columbia in collaboration with Tokyo Gas Company [13,14,55]. It was experimentally demonstrated the sorption-enhanced membrane reactor for low-temperature WGS reaction [11], using commercial layered double hydroxides (MG 50 Sasol) for high capacity high temperature CO_2 adsorption; the sorbent regeneration was accomplished using a purge gas. The concept considers the sorbent mixed with a $CuO/ZnO/Al_2O_3$ commercial catalyst placed on the retentate side, and in another embodiment the sorbent was placed in the permeate side and the catalyst in the retentate side. A CMS membrane presenting high H_2 selectivity was used for hydrogen separation in situ. The HSEMR was investigated for a range of temperatures, pressures, and other experimental conditions showing potential interest to produce a purified hydrogen stream. In this study [11], it was evaluated experimentally the ability of the membrane, catalyst and adsorbent to run a cyclic operation in a four-cycle experiment. The results showed that the HSEMR can operate smoothly without any apparent performance degradation.

A four-bed HSEMR system for running the WGS reaction was designed on the basis of the principles of a PSA and then modelled and simulated [52]. The reaction mixture of CO, H_2 and H_2O with a molar ratio of (1:4:1.1) was fed at a pressure of 3 bar, and the reaction was carried at 250 °C. Changing the space–time-ratio and the reaction cycle duration, the operating conditions that originate the highest CO conversions were assessed. For desorption (purge) step air was used, but this is not a practical solution, as Cu-based catalyst oxidizes getting immediately deactivated. Nevertheless, it was shown that this technology can deliver a hydrogen stream ready to use directly in PEM fuel cells without the need for any additional downstream. Tsotsis *et al.* [12] also simulated the use of a HSEMR for the SMR process. They considered a fixed bed reactor consisting of two concentric tubes separated by a membrane with the outer (feed) shell containing the SMR catalyst and the CO_2 sorbent and the inner shell carrying the sweep gas and permeated H_2.

A more detailed application of HSEMR for WGS reaction was carried experimentally by Soria *et al.* [50]. These authors used a HSEMR to produce hydrogen via low temperature WGS reaction combining both CO_2 and H_2 removal from the reaction bulk. They inserted the catalyst ($CuO/ZnO/Al_2O_3$) and the adsorbent (K_2CO_3-promoted hydrotalcite) in the inner side of a self-supported Pd–Ag membrane tube, Figure 13.5. They compared the performance of the HSEMR with that obtained with a traditional and a sorption-enhanced reactor (only CO_2 is removed) operating in the same operational conditions. The performance of the HSEMR showed to be clearly higher than the traditional WGS or of the sorption-enhanced WGS (Figure 13.6), allowing to overcome the equilibrium limitations and producing a highly pure hydrogen stream; in the pre-breakthrough zone full CO conversion was obtained at 5.5 bar and 300 °C. This suggests that combining CO_2 and H_2 removal in a single unit allows producing a CO-free high-purity hydrogen stream. Actually, the authors observed that the stream is CO and CO_2 free in the pre-breakthrough zone, as the CO_2 produced was completely sorbed on the hydrotalcite, and the CO was completely converted. This result is favourable to the Pd–Ag membrane stability, as CO_2 and mainly CO are responsible for the membrane poisoning. The potassium-promoted hydrotalcite in breakthrough experiments showed an increase in the CO_2 sorption capacity in the presence of water vapour. After several sorption–desorption cycles with a dry stream, the sorbent capacity decreased, which should be partially compensated if the feed stream is humidified. The adsorbent regeneration was accomplished using a nitrogen stream.

13.3.3.1 Hybrid sorption-enhanced membrane reactor with a fluidized bed

The University of British Columbia in collaboration with the Tokyo Gas Company achieved one of the most important practical results by building a fluidized bed reactor pilot for hydrogen production combining a hydrogen selective membrane separation with CO_2 selective adsorbent [55]. A reformer comprising a fluidized

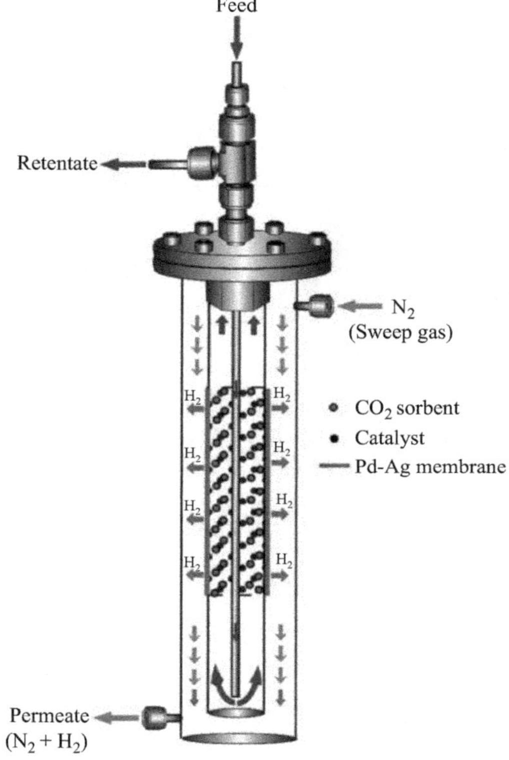

Figure 13.5   Scheme of hybrid sorbent-membrane reactor. Reprinted, with permission, from Reference 50

bed membrane reactor (FBMR) and a calciner (regenerator) was conceived [55]. The FBMR consisted of a 2 m long stainless steel vessel with rectangular cross-sectional area of 48.4 cm^2, able to hold up to four double-sided membrane modules, each with a nominal permeation area of 300 cm^2. No sweep gas was introduced into the permeate side, and the permeation was assured using a vacuum pump. In their experiments, the authors used a mixture of solid particles composed of limestone (Strassburg) and alumina Ni-based SMR commercial catalyst. Experimental tests pointed out that the screened limestone/catalyst fluidized in a little extension and was prone to channelling. To overcome these issues, significant amounts of alumina were added to the solid mixture making, on the other hand, the CO_2 sorption capacity of the FBMR to decrease. The bed was operated within the bubbling fluidization flow regime at much lower temperature (550 °C) than in conventional reformers (850 °C), using a steam-to-carbon molar ratio of 3.0, reaction pressure of 300 kPa and a permeate pressure of 30 kPa. The temperature in the FBMR was assured by two electric heaters, one internal and the other external to the reactor.

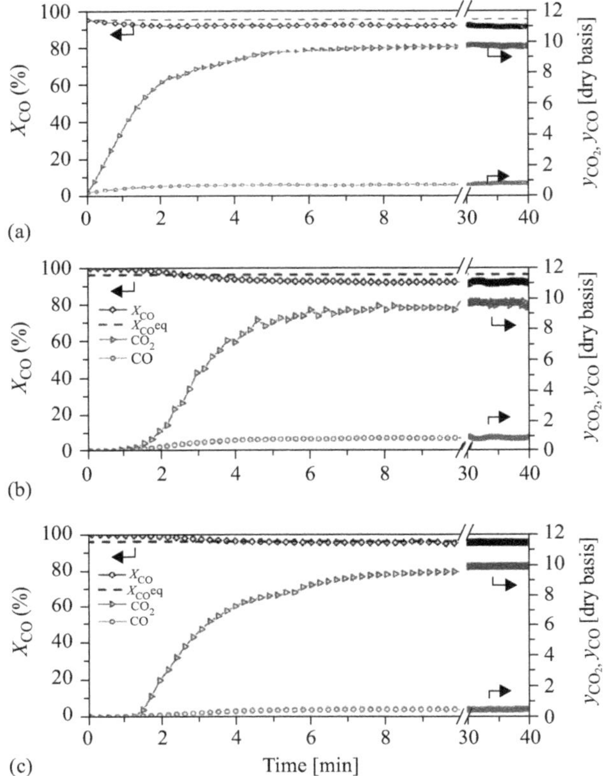

Figure 13.6   *CO conversion and CO and CO_2 composition (vol. per cent) history during WGS at 300 °C and 3 bar in a traditional reactor (a); sorption-enhanced reactor (b); hybrid sorbent-membrane reactor (c). Reprinted, with permission, from Reference 50*

Although the carbonation reaction is exothermic and in terms of heat management could provide enough heat to sustain the endothermic MSR reaction (see (13.2)), the large surface area to volume ratio of the reactor originated considerable heat losses forcing the introduction of an additional heating. The membranes used intended to shift the equilibrium in the forward direction (see (13.1) and (13.2)), to enhance hydrogen production and also purifying the product in order to obtain ultra-pure hydrogen for fuel cell applications. Each membrane panel had an active area of 300 cm^2 (Figure 13.7), and it was developed by Membrane Reactor Technologies, Ltd. The panels were made of a 25 μm thick PdAg25 foil bonded on both sides to a stainless steel substrate. The membrane foil was supported in a porous metal plate that could tolerate transmembrane pressures of *ca.* 2.5 MPa. To assure nearly infinite selectivity, the authors used considerably thick membranes, guaranteeing that only atomic hydrogen can diffuse through the metal foil.

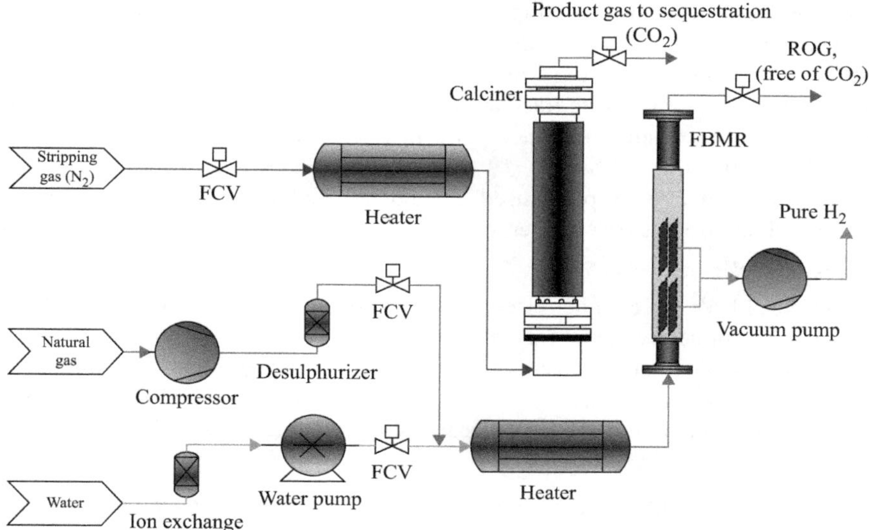

*Figure 13.7   Schematic of a fluidized-bed membrane reformer with CO$_2$
separation and the regenerator. Reprinted, with permission, from
Reference 55*

The unit also included a calciner with a nominal volume of 33 dm^3 used for
regenerating the adsorbent from CaCO$_3$ back to CaO. The CO$_2$ produced in the
FBMR is immediately adsorbed by the CaO particles (sorbent), producing CaCO$_3$
particles, which are then transferred to the calciner. The reactors were operated
batch wise, but continuous (or periodic) sorbent transfer between the reformer and
the regenerator is required for the industrial implementation of the process. CaCO$_3$
particles in the regenerator were heated up to 850 °C (supplied by a 17 kW external
electric heater) at 0.2 MPa using nitrogen as the stripping/fluidization gas to release
CO$_2$. The drop of the CO$_2$ concentration on the regenerator off-gas was used as
an indication of the calcination process end. To determine the CO$_2$ sorption
capacity, the authors performed preliminary calcination experiments with fresh
Strassburg limestone at 850 °C and 0.2 MPa using a nitrogen stream and obtained a
weight loss of 25% compared with a full calcination weight loss of 44%. It was
concluded that the limestone could adsorb 0.31 kgCO$_2$ kgCaO^{-1} (0.39 molCO$_2$
molCaO^{-1}). However, cyclic calcination/carbonation experimental tests showed a
rapid decrease of the CO$_2$ sorption capacity to values of 10% to 25% of the initial
capacity.

The pilot plant (Figure 13.7) produced a hydrogen stream of +99.99% in
the permeate side with a hydrogen-permeate-to-feed methane molar ratio of 1.9,
which gives a hydrogen recovery of *ca.* 50%. According to the authors, the
hydrogen recovery could increase if additional membrane area is installed or by
purifying the reactor off-gas. The maximum CO$_2$ adsorption capacity reported was

0.19 mol CO_2 mol CaO^{-1} during the carbonation steps, with a carbon capture efficiency of 87%. This CO_2 sorption capacity is lower than the obtained for initial tests with the fresh Strassburg limestone adsorbent. The batch-wise sorbent was able to provide a stream with trace amounts of CO and CO_2 during up to 30 min, indicating the maximum residence time in a continuous operation mode with recirculation of solids (sorbent).

The developed pilot plant used a fluidized bed instead of a packed bed reactor, which despite the higher complexity offers several potential benefits, namely [14,56]:

1. fluidized beds have higher effective heat transfer coefficients than fixed beds, operating in nearly isothermal conditions;
2. fluidized beds impose considerably lower pressure drops than fixed beds;
3. small catalyst particles, *ca.* 100 μm, can be used in the fluidized bed, eliminating internal diffusion resistances and leading to catalyst effectiveness factors close to the unity.

The possibility of continuous operation with recirculation of solids in the pilot scale was studied experimentally and by simulation [55] for assessing the industrial viability of this new reactor (Figure 13.8). The CaO adsorbent particles were removed from the top of the regenerator using a standpipe inserted through the bottom of the reformer and using a fluidization gas to assist the solids transfer. In the reformer, the calcium particles showed an upward movement and were removed through a flange placed in an upper side position. The regenerator was placed in a lower position, allowing a gravitational driving force, and a nitrogen stream was used to assist the $CaCO_3$ transfer. Using clear polymer tubes to make the connections, the authors observed reproducible recirculation from the regenerator to the reformer through the *J*-valve, but reverse flow from the reformer to the regenerator was occasionally observed. They concluded that the solids circulation was a very challenging task for the current configuration.

The energy efficiency for an industrial scale plant with recirculation of solids was estimated by simulation. The overall process thermal efficiency was defined as the ratio of total heating value of product hydrogen to the higher heat value of methane feed and process net energy input without considering the heat losses. The results showed a thermal efficiency of 78% [55], highlighting the potential of this technology to produce high-purity hydrogen and a separate CO_2 stream at high-energy efficiencies.

13.3.3.2 CO_2/H_2 active membrane piston reactor with CO_2 adsorption (CHAMP-SORB)

Despite the great advances made by University of British Columbia and Tokyo Gas Company together, the pilot reformer still operates at rather high temperatures (550 °C) and utilized an excess of steam (S/C = 3) to achieve high methane conversion. Anderson *et al.* [51,57] used the concept of CO_2/H_2 active membrane piston (CHAMP) reactor developed for on-vehicle methanol reforming [58,59]

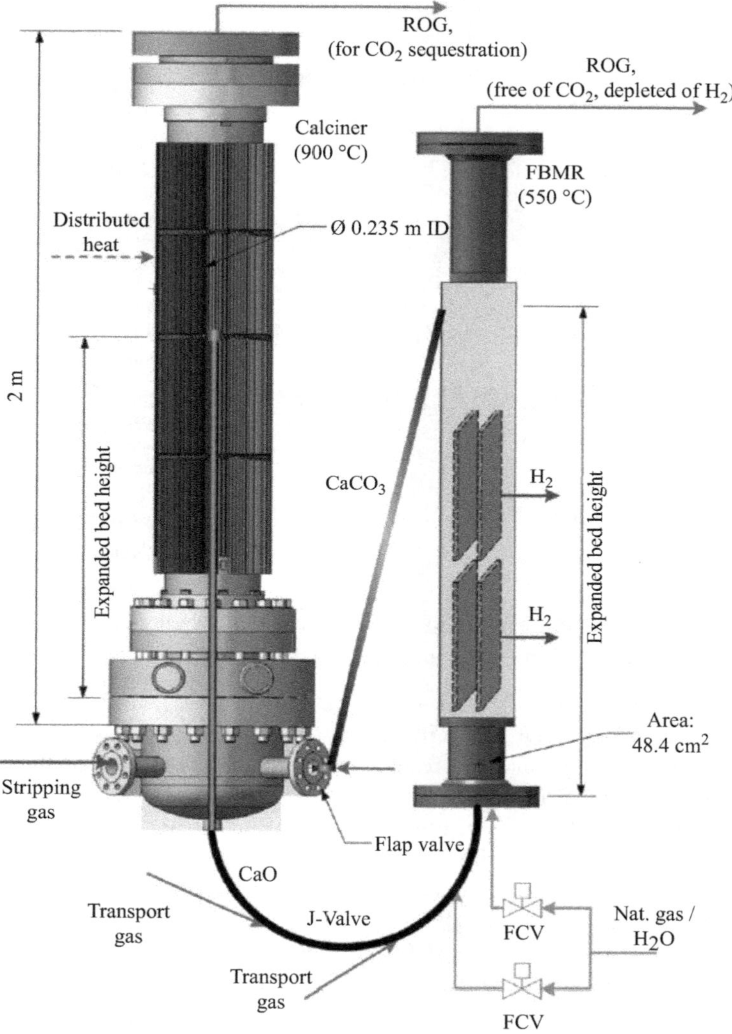

Figure 13.8　Solid circulation scheme between the fluidized-bed membrane reformer with CO$_2$ separation and the regenerator. Reprinted, with permission, from Reference 55

and combined it with CO$_2$ adsorption and H$_2$ production in situ via MSR at temperatures below 500 °C. These authors termed this approach as CHAMP-SORB; the technology is essentially a hybrid sorption-membrane reactor with variable reaction volume operation using a four-stroke cycle, as illustrated in Figure 13.9.

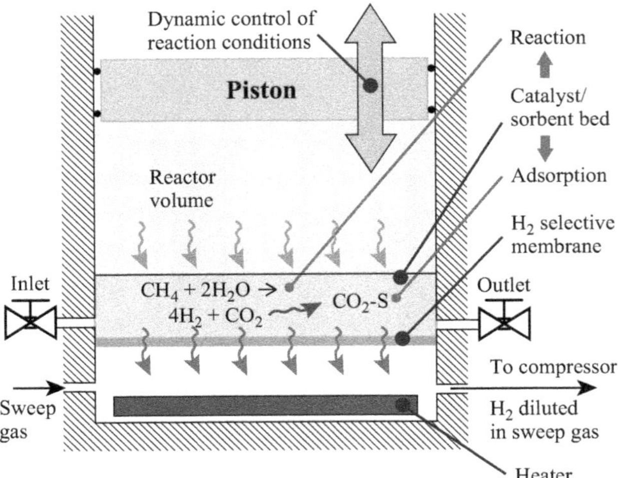

Figure 13.9   Schematic of the CHAMP-SORB reactor concept for steam methane reforming. Reprinted, with permission, from Reference 51

The CHAMP-SORB operation is based on the principles of a four-stroke internal combustion reciprocating engine:

1. Intake – in this stroke, the piston begins at bottom dead centre (BDC) and ends at top dead centre (TDC). The intake valve is opened, whereas the piston pulls a CH_4/H_2O mixture into the reactor by producing vacuum into the cylinder through its upward motion.
2. Compression – this stroke begins at TDC and ends at BDC. Piston compresses the CH_4/H_2O mixture initializing the SMR reactions, CO_2 adsorption, and H_2 membrane permeation to produce a pure H_2 product. At this point, both intake and exhaust valves are closed.
3. Reaction – while piston is at BDC, the SMR reactions, CO_2 adsorption and H_2 membrane permeation occur until a sufficient level of CH_4 conversion and/or a target H_2 yield is achieved. As piston returns to the TDC position, vacuum is produced creating the favourable conditions for the CO_2 desorption.
4. Exhaust – during this exhaust stroke, piston returns to BDC position from the TDC position, expelling the gases through the exhaust valve.

The CHAMP-SORB presents several features that favour the forward direction of the MSR and WGS reactions: (i) H_2 is selectively removed from the reactor bulk through the membrane, limiting the increase of H_2 concentration and partially suppressing the MSR backward reaction; (ii) CO_2 is adsorbed as it is produced simultaneously with the hydrogen removal via membrane permeation, this keeps the fuel concentrations (CH_4 and H_2O) higher than the products; (iii) high pressure is kept as the piston moves downward, maintaining the driving force for SMR reaction, H_2 permeation and CO_2 adsorption.

In the first CHAMP-SORB report, Anderson *et al.* [57] presented a thermodynamic study and concluded that (i) high conversion of CH_4 at low temperature and low stoichiometric S/C ratio levels could only be achieved by removing both CO_2 and H_2; (ii) the CO_2 removal from the reaction bulk as it is produced makes the coke formation by the Boudouard reaction ($2CO \rightleftharpoons CO_2 + C$) thermodynamically unfavourable; (iii) the reactor volume and piston displacement can be adjusted to operate at the maximum pressure allowed by the membrane, in order to maximize the performance enhancement. In the second report, Anderson *et al.* [51] made the first experimental and modelling demonstration of the steam methane reforming reaction carried out in a CHAMP-SORB reactor at a S/C molar ratio of 2:1 and 400 °C. The authors used Ni-based SMR catalyst and potassium-promoted hydrotalcite as CO_2 adsorbent and a 50 μm thick Pd/Ag (77/23% w/w) H_2 selective membrane. The results showed that the SMR process could be accomplished at temperatures as low as 400 °C and at the stoichiometric S/C molar ratios of 2:1 with a very low CO production due to a shift in equilibrium of the WGS reaction. The simulations showed that CHAMP-SORB technology is able to keep CH_4 and H_2O at high concentrations during the reaction step, due to simultaneous H_2 and CO_2 removal. In addition, high H_2 permeation was observed when included a CO_2 sorbent due to lower CO_2 dilution. A time scale analysis showed that the appropriate quantity of sorbent must have the capacity to sorb all CO_2 produced assuming full CH_4 conversion. Moreover, the ability to carry out the MSR at lower temperatures depends on the ability to remove H_2 effectively. Despite the experimental demonstration of the CHAMP-SORB concept, numerous practical challenges still need to be addressed, such as the effect of the cycles on the sorbent, interactions catalyst/membrane/adsorbent, membrane longevity, integration of heat sources other than electric resistances and material endurance [51].

13.3.3.3   Modelling the hybrid sorption-membrane reactor

For designing a hybrid sorption-membrane reactor for commercial applications, it is mandatory to have accurate information on all variables that influence the reactor performance, namely reaction kinetics, membrane permeation, adsorption equilibrium, adsorption and desorption kinetics, in order to describe as accurately as possible the governing model equations. Few articles describe models and simulators of sorption-enhanced reactors and membrane-enhanced steam reforming. Ding and Alpay [60] proposed a non-isothermal and non-isobaric dynamic reactor model to predict the concentration histories and concentration profiles in an adsorptive reactor. Xiu *et al.* [7] developed a mathematical model that takes into account multi-component mass balances, pressure drop and energy balance to describe the SE-SMR cyclic process. Kim *et al.* [61] simulated the performance of a packed bed membrane reactor assuming hydrogen permeable palladium membranes. Gallucci *et al.* [62] modelled the steam reforming of methane in a membrane reactor using a 1D approach and focusing on the influence of the different retentate and permeate flow configurations on both methane conversion and hydrogen recovery. Tsuru *et al.* [63] used two dimensionless numbers, Damkohler (Da) and the permeation number (θ), to characterize the performance of an isothermal catalytic membrane reactor operating in concurrent and exhibiting plug-flow pattern; these authors assumed microporous hydrogen selective membranes.

Despite the mentioned studies, only few authors studied hybrid sorption and membrane reforming reactors. Tsotsis group simulated hybrid sorption and membrane reactors for methane-steam-reforming assuming a porous ceramic hydrogen selective membrane coupled with a CO_2 adsorption system [12] and for WGS reaction in a continuous process using four-bed HSEMR [52]. Chen *et al.* [14] simulated a FBMR with in situ and ex situ hydrogen and/or CO_2 removal for producing purified hydrogen by steam methane reforming. Chen *et al.* [13] also developed a kinetic two-phase model to study the effect of the operating parameters in a sorption-enhanced steam reforming.

The carbonation reaction of CaO is controlled by two rate regimes. In the initial stage, the reaction occurs rapidly by heterogeneous surface chemical reaction. As the compact layer of $CaCO_3$ is formed on the CaO particles surface, it limits the diffusion of the reactive species decreasing the reaction rate and, eventually, stopping it. Kinetic studies have been developed by Bhatia and Perlmutter [64] and Gupta and Fan [65] reporting that the reaction does not achieve full conversion, with maximum conversions in the range of 70%–80% using 80–137 μm CaO particles, reaching up to 90% conversion in specific conditions [65].

To describe such gas–solid reaction kinetics, various models have been proposed. Ideally, one would like to account explicitly for both external and internal mass transport and finite rates of sorption. Yet, there are currently no experimental data concerning high-temperature CO_2 adsorption to explain this level of mathematical detail; instead, concerning the sorption processes, simple models have been used [7,60]. The most used models are as follows: (i) continuous model or dynamic model based on the assumption that the diffusion of the reactant gas into a particle is rapid enough compared with the chemisorption; this model is not adequate for representing the CaO carbonation when in diffusion control regime [60,66]; (ii) the unreacted core model, known as shrinking core model, assumes that the reaction zone is restricted to a thin front advancing from the outer surface into the particle core; (iii) the linear driving force (LDF) model. The shrinking core model considers regimes controlled by chemical reaction (see (13.14)) and controlled by diffusion (see (13.15)):

$$\frac{t}{\tau} = 1 - (1 - X)^{1/3} \tag{13.14}$$

$$\frac{t}{\tau} = 1 - 3(1 - X)^{2/3} + 2(1 - X) \tag{13.15}$$

where t is the time, X the fractional conversion of CaO to $CaCO_3$ and τ is the time required to full conversion. The shrinking core model, developed by Levenspiel [67], is not practical, since the conversion is implicitly given as a function of time

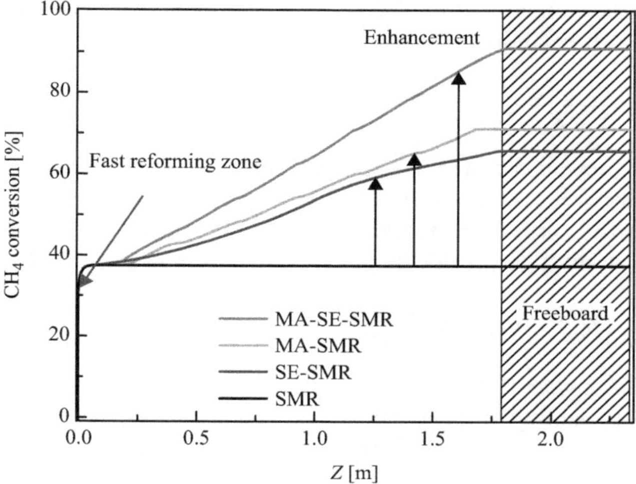

Figure 13.10   Predicted CH$_4$ conversion at different heights above the distributor for traditional SMR, membrane enhanced SMR, sorption enhanced SMR and hybrid sorption membrane reactor. Reprinted, with permission, from Reference 13

and does not describe the actual kinetic behaviour in the diffusion control regime of CaO-carbonation. To overcome these limitations, several authors reported modifications to the original shrinking core model. Bhatia and Perlmutter [64] correlated the chemisorption behaviour with the internal pore structure. Johnsen [68] modified the shrinking core model to incorporate three resistances for the gas–solid reaction, namely external mass transfer, intra-particle diffusion and chemical reaction and analysed the importance of these individual resistances in the reaction as in the conversion progresses. The model developed by Johnsen [68] was used by Chen *et al.* [13] in a kinetic two-phase model for sorption-enhanced steam reforming (Figure 13.10).

Due to its simplicity and suitable representation of the experimental data, LDF models are the most used:

$$\frac{dC_{CO_2}}{dt} = k_a \left(C_{CO_2}^{eq} - C_{CO_2} \right) \tag{13.16}$$

where $C_{CO_2}^{eq}$ is the adsorption equilibrium concentration of CO_2 on the adsorbent (mol kg^{-1}) corresponding to the prevailing gas-phase concentration, C_{CO_2} is the CO_2 concentration in solid-phase, and k_a is the mass-transfer parameter, which incorporates effects of external and intra-particle mass transport and the sorption processes. The mass-transfer parameter, k_a, is temperature dependent, but it is often assumed temperature and pressure independent [66]. For potassium promoted hydrotalcite, the CO_2 adsorption follows Langmuir [66] adsorption isotherm under both dry and wet conditions:

$$C_{CO_2}^{eq} = \frac{m_{CO_2} K_{CO_2} p_{CO_2}}{1 + K_{CO_2} p_{CO_2}} \tag{13.17}$$

where m_{CO_2} is the total adsorbent capacity, and K_{CO_2} the adsorption equilibrium and p_{CO_2} is the partial pressure of CO_2.

In the hybrid sorption–membrane reactor, besides the CO_2 adsorption, hydrogen permeation must be considered in the model. The permeation flow rate is:

$$J_i = \frac{P_e}{A\delta} \left(p_{i,\text{ret}}^{1/n} - p_{i,\text{perm}}^{1/n} \right) \tag{13.18}$$

where J_i is the molar flow rate of species i, A is the membrane area, P_e is the membrane permeability, δ is the membrane thickness, and $p_{i,\text{ret}}$ and $p_{i,\text{perm}}$ are the partial pressure in the retentate side and permeate side, respectively, of species i. P_e follows the Arrhenius equation (see (13.19)):

$$P_e = P_{e0} e^{-E_a/RT} \tag{13.19}$$

where P_{e0} is the pre-exponential factor, E_a is the activation energy, R is the ideal gas constant, and T is the temperature. Depending on the membrane and type of mechanism, n can assume different values. For defect-free membranes of Pd and Pd alloys, n is equal to 2, due to the hydrogen dissociative sorption mechanism on the palladium surface; with n equal to 2, (13.1) renders the Fick–Sieverts equation [69]. Thanks to the hydrogen dissociative sorption mechanism, Pd membranes present nearly infinite selectivity. Microporous membranes exhibit a permeation mechanism based on molecular sieving, in which gases with larger kinetic diameters than hydrogen can permeate through membrane pinholes and cracks by Knudsen diffusion. This fact justifies the lower selectivity normally shown by microporous membranes compared with Pd membranes. For microporous membranes n is equal to 1; however, n can assume other values that fit better to the experimental data; actually, the permeation in molecular sieve membranes is driven by the sorbed concentration gradient [70]. Considerable efforts are being made for a better understanding of the phenomena that occur during molecular transport through microporous systems [70].

13.4 Conclusions

The hydrogen market is growing from an estimated value of \$103.5 billion in 2014 to \$138.2 billion by 2019, with a compound annual growth rate of 5.9% [71]. Hydrogen has been suggested as the best energy vector to replace fossil fuels in transports, either burned as fuel in the internal combustion engines or used in fuel cells to produce electricity. But the main growth in the hydrogen consumption is related to its use in refineries to improve the quality of petroleum products [71]. As the hydrogen market grows, the development of more efficient industrial processes is required. Hydrogen can be obtained from renewable or nuclear sources, using electricity as an intermediate energy carrier. Industrially hydrogen is presently obtained by steam reforming of methane or other hydrocarbons or by gasification of coal and oil refining residues; hydrogen can also be recovered from the refinery off-gases or syngas. MSR process is being used for decades as the main industrial process for hydrogen production, despite its low-energy efficiency, which is a challenge for the development of more efficient and reliable processes.

Reactors integrating synergistically membrane and/or adsorptive processes are reviewed. Combining reaction with product separation can provide significant improvements to the conventional MSR process. Membrane reactors allow higher MSR reaction conversions due to the H_2 selective removal from the reaction bulk, which shifts the equilibrium towards the reaction products and increases the reactants residence time. They also allow lower operating temperature and have lower downstream purification requirements. The main disadvantage of this technology is related to the long-term operation stability due to the irreversible damages caused by impurities to the membrane, such as CO and sulphur compounds. In adsorptive reactors, the reforming catalyst is combined with a CO_2 sorbent for allowing CO_2 to be removed in situ, as it is formed from the reaction medium. Higher conversions for reforming and WGS reactions are achieved, making possible to produce hydrogen with a purity of *ca.* 90%–95% in a single-step process. CO_2 sorption on CaO is a highly exothermic process releasing enough heat to run the MSR reaction. The adsorptive reactors present, however, few drawbacks: they require extra energy to regenerate the CO_2 sorbent, and the sorption capacity reduces significantly during the carbonation/calcination cycles. As most of the hydrogen applications require hydrogen purities >99.99%, further downstream purification is required. Hybrid adsorptive membrane reactors incorporating synergistically adsorption and permeation processes, allow higher conversions, higher selectivity, higher hydrogen purity and lower operating temperature than other reactors.

This technology is still in its early stage of development. It has been investigated for hydrogen production, CO_2 capturing and water treatment showing very promise results. In the field of hydrogen production, the most noticeable achievement was the development of a fluidized bed sorption-enhanced membrane reactor in-line with a calciner (sorbent regenerator) by the join collaboration between University of British Columbia and Tokyo Gas Company. The pilot-plant produced a hydrogen stream with a purity of +99.99% and a hydrogen recovery of *ca.* 50%, at 550 °C and 300 kPa. Hybrid adsorptive membrane reactors still exhibiting few challenges but show a promising future for running some industrially challenging reactions.

Nomenclature

Acronyms

BDC   bottom dead centre
CHAMP CO_2/H_2 active membrane piston reactor

CMS	carbon molecular sieve
FBMR	fluidized bed membrane reactor
HSEMR	hybrid sorption-enhanced membrane reactor
MSR	methane steam reforming
PEMFC	Polymer electrolyte membrane fuel cell
PROX	partial oxidation
PSA	pressure swing adsorption
S/C	steam-to-carbon ratios
SERP	sorption-enhanced reaction process
TDC	top dead centre
TSA	thermal swing adsorption
WGS	water-gas shift

Symbols

A	membrane area
$C_{CO_2}^{eq}$	adsorption equilibrium concentration of CO_2 on the adsorbent
C_{CO_2}	CO_2 concentration in solid-phase
Da	Damkohler
E_a	activation energy
J_i	molar flow rate of species i
k_a	mass-transfer parameter
$K_{eq,i}$	equilibrium constants of the reaction i
k_i	reaction rate constant of the reaction i at a defined temperature
K_j	adsorption constants of species j at a defined temperature
m_{CO_2}	total adsorbent capacity
P_e	membrane permeability
P_{e0}	pre-exponential factor
p_j	partial pressure of specie j
R	ideal gas constant
r_i	reaction rate of the reaction i
t	time
T	temperature
X	fractional conversion

Greek

ΔH_{298}°	standard enthalpy of formation
θ	permeation number
τ	time required to full conversion
δ	membrane thickness

References

[1]   Darwish N. A., Hilal N., Versteeg G., Heesink B. 'Feasibility of the direct generation of hydrogen for fuel-cell-powered vehicles by on-board steam reforming of naphtha.' *Fuel*. 2004;**83**(4–5):409–417.

[2]   Semelsberger T. A., Brown L. F., Borup R. L., Inbody M. A. 'Equilibrium products from autothermal processes for generating hydrogen-rich fuel-cell feeds.' *International Journal of Hydrogen Energy*. 2004;**29**(10): 1047–1064.

[3]   Xu J. G., Froment G. F. 'Methane-steam reforming, methanation and water gas shift – I: intrinsic kinetics.' *American Institute of Chemical Engineers Journal*. 1989;**35**:88–96.

[4]   Adris A. M., Pruden B. B., Lim C. J., Grace J. R. 'On the reported attempts to radically improve the performance of the steam methane reforming reactor.' *Canadian Journal of Chemical Engineering*. 1996;**74**:177–182.

[5]   Criscuoli A., Basile A., Drioli E., Loiacono O. 'An economic feasibility study for water gas shift membrane reactor.' *Journal of Membrane Science*. 2001;**181**:21–27.

[6]   Saracco G., Specchia V. 'Catalytic inorganic-membrane reactors: present experience and future opportunities.' *Catalysis Reviews: Science and Engineering*. 1994;**36**:305.

[7]   Xiu G. H., Soares J. L., Li P., Rodrigues A. E. 'Simulation of five-step one-bed sorption-enhanced reaction process.' *American Institute of Chemical Engineers Journal*. 2002;**48**(12):2817–2832.

[8]   Ding Y., Alpay E. 'Equilibria and kinetics of CO_2 adsorption on hydrotalcite adsorbent.' *Chemical Engineering Science – Journal*. 2000;**55**:3461.

[9]   Xiu G., Li P., Rodrigues A. E. 'Sorption-enhanced reaction process with reactive regeneration.' *Chemical Engineering Science – Journal*. 2002; **57**:3893.

[10] Park B., Tsotsis T. T. 'Models and experiments with pervaporation membrane reactors integrated with an adsorbent system.' *Chemical Engineering and Processing*. 2004;**43**:1171.

[11] Harale A., Hwang H. T., Liu P. K. T., Sahimi M., Tsotsis T. T. 'Experimental studies of a hybrid adsorbent-membrane reactor (HAMR) system for hydrogen production.' *Chemical Engineering Science – Journal*. 2007; **62**:4126.

[12] Fayyaz B., Harale A., Park B. G., Liu P. K. T., Sahimi M., Tsotsis T. T. 'Design aspects of hybrid adsorbent-membrane reactors for hydrogen production.' *Industrial & Engineering Chemistry Research*. 2005;**44**:9398.

[13] Chen Y., Mahecha-Botero A. S., Lim C. J., *et al.* 'Hydrogen production in a sorption-enhanced fluidized-bed membrane reactor: operating parameter investigation.' *Industrial & Engineering Chemistry Research*. 2014;**53**:6230.

[14] Chen Z., Po F., Grace J. R., *et al.* 'Sorbent-enhanced/membrane-assisted steam-methane reforming.' *Chemical Engineering Science – Journal*. 2008;**63**:170.

[15]   Iulianelli A., Ribeirinha P., Mendes A., Basile A. 'Methanol steam reforming for hydrogen generation via conventional and membrane reactors: a review.' *Renewable and Sustainable Energy Reviews*. 2014;**29**:355–368.

[16]   Katiyar N., Kumar S., Kumar S. 'Comparative thermodynamic analysis of adsorption, membrane and adsorption-membrane hybrid reactor systems for methanol steam reforming.' *International Journal of Hydrogen Energy*. 2013;**38**:1363–1375.

[17]   Baker E.H. 'The calcium oxide–carbon dioxide system in the pressure range 1–300 atmospheres.' *Journal of the Chemical Society*. 1962;**70**: 464–470.

[18]   Ding Y., Alpay E. 'Equilibria and kinetics of CO_2 adsorption on hydrotalcite adsorbent.' *Chemical Engineering Science – Journal*. 2000;**55**:3461–3474.

[19]   Yi K. B., Eriksen D. 'Low temperature liquid state synthesis of lithium zirconate and its characteristics as a CO_2 sorbent.' *Separation Science and Technology*. 2006;**41**:283–296.

[20]   Tessié du Motay C. R., Maréchal C. R. 'Industrial preparation of hydrogen.' BullMensuel de La SociétéChimique de Paris. 1868; **9**:334–334.

[21]   Minet R. G., Desai K. 'Cost-effective methods for hydrogen production.' *International Journal of Hydrogen Energy*. 1983;**8**(4):285–290.

[22]   Steinberg M., Cheng H. C. 'Modern and prospective technologies for hydrogen production from fossil fuels.' *International Journal of Hydrogen Energy*. 1989;**14**:797.

[23]   Kirk-Othmer, *Concise Encyclopedia of Chemical Technology*, 4th ed. New York: Wiley; 1999. Vol. 12, p. 950.

[24]   International Organisation for Standardisation, ISO 14687-2:2012, *Hydrogen Fuel-Product Specification—Part 2: Proton Exchange Membrane Fuel Cell Applications for Road Vehicles*. Published International, 2012. Available from http://www.iso.org/iso/catalogue_detail.htm?csnumber=55083. [Accessed 14 November 2016].

[25]   Akers W. W., Camp D. P. 'Kinetics of the methane-steam reaction.' *American Institute of Chemical Engineers Journal*. 1955;**1**:471.

[26]   Allen D. W., Gerhard E. R., Likins Jr. M. R. 'Kinetics of the methane-steam reaction.' *Industrial & Engineering Chemistry Process Design and Development*. 1975;**14**(3):256.

[27]   Bodrov I. M., Apel'baum L. O., Temkin M. I. 'Kinetics of the reaction of methane with steam on the surface of nickel', *Kinetics and Catalysis*. 1964; **5**(4):696.

[28]   Froment G. F., Bischoff K. B. *Chemical Reactor Analysis and Design*. New York: Wiley Publishers; 1990.

[29]   Ruthven D., Farooq S., Knaebel K. *Pressure Swing Adsorption*. New York: VCH Publishers; 1994.

[30]   Tomita T., Sakamoto T., Ohkamo U., Suzuki M. *Fundamentals of Adsorption II*. New York: Engineering Foundation; 1987, p. 89.

[31]   Sircar S., Golden T. C. 'Purification of hydrogen by pressure swing adsorption separation.' *Science and Technology*. 2000;**35**:667–687.

[32] Adris A. M., Pruden B. B., Lim C. J., Grace J. R. 'On the reported attempts to radically improve the performance of the steam methane reforming reactor', *Canadian Journal of Chemical Engineering*. 1996;**74**:177.

[33] Bird R. B., Stewart W. E., Lightfoot E. N. *Transport Phenomena*, 2nd ed. New York: John Wiley & Sons; 2001.

[34] Trimm D. 'Catalysts for the control of coking during steam reforming.' *Catalysis Today*. 1999;**49**:3–10.

[35] Gluud W., Keller K., Schonfelder R., Klempt W. 'Production of hydrogen.' US Patent 1 816 523, 1931.

[36] Gorin E, Retallick W. B. 'Method for the production of hydrogen.' US Patent 3 108 857, 1963.

[37] Brun-Tsekhovoi A. R., Zadorin A. N., Katsobashvili Y. R., Kourdyumov S. S. 'The process of catalytic steam-reforming of hydrocarbons in the presence of a carbon dioxide acceptor.' In *Proceedings of the Seventh World Hydrogen Energy Conference*, Moscow, Russia, Sep 1988. New York: Pergamon Press; p. 885.

[38] Kumar R., Cole J., Lyon R. 'Unmixed reforming: an advanced steam reforming process.' In *Presented at the Fuel Cell Reformer Conference*, South Coast Air Quality District, Diamond Bar, CA, 1999.

[39] Barelli L., Bidini G., Corradetti A., Desideri U. 'Study of the carbonation–calcination reaction applied to the hydrogen production from syngas.' *Energy*. 2007;**32**:697–710.

[40] Hufton J., Weigel S., Waldron W., Nataraj S., Rao M., Sircar S. 'Sorption enhanced reaction process (SERP) for the production of hydrogen.' In *Proceeding of the 1999 U.S DOE Hydrogen Program Review*, NREL/CP-570-26938.

[41] Han C., Harrison D. P. 'Simultaneous shift reaction and carbon dioxide separation for the direct production of hydrogen.' *Chemical Engineering Science*. 1994;**49**:5875–5883.

[42] Goto S., Tagawa T., Oomiya T. 'Dehydrogen of cyclohexane in a PSA reactor using hydrogen storage alloy.' *Chemical Engineering Essays (Japan)*. 1193;**19**:978–983.

[43] Carvil B. T., Hufton J. R., Sircar S. 'Sorption-enhanced reaction process.' *American Institute of Chemical Engineer Journal*. 1996;**42**:2765–2772.

[44] Hufton J. R., Mayorga S., Sircar S. 'Sorption-enhanced reaction process for hydrogen production.' *American Institute of Chemical Engineer Journal*. 1999;**45**:248–256.

[45] Johnsen K., Ryu H. J., Grace J. R., Lim C. J. 'Sorption-enhanced steam reforming of methane in a fluidized bed reactor with dolomite as CO-acceptor.' *Chemical Engineering Science*. 2006;**61**:1195–1202.

[46] Shu J., Grandjean B. P. A., Van Neste A., Kaliaguine S. 'Catalytic palladium-based membrane reactors: a review.' *Canadian Journal of Chemical Engineering*. 1991;**69**(5):1036–1060.

[47] Bischoff B. L., Judkins R. R. 'Development of inorganic membranes for Hydrogen separation.' Oak Ridge National Laboratory Report, 2008.

[48]   Shah M. M., Drnevich R. F., Balachandran U. 'Integrated ceramic membrane system for hydrogen production.' In *Proceedings of the 2000 Hydrogen Program Review*, NREL/CP-570-28890.

[49]   Park B. G. 'A hybrid adsorbent-membrane reactor (HAMR) system for hydrogen production.' *Korean Journal of Chemical Engineering*. 2004; **21**(4):782–792.

[50]   Soria M. A., Tosti S., Mendes A., Madeira L. M. 'Enhancing the low temperature water–gas shift reaction through a hybrid sorption-enhanced membrane reactor for high-purity hydrogen production.' *Fuel*. 2015;**159**: 854–863.

[51]   Anderson D. M., Mohamed H. N., Thomas M. Y., Kottke P. A., Andrei G. F. 'Sorption-enhanced variable-volume batch−membrane steam methane reforming at low temperature: experimental demonstration and kinetic modeling.' *Industrial & Engineering Chemistry Research*. 2015;**54**: 8422−8436.

[52]   Harale A., Hwang H. T., Liu P. K. T., Sahimi M., Tsotsis T. T. 'Design aspects of the cyclic hybrid adsorbent-membrane reactor (HAMR) system for hydrogen production'. *Chemical Engineering Science*. 2010;**65**:427.

[53]   Yang H., Xu Z., Fan M., *et al.* 'Progress in carbon dioxide separation and capture: a review.' *Journal of Environmental Sciences*. 2008;**20**:14–27.

[54]   Stoquarta C., Servaisb P., Bérubéc P. R., Barbeaua B. 'Hybrid membrane processes using activated carbon treatment for drinking water: a review.' *Journal of Membrane Science*. 2012:**411–441**:1–12.

[55]   Andrés M. B., Boyd T., Grace J. R., *et al.* '*In-situ* CO_2 capture in a pilot-scale fluidized-bed membrane reformer for ultra-pure hydrogen production.' *International Journal of Hydrogen Energy*. 2011;**36**:4038.

[56]   Deshmukh S., Heinrich S., Mörl L., Annaland M., Kuipers J. 'Membrane assisted fluidized bed reactors: potentials and hurdles'. *Chemical Engineering Science*. 2007;**62**:416–436.

[57]   Anderson D. M., Kottke P. A., Fedorov A. G. 'Thermodynamic analysis of hydrogen production via sorption-enhanced steam methane reforming in a new class of variable volume batch-membrane reactor.' *International Journal of Hydrogen Energy*. 2014;**39**:17985.

[58]   Damm D. L., Fedorov A. G. 'Batch reactors for hydrogen production: theoretical analysis and experimental characterisation.' *Industrial & Engineering Chemistry Research*. 2009;**48**:5610.

[59]   Yun T. M., Kottke P. A., Anderson D. M., Fedorov A. G. 'Power density assessment of variable volume batch reactors for hydrogen production with dynamically modulated liquid fuel introduction.' *Industrial & Engineering Chemistry Research*. 2014;**53**:18140.

[60]   Ding Y., Alpay E. 'Adsorption-enhanced steam methane reforming.' *Chemical Engineering Science*. 2000;**55**:3929–3940.

[61]   Kim J. H., Choi B. S., Yi J. 'Modified simulation of methane steam reforming in Pd-membrane/packed-bed type reactor.' *Journal of Chemical Engineering of Japan*. 1999;**32**:760.

[62] Gallucci F., Comite A., Capannelli G., Basile A. 'Steam reforming of methane in a membrane reactor: an industrial case study.' *Industrial & Engineering Chemistry Research*. 2006;**45**:9.

[63] Tsuru T., Yamaguchi K., Yoshioka T., Asaeda M. 'Methane steam reforming by microporous catalytic membrane reactors.' *American Institute of Chemical Engineers*. 2004;**50**:11.

[64] Bhatia S. K., Perlmutter D. D. 'Effect of the product layer on the kinetics of the CO_2-lime reaction.' *American Institute of Chemical Engineers*. 1983;**29**:79–86.

[65] Gupta H., Fan L.S. 'Carbonation-calcination cycle using high reactivity calcium oxide for carbon dioxide separation from flue gas.' *Industrial & Engineering Chemistry Research*. 2002;**41**:4035.

[66] Ding Y., Alpay E. 'Equilibria and kinetics of CO_2 adsorption on hydrotalcite adsorbent.' *Chemical Engineering Science*. 2000;**55**:3461.

[67] Levenspiel O. *Chemical Reaction Engineering*. New York: Wiley; 1972.

[68] Johnsen K. 'Sorption-enhanced steam methane reforming in fluidized bed reactor.' Ph.D. Thesis, Norwegian University of Science and Technology, Norway, 2006.

[69] Sieverts A., Zapf G. 'The solubility of deuterium and hydrogen in solid palladium.' *Zeitschrift für Physikalische Chemie*. 1935;**174**:359–364.

[70] Li H., Haas-Santo K., Schygulla U., Dittmeyer R. 'Inorganic microporous membranes for H_2 and CO_2 separation – review of experimental and modeling progress'. *Chemical Engineering Science*. 2015;**127**:401–417.

[71] MarketsandMarkets Market Report. 2014. Available from http://www.marketsandmarkets.com/PressReleases/hydrogen.asp

Chapter 14

Carbon-based membranes

Jon Arvid Lie[1], Xuezhong He[1], Izumi Kumakiri[2],
Hidetoshi Kita[2] and May-Britt Hägg[1]

Abstract

Carbon membranes as a promising candidate for energy-efficient gas separation processes have been studied for more than 20 years. This chapter describes the status and perspectives of both self-supported and supported carbon membranes. The key steps on the development of high performance hollow-fiber carbon membranes are discussed, including precursor selection, tuning carbon membrane structure, and regeneration. The module design and continuous carbonization process are pointed out to be the main challenges related to upscaling. Supported carbon membranes open new opportunities for high-temperature and high-pressure applications. The main challenges of supported carbon membranes are the lower packing density and relatively high production cost compared to the self-supported hollow-fiber carbon membranes – this directs their applications more towards the medium to small gas volume processes. Finally, the potential applications of carbon membranes are also briefly mentioned. The recovery of hydrogen from various gas streams may become a major application, as well as olefin–paraffin separation, but also removal of CO_2 from natural gas or biogas (CO_2–CH_4 separation) has a very nice potential. The carbon membranes show great potentials in gas separation applications with the possibility of tailoring/controlling the membrane pore size on a molecular sieving level.

14.1   Introduction

This chapter describes carbon-based membranes: from material to application. Carbon membranes have been studied in more than 20 years as a promising candidate for energy-efficient gas separation processes. Strong interests have been focused on preparation of carbon membranes for gas separations such as H_2/CH_4, H_2/CO_2, CO_2/CH_4 and olefin/paraffin. The first carbon membranes were prepared

[1]Norwegian University of Science and Technology, Trondheim, Norway
[2]Yamaguchi University, Yamaguchi, Japan

from the carbonization of cellulose hollow fibers by Koresh and Soffer [1] and has since then been developed further following different routes of pyrolysis by numerous researchers around the world – some examples are References 2–6. There are several polymeric materials which are suitable as precursors, but the main groups of material used are cellulose derivatives (in Memfo group at NTNU [3,4,7,8]), polyimide (mainly in Koros group) [2,9–12], and polyacrylonitrile [13].

Carbon membranes can be divided into two categories: unsupported and supported carbon membranes [14]. Unsupported membranes have three different configurations: flat-sheet film, hollow fiber and capillary tubes, whereas supported carbon membranes involve two configurations: flat and tube. The unsupported hollow-fiber carbon membranes are prepared from hollow-fiber precursors (see details in Section 14.2) which could be the only viable module configuration in which large areas are needed in industrial applications due to high packing density. However, the supported carbon membranes have better mechanical stability and are typically prepared by coating a support with a thin, uniform polymeric layer and carbonization afterwards (see details in Section 14.3).

The intriguing idea is that by careful control of the pyrolysis conditions, the pores in the resulting carbon material skeleton can be tailored on nanoscale and thus be able to separate gas molecules which are much alike both in size and physical properties (i.e., carbon molecular sieve (CMS) membranes). This has also proven to be possible by reviewing the many published papers on the topic [6,11,15]. Special challenging separations are typically olefin/paraffin (e.g., C_3H_6/C_3H_8, C_2H_4/C_2H_6). Some of the techniques used for pore tailoring are using purge gas (e.g. CO_2, N_2), control final carbonization temperature and soak time and post-treatment (post-oxidation (PO) and chemical vapor deposition (CVD)). These techniques are described in more detail in Sections 14.2 and 14.3.

Carbon membranes have great advantages on mechanical and chemical stability when exposed to high pressure and temperature in processes. They can easily exceed the Robeson upper bound (2008) based on their molecular sieve mechanism and thus reach the industrial attractive region as illustrated in Figure 14.1 [16]. By further improvement of membrane performance, it can potentially offset the relatively high production cost compared to polymeric membranes. However, there are still some challenges related to module making for upscaling, especially on carbon membranes mounting, potting and sealing for high-pressure applications see Figure 14.2. There is a need to find a suitable potting material that can easily penetrate a bundle of fibers even at a high packing density, and which should cure relatively fast, without generating excessive heat (see details in Section 14.2). Ceramic supported carbon membranes provide good mechanical strength and are easier on module construction, which has been widely investigated for high temperature applications (see Section 14.3). However, the challenges related to the fabrication of a thin, defect-free carbon selective layer on the top of the support and the compatibility between support and carbon layer as well as high cost of ceramic tubes still hinder its upscaling [17]. The recently developed graphene oxide (GO) membranes with an ultrathin GO layer (<20 nm) shows great potential for H_2/CO_2 separation with selectivity >3,400 [18] – this graphene-based new carbon

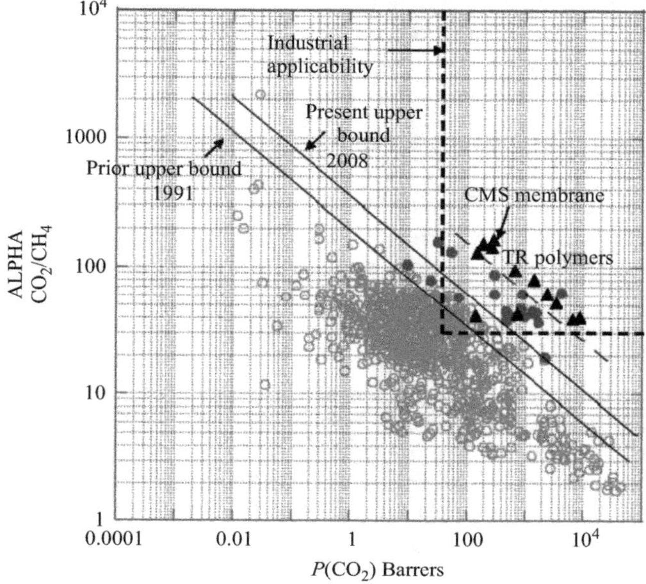

Figure 14.1   Comparison of carbon membranes with other polymeric and thermally rearranged (TR) membranes for CO_2/CH_4 separation based on Robeson upper bound (2008) adapted from Reference 16

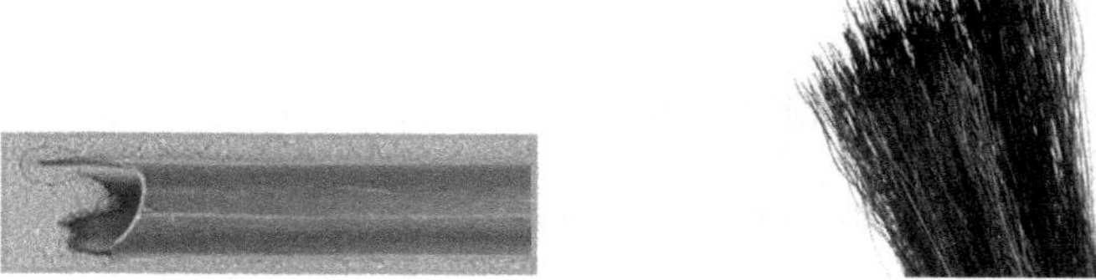

Figure 14.2   Left: SEM-picture of a CMS fiber; wall thickness 16 μm. Right: a bundle of CMS hollow fibers ready for mounting in a pilot module in our group (1 m^2 area, OD = 200 μm)

membranes are a promising research area for the future. Also self-supported GO membranes have been successfully investigated for gas separation [19] and may play an important role in future applications.

There are several potential industrial applications for carbon membranes that can potentially overcome the trade-off of permeability/selectivity as well as the limitation of operation temperature, pressure and adverse conditions such as the presence of acid gases SO_2 and NO_x for polymeric membranes (e.g., H_2 separation from syngas, H_2/CH_4 separation, natural gas processing, biogas upgrading and more). A few companies have challenged the difficulties of module making and scaling up but not yet fully succeeded to make large-scale modules for industrial applications (as per 2016).

14.2 Hollow-fiber carbon membranes

The self-supported hollow-fiber carbon membranes have been intensively investigated for gas separation processes. The carbon membrane production cost and performance significantly depend on precursor and carbonization procedure. Membrane aging due to the adsorption of gas molecules inside carbon matrix exists in most carbon membranes, and suitable regeneration methods should be employed to recover the membrane performance over time. The main challenges on carbon membrane upscaling are firstly to establish a low cost continuous carbonization process and secondly to secure that the module design will be suitable for high-pressure and high-temperature applications. This section on hollow-fiber carbon membranes is thus divided into four subsections due to the complex preparation of these membranes. Finally, some promising potential applications on H_2 purification as well as olefin/paraffin and CO_2/CH_4 separation are described.

14.2.1 Preparation of precursor

Choice of precursor is one of the main parameters deciding the performance of carbon membranes, together with temperature and atmosphere during carbonization, as well as post-fabrication conditions. Current research and pilot scale work has mainly focused on two precursors: cellulose and polyimide [7,20,21]. Table 14.1 lists some pros and cons for the two types of polymer. The difference in costs (estimated) is basically due to the availability of the material.

The only viable configuration for industrial carbon membranes is hollow fibers, mainly for mechanical reasons and the need for a high membrane area per volume of separation unit. A carbon membrane fabricated in a flat-sheet configuration needs a support to withstand high pressure differences; hence, the cost will increase and the packing density of the unit will decrease dramatically. For smaller gas volumes, supported carbon membranes on ceramic tubes may, however, be a

Table 14.1   Comparison of cellulose and polyimide as precursors for carbon membranes

Property	Cellulose	Polyimide
Availability	High (abundant)	Limited
Sustainability	Renewable biopolymer	Synthetic polymer
Processability	Moderate	High
Free volume	Low	Moderate to high
Fusing risk in carbonization	Moderate	Very high
Material cost ratio	1	10

choice if the permeance is sufficiently high. To obtain perfectly a flat carbon sheet is another challenge due to the reduction in surface energy during carbonization which will twist the sheet in different directions, compared to the self-supported hollow fibers which upon carbonization will keep their original shape as long as the cross section is symmetric.

The hollow fibers are normally made by a dry/wet spinning method, that is, the polymer solution is extruded through a nozzle, then passing an air gap before entering a quench bath of non-solvent. In this work, cellulose acetate (CA) was used as starting polymer since solvents for cellulose are rare or hazardous. CA and polyvinylpyrrolidone (PVP) were dissolved in *N*-methyl-2-pyrrolidone (NMP). The role of PVP is to reduce macrovoid formation and hence improves the mechanical properties of the fiber and gives a more stable spin line. PVP will decompose during carbonization and may work as a porogen as well (see Section 14.2.2). The bore liquid used is usually a mixture of deionized water and NMP. Optional ingredients include metal salts, which will be explained in Section 14.2.2. To obtain an effective cleaning, procedure for nozzles is critical to make symmetric and reproducible fibers. It is also important to keep a small dope flow rate through the nozzle (preferentially small nozzle dimensions) in order to avoid high shear rates and keep the fiber dimensions small. The rate of coagulation of the fiber is another critical parameter during spinning, because on the one hand the fiber has to be stiff enough to avoid deformation on its way out of the coagulation bath, whereas on the other hand if coagulation is too fast, polymer particles may precipitate on the bore side and plug the fiber. The last phenomenon may create defect fibers in the later formation steps. A solution to this issue was found in lowering the coagulation temperature. A brief flow sheet of the typical production process of carbon membranes from CA is given in Figure 14.3 [16].

After the CA fibers were washed and treated with glycerol (to keep the pore structure open), the fibers were immersed in a sodium hydroxide solution to remove the acetate groups. In general, polyesters undergo other pyrolysis mechanisms and

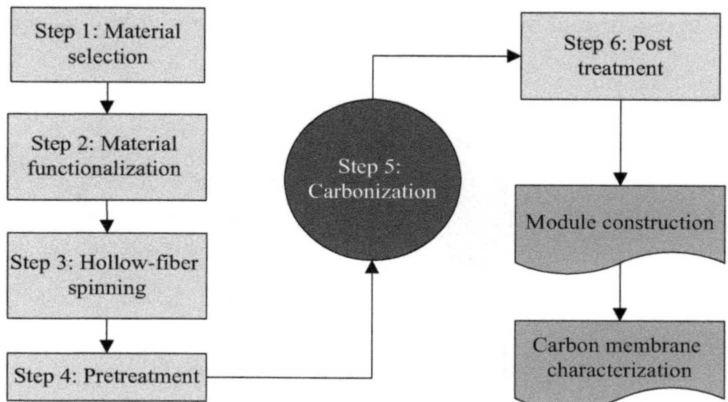

Figure 14.3   Steps in the production of carbon hollow fibers from CA [16]

cannot be transformed to carbon membranes due to fracturing and extremely high weight loss.

The cellulose fibers were then washed in tap water before drying in a controlled environment. Due to the intrinsic chirality of cellulose chains, it is essential to dry the fibers at a slow speed. Otherwise, the fibers will curl up and destroy the packing density of the final separation module. The drying protocol needs to start at high humidity and end at the same conditions as in the fiber storage environment.

14.2.2   Carbonization and regeneration

One of the major advantages of carbon membranes is that the pore size distribution can be finely adjusted to separate a variety of gas mixtures. In order to choose the right fabrication conditions, the gas components and operational conditions must be known. Carbon membranes can achieve high selectivity between gases very similar in size, due to the shape and size discrimination that occurs in the many stages (pore constrictions) from feed to permeate side of the carbon matrix. A challenge with carbon membranes has been loss of permeability or capacity after fabrication, either during storage or during operation. This aging may be divided into two categories [22]:

1.   Physical aging
2.   Sorption-induced aging

Polyimide films with high free volume results in carbon membranes with initially high permeability, but capacity will decrease significantly during the first days of operation [22]. The structure seems not to be thermodynamically stable, and a denser packing of the carbon structure is believed to occur just after fabrication. Especially the use of vacuum seems to make parts of the structure collapse.

Figure 14.4 illustrates different ways of modifying the carbon structure [23]. *Porogens* are compounds added to the precursor in order to increase the pore volume or porosity of the resultant carbon. An example of a porogen is PVP, which will decompose during carbonization. When the content of porogens increases, the mechanical properties of the carbon may deteriorate and the risk of physical aging increases.

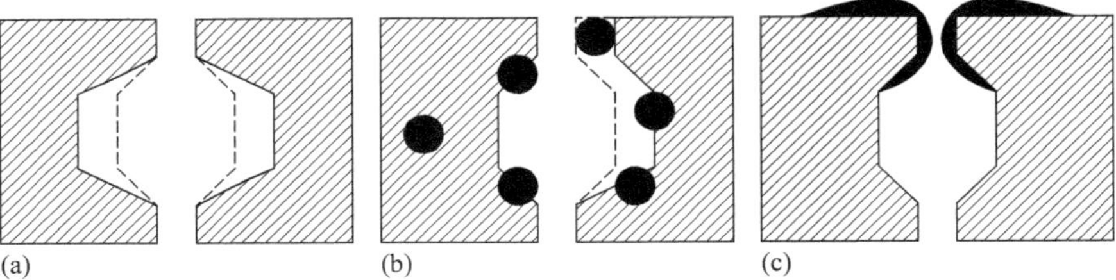

Figure 14.4   Three methods for carbon modification: (a) porogens, (b) doping/ spacers, and (c) surface modifiers [23]

Dopants or spacers like lithium salts or other small interstitial agents are very interesting and should be explored in order to prevent physical aging. Since carbon hollow fibers can withstand high pressure differences also when the bore side is pressurized and the shell side is at atmospheric pressure, this inside-out configuration will probably create a slight expansion of the carbon matrix during operation and may be another way of preventing physical aging and achieving a stable membrane capacity. Long-term experiments in our research group support this phenomenon.

When it comes to the sorption-induced aging, it can be further divided into chemisorption and adsorption of different gases. Chemisorption of oxygen on carbon has been discussed by several research groups, for example Boehm [24]. Oxygen gas may react slowly with the edge of graphene sheets or other defects in the carbon structure. The carbon microvoids are organophilic in nature and tend to adsorb organic gases or vapors during storage and operation. Unless dealt properly with, both types of sorption leads to undesired occupation of pore volume, and hence loss of membrane capacity. However, strong adsorption of gases may also be exploited through the selective surface flow (SSF) mechanism [25]. As this mechanism increases the permeability by one order of magnitude, this may be a solution to the capacity limitation, if a strongly adsorbing gas is the permeant.

The third type of modification (see Figure 14.4(c)) is *surface modification*. Jones and Koros [26] successfully coated the surface of carbon with a Teflon layer to protect against water sorption. A more common technique is CVD, in order to tune the selectivity of the carbon membrane. This may often be necessary if oxidation is used to open the structure after carbonization (PO).

The effect of the different parameters during carbonization has been studied by many researchers [8,27,28]. Table 14.2 lists the most common parameters and how they typically affect the properties of carbon. A crystalline precursor like polyacrylonitrile (low free volume) results in carbon with very low gas permeability. However, there is not always a correlation between the permeability of precursors and the permeability of the resulting carbon membranes [6]. Oxidation causes C-atoms in the pore walls to burn off, shifting the pore size distribution to a higher value. Increasing the final temperature will first open the carbon structure until a certain temperature when reorganization of the structure starts to densify the structure. Keeping the final temperature for some time (soak) will also make the carbon somewhat denser but may eliminate dangling bonds at the edge of graphene sheets; hence, stability improves. Therefore, it is crucial to control carbonization condition in order to tailor pore size and structure and prepare a high performance carbon membrane. Orthogonal experimental design (OED) and conjoint analysis was reported by He and Hagg [8] to optimize the carbonization procedure, and the influence of different carbonization parameters on carbon membrane performance was also systematically investigated.

Most self-supported carbon hollow fibers will need post-treatment to increase their capacity to the same range as polymeric fibers. Supported carbon membranes (Section 14.3) usually have higher capacity due to a thinner selective layer. PO was carried out on a module containing 2,500 fibers according to a given temperature

Table 14.2   Carbonization parameters and their typical correlation with permeability, selectivity and stability of the resulting carbon membranes

Carbonization parameters	Permeability	Selectivity	Stability
Precursor free volume	+	−	−
Oxidative atmosphere	+	−	−
Heating rate	+	−	−
Final temperature	+, then −	−, then +	+
Final isotherm	−	+	+

\+ Parameter has positive contribution on membrane performance.
− Parameter has negative contribution on membrane performance.

Table 14.3   Effect of post-oxidation on permeability and selectivity. Mixed gas performance at 20 °C and 10 bar feed (unpublished pilot scale data).

Module	No. of fibers	Permeance CO_2 [GPU*]	Selectivity CO_2/CH_4
As carbonized	2,500	6.0	100
Post-oxidized	2,500	40	45

*GPU = Gas permeance unit [1,000 GPU = 2.76 m^3(STP)/(m^2 h bar)].

protocol. The effect of PO on permeance and selectivity is given in Table 14.3. This work was carried out in our group.

Loss of permeability can be offset by increasing the operating temperature. High thermal resistance is one of the advantages of carbon membranes and should be exploited. This brings us to the last topic of this section – possible regeneration methods. Several methods are relevant:

1. Thermal
2. Thermochemical
3. Chemical
4. Electrothermal

Thermal regeneration involves heating the module with an inert gas inside, to desorb gases or vapors from the pores and voids. Thermochemical regeneration means heating in a reactive atmosphere like air, to increase the actual pore size distribution. Reactive surface groups can be made passive by reduction in, for example, hydrogen. This method can restore or even increase the capacity of a used membrane module compared to its initial value and offers a fast route to a new module (compared to a polymeric module which in most cases cannot be reused). The chemical method exposes the module to an agent which will dissolve some of the adsorbed gases or vapors, like the use of propylene [29]. Finally, electrothermal regeneration applies a low-voltage current across the carbon to release sorbed molecules. This requires a certain conductivity of the carbon matrix, and membranes doped with metals are suitable (Figure 14.4). For safety reasons, this method should be performed in non-oxidizing gas streams.

14.2.3   Challenges of scaling up

When a membrane is going to be manufactured on a large scale, it is important to secure a steady and reliable supply of raw materials. This is an important advantage of cellulosic materials, in addition to its sustainable character (revisit Table 14.1). The environmental impact of each raw material and any side products should be considered. The cost of cellulose is about 10% of the cost of most polyimides. This is of economic interest, but it is not a dominant cost in the membrane production.

This section contains results and discussions on carbon membranes produced in a batch-wise manner (except for the spinning of fibers) on pilot-scale. Some thoughts about fabrication in a continuous way are given at the end of the section. A general challenge for batch-wise production is to obtain equal conditions for each fiber inside a chamber or container. The parameters involved are usually concentration of different compounds and temperature. Equal conditions can be obtained by proper stirring or cross-flow of the medium onto the fibers.

When scaling up, the fiber yield has to be thoroughly evaluated in each process step since it is of economic importance. The fiber yield in this context is the number of fibers that meet certain agreed specifications compared to the total number of fibers that initially were produced. Obtaining a high yield is an exercise in perfection of each process step. To have adequate and calibrated measuring equipment is important in order to monitor the conditions in each process over time. When a process step is optimized, a detailed operating procedure has to be made to secure that the same conditions will be repeated or present for the next batch of fibers. Figure 14.5 illustrates some factors which have an impact on the fiber yield in the manufacturing process.

When the fiber length increases, the risk of a broken fiber increases due to handling and mechanical properties of the fibers, especially after carbonization.

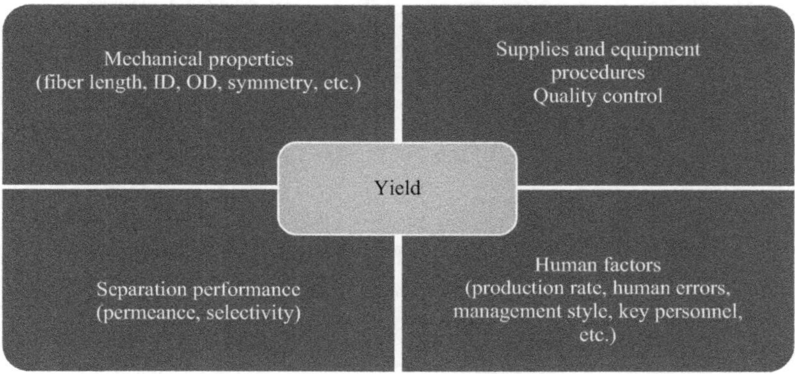

Figure 14.5   Factors affecting yield, that is, fraction of fibers that meet the product targets

Also, an asymmetric cross-section will make the fiber curl more upon drying. Such fibers have to be removed during the manufacturing process and hence yield is reduced. A low fiber diameter increases the flexibility of the fiber, which is important during handling and module making. However, the inner diameter (ID) has a lower limit due to pressure loss across the module in operation. Fibers made in our group typically have an outer diameter of about 180 μm, and a wall thickness of about 25 μm. A carbon fiber does not need to be free of small bends; they may be an advantage if temperature changes (expansion/contraction) and may improve the flow pattern on the shell side of the module. However, weak points in the fiber *have to* be eliminated; otherwise, the fiber will break at some point during operation. This is an important part of the quality control. Carbon fiber modules which do not meet the separation performance also have to be removed; otherwise, the purity and recovery demands in the separation process cannot be met.

Karvan *et al.* [21] found that if two polyimide fibers were in contact with each other during carbonization, they fused together at high temperatures, rendering them unusable. Polyimide and cellulose undergo different carbonization mechanisms and have different surface properties. For these reasons, cellulose fibers are less prone to fusing, and thousands of fibers may be carbonized in the same batch and in contact with each other. However, it is crucial to drain the tars and vapors as the carbonization progresses to higher temperatures. Initially, we used CO_2 as furnace atmosphere, since the carbon fibers showed higher permeability compared to inert sweep gas. We found however that the use of N_2 as sweep gas instead of CO_2 produces other or less amount of tar compounds, and a lower number of fibers were fused together after carbonization. As CO_2 is a major pyrolysis by-product, it is speculated that when lowering the CO_2 partial pressure in the furnace atmosphere, more gas and less tar is produced due to a shift in the equilibrium of the reactions. Tar and other by-products were drained using sweep gas flow in a cross-flow pattern, with multiple inlets of fresh gas along the fiber length.

In order to drain the tar and water produced during carbonization, a small tilt of the furnace (or worktube) was done to have an angle between the floor and the furnace. But this angle may become too high, as is shown in Table 14.4, in which the angle was halved from Y to $Y/2$.

Lowering the angle between the furnace tube and the floor significantly increased the yield of carbon fibers. Keeping a low angle probably prevents buoyancy of tars and vapors; hence, better draining or sweeping of these by-products during carbonization is achieved. The yield of fibers upstream carbonization

Table 14.4   Effect of furnace angle on yield during carbonization

Angle	Atmosphere	No. of fibers	Carbonization yield (%)	Total yield (%)
Y	N_2	1,000	52	47
$Y/2$	N_2	1,000	80	71

was the same for the two batches. A total yield of more than 70% is acceptable for a production process involving such a high number of steps.

Another important factor affecting bundle quality is a clean fiber surface when carbonization starts. In the washing step prior to drying, we compared using a surface active compound (quaternary ammonium salt), a sugar solution and an alcohol in the washing solution. We found that the degree of fused fibers after carbonization were in the order salt > sugar > alcohol.

High content of non-solvent in bore liquid is preferred to reduce the need for washing of fibers, and prevent the polymer fibers from re-dissolving. To secure high quality of fibers, the extrusion rate should be low, with many fiber lines in parallel. Low take-up rates also allow the contact time in each bath to be sufficient without having to build equipment with high space demand. A continuous line should be easier to automatize compared to a batch-wise production which is more dependent on human factors. However, it will be critical to control the tension of the fibers carefully.

Another challenge on upscaling of carbon membranes is related to module design and construction, typically on carbon membranes mounting, potting, and sealing for high temperature/pressure applications, which need to find a suitable potting material that can easily penetrate a bundle of fibers even at a high packing density, and should cure relatively fast, without generating excessive heat.

14.2.4   Sample applications

Hydrogen separation from different gas mixtures is one of the most attractive applications for carbon membranes, as its permeability is high. This application is probably the only one in which oxidation during carbonization or PO may not be needed in order to achieve a competitive permeance. Alternatively, making a carbon membrane separating according to the SSF mechanism reduces the need for recompression of hydrogen after separation/recovery.

Other promising applications include separation of molecules which differ in shape but otherwise have similar properties. Some examples are given below.

14.2.4.1   Hydrogen separation from different gas mixtures

Industrially, most of the hydrogen is produced by steam reforming, for instance in large multi-tubular fixed-bed reactors. The reactions are equilibrium limited and produce a hydrogen-rich gas mixture containing carbon oxides and other by-products. To improve purity and yield of hydrogen, a membrane reactor system could be a solution. Carbon membranes are suitable for membrane reactors due to their thermal and chemical resistance. Several researchers are working on carbon membranes for, for example, the water gas shift reactions [30].

The produced hydrogen may be distributed in the existing natural gas grid and separated at the end user sites. The NATURALHY project (sixth EU framework program) was launched in 2002 and aimed to test all critical components in a mixed network by adding hydrogen to existing natural gas networks (illustrated in Figure 14.6 [31]). Grainger and Hägg [32] studied the separation performance of H_2/CH_4 with CMS membranes based on experiments and process simulation in the

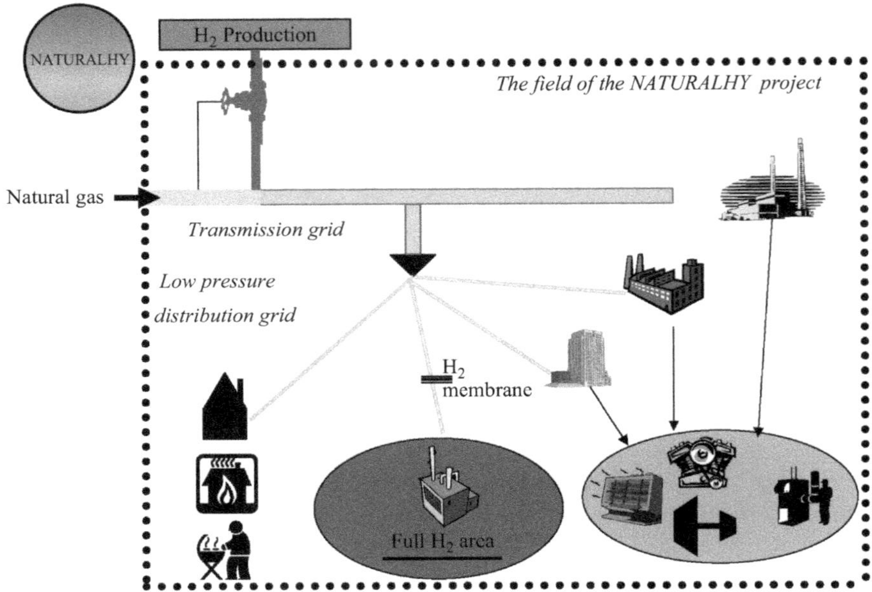

Figure 14.6   The field of NATURALHY project [31]

mentioned NATURALHY project. Their techno-economic evaluation indicated that CMS membranes can offer a great potential for hydrogen separation from hydrocarbon, and high purity hydrogen can be recovered from leaner streams in natural gas networks at low energy consumption.

14.2.4.2 Olefin/paraffin separation

Separation of gases according to shape discrimination is perhaps the most important feature or advantage of carbon membranes. For example: alkanes – alkenes are chemically and physically quite similar compounds with almost identical critical properties, it must be separated on the basis of molecular size difference with the Lennard-Jones diameter of 4.7 and 5.1 Å for propene and propane, respectively. There are few competing technologies in this market, perhaps only distillation, which is highly energy demanding.

Xu *et al.* [22] carbonized polyimide membranes and performed mixed gas tests with ethylene/ethane at 35 °C, at feed pressure 8 bar and permeate pressure 1 bar. After about 40 h the ethylene permeance stabilized at 13 GPU, and the selectivity at 4.

Hägg *et al.* [33] investigated carbon hollow-fiber lab-scale modules produced by Carbon Membranes Ltd. and performed mixed gas tests with propylene/propane at 50 °C, at feed pressure 3 bar and permeate pressure 1 bar. The permeance of propylene was 0.07 m^3 (STP)/(m^2 h bar) (i.e., 27 GPU), with a selectivity of 50.

14.2.4.3 CO_2/CH_4 separation

Fu *et al.* [6] prepared carbon membranes from four different polyimide precursors. The carbon with the highest initial permeability showed the highest permeability after aging (stabilization) in the feed gas (50/50 CO_2/CH_4 at 3.4 bar) for one month (although it showed the highest percentage of aging). The CO_2 permeability stabilized after about 10 days and was still more than 10,000 Barrer, and the CO_2/CH_4 selectivity was 35.

In the Memfo group at NTNU a module with 2,000 fibers was exposed to a feed of 5.5% CO_2 and balance CH_4 at a rate of 1 N d m^3/h, and found no change after 16 days (room temperature, 10 bar feed, 1 bar on permeate side). However, with the addition of 1,000 ppm *n*-heptane the CO_2 permeance was reduced by 23% for the same conditions. The gas was fed on the shell side, and bore side feed should therefore be investigated. We hypothesize that the extent of aging is then reduced (see Section 14.2.2).

Several modules of the same type as above were exposed to real biogas (63% CH_4, 1 ppm H_2S, balance CO_2) at a waste treatment plant in Southern Norway. The modules processed about 1 N m^3/h of biogas at 15–20 °C and 20 bar feed pressure. After 200 days in operation, the CO_2 permeance was reduced by 30%. Operating with an inside-out feed and at higher temperatures could be done to restore the capacity of the modules.

The pilot scale modules were also tested in lab for high pressures ($\rightarrow$ 50 bar) for potential application in natural gas sweetening. The separation performance was basically maintained at high pressures, which show an extraordinary advantage for the carbon membranes related to this application. The membrane showed no signs of plasticization by CO_2 which has been a drawback for polymeric membranes.

14.3   Supported carbon membranes

Although CMS membranes show significantly high separation performances, the major drawbacks are their rather small permeation rate and brittleness. Fabricating carbon membranes in the form of hollow fibers is one way to overcome the permeation limitations by applying a module of high packing density. Preparing CMS membranes on porous supports is another approach. As supports provide mechanical strength, the thickness of CMS membranes can be less than 1 μm, reducing the resistance for permeation and, thus, improving the flux. In addition, supported membranes give more variety in the choice of precursors. Furthermore, supported CMS membranes open new opportunities for high-temperature and high-pressure applications. One of the early studies of supported CMS membranes is reported by Rao and Sircar [26] at Air Products and Chemicals, called SSF membranes. Recent developments of supported CMS membranes in laboratory and in larger-scale applications are described in the following sections with some examples.

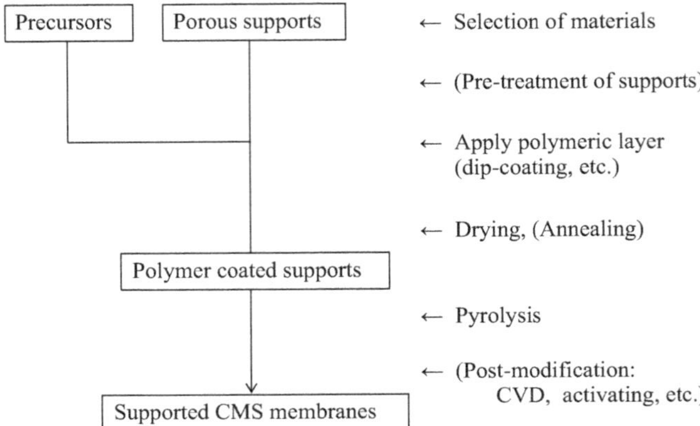

Figure 14.7   Schematic of a supported membrane preparation procedure

14.3.1   Preparation

Typical fabrication process of a supported CMS membrane is illustrated in Figure 14.7. The surface of a porous support is coated with a polymer thin layer and then pyrolyzed. The coating and the carbonizing processes are sometimes repeated to eliminate defects in the supported CMS membranes.

Various types of supports have been used that include polymeric [34], ceramic [35], carbon [36,37], and metallic [38] materials shaped as flat sheets, disks, tubulars, fibers, and monoliths. Dip-coating [39–41], spin coating [42], spray coating [43], ultrasonic deposition [38,44], vapor deposition [45], and other techniques [46] are employed to apply precursor layers on the supports.

Dip-coating is a simple method to apply polymeric layer on a porous support. Similar to the preparation of sol–gel delivered membranes [47], pore size of the support, properties of a precursor solution, withdrawal speed of supports from the precursor solution and drying conditions affect the morphology of a polymeric coating layer.

Same as self-supported CMS membranes, choice of the precursors, and the pyrolysis conditions affect strongly the properties of supported CMS membranes. Typically, supported CMS membranes have thickness of ca. 1–5 µm as shown in Figure 14.8.

Koga *et al.* [48] examined the influence of pyrolysis conditions by using two different heating methods: a conventional tube furnace with a heating rate of 5 °C/min and a high frequency induction heating with a heating rate of 500 °C/min. CMS membranes prepared under a rapid pyrolysis showed about ten times faster permeation rates than the membranes prepared in a conventional oven as shown in Figure 14.9. Placing position of membranes during the pyrolysis also affected the membrane properties. CMS membranes placed vertically during the pyrolysis showed higher separation ability and lower permeation flux than those placed horizontally. The difference is explained by the deposition of organic

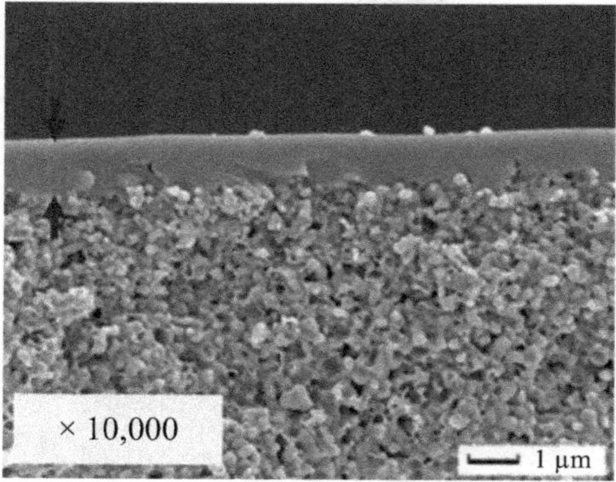

Figure 14.8   Cross-sectional view of a supported CMS membrane formed on a porous alumina tube

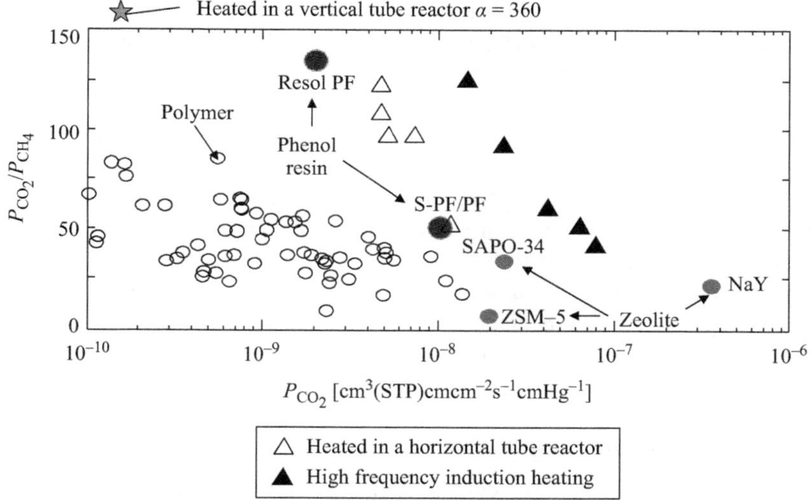

Figure 14.9   Influence of pyrolysis conditions on the membrane performance [48]

molecules released from the precursor during the carbonization. Membranes placed vertically got more deposition of organics, resulting in the formation of denser CMS layers [49].

As the CMS layer can be thin as 1 μm or less, the surface roughness of a support can be transferred to the CMS membrane formed over it. Therefore, surface modifications are sometimes applied to reduce the pore size and the surface roughness of a support, which helps a formation of thin continuous layer. Applying γ-Al_2O_3 layer on a commercial α-Al_2O_3 porous support is one of the major

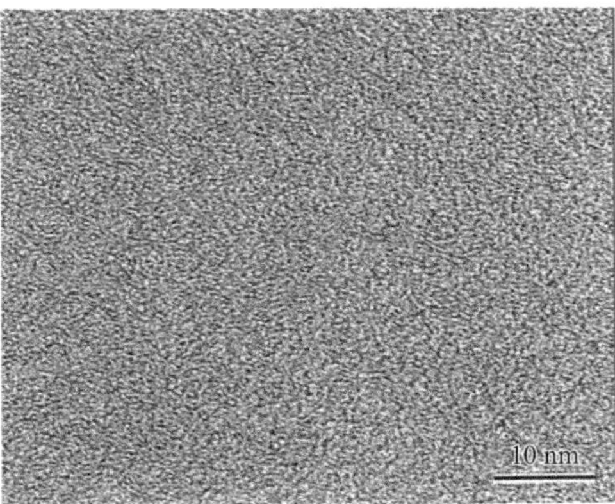

*Figure 14.10   TEM observation of a CMS membrane derived from lignophenol
with pyrolysis at 600 °C for 10 min in nitrogen [49]*

modification methods that can improve separation performance of resulting supported CMS membranes [35]. Pre-coating ceramic supports with a pencil is an interesting method that can reduce the surface roughness by a simple dry method [50].

Figure 14.10 shows a TEM observation of supported CMS membrane prepared from lignophenol. The CMS membrane was consisting of mostly amorphous phase. No significant differences were observed when the membrane morphologies at the membrane surface, at the center of carbon layer and at the interface to the support were compared [49]. The micro-structure of supported CMS membranes can be more compact than that of self-standing CMS membranes due to the interaction between the precursor and the support [51]. However, the influence of supports on perm-selectivity is not clear due to various other factors affecting more strongly on the membrane formation.

14.3.2   Status of development

Various types of polymers have been used as precursors of supported CMS membranes. Some examples are shown in Table 14.5. Permeation rate is expressed with a unit of $mol/m^2/s/Pa$ in the table. In some cases, post-modifications are applied on CMS membranes to plug defects by CVD [46] and to improve the permeating flux by activation with H_2, CO_2, and other gases that reduces the thickness of the CMS membrane and/or increases the pore size [52].

Figure 14.11(a) and (b) shows examples of supported CMS membrane properties in H_2/CH_4 and CO_2/CH_4 separations. Self-supported hollow CMS membranes [4,53] and supported membranes [37,38,45,51,54] show comparable performances in these systems. Olefin/paraffin separations are other separation processes in which membrane can play an important role. Although not many results can be found with supported CMS membranes for these applications, recently reported results showed higher performance than the trade-off curve [35]

Table 14.5   *Examples of supported CMS membrane preparation conditions and gas permeation results*

Supported Support/geometry	Modification	Pore size (µm)	Precursor	Method*	Pyrolysis	Gas test temp.	Permeance $(10^{-9}\,\text{mol/m}^2/\text{s/Pa})$**			Ideal selectivity (separation factor)			Reference
							H_2 (He)	CO_2	O_2	H_2/N_2	O_2/N_2	CO_2/CH_4	
Carbon/Disk Ø 35 mm	–		Matrimid	S	700 °C–2 h (vacuum)	25 °C	(27)	36.6	8.1	–	4?	8.1 (23)	[36]
Carbon/Disk Ø 35 mm			Phenolic resin	S	700 °C (vacuum)	25 °C	(8.2)	2.0	1.2	265		87 (165)	[37]
α-Al$_2$O$_3$/Tube (o.d. 2.3 mm)	–	0.14	Phenolic resin + Sulfonated phenolic resin	D	500 °C–1 h (N$_2$)	35 °C	56 134	10 40	2.3 10	160 –	10.8 12	170 54	[41]
α-Al$_2$O$_3$/Tube (o.d. 10 mm, i.d. 7 mm)	–	0.2	Novolac resin + Boehmite	D	550 °C–2 h (N$_2$)	RT†	145 (79.5)	–	3.0	725	15	–	[40]
α-Al$_2$O$_3$/Tube (o.d. 0.9 mm, i.d. 0.6 mm)	γ-Al$_2$O$_3$ Glass	0.005 0.0035	Furfuryl alcohol	V	600 °C–1 h 600 °C–1 h	25 °C 25 °C	25.5 6.04	5.82 2.67	0.775 0.845	347 91	10.6 12.7	92 82	[45]
α-Al$_2$O$_3$/o.d. 2.25 mm, i.d. 1.8 mm	–		Lignocresol	D	600 °C–1 h (N$_2$)	35 °C 105 °C	56 (29) 82 (43)	17 2.3	2.7 8.2	167 44	8.0 4.5	87 17	[39] [18]
Anodic alumina/4 cm^2	–	0.020	Graphene oxide	V	–	20 °C	100	0.03?	–	~900	–	–	[18]
Coal disk/Ø 40 mm × 2 mm thick	Meso-porous carbon	0.71 (largest size)	PMDA-ODA	S	800 °C–2 h		54.55	8.80	7.45	76.3	10.4	–	[42]
α-Al$_2$O$_3$/o.d. 13 mm i.d. 8 mm	2B pencil	3.0	Polyfurfuryl alcohol (PFA)	D	700 °C–4 h (Ar)		35	10	7.5	58	12.5	–	[50]

*C: cast coating, D: dip coating, S: spin coating, V: vapor phase deposition.

**: values were read from figures.

†: after air exposure.

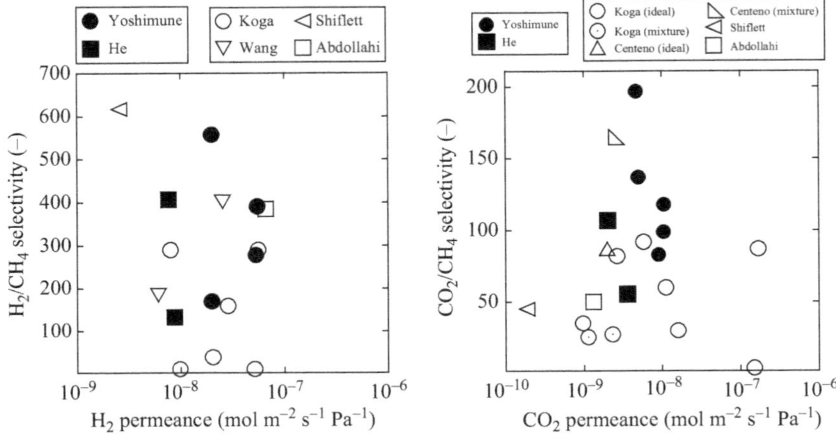

Figure 14.11   *Permeation performances of supported CMS membranes measured at 298–308 K (a) H₂/CH₄ selectivity as a function of H₂ permeance, (b) CO₂/CH₄ selectivity as a function of CO₂ permeance (Koga [48, 49], Wang [45], Shiflett [38], Centeno [37], Abdollahi [54], He [4] (thickness of 25 mm is used to calculate Barrer to permeance), Yoshimune [53]) (open keys: supported CMS membranes, closed keys: self-standing CMS membranes)*

(Table 14.6). Further developments of supported CMS membranes, such as applying variety of precursors and post-modifications will improve the membrane performance further.

As briefly mentioned in the introduction part, new types of carbon materials have recently been researched. These are typically carbon nanotube, graphene and GO, and diamond-like carbon (DLC), which are shaped into membranes and their interesting permeation properties are reported. For example, as graphene has a single atomic layer, the thickness of the graphene membranes can be quite small and, thus, higher flux can be expected. Li *et al.* covered an anodic alumina oxide support with single-layered GO flakes by filtration [44]. GO membranes with thickness varied from 1.8 to 18 nm were prepared. These membranes showed significantly high hydrogen/carbon dioxide separation factor over 2,000 at 20 °C with hydrogen permeance of ca. 1×10^{-7} mol/m^2/s/Pa.

Karan *et al.* [55] prepared DLC membranes with a thickness ranging from 10 to 40 nm by plasma-enhanced CVD on anodized alumina flat supports pre-coated with a sacrificial layer. The DLC membranes have pores of ca. 1 nm and showed high solvent flux of, for example, ethanol flux of 67.3 L/m^2/h at 80 kPa pressure difference with 100% rejection of molecules with ca. 1.5 nm size. Cortese *et al.* [56] coated the surface of cotton textiles with DLC films by plasma-enhanced CVD and examined their oil–water separation performance. Anti-fouling properties

Table 14.6   Olefin/paraffin separation examples with CMS membranes

Support (pore size in μm)	Precursor	Pyrolysis temp.-time	Perm. Temp. (°C)	Permeance $(10^{-9}\ mol/m^2/s/Pa)^*$		Permselectivity (separation factor)		Reference
				C_2H_4	C_3H_6	$C_2H_4/$ C_2H_6	$C_3H_6/$ C_6H_8	
α-Al$_2$O$_3$ (0.14)	BPDA-ODA	700 °C–0	35		0.87 (0.79)	4.4	54 (46)	[57]
			100		1.7 (1.5)	4.5	33 (29)	
γ-Al$_2$O$_3$ (–)	6FDA	550 °C–2 h	25	–	3 (3.2)**	–	– (36)**	[35]

*Values are read from figures.
**50:50 mixture.

of, for example, silicon- and fluorine-doped DLC [58,59] are also reported. These results show various new potential applications of carbon membranes.

14.3.3   Sample applications

Membrane module design is one of the keys to bring the membranes from laboratory scale to pilot scale. Scalable, reproducible, and cost-effective preparation of CMS membranes and modules are other important factors. Several module configurations have been reported for the supported CMS membranes. Some examples are shown in Figure 14.12. Rao and Sircar at Air Products and Chemicals formed CMS membranes on flat porous sheets having 0.7 μm pore size [60,61]. CMS membrane sheets were assembled to form a module with a total surface area of 0.356 m^2 (Figure 14.12(a)). They have tested the module in separating hydrogen–hydrocarbon mixtures, such as hydrogen recovery from a fluid catalytic cracker off-gas stream, and from the PSA waste gas [62]. According to R. Baker, "the process was tried at the pilot-plant scale, but eventually abandoned in part because of blocking of the membranes by permanently adsorbed higher hydrocarbons."

A honeycomb configuration is reported by Blue Membrane GmbH (Germany) (Figure 14.12(b)) [46]. The module is produced by applying thin precursor film on a flat paper modified with ceramic fibers, stamping wavy pattern, applying sealing to the edges and then carbonized. CVD was applied to tune the membrane property. The maximum packing density was up to 2,500 m^2/m^3 with 10 m^2 per module [46]. NGK (Japan) reported a multi-channel configuration (Figure 14.12(c)) [63]. Asymmetric multi-channel supports made of Al$_2$O$_3$ and TiO$_2$ with the finest pore size of 0.1 μm were coated with phenol resin and heated at 700 °C. CMS membranes showed water selective permeation with a separation factor of ca. 300 and flux of 1.2 kg/m^2/h when 70% acetic acid solution was applied at 70 °C. Dehydration of acidic fluids is a potential application of CMS membranes in which commercialized A-type zeolite membranes cannot be employed due to their insufficient acid stabilities. For industrialization, flux may need to be improved with a factor of 2 or more.

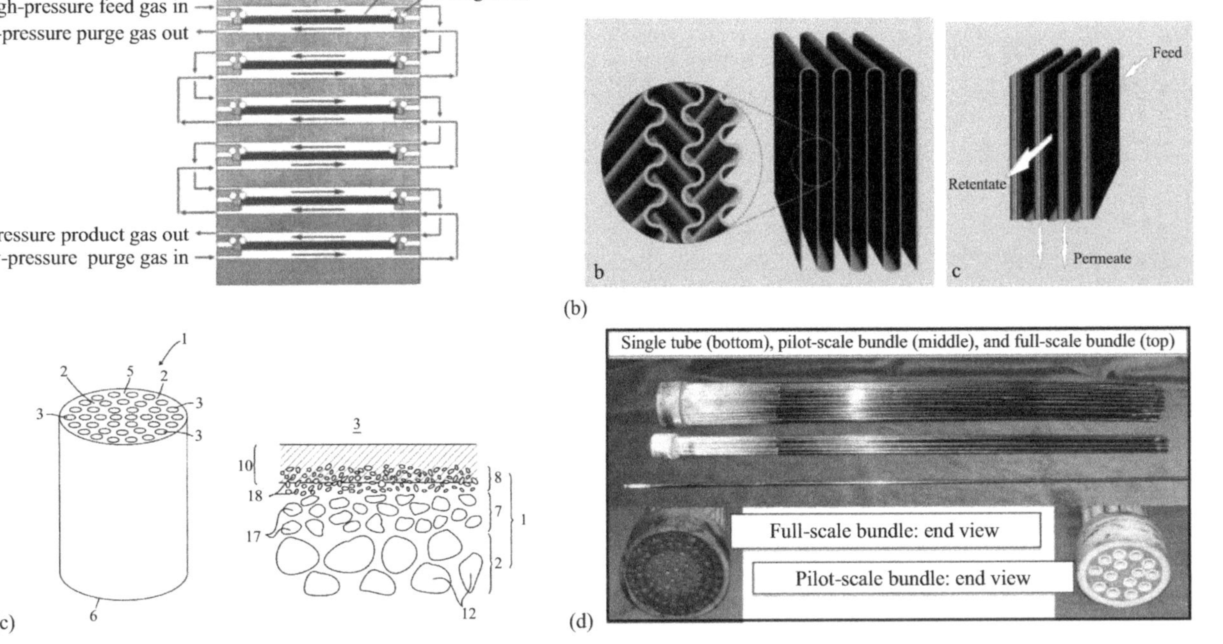

Figure 14.12   Examples of module design with supported CMS membranes (a) [25], (b) [64], (c) [63], and (d) [65]

Media and Process Technology Inc., and Tsotsis *et al.* at University of Southern California have been investing the application of CMS membranes to water–gas shift (WGS) reaction [54,65]. They have demonstrated a membrane module consisting of tubular CMS membranes with a surface area of 0.76 m^2 and a packing density of 222 m^2/m^3 [65] (Figure 14.12(d)). A bundle of supports was fabricated beforehand the application of CMS layers via dip-coating and carbonization. The performance of CMS membrane modules was tested in hydrogen recovery from syngas. Stable hydrogen permeation from raw coal-delivered and biomass-delivered syngas was demonstrated through field-tests for more than 500 cumulative hours. Integration of CMS membranes to WGS reaction as membrane reactors improved CO conversion [54]. Palladium and palladium-alloy membranes are alternative choice for hydrogen recovery and have been intensively studied in the application of WGS reactions as a membrane reactor. However, surfer poisoning is reported with palladium-based membranes that reduce the permeation rate [66,67]. CMS membranes, on the contrary, showed high stability against H_2S, NH_3, and other contaminants in the syngas [54,65].

Hydrogen carrier system is another example in which CMS membranes can be applied. Transporting and storing hydrogen in the form of liquid hydrocarbon is considered through, for example, Euro-Quebec project [68] and ALCA/SIP project in Japan (Council for Science, Technology and Innovation (CSTI), Cross-ministerial Strategic Innovation Promotion Program (SIP), "energy carrier" (Funding agency: JST), 2013–2017). For example, methyl-cyclohexane releases hydrogen with a reaction as methyl-cyclohexane $\rightleftarrows$ toluene $+ 3H_2$. The dehydrogenation is an endothermic reaction and requires over 573 K to reach 100% conversion. Combining membranes to the dehydrogenation reaction for a simultaneous removal of formed hydrogen is studied with various types of membranes [52,69,70]. Itoh and Haraya [70] applied carbon fibers, prepared by carbonizing commercial polyimide hollow fibers, to the dehydrogenation of cyclohexane. Hirota *et al.* [52] prepared supported CMS membranes on α-Al_2O_3 tubes from furfuryl alcohol by a vapor-phase synthesis and applied them in dehydrogenation of methyl-cyclohexane. In both cases, CMS membrane reactors improved the conversion rate from the equilibrium value; for example ca. 20% to ca. 50% at 473 K [52]. The results suggest a possibility to reduce the dehydrogenation temperature by integrating membrane separation to the catalytic reaction. Lower temperature will be able to improve the efficiency of hydrogen carrier system, and also to prevent carbon formation on the catalysts.

Integrating membrane separation process to photo-catalysis for a production of hydrogen via photo-catalytic water splitting is under investigation through a Japanese national project financially supported by New Energy and Industrial Technology (Japan Technological Research Association of Artificial Photosynthetic Chemical Process (ARPChem), 2011–2016). Membranes having hydrogen/oxygen separation factor over 45 is required to obtain hydrogen purity over 96% in the permeate with a hydrogen recovery over 98% [71]. CMS membranes are one of the promising candidates due to their molecular-sieving properties and the hydrophobic nature.

Supported CMS membranes, especially when ceramic supports are used, have less economic competitiveness today due to the rather expensive support price. However, the uniqueness of supported CMS membranes compared to the conventional polymeric membranes is obvious in high-temperature and high-pressure applications, in which polymeric membranes cannot be employed. Integrating CMS membranes to dehydrogenation reactions is one example of such. Solvent filtration is another interesting potential application. Several module designs and fabrication methods have also been proposed and demonstrated. Resent results confirmed high resistance of CMS membranes against H_2S, NH_3, and other chemicals. Further improvements on supported membrane properties will strengthen the competitiveness of CMS membranes and will contribute realizing environmental and economic friendly processes.

14.4   Conclusions

In this current chapter, the status of carbon membranes has been reviewed with respect to preparation and the level of development with respect to scaling up and their potential use for industrial applications. Different designs exist, such as self-supported hollow fibers, supported flat sheets or sheets of honeycomb structure, as well as carbon membranes supported on ceramic tubes. A small number of companies (5–6) have tried to bring these membranes to the market on industrial scale, but not yet fully succeeded. A main challenge for all has been the module design and the costly carbon membrane production which is very difficult to run as a continuous process. The carbonization step must be very fine-tuned to have a high yield of perfect membranes. However, the excellent separation performance these membranes usually have indicates that the research for continuous more robust membranes and the problem of module design will be solved. The self-supported hollow fibers have a potential of lower costs and application for larger gas volumes, whereas the supported carbon membranes which have a lower packing density will probably also be a good choice for medium to small gas volumes. The recovery of hydrogen from various gas streams may become a major application, as well as olefin–paraffin separation, but also removal of CO_2 from natural gas or biogas (CO_2–CH_4 separation) has a very nice potential. For some gas streams containing water vapor or higher hydrocarbons, a system for regular regeneration of the carbon membranes will have to be put up – there are various techniques documented for this. With the possibility of tailoring, the membrane pore size which these membranes have, there are a huge number of potential gas separation applications for the carbon membranes.

List of acronyms

CMS   carbon molecular sieve
DLC   diamond-like carbon

GO   graphene oxide
GPU   gas permeance unit
PSA   pressure swing adsorption
SEM   scanning electron microscopy
STP   standard temperature and pressure
SSF   selective surface flow
TEM   transmission electron microscopy

References

[1] Koresh JE, Soffer A. Molecular sieve carbon membrane. Part I: presentation of a new device for gas mixture separation. Separ Sci Technol. 1983;18: 723–34.

[2] Steel KM, Koros WJ. Investigation of porosity of carbon materials and related effects on gas separation properties. Carbon. 2003;41(2):253–66.

[3] Lie JA, Hagg M-B. Carbon membranes from cellulose: synthesis, performance and regeneration. J Membr Sci. 2006;284(1–2):79–86.

[4] He X, Hägg M-B. Structural, kinetic and performance characterization of hollow fiber carbon membranes. J Membr Sci. 2012;390–391(0):23–31.

[5] Zhang B, Wu Y, Lu Y, Wang T, Jian X, Qiu J. Preparation and characterization of carbon and carbon/zeolite membranes from ODPA–ODA type polyetherimide. J Membr Sci. 2015;474:114–21.

[6] Fu S, Sanders ES, Kulkarni SS, Koros WJ. Carbon molecular sieve membrane structure–property relationships for four novel 6FDA based polyimide precursors. J Membr Sci. 2015;487:60–73.

[7] He X, Lie JA, Sheridan E, Hägg M-B. Preparation and characterization of hollow fiber carbon membranes from cellulose acetate precursors. Ind Eng Chem Res. 2011;50(4):2080–7.

[8] He X, Hagg M-B. Optimization of carbonization process for preparation of high performance hollow fiber carbon membranes. Ind Eng Chem Res. 2011;50(13):8065–72.

[9] Suda H, Haraya K. Gas permeation through micropores of carbon molecular sieve membranes derived from Kapton polyimide. J Phys Chem B. 1997; 101(20):3988–94.

[10] Barsema JN, van der Vegt NFA, Koops GH, Wessling M. Ag-functionalized carbon molecular-sieve membranes based on polyelectrolyte/polyimide blend precursors. Adv Funct Mater. 2005;15(1):69–75.

[11] Rungta M, Xu L, Koros WJ. Carbon molecular sieve dense film membranes derived from Matrimid® for ethylene/ethane separation. Carbon. 2012;50(4): 1488–502.

[12] Bhuwania N, Labreche Y, Achoundong CSK, *et al.* Engineering substructure morphology of asymmetric carbon molecular sieve hollow fiber membranes. Carbon. 2014;76:417–34.

[13]   David LIB, Ismail AF. Influence of the thermastabilization process and soak time during pyrolysis process on the polyacrylonitrile carbon membranes for O_2/N_2 separation. J Membr Sci. 2003;213(1–2):285–91.

[14]   Fuertes AB, Centeno TA. Preparation of supported asymmetric carbon molecular sieve membranes. J Membr Sci. 1998;144(1–2):105–11.

[15]   Xu L, Rungta M, Brayden MK, *et al.* Olefins-selective asymmetric carbon molecular sieve hollow fiber membranes for hybrid membrane-distillation processes for olefin/paraffin separations. J Membr Sci. 2012;423–424:314–23.

[16]   Hägg MB, He X. Chapter 15: Carbon molecular sieve membranes for gas separation. In Drioli E, Barbieri G., eds. Membrane Engineering for the Treatment of Gases: Gas-separation Problems Combined with Membrane Reactors, The Royal Society of Chemistry, ISBN:978-1-84973-239, 2011. p. 162–191.

[17]   Salleh WNW, Ismail AF. Carbon membranes for gas separation processes: recent progress and future perspective. J Membr Sci Res. 2015;1:2–15.

[18]   Li H, Song Z, Zhang X, *et al.* Ultrathin, molecular-sieving graphene oxide membranes for selective hydrogen separation. Science. 2013;342(6154): 95–8.

[19]   Romanos G, Pastrana-Martínez LM, Tsoufis T, *et al.* A facile approach for the development of fine-tuned self-standing graphene oxide membranes and their gas and vapor separation performance. J Membr Sci. 2015;493: 734–47.

[20]   Soffer A, Gilron J, Suguee S, Hed-Ofek R, Cohen H. Process for the production of hollow carbon fiber membranes. EP Patent 0671202. 1995.

[21]   Karvan O, Johnson JR, Williams PJ, Koros WJ. A pilot-scale system for carbon molecular sieve hollow fiber membrane manufacturing. Chem Eng Technol. 2013;36(1):53–61.

[22]   Xu L, Rungta M, Hessler J, *et al.* Physical aging in carbon molecular sieve membranes. Carbon. 2014;80:155–66.

[23]   Lie JA. Synthesis, performance and regeneration of carbon membranes for biogas upgrading-a future energy carrier. PhD Thesis, Norwegian University of Science and Technology, 2005.

[24]   Boehm HP. Surface oxides on carbon and their analysis: a critical assessment. Carbon. 2002;40(2):145–9.

[25]   Rao MB, Sircar S. Nanoporous carbon membranes for separation of gas mixtures by selective surface flow. J Membr Sci. 1993;85(3):253–64.

[26]   Jones CW, Koros WJ. Carbon composite membranes: a solution to adverse humidity effects. Ind Eng Chem Res. 1995;34(1):164–7.

[27]   Geiszler VC, Koros WJ. Effects of polyimide pyrolysis conditions on carbon molecular sieve membrane properties. Ind Eng Chem Res. 1996;35(9): 2999–3003.

[28]   Su J, Lua AC. Influence of carbonisation parameters on the transport properties of carbon membranes by statistical analysis. J Membr Sci. 2006; 278(1–2):335–43.

[29] Jones CW, Koros WJ. Carbon molecular sieve gas separation membranes-II. Regeneration following organic exposure. Carbon. 1994;32(8):1427–32.

[30] Liu PKT, Sahimi M, Tsotsis TT. Process intensification in hydrogen production from coal and biomass via the use of membrane-based reactive separations. Curr Opin Chem Eng. 2012;1(3):342–51.

[31] Huizing R. The NATURALHY project: preparing for the hydrogen economy by using the existing natural gas system as a catalyst [cited 16 October 2015]. Available from: https://www.coleurope.eu/content/study programmes/eco/conferences/Files/Papers/Florisson_The_NATURALHY_ Project.pdf

[32] Grainger D, Hägg M-B. The recovery by carbon molecular sieve membranes of hydrogen transmitted in natural gas networks. Int J Hydrogen Energy. 2008;33(9):2379–88.

[33] Hagg M-B, Lie JA, Lindbrathen A. Carbon molecular sieve membranes. A promising alternative for selected industrial applications. Ann Ny Acad Sci. 2003;984(1):329–345.

[34] Wei W, Qin G, Hu H, You L, Chen G. Preparation of supported carbon molecular sieve membrane from Novolac phenol–formaldehyde resin. J Membr Sci. 2007;303(1–2):80–5.

[35] Ma X, Lin BK, Wei X, Kniep J, Lin YS. Gamma-alumina supported carbon molecular sieve membrane for propylene/propane separation. Ind Eng Chem Res. 2013;52(11):4297–305.

[36] Fuertes AB, Nevskaia DM, Centeno TA. Carbon composite membranes from Matrimid® and Kapton® polyimides for gas separation. Micropor Mesopor Mater. 1999;33(1–3):115–25.

[37] Centeno TA, Fuertes AB. Supported carbon molecular sieve membranes based on a phenolic resin. J Membr Sci. 1999;160(2):201–11.

[38] Shiflett MB, Foley HC. On the preparation of supported nanoporous carbon membranes. J Membr Sci. 2000;179(1–2):275–82.

[39] Koga T, Kita H, Uemura K, Tanaka K, Kawafune I, Funaoka M. Structure and separation performance of carbon membranes from lignin-based materials. Trans Mater Res Soc Japan. 2009;34(3):371–4.

[40] Llosa Tanco MA, Pacheco Tanaka DA, Mendes A. Composite–alumina–carbon molecular sieve membranes prepared from Novolac resin and Boehmite. Part II: effect of the carbonization temperature on the gas permeation properties. Int J Hydrogen Energy. 2015;40(8):3485–96.

[41] Zhou W, Yoshino M, Kita H, Okamoto K-i. Preparation and gas permeation properties of carbon molecular sieve membranes based on sulfonated phenolic resin. J Membr Sci. 2003;217(1–2):55–67.

[42] Li L, Song C, Jiang H, Qiu J, Wang T. Preparation and gas separation performance of supported carbon membranes with ordered mesoporous carbon interlayer. J Membr Sci. 2014;450:469–77.

[43] Acharya M, Foley HC. Spray-coating of nanoporous carbon membranes for air separation. J Membr Sci. 1999;161(1–2):1–5.

[44] Li H, Song Z, Zhang X, *et al.* Ultrathin, Molecular-Sieving Graphene Oxide Membranes for Selective Hydrogen Separation. Science. 2013;342 (6154):95–8.

[45] Wang H, Zhang L, Gavalas GR. Preparation of supported carbon membranes from furfuryl alcohol by vapor deposition polymerization. J Membr Sci. 2000;177(1–2):25–31.

[46] Lagorsse S, Leite A, Magalhães FD, Bischofberger N, Rathenow J, Mendes A. Novel carbon molecular sieve honeycomb membrane module: configuration and membrane characterization. Carbon. 2005;43(4):809–19.

[47] Brinker CJ, Sehgal R, Hietala SL, *et al.* Sol–gel strategies for controlled porosity inorganic materials. J Membr Sci. 1994;94(1):85–102.

[48] Koga T, Kita H, Tanaka K, Funaoka M. Rapid pyrolysis of lignin-based materials for the synthesis of molecular sieve carbon membrane, Trans Matter Res Soc. Japan. 2006;31(2):287–290.

[49] Koga T. Studies on carbon membranes from wood materials. PhD Thesis. Yamaguchi University, Graduate School of Science and Engineering, Ube, Japan. 2013.

[50] Wang C, Ling L, Huang Y, Yao Y, Song Q. Decoration of porous ceramic substrate with pencil for enhanced gas separation performance of carbon membrane. Carbon. 2015;84:151–9.

[51] Briceño K, Montané D, Garcia-Valls R, Iulianelli A, Basile A. Fabrication variables affecting the structure and properties of supported carbon molecular sieve membranes for hydrogen separation. J Membr Sci. 2012;415–416:288–97.

[52] Hirota Y, Ishikado A, Uchida Y, Egashira Y, Nishiyama N. Pore size control of microporous carbon membranes by post-synthesis activation and their use in a membrane reactor for dehydrogenation of methylcyclohexane. J Membr Sci. 2013;440:134–9.

[53] Yoshimune M, Haraya K. Flexible carbon hollow fiber membranes derived from sulfonated poly(phenylene oxide). Sep Purif Technol. 2010;75(2): 193–7.

[54] Abdollahi M, Yu J, Liu PKT, Ciora R, Sahimi M, Tsotsis TT. Hydrogen production from coal-derived syngas using a catalytic membrane reactor based process. J Membr Sci. 2010;363(1–2):160–9.

[55] Karan S, Samitsu S, Peng X, Kurashima K, Ichinose I. Ultrafast viscous permeation of organic solvents through diamond-like carbon nanosheets. Science. 2012;335(6067):444–7.

[56] Cortese B, Caschera D, Federici F, Ingo GM, Gigli G. Superhydrophobic fabrics for oil–water separation through a diamond like carbon (DLC) coating. J Mater Chem A. 2014;2(19):6781–9.

[57] Hayashi J-i, Mizuta H, Yamamoto M, Kusakabe K, Morooka S, Suh S-H. Separation of ethane/ethylene and propane/propylene systems with a car-bonized BPDA−pp'ODA polyimide membrane. Ind Eng Chem Res. 1996;35(11):4176–81.

[58] Su XJ, Zhao Q, Wang S, Bendavid A. Modification of diamond-like carbon coatings with fluorine to reduce biofouling adhesion. Surf Coat Technol. 2010;204(15):2454–8.

[59] Zhao Q, Liu Y, Wang C, Wang S. Bacterial adhesion on silicon-doped diamond-like carbon films. Diamond Relat Mater. 2007;16(8):1682–7.

[60] Rao MB, Sircar S. Performance and pore characterization of nanoporous carbon membranes for gas separation. J Membr Sci. 1996;110(1):109–18.

[61] Liu PKT. Carbon molecular sieve membrane as a true one box unit for large scale hydrogen production, Final technical report, Vol. Department of energy, The United States, DE-FC26-07NT43057, 2012.

[62] Sircar S, Waldron WE, Rao MB, Anand M. Hydrogen production by hybrid SMR–PSA–SSF membrane system. Sep Purif Technol. 1999;17(1):11–20.

[63] Ichikawa A, Sakai T, Kinoshita N, Isoda Y, Kimata T, inventors; Method for producing carbon film. 2013: Vol. WO/2013/042262.

[64] Lagorsse S, Campo MC, Magalhaes FD, Mendes A. Water adsorption on carbon molecular sieve membranes: experimental data and isotherm model. Carbon. 2005;43(13):2769–79.

[65] Parsley D, Ciora Jr RJ, Flowers DL, *et al.* Field evaluation of carbon molecular sieve membranes for the separation and purification of hydrogen from coal- and biomass-derived syngas. J Membr Sci. 2014;450:81–92.

[66] Kajiwara M, Uemiya S, Kojima T. Stability and hydrogen permeation behavior of supported platinum membranes in presence of hydrogen sulfide. Int J Hydrogen Energy. 1999;24(9):839–44.

[67] Morreale BD, Ciocco MV, Howard BH, Killmeyer RP, Cugini AV, Enick RM. Effect of hydrogen-sulfide on the hydrogen permeance of palladium–copper alloys at elevated temperatures. J Membr Sci. 2004;241(2):219–24.

[68] Gretz J, Baselt JP, Ullmann O, Wendt H. The 100 MW Euro-Quebec hydro-hydrogen pilot project. Int J Hydrogen Energy. 1990;15(6):419–24.

[69] Oda K, Akamatsu K, Sugawara T, Kikuchi R, Segawa A, Nakao S-i. Dehydrogenation of methylcyclohexane to produce high-purity hydrogen using membrane reactors with amorphous silica membranes. Ind Eng Chem Res. 2010;49(22):11287–93.

[70] Itoh N, Haraya K. A carbon membrane reactor. Catal Today. 2000;56(1–3):103–11.

[71] Tanaka K, Sakata Y. Present and future prospects of hydrogen production process constructed by the combination of photocatalytic H_2O splitting and membrane separation process. Membrane. 2011;36(3):113–21.

Chapter 15

Separation of hydrogen isotopes by cryogenic distillation

Gianluca Valenti[1]

Abstract

The hydrogen element exists naturally in the form of three isotopes, sharing the same number of proton and electron, which is equal to 1, but not that of neutrons, which ranges from 0 to 2. In order, these isotopes are protium, commonly said light hydrogen and indicated with $_1^1H$ or simply H; deuterium, commonly heavy hydrogen indicated with $_1^2H$ or D; and tritium, $_1^3H$ or T. Naturally, deuterium abundance is 0.0115%, whereas tritium is rare and radioactively unstable. Protium, deuterium and tritium form diatomic molecules bonding together, which can be homonuclear, H_2, D_2 and T_2, or heteronuclear, HD, HT and DT. Homonuclear molecules can exist in either an *ortho* modification, oH_2, oD_2, oT_2, or a *para* modification, pH_2, pD_2, pT_2. Hydrogen has the largest isotope effects principally due to the largest differences in the relative mass of its isotopes. Isotope effects are differences in chemical and physical properties arising from differences in the nuclear mass. In particular, lighter hydrogen molecules are characterized by higher vapour pressures than heavier ones; in other words, lighter molecules are more volatile. Among the isotope separation techniques, distillation is adopted in industrial applications because of the advantages of achieving high separation degrees and of processing large quantities of fluids. Distillation is based on the different vapour pressures of the components to be separate and; hence, it requires the coexistence of liquid and vapour phases. Coexistence occurs in the cryogenic range of 10–40 K for molecular hydrogen. The number of cryogenic distillation plants constructed for deuterium and tritium separation is small due to their limited market. One example is the deuterium plant built in Germany in the late 1960s, and another the tritium plant in Canada in the late 1980s. Both plants proved the possibility to achieve high purities, exceeding 99.8%, as well as high separation factors. Today, deuterium is employed mostly as constituent of heavy water as neutron moderator for a number of nuclear fission reactors; it is also utilized for the preparation of nuclear

[1]Politecnico di Milano, Via Lambruschini 4A, 20156 Milano, Italy

weapons or as a non-radioactive tracer in chemical and metabolic reactions. Tritium is used instead as a radioactive tracer in chemistry and biology. Both deuterium and tritium are adopted for the research on the physics of matter and, notably, they have been selected for the future International Thermonuclear Experimental Reactor nuclear fusion reactor.

15.1   Introduction to the rationale of separating hydrogen isotopes

The origin of hydrogen isotopes, as after all the origin of everything, dates back to about 13.7 billion years, at the time of the Big Bang. The instant after the bang, the Universe was filled with elementary particles including electrons, whereas protons and neutrons were formed shortly after, but within 1 s. Few minutes later, protons and neutrons combined into the first nuclei of deuterium and helium, whereas most protons remained separated as hydrogen nuclei. It took hundreds of thousands of years for electrons and nuclei to unite into the first atoms, which were and have remained mostly hydrogen [1,2]. Today, the hydrogen element constitutes 90% of the total atoms or 75% of the total matter preceding by far helium, the second most abundant element. In Space, hydrogen exists predominantly in the atomic and in the ionic state, while partly in the molecular form. On Earth, hydrogen is bonded mostly to other atoms in chemical compounds, such as water and hydrocarbons.

Hydrogen exists naturally as three *isotopes*, which are called commonly (light) hydrogen (H), deuterium (D) and tritium (T). (The complete terminology is presented in Section 15.2.) They share the same number of proton and electron, which is simply equal to 1, but not that of neutrons, which ranges from 0 to 2, respectively. The difference in the neutron number leads to a large relative difference in the atom mass and, ultimately, in physical and chemical properties of the isotopes. Above all, hydrogen and deuterium are stable isotopes, whereas tritium is a radioactive unstable isotope with a relatively long decay time.

In most hydrogen processes, a distinction among its isotopes is unnecessary being the concentration of the light isotope larger by far than that of the heavier isotopes. However, these heavier isotopes have peculiar applications, either civil or military, scientific or industrial, that place a great attention upon them. Deuterium is employed as constituent of the so-called heavy water (water molecules containing mostly one or two deuterium atoms, HDO and more importantly D_2O, instead of hydrogen atoms, H_2O), which is employed in types of nuclear fission reactors for stationary power generation as *neutron moderator* [3]. It is also employed for the preparation of nuclear weapons or as a non-radioactive tracer in chemical and metabolic reactions. Tritium is used instead as a radioactive tracer in chemistry and biology. Both deuterium and tritium are adopted as substrates for the research on the physics of matter and, notably, they have been selected as the best combination for fuelling the future International Thermonuclear Experimental Reactor [widely known as International Thermonuclear Experimental Reactor (ITER)] for power generation [4,5]. Today, the main market of deuterium remains the production of heavy water, whereas tritium is still a niche product.

In general, the techniques for *isotope separation* from their mixture are as follows:

1. dependent directly on their diverse atom mass;
2. dependent indirectly on the mass;
3. independent from the mass.

The first kind comprises gravity sedimentation, thermal diffusion, centrifugation, permeation through membranes, electromagnetic deflection, and laser excitation coupled to electric attraction. The second include physical and chemical methods based on the fact that lighter isotopes evaporate at lower temperatures and react at higher rates than heavier ones; in particular, the change in the rate of a chemical reaction upon substitution of one isotope is named *kinetic isotope effect*. This second class includes *distillation* and chemical processes. The third one embraces gas chromatography, absorption and adsorption. The alternative to isotope separation is *isotope synthetization*, which allows obtaining the desired isotope in a pure form by irradiating with neutrons a suitable substrate.

In the case of the hydrogen element, deuterium is produced conventionally by isotopic separation from light hydrogen in an either direct or indirect method. A *direct method* processes a mixture containing exclusively the hydrogen isotopes, whereas an *indirect method* requires a *third body*. The deuterium direct separation is achieved mostly by way of the distillation of a light hydrogen–deuterium mixture at cryogenic temperatures, whereas the indirect by way of the deuterium-enrichment of conventional water to heavy water coupled to the electrolysis of the produced heavy water (water is the third body) [6]. Due to its instability and low natural concentration, tritium is produced conventionally by synthetization irradiating lithium or heavy water with neutrons. The neutron irradiation of heavy water occurs intrinsically in those nuclear reactors employing heavy water as neutron moderator. In such a case, tritium is produced by the *detritiation* of the irradiated heavy water by way of a catalytic transfer of tritium from the heavy water to a light hydrogen–deuterium stream coupled to the cryogenic distillation of that stream [7,8]. In the future, fusion reactors will also employ a system for the separation of hydrogen isotope based on cryogenic distillation [9,10].

The present chapter focuses on the cryogenic distillation because, on one side, it is a mature process adapt for industrial-size applications and, on the other, it is a direct method for separating the hydrogen isotopes from their mixture in order to 'purify' a hydrogen stream into ideally three streams: light hydrogen, deuterium and tritium streams. The following paragraphs illustrate the physical characteristics of the hydrogen isotopes and of the molecules that they form bonding together. They also explain both the basics of the cryogenic distillation as well as of the cryogenic liquefaction. Lastly, they describe two large-scale plants built during the years of the growing interests in the nuclear energy. All paragraphs aim at reporting general concepts, while providing a number of references that can be referred to for retrieving specific notions.

15.2 Hydrogen isotopes

Hydrogen is the most diffuse element in the Universe, one of the most common on Earth and a very vital one for industrial processes and, not less importantly, for life. The following sections are dedicated to the description of hydrogen as an element and as a molecule, highlighting in particular their thermodynamic behaviours that have a direct effect on the design of the liquefaction cycle and the cryogenic distillation. For completeness, a first section reports the general terminology about isotopes.

15.2.1 General terminology

The general terminology about isotopes is provided accordingly to the International Union of Pure and Applied Chemistry (IUPAC) [11]. Terms are listed in a logical sequence rather than in an alphabetical one.

Atom. The smallest particle still characterizing a chemical element. It consists of a nucleus of a positive charge (Z is the proton number and e the elementary charge) carrying almost all its mass (more than 99.9%) and Z electrons determining its size.

Element or *chemical element*. A species of atoms; all atoms with the same number of protons in the atomic nucleus.

Nuclide. Species of atom, characterized by its mass number, atomic number and nuclear energy state, provided that the mean life in that state is long enough to be observable.

Isotopes. Nuclides having the same atomic number but different mass numbers.

Atomic number or *proton number* (Z). The number of protons in the atomic nucleus.

Mass number or *nucleon number* (A). Total number of heavy particles (protons and neutrons jointly called nucleons) in the atomic nucleus.

Elementary charge or *proton charge* (e). Electromagnetic fundamental physical constant equal to the charge of a proton and used as atomic unit of charge, equal to $1.602\ 176\ 487(40) \times 10^{-19}\ °C$.

Isotope exchange. A chemical reaction in which the reactant and product chemical species are chemically identical but have different isotopic composition.

Nuclear decay. A spontaneous nuclear transformation.

Nuclear transformation. The change of one nuclide into another with a different proton number or nucleon number.

Half-life. Time needed for the concentration of an entity to decrease to one half of its original value.

In addition, an isotope can be *stable* or *unstable*, whether it is subjected to a nuclear decay or not. Moreover, they can be *natural*, whether occurring naturally, as opposed to *synthetic*. In general, natural isotopes may be either stable or unstable, whereas synthetic isotopes are typically unstable.

15.2.2   Hydrogen element

Hydrogen is the first element in the periodic table having the simplest atomic structure of all the elements; its nucleus comprises a single proton, and it has only one outer electron. There are *three isotopes* of hydrogen occurring naturally and having none, one or two neutrons in the nucleus respectively [12–14]:

- $^1_1$H, also called protium,[a] light hydrogen, ordinary hydrogen, hydrogen-1 or H;
- $^2_1$H, also deuterium, heavy hydrogen, hydrogen-2 or D;
- $^3_1$H, also tritium, super heavy hydrogen, heavy-heavy hydrogen, hydrogen-3 or T.

In 1988, Bunnett and Jones [15] established the nomenclature for hydrogen atoms, ions and groups in agreement with the IUPAC rules. The general name of the atom is thus *hydrogen*, whereas the specific names for the isotopes are *protium*, *deuterium* and *tritium*. Accordingly, the cations, positively charged, are *proton*, *deuteron* and *triton*, whereas the anions *protide*, *deuteride* and *tritide*, respectively (these latter ones will be recalled in Section 15.2.3). Figure 15.1 visualizes schematically the structure of the three isotopes of hydrogen. On top of the isotopes occurring naturally, even heavier ones can be synthetized in laboratory: Korsheninnikov *et al.* [16] proved the possibility of manufacturing isotopes up to $^7_1$H (hydrogen-7).

Historically, Birge and Menzel suggested the existence of the two stable hydrogen isotopes in mid-1931, whereas Urey, Brickwedde and Menzel proved it in late 1931. Urey *et al.* proposed the special names of *protium* and *deuterium*, which derive from the ancient Greek words *first* and *second*. Rutherford, Oliphant and Harteck observed an unstable isotope in 1934 and gave it the name *tritium*, meaning *third*. In recent years, Berglund and Wieser reported – on behalf of the Commission on Isotopic Abundances and Atomic Weights (CIAAW) of IUPAC – the table of isotopic compositions of the elements. According to the table, the *representative isotopic composition* of protium is 0.999 885(70), whereas that of deuterium is 0.000 115(70);[b] their observed range of natural variations are

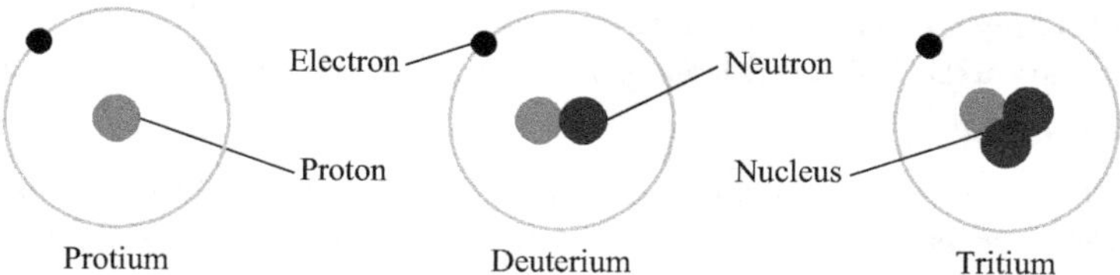

Figure 15.1   Schematic illustration of the three isotopes of hydrogen element that occur naturally: protium (or ordinary hydrogen, which is the most abundant by far), deuterium (heavy hydrogen) and tritium

[a]For the generic element X, the nomenclature ^A_ZX indicates the mass number A and the atomic number Z, thus identifying univocally the specific isotope of X.

[b]The representative isotopic composition is, according to CIAAW, "the isotopic composition of chemicals and/or natural materials that are likely to be encountered in the laboratory. [. . .] for elements with known isotope-abundance variations, they may not necessarily correspond to the best measurements."

Table 15.1   Atomic properties of the three isotopes of hydrogen that occur naturally: protium, deuterium and tritium (extracted from Reference 12)

Properties	Units	Protium, $_1^1H$	Deuterium, $_1^2H$	Tritium, $_1^3H$
Relative atomic mass	a.m.u.	1.007 825	2.014 102	3.016 049
Nuclear spin quantum number	–	1/2	1	1/2
Radioactive stability	–	Stable	Stable	Unstable

0.999 816 to 0.999 974 and 0.000 026 to 0.000 184, respectively [17]. In contrast, tritium occurs naturally only in traces beings formed by the interaction of cosmic rays with atmospheric gases; its atmospheric content is estimated to be 1 tritium atom every 10^{18} protium atoms [12].

Differences in chemical and physical properties arising from differences in the nuclear mass of an element are called *isotope effects*. Among all the elements, hydrogen has the largest isotope effects principally due to the largest differences in the relative mass of the isotopes. (This fact justifies the use of distinctive names for hydrogen isotopes in contrast to all other elements that have no separate names for their isotopes.) In the case of hydrogen, the chemical properties of protium, deuterium and tritium are essentially equal; exception made in matters such as equilibrium constants and reaction rates. Moreover, protium and deuterium are stable isotopes, whereas tritium is radioactive unstable isotope, subjected to a nuclear decay with a half-life of 12.4 years. Table 15.1 shows the main atomic properties of the three isotopes [12].

15.2.3   *Hydrogen molecules: isotopic forms and* ortho/para *modifications*

Protium, deuterium and tritium can combine and form *diatomic molecules*. If the molecule is made by two atoms of the same isotope, as for the case of H_2, D_2 and T_2, it is said *homonuclear diatomic molecule* (or also *isotopic diatomic molecules*); on the contrary, HD, HT and DT, *heteronuclear diatomic molecule*. Thus, there are six *isotopic forms*:

- H_2, *diprotium* (generally said molecular hydrogen or simply hydrogen),
- D_2, *dideuterium* (generally molecular deuterium or simply deuterium),
- T_2, *ditritium* (molecular tritium or tritium),
- HD, *hydrogen deuteride*,
- HT, *hydrogen tritide*,
- DT, *deuterium tritide*.

Being deuterium scarcely abundant and tritium rare, molecular hydrogen is approximated to be solely H_2 for most industrial applications. Nevertheless, this approximation is not applicable for peculiar applications, like the isotope separation covered in this chapter. Figure 15.2 visualizes schematically the structure of the six isotopic forms of diatomic molecular hydrogen.

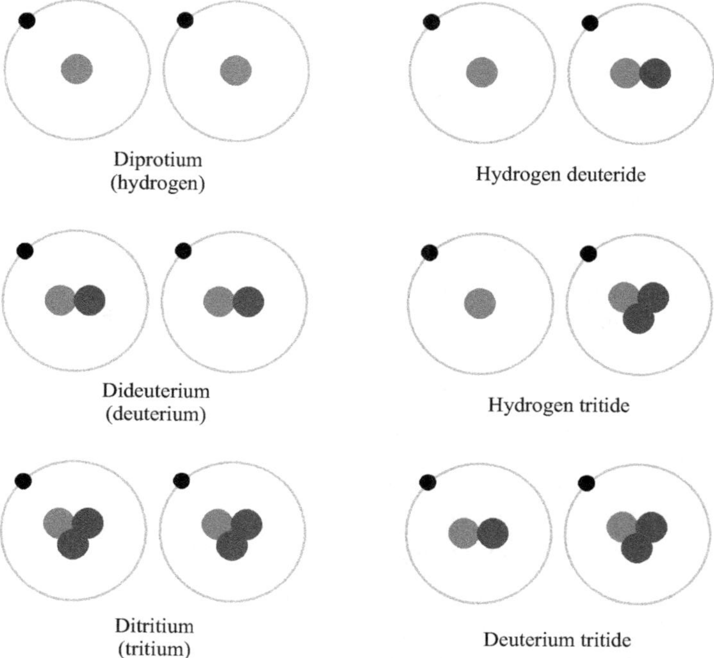

Figure 15.2   Schematic illustration of the six isotopic forms of the diatomic molecules of hydrogen that occur naturally: diprotium (or simply hydrogen, which is the most abundant by far), dideuterium (or hydrogen), ditritium (or tritium), hydrogen deuteride, hydrogen tritide and deuterium tritide

Molecular hydrogen was observed accidentally for the first time by Theophrastus von Hohenheim, also known as Paracelsus, who noted that the attack of strong acids against metals generated a flammable gas. Other chemists and physicists repeated his experience, including Robert Boyle who described the properties of this flammable gas in 1671. The credit of the discovery that the generated gas was made of a new element is attributed commonly to Henry Cavendish because he was able to isolate it and measure its relevant properties in 1776. Shortly later, in 1783, Antoine Lavoisier gave the new element the name *hydrogen* after he discovered, in collaboration with Pierre-Simon Laplace, that burning it in air produced water. Etymologically, hydrogen means forming, *genes*, water, *hydro*, both words belonging to ancient Greek. Lastly, peculiar properties of the homonuclear hydrogen molecules, as opposed to the heteronuclear ones, were discovered spectroscopically and interpreted via quantum mechanics by Heisenberg in 1927.

Molecular hydrogen, as any other substance, can be described by *Statistical thermodynamics*, the branch of Physics that studies the matter combining the probability theory and the quantum mechanics with a microscopic mechanical model of the constituents of the matter itself. Its aim is predicting the properties that are measurable at the macroscopic level based on the microscopic behaviour of the measured substance [18].

According to the Statistical thermodynamic theory, for the homonuclear diatomic molecules, such as H_2, D_2 and T_2, there is a division of the *quantized rotational energy levels* of their nuclear spins into two groups that are referred to as the *ortho* and the *para* modifications. One series is composed of the even numbered and the other of the odd numbered levels. For the case of H_2 and T_2, *ortho* modification is composed of odd numbered, whereas *para* of even ones; on the contrary, for the case of D_2, the *ortho* modification is composed of even numbered, whereas *para* of odd ones (the motivation is provided later in this section). Thus, there are six *ortho/para modifications*:

- oH_2, *orthodiprotium* (generally said orthohydrogen),
- pH_2, *paradiprotium* (generally parahydrogen),
- oD_2, *orthodideuterium* (or simply orthodeuterium),
- pD_2, *paradideuterium* (simply paradeuterium),
- oT_2, *orthoditritium* (orthotritium),
- pT_2, *paraditritium* (paratriutium).

In simpler words, the distinction between the two modifications of homonuclear diatomic molecules is the relative spin of the two nuclei. Figure 15.3 visualizes schematically the structure of the six *ortho/para* modifications of homonuclear diatomic hydrogen.

In few words, molecular hydrogen exists in various isotopic forms and *ortho/para* modifications, yielding a general mixture of nine molecules: oH_2, pH_2, oD_2, pD_2, oT_2, pT_2, HD, HT, and DT.

The relative composition between *ortho* and *para* modifications of homonuclear diatomic molecules varies with temperature. The equilibrium composition at room temperature is named the *normal composition*, and the abundant component is indicated as the *ortho* modification. In addition, the equilibrium composition at any temperature is named simply *equilibrium composition* and, in particular, at ambient temperature, normal and equilibrium compositions are the same mixture. Thus, there are six important compositions:

- n-H_2, *normal-composition diprotium* (generally said normal-hydrogen),
- e-H_2, *equilibrium-composition diprotium* (generally equilibrium-hydrogen),
- n-D_2, *normal-composition dideuterium* (normal-deuterium),
- e-D_2, *equilibrium-composition dideuterium* (equilibrium-deuterium),
- n-T_2, *normal-composition ditritium* (normal-tritium),
- e-T_2, *equilibrium-composition ditritium* (equilibrium-tritium).

Figure 15.4 visualizes how the *ortho–para* composition at equilibrium varies for H_2, D_2 and T_2 in the temperature range from 0 to 300 K.

In 1948, Woolley *et al.* [19] reported the computation of thermal data for H_2, HD and D_2 in solid, liquid and gaseous states, including the distinctive properties of *ortho* and *para* modifications of H_2 and D_2. Hoge and Arnold [20] presented in 1951

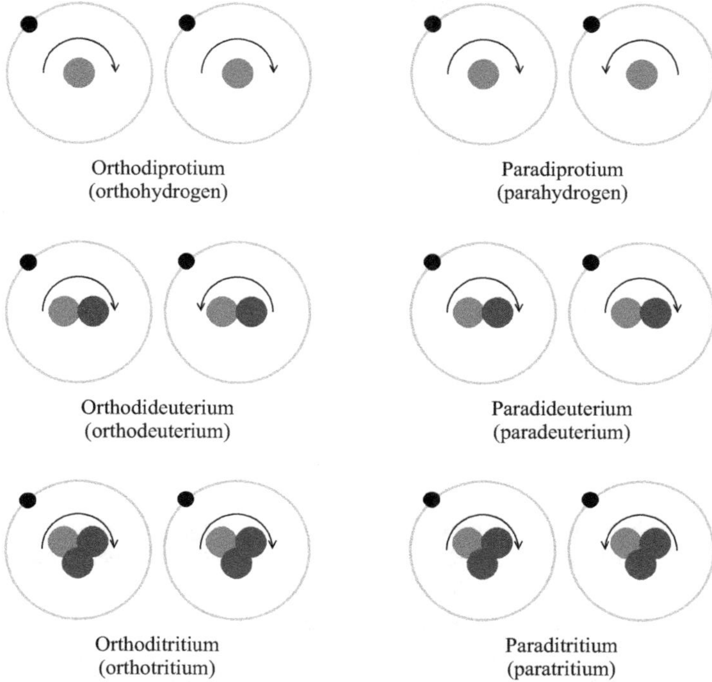

Figure 15.3   *Schematic illustration of the six ortho/para modifications of the homonuclear diatomic hydrogen that occur naturally showing the relative spin of the two nuclei: ortho- and paradiprotium (or simply ortho- and parahydrogen, which are the most abundant by far), ortho- and paradideuterium (or ortho- and paradeuterium), ortho- and paraditritium (or ortho- and paratritium)*

the measured vapour pressures of e-H_2, HD and e-D_2 in the wide range from near their triples points up to their critical points. Moreover, they also provided the dew-point pressures of several binary mixtures, indicating a difference of about 3% above those predicted by the law of ideal solutions, also known as the *Raoult's Law*. In other words, the *ideal solution model* is appropriate for the H_2, HD and D_2 mixture for engineering purposes. In the same year, Hoge and Arnold [21] reported the measurements of their critical temperatures, pressures and volumes. Given the growing role of hydrogen as research tool in chemistry, physics, biology and nuclear engineering, Haar *et al.* [22] recounted in 1961 the results of a decade-long research on the isotope effects in hydrogen compounds undertaken by the United States National Bureau of Standards, now National Institute of Standards and Technology (NIST). The research covered data of state, reaction kinetics, thermodynamic functions and a variety of other physico-chemical phenomena of diatomic hydrides, deuterides and tritides. These molecules are diatomic compounds *comprising protide, deuteride and tritide* and eventually another element. Roder *et al.* [23]

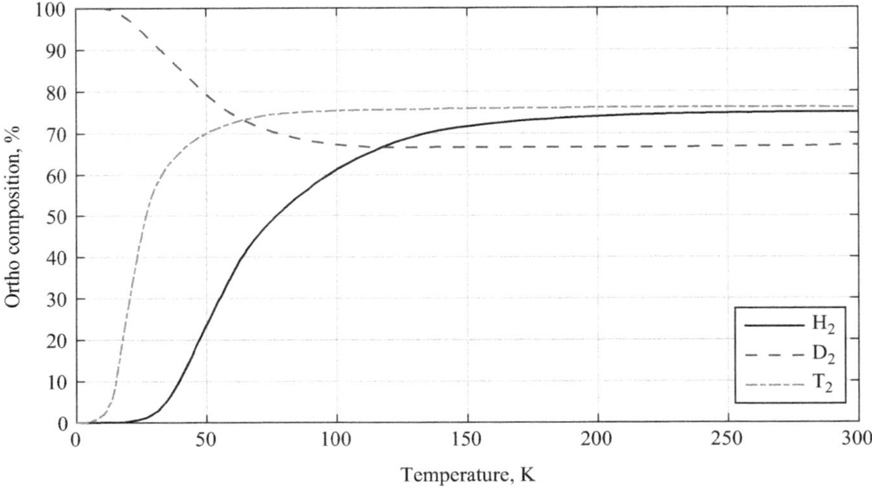

Figure 15.4   *Ortho–para composition for H₂, D₂ and T₂ at equilibrium from 0 to 300 K. However, the ortho–para conversion is a very slow process. Therefore, the equilibrium composition at room temperature, called normal composition, is of interest for the cryogenic distillation of hydrogen isotope because it is fast process compared to that conversion (redrawn from Figure 6 of Reference 12)*

completed a survey, as of the year 1972, about volumetric, thermodynamic, thermal, transport, electrical and mechanical properties of all hydrogen isotopes forms and *ortho/para* modifications below their critical points. Shortly later, in 1981, McCarty *et al.* [24] replicated the survey focusing mainly on the light isotope of hydrogen. Today, the studies by Jacobsen, Leachman, Richardson, Lemmon and colleagues are likely the most advanced analyses of thermodynamic and thermophysical properties of hydrogen in all isotopic forms and *ortho/para* modifications [25–29]. All these cited publications are still of large use for the investigation about the properties of the hydrogen. As a rule, isotopic forms and *ortho/para* modifications yield large relative differences in the thermodynamic properties of the diverse molecules at same temperature and pressure only within the cryogenic region.

The *ortho/para conversion reaction* from the *ortho* to the *para* modification, as well as the reversed reaction, is a very slow reaction, unless it is promoted catalytically. This kind of reactions describes the equilibrium among the *ortho/para* modification as follows:

- $oH_2 \rightleftarrows pH_2$
- $oD_2 \rightleftarrows pD_2$
- $oT_2 \rightleftarrows pT_2$

The conversion reactions are exothermic as temperature decreases and characterized by values of the *enthalpy of conversion* of similar magnitude to those of the *enthalpy of evaporation*. Therefore, the analysis of *ortho–para* composition and the catalytic promotion of their conversion are of utmost importance for the storage of the liquefied hydrogen, which would evaporate otherwise in the tank during the slow conversion from one modification to the other.

Similarly to the conversion reaction, also the *isotope exchange* (Section 15.2.1) between any two hydrogen isotopes is a slow reaction, unless promoted catalytically. This kind of reactions describes the equilibrium among heteronuclear and homonuclear hydrogen molecules as follows:

- $2HD \rightleftarrows H_2 + D_2$
- $2HT \rightleftarrows H_2 + T_2$
- $2DT \rightleftarrows D_2 + T_2$

The isotope exchanges shift rightward, which is the equilibrium shifts from heteronuclear molecules to homonuclear molecules, at lower temperatures. This characteristic is exploited favourably in the distillation processes to concentrate the isotopes in their homonuclear diatomic molecules.

As will be motivated in Section 15.3, the working principle of distillation is the difference among *vapour pressures* of the components to be separated. Table 15.2 reports the fixed points, which are the triple and critical points, for the diatomic molecules of interest, whereas Figure 15.5 visualizes their vapour pressures between the fixed points. As a general rule, lighter hydrogen molecules are characterized by higher vapour pressures with respect to heavier molecules at the same temperature. In other words, lighter molecules are more volatile than heavier ones.

Table 15.2   Fixed points for (most of) the isotopic forms and ortho/para *modifications of molecular hydrogen (elaborated from Table 11.1 of Reference 23)*

Form and modification	Triple point temperature (K)	Triple point pressure (bar)	Normal boiling point (K)	Critical point temperature (K)	Critical point pressure (bar)
Homonuclear diprotium					
oH_2	14.05	0.0735	20.454	NA	NA
pH_2 or $e\text{-}H_2$	13.803	0.07042	20.268	32.976	12.928
$n\text{-}H_2$	13.957	0.07205	20.39	33.19	13.15
Homonuclear dideuterium					
oD_2 or $e\text{-}D_2$	18.691	0.1713	23.63	38.262	16.450
pD_2	18.78	0.1713	23.66	NA	NA
$n\text{-}D_2$	18.71	0.1713	23.66	38.34	16.650
Ditritium					
T_2	20.62	0.2160	25.04	40.44	18.502
Heteronuclear diatomic hydrogen					
HD	16.60	0.124	22.143	35.908	14.84
HT	17.62	0.1460	22.92	37.13	15.70
DT	19.71	0.1942	24.38	39.42	17.70

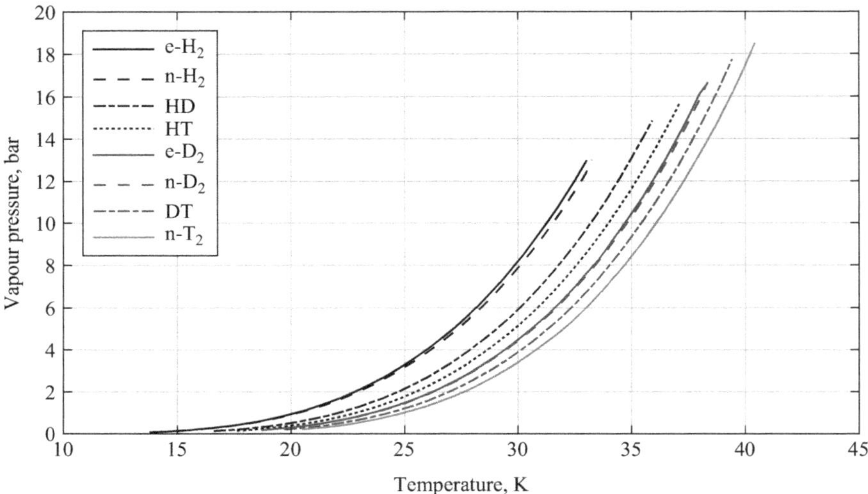

Figure 15.5   Vapour pressures for (most of) the isotopic forms and ortho/para
*modifications of molecular hydrogen from triple point to critical
point (elaborated and drawn from Table 10.2 of Reference 23)*

15.3   Basics of cryogenic distillation

Distillation is a process for separating two or more components from a mixture
based on their different vapour pressures. It is adopted commonly in industrial
applications because of the advantages of achieving high separation degrees and of
processing large quantities of fluids. Green and Perry [30] as well as McCabe *et al.*
[31] describe the general process in two well-known textbook, whereas Andreev
et al. [6] focus on the process for separating biogenic isotopes. Flynn *et al.* [32] as
well as Benedict *et al.* [33] illustrate specifically the deuterium distillation. Lastly,
Busigin and Sood [7] as well as Cristescu *et al.* [34] specifically illustrate the
tritium separation. Distillation requires the coexistence of liquid and vapour phases,
a condition that occurs exclusively at cryogenic conditions for molecular hydrogen.
Cryogenics means literally the production of icy cold from ancient Greek and it is
used today as synonym for temperatures lower than, indicatively, $-150\,^{\circ}C$ (123 K).
As seen in Section 15.2.3, triple and critical temperatures of the various isotopic
forms and *ortho/para* modifications of molecular hydrogen fall in the range of
10–40 K. This section describes first the fundamental working principle of the distillation process and outlines after its application to the hydrogen mixtures.

15.3.1   Fundamental working principle

In general terms, the *separation operation* of components from their mixture is
achieved by the creation of two or more adjacent zones that differ gradually in

temperature, pressure, composition and/or phase state [30]. Distillation is a separation operation exploiting the physical equilibria between the vapour and the liquid phase within each zone. It is accomplished by vertical columns in which these zones are created by stacked trays or by packings. The *feed* to be separated is introduced at one or more points along the column height. Because of the difference in density between vapour and liquid phases, liquid runs down the column, cascading from zone to zone, whereas vapour flows up the column, contacting liquid at each zone. Liquid reaching the bottom of the column is partially vaporized by a heating mean to provide *boil-up*, which flows up the column. The remainder of the bottom liquid is withdrawn as *bottom product*. Vapour reaching the top of the column is condensed by a cooling mean. Part of this liquid is returned to the column as *reflux*, which runs down the column. The remainder of the liquid is withdrawn as *overhead product*. In some cases, only part of the vapour is condensed so that the overhead product is a vapour. This overall flow pattern in a distillation column provides countercurrent contacting of vapour and liquid streams at all zones along the column, whereas the heating of the bottom and the cooling of the top support a temperature gradient. The lighter components tend to concentrate in the vapour phase at the lower temperatures, whereas the heavier components in the liquid phase at higher temperatures. Lighter components are those characterized by *higher vapour pressures*, hence higher volatility and lower boiling points; vice versa, heavier by *lower vapour pressures*, hence lower volatility and higher boiling points. Therefore, the overall result is a vapour phase that becomes richer in lighter components as it flows up the column and a liquid phase that becomes richer in heavier components as it runs down.

A distillation column is assessed by its *separation factor*, which may be defined conceptually as the ratio of the concentration of the desired component in the product in which it concentrates (either bottom or overhead) with respect to its concentration in the other product. If the separation factor of a single column does not allow achieving the targeted purity of the desired component, more columns can be arranged in a proper sequence. Each column is referred to as a *distillation stage*.

15.3.2   Application to hydrogen isotopes

In *hydrogen distillation*, the desired components are typically the heavier isotopes, deuterium and tritium, which are characterized by lower vapour pressures than protium, as reported in Section 15.2.3. Moreover, typical targeted purities require three or four distillation stages. Hence, they concentrate progressively on the liquid bottom products, whereas protium concentrates on the gaseous overhead products.

Furthermore, hydrogen mixtures do not contain only homonuclear diatomic molecules, H_2, D_2, T_2, but primarily heteronuclear ones, HD, HT and DT. The isotope exchange reactions describe the equilibrium among heteronuclear and homonuclear molecules (Section 15.2.3). The equilibrium is reached slowly unless catalysed but shifts favourably towards the homonuclear molecules at lower temperatures. Hence, *catalytic reactors* (also said *catalytic equilibrators*) are placed

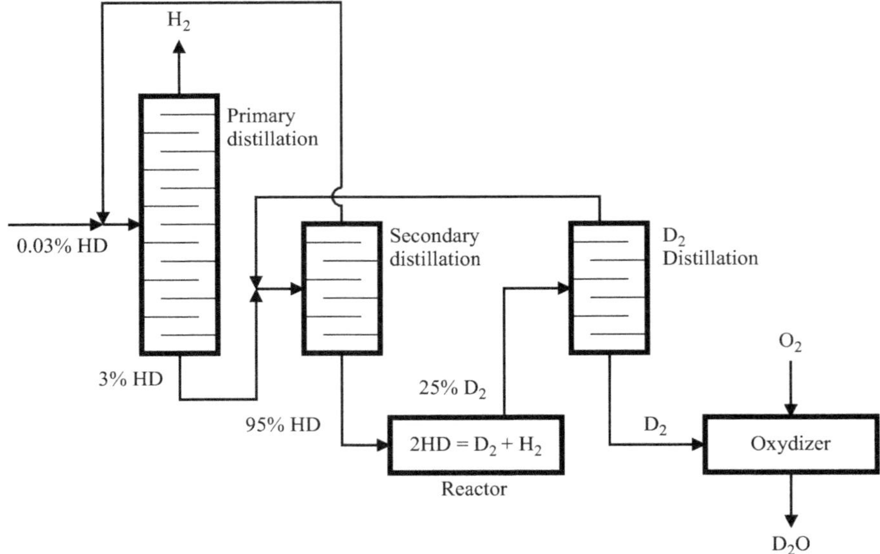

Figure 15.6   Process flow diagram of deuterium separation from an H_2, HD, D_2 mixture by way of a three-stage distillation and one catalytic reactor that promotes the formation of the homonuclear diatomic molecules at low temperature (redrawn from Flynn et al. [32])

generally between the last two distillation stages to promote the formation of H_2, D_2 and T_2 and ease the separation of the isotopes, D and T, in their homonuclear diatomic molecules.

Figure 15.6 depicts the conceptual design of deuterium separation as proposed by Flynn *et al.* [32]. Since deuterium occurs in the form of HD rather than D_2, as recalled above, the first and second distillation stages aim at concentrating HD in their bottom products (respectively 3% and 95%). Then, the bottom product of the second column is enriched in D_2 by a catalytic reactor. A third distillation stage allows for a high purity D_2 in a liquid phase. (In this design, the separated deuterium is utilized along with oxygen to produce heavy water.) A similar concept could be implemented for the separation of tritium.

15.4   Basics of cryogenic liquefaction

Cryogenic liquefaction is the process of turning a gaseous component at atmospheric conditions into a liquid at atmospheric pressure yet cryogenic temperature. Cryogenic liquefaction applies to the so-called permanent gases: helium, hydrogen, neon, nitrogen and oxygen (as well as air in a more general sense). The following sections outline briefly the fundamental cooling effects, the fundamental liquefaction cycles and the current industrial technology. The textbooks by Barron [35] or by Flynn [36] shall be referred to for the general topic of cryogenic systems, whereas the review by Valenti [37] for the specific topic of hydrogen liquefaction.

15.4.1   Fundamental cooling effects

Cryogenic liquefiers are based on the thermodynamic concept that the cooling effect is obtained by expanding adiabatically a fluid from a proper initial condition. The adiabatic expansion can be executed either with or without extracting energy from the fluid being expanded. The expansion without energy extraction, which is an isenthalpic process, is realized by a throttling valve; the expansion with energy extraction, which is ideally an isentropic process, by an expanding fluid machine. Because liquefiers are based on the expansion of a fluid, they require a compression process to complete the cycle. This compression is executed at ambient temperature and, commonly, in an intercooled manner. Thus, there are two methods for achieving a cooling effect within a liquefaction plant:

- *throttling through a valve*, which is described from a thermodynamic point of view by the Joule-Thomson coefficient and the inversion curve;
- *expanding via a fluid machine*.

15.4.2   Fundamental liquefaction cycles

The most fundamental liquefaction scheme is the (simple or single-pressure) *Linde–Hampson cycle*, illustrated in Figure 15.7(a), which employs exclusively a

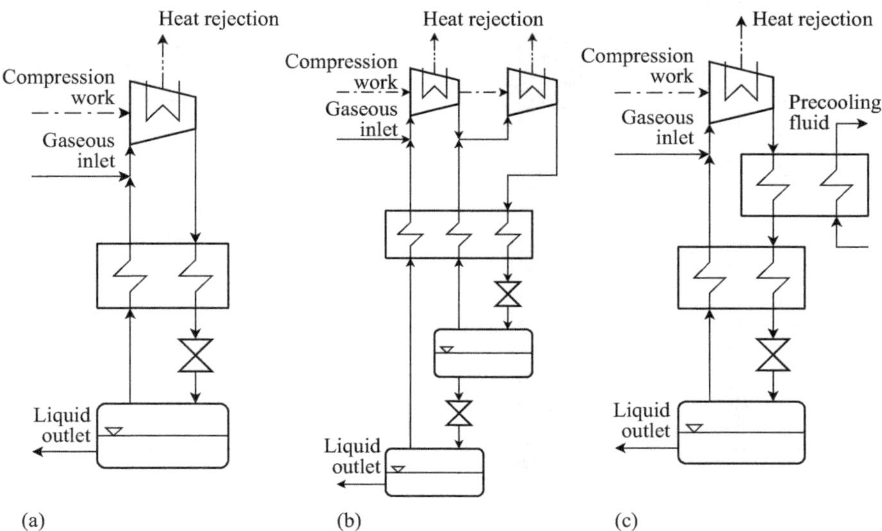

Figure 15.7   Linde–Hampson cycle in the simple or single-pressure configuration (a), dual-pressure (b) and single-pressure with precooling (c). In these cycles, the processed fluid is also the working fluid

throttling valve to achieve the cooling effect. The performance of the plant, which is however never high, can be improved by the splitting of the throttling over two pressure levels. This new scheme is named *dual-pressure Linde–Hampson cycle* and depicted by Figure 15.7(b). In case the fluid to be liquefied has an inversion curve below the ambient temperature for any pressure, like hydrogen, the process requires a precooling system, which cools the stream before the recuperative heat exchanger and to the throttling valve (Figure 15.7(c)).

The performance of the liquefier can be increased remarkably by the adoption of an expanding fluid machine. The fundamental scheme comprising the expander is the *Claude cycle* (Figure 15.8(a)). Likewise the Linde–Hampson, the Claude cycle can operate over an additional pressure level. In this case, the throttling valve works between maximum and minimum pressures, whereas the machine between mid and minimum (Figure 15.8(b)). Moreover, the Claude cycle can adopt a precooling system, despite it is not strictly necessarily also for those fluids that, like hydrogen, have a low inversion curve (Figure 15.8(c)).

Furthermore, there are few other fundamental cycles derived from the Claude one as well as additional liquefaction cycles operating with working fluids different from hydrogen. However, these cycles are not reported here because not so pertinent to the topic of current hydrogen liquefaction and distillation. In any of these cycles, the stream of hydrogen to be liquefied is always at a pressure higher than its critical pressure. At supercritical pressure, indeed, there is not a phase change

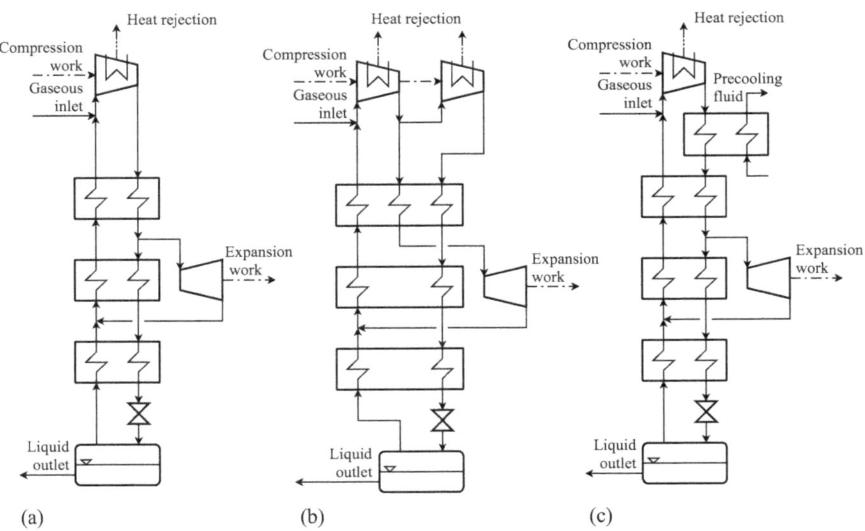

Figure 15.8 *Claude cycle in the simple or single-pressure configuration (a), dual-pressure (b) and single-pressure with precooling (c). In these cycles, the processed fluid is also the working fluid*

within any of the heat exchangers, which would lead to small temperature variations upon large energy transfers. In contrast, a subcritical (pure) fluid would require extracting the condensation energy at a constant temperature, which is an adverse situation for achieving high effectiveness of the heat exchangers (or low-entropy generation from a second-law perspective).

15.4.3　Current hydrogen liquefaction plants

A hydrogen liquefier may be classified as small if its capacity is up to 0.5 t day^{-1} and large if greater than 5.0 t day^{-1}, whereas the World hydrogen liquefaction capacity is estimated to fall within the range of 300–350 t day^{-1}. Small- and large-scale hydrogen liquefiers do not differ just in the capacity but also in the manner by which liquefaction is accomplished. Small-scale hydrogen liquefaction is realized by cryogenic refrigerators, which are systems operating with a refrigerant in a closed-loop thermodynamic cycle that cools and liquefies the hydrogen stream. In the case of hydrogen, the adopted refrigerant is helium. In contrast, large-scale liquefaction is accomplished by open-loop thermodynamic cycles in which hydrogen is both the processed fluid and the working fluid. This distinction is only technological because from a theoretical perspective closed-loop refrigerators can be resized to large scales, as well as open-loop plants downsized. In addition, for very small hydrogen liquefiers, magnetic refrigerators may be employed.

Large-scale hydrogen liquefiers must employ a *catalytic ortho-to-parahydrogen conversion reactors* if the liquid has to be stored over time because, as seen in Section 15.2.3, the reaction is highly exothermic and otherwise very slow. The reaction is so exothermic that the associated enthalpy of conversion is greater than the enthalpy of vaporization at storage tank pressure. In brief, if orthohydrogen were not converted during liquefaction, the conversion would occur slowly in the storage tank, and it would release an amount of energy large enough to vaporize the entire content of that tank over a timespan of about 2 months.

Today's large-scale hydrogen liquefiers are based on the modification of the precooled Claude cycle shown in Figure 15.9. Precooling is obtained by a dedicated refrigeration cycle or by liquid nitrogen. Catalytic conversion of orthohydrogen to parahydrogen can be executed in a *batch mode* or in a *continuous mode*. If in batch mode, reactors can be either adiabatic or isothermal. Isothermal reactors are immersed in liquid nitrogen or in liquid hydrogen baths. If in continuous mode, the catalyst is placed within the heat exchangers. From a point of view of liquefaction work requirement, continuous conversion is definitely more efficient. Typically, hydrogen liquefiers are designed to produce a liquid with a parahydrogen content greater than 95%. For reference, Bracha *et al.* give an accurate description of a recent hydrogen liquefier that, despite not any more in operation, is still representative for the current technology.

The hydrogen liquefaction work depends strongly on inlet conditions (pressure and temperature), outlet conditions (saturation pressure and parahydrogen content) and ambient temperature. For an inlet at 1 bar and 288 K, an outlet of saturated liquid at 1 bar and equilibrium composition, and an ambient at 288 K, the *ideal liquefaction*

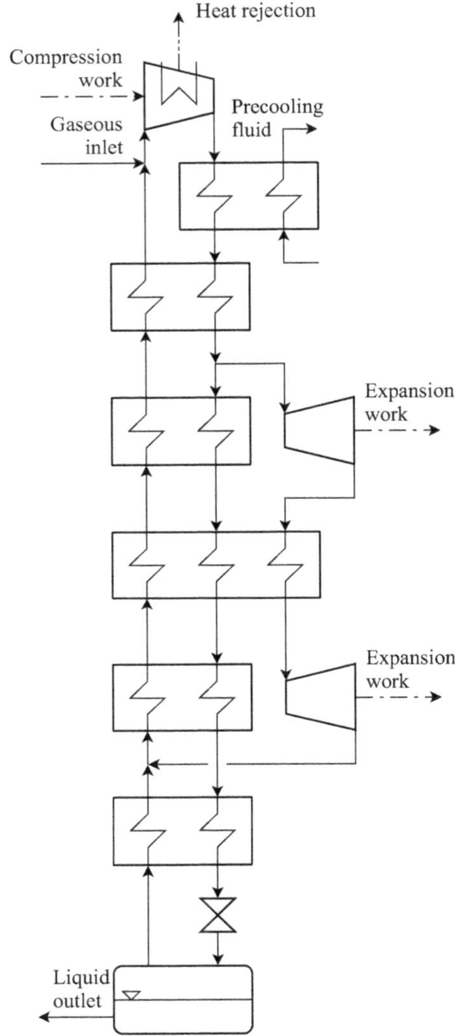

Figure 15.9   Modern large-scale hydrogen liquefiers are based on modifications of the fundamental Claude cycle and employ catalytic orthohydrogen to parahydrogen conversion reactors, either in a batch mode or in continuous mode

work is 13.6 MJ kg^{-1} (3.78 kW h kg^{-1}). If the inlet pressure is 20 bar instead of 1 bar, as typical for a number of hydrogen production processes, the ideal liquefaction work is 9.97 MJ kg^{-1} (2.77 kW h kg^{-1}). The *real liquefaction work of current large-scale liquefiers* is in the range of 30–45 MJ kg^{-1} (about 8–12 kW h kg^{-1}), for an inlet at about 20 bar and a parahydrogen content of at least of 95%.

15.5   Reference plants

Despite being adapt for industrial-size applications, the number of cryogenic distillation plants constructed for deuterium and tritium separation is small due to the *limited market* of these two isotopes. The following two sections provide an example of a separation plant for deuterium and one for tritium.

15.5.1   Deuterium separation

A cryogenic distillation facility for the separation of deuterium from a hydrogen stream was erected in Hoechst, Germany, during the late 1960s [38]. The scope of the facility, which was operated for a limited number of years, was producing heavy water for fission nuclear reactors. The facility comprised two distillation sections:

- the *first distillation section* concentrated on HD at a purity of 95%, because molecular hydrogen occurs naturally in H_2 and HD forms rather than D_2 (there are approximately 10^4 molecules of HD for 1 molecule of D_2),
- the *second distillation section* promotes catalytically the isotope exchange reaction to yield H_2 and D_2 from HD, and then it separates D_2 at a purity of 99.8%.

Figure 15.10 depicts a schematic process flow diagram of the facility. The first section includes two columns, the first one of which concentrates on HD at the

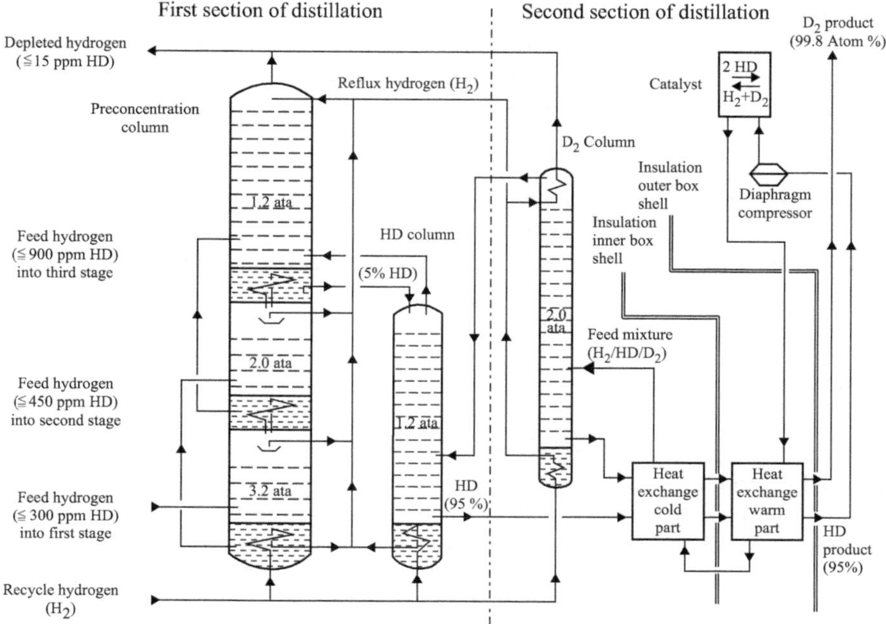

Figure 15.10   Process flow diagram of the Linde cryogenic distillation plant erected in the late 1960s (by Scott et al. [38])

intermediate purity of 5%; the second section a single column. Both sections operate in the 20–25 K range, which is reached by way of a single-pressure Linde–Hampson cycle with liquid nitrogen precooling at around 80 K. Overall, the distillation facility enriched deuterium from just 150 ppm in the hydrogen inlet stream to a purity 99.8% in the deuterium outlet stream. In addition, 95% of the deuterium entering the distillation facility was separated, whereas the total annual production was 1,080 m^3 (at standard conditions). Shortly later, Linde built a similar distillation plant in Nangal, India, that was in service until recent years.

15.5.2   Tritium separation

A *tritium removal facility* for the detritiation of the heavy water from nuclear fission reactors was erected in Darlington, Canada, during the late 1980s [7,39]. The scope of the facility, still in operation, is reducing the radioactivity environmental emissions and worker radiation doses caused by heavy water leakages from the equipment of the nuclear power plant. The facility comprises two sections:

- a *vapour phase catalytic exchange section*, which extracts tritium from the irradiated water into a deuterium-rich stream by way of eight stages,
- a *cryogenic distillation section*, which concentrates tritium by way of four columns with several hundred theoretical stages and a catalytic equilibrator.

Figure 15.11 depicts a schematic process flow diagram of the facility. The four columns are arranged in a simple cascade such that the (liquid) bottom product of each column, richer in tritium than that column feed, enters the next column, whereas its (vapour) overhead product, richer in the lighter isotopes, enters the previous column. The overhead product of the fourth and last column, prior to entering the third column, flows into a catalytic equilibrator, which is a catalysed reactor that favours the isotopic exchange of heteronuclear molecules to homonuclear ones (Section 15.2.3) and, hence, promotes the formation of T$_2$ to be separated. Overall, the facility enriches tritium from just parts per millions in the heavy water inlet stream to a purity of 99.9% in the tritium outlet stream, achieving a remarkable separation factor. Ultimately, tritium is stored as a *metal hydride*.

15.6   Further reading

The interested reader may refer first to the two well-known textbooks by Barron [35] and Flynn [36]. In addition, the reader can locate the other reports and articles cited in this work. More in general, detailed information may be derived from two reference journals of the sector: (i) the *International Journal of Hydrogen Energy* and (ii) *Cryogenics*. Furthermore, the proceedings of Romanian conference *Progress in Cryogenics and Isotopes Separation* may be also valuable sources of data, as well as the websites of IUPAC, CIAAW, NIST, ITER and the International Atomic Energy Agency.

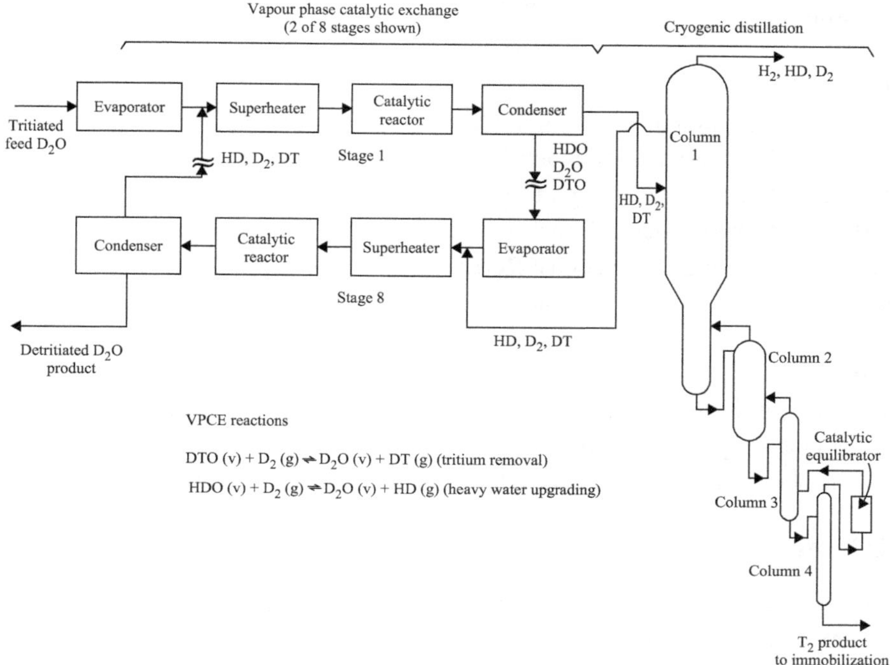

Figure 15.11   Process flow diagram of the Darlington Tritium Removal Facility erected in the late 1980s (by Busigin and Sood [7,39])

15.7   Conclusions

In most hydrogen processes, a distinction among its isotopes is unnecessary being the concentration of protium larger by far than that of deuterium and tritium. However, these two heavier isotopes have peculiar applications, such as those related with nuclear fission as well as fusion reactors, that place a great attention upon them. There exist a number of techniques for isotope separation and, among them, cryogenic distillation has been proved a mature process for the industrial-scale processing of hydrogen. Two reference plants were built in the 1960s and 1980s, respectively, during the years of growing interest for nuclear energy in the civil and military sectors. Today, such interest has decreased, but it is expected to increase more than ever before in view of the future nuclear fusion reactors.

Acknowledgements

The author is very grateful to Prof. Hans Quack for sharing his knowledge and to Michela Capoccia as well as Andrea Seghezzi, from SAPIO Produzione Idrogeno Ossigeno S.r.l., for exchanging information.

Nomenclature

Acronyms

CIAAW	Commission on Isotopic Abundances and Atomic Weights
ITER	International Thermonuclear Experimental Reactor
IUPAC	International Union of Pure and Applied Chemistry
NASA	United States National Aeronautics and Space Administration
NBS	United States National Bureau of Standards
NIST	United States National Institute of Standards and Technology

Symbols

A mass (or nucleon) number

Z atomic (or proton) number

e elementary (or proton) charge

References

[1]   Hawking S., Mlodinow L. *A briefer history of time*. Bantam Press, New York (NY), USA. 2005.

[2]   Ridgen J. *Hydrogen: The essential element*. Harvard University Press, Cambridge (MA), USA. 2003.

[3]   Murray R.L., Holbert K.E. *Nuclear energy: An introduction to the concepts, systems, and applications of nuclear processes*. Butterworth-Heinemann, Amsterdam, The Netherlands. 2014.

[4]   Horton C.W., Benkadda S. *ITER physics*. New Jersey, World Scientific. 2015.

[5]   http://www.iter.org

[6]   Andreev B.M., Magomedbekov E.P., Raitman A.A., Pozenkevich M.B., Sakharovsky Yu.A., Khoroshilov A.V. *Separation of isotopes of biogenic elements in two-phase systems*. Elsevier. 2006.

[7]   Busigin A., Sood S.K. 'Optimization of Darlington Tritium Removal Facility performance: Effects of key process variables'. *Nuclear Journal of Canada*. 1987;**1**(4), 368–371.

[8]   Ahn D.H., Lee H.S., Chung H.S., Song M.-J., Son S.-H. 'Optimum design of the Wolsong Tritium Removal Facility'. *Journal of the Korean Nuclear Society*. 1996;**28**(4), 415–423.

[9]   Bartlit J.R., Sherman R.H., Stutz R.A., Denton W.H. 'Hydrogen isotope distillation for fusion power reactors'. *Cryogenics*. 1979;**19**(5), 275–279. doi:10.1016/0011-2275(79)90142-5.

[10]   Busigin A., Sood S.K., Kveton O.K., Dinner P.J., Murdoch D.K., Leger D. 'ITER hydrogen isotope separation system conceptual design description'. *Fusion Engineering and Design*. 1990;**13**(1), 77–89. doi:10.1016/0920-3796(90)90035-5.

[11]   *International Union of Pure and Applied Chemistry (IUPAC). Golden Book.* 2014. https://goldbook.iupac.org/PDF/goldbook.pdf

[12]   Kakiuchi M. *Hydrogen: Inorganic chemistry.* In *Encyclopaedia of Inorganic and Bioinorganic Chemistry.* John Wiley & Sons, Online. 2011. doi:10.1002/9781119951438.eibc0086.

[13]   *Deuterium.* In *Encyclopaedia of Inorganic and Bioinorganic Chemistry.* John Wiley & Sons, Online. 2011. doi:10.1002/9781119951438.eibd0219.

[14]   Galeriu D., Melintescu A. *Tritium: Radionuclide.* In *Encyclopaedia of Inorganic and Bioinorganic Chemistry.* John Wiley & Sons, Online. 2011. doi:10.1002/9781119951438.eibc0413.

[15]   Bunnett J.F., Jones R.A.Y. 'Names for hydrogen atoms, ions, and groups, and for reaction involving them'. *Pure and Applied Chemistry.* 1988; **60**(7), 1115–1116. doi:10.1351/pac198860071115.

[16]   Korsheninnikov A.A., Nikolskii E.Yu., Kuzmin E.A., *et al.* 'Experimental evidence for the existence of ^{7}H and for a specific structure of ^{8}He'. *Physical Review Letters.* 2003;**90**(8), 082501-1-4. doi:10.1103/PhysRevLett.90.082501.

[17]   Berglund M., Wieser M.E. 'Isotopic composition of the elements 2009 (IUPAC technical report)'. *Pure and Applied Chemistry.* 2011;**83**(2), 397–410. doi:10.1351/PAC-REP-10-06-02.

[18]   Tien, C.L., Lienhard, J.H. *Statistical thermodynamic.* Revised printing, Hemisphere Publishing, Oxford, UK. 1988.

[19]   Woolley H.W., Scott R.B., Brickedde F.G. 'Compilation of thermal properties of hydrogen in its various isotopic and ortho–para modifications'. Research paper RP1932, *Journal of Research of the National Bureau of Standards.* 1948;**41**(5), 379–475.

[20]   Hoge H.J., Arnold R.D. 'Vapor pressures of hydrogen, deuterium, and hydrogen deuteride and dew-point pressures of their mixtures'. Research paper 2228, *Journal of Research of the National Bureau of Standards.* 1951;**47**(2), 63–74.

[21]   Hoge H.J., Arnold R.D. 'Critical temperatures, pressures, and volumes of hydrogen, deuterium, and hydrogen deuteride'. Research paper 2229, *Journal of Research of the National Bureau of Standards.* 1951;**47**(2), 75–79.

[22]   Haar L., Friedman A.S., Beckett C.W. *Ideal gas thermodynamic functions and isotope exchange functions for the diatomic hydrides, deuterides, and tritides.* United States National Bureau of Standards Monograph 20, 1961.

[23]   Roder H.M., Childs G.E., McCarty R.D., Angerhofer P.E. *Survey of the properties of the hydrogen isotopes below their critical temperatures.* United States National Bureau of Standards Technical Note 641, 1973.

[24]   McCarty R.D., Hord J., Roder H M. *Selected properties of hydrogen (engineering design data),* National Bureau of Standards Monograph 168, 1981.

[25]   Jacobsen R.T., Leachman J.W., Penoncello S.G., Lemmon E.W. 'Current status of thermodynamic properties of hydrogen'. *International Journal of Thermophysics.* 2007;**28**(3), 758–772. doi:10.1007/s10765-007-0226-7.

[26] Leachman, J.W., Jacobsen R.T., Penoncello S.G., Huber M.L. 'Current status of transport properties of hydrogen'. *International Journal of Thermophysics.* 2007;**28**(3), 773–795. doi:10.1007/s10765-007-0229-4.

[27] Leachman J.W., Jacobsen R.T., Penoncello S.G., Lemmon E.W. 'Fundamental equations of state for parahydrogen, normal hydrogen, and orthohydrogen'. *Journal of Physical and Chemical Reference Data.* 2009;**38**(3), 721–748. doi:10.1063/1.3160306.

[28] Richardson I.A., Leachman J.W. 'Thermodynamic properties status of deuterium and tritium'. *AIP Conference Proceedings.* 2012;**1434**, 1841–1848. doi:10.1063/1.4707121.

[29] Richardons I.A., Leachman J.W., Lemmon E.W. 'Fundamental equation of state for deuterium'. *Journal of Physical and Chemical Reference Data.* 2014;**43**(1), 013103-1-013103-13. doi:10.1063/1.4864752.

[30] Green D., Perry R. *Perry's chemical engineer's handbook.* Eighth edition. McGraw-Hill Education, New York (NY), USA. 2007.

[31] McCabe W., Smith J.C., Harriott P. *Unit operations of chemical engineering.* Seventh edition. McGraw-Hill Education, New York (NY), USA. 2004.

[32] Flynn T.M., Weitzel D.H., Timmerhaus K.D., Vander Arend P.C., Draper J.W. 'Distillation of hydrogen–deuterium mixtures'. *Advances in Cryogenic Engineering.* 1960;**2**, 39–44. doi:10.1007/978-1-4684-3102-5_6.

[33] Benedict M., Pigford T.H., Levi H.W. *Nuclear chemical engineering.* Second edition. McGraw-Hill, New York (NY), USA. 1981.

[34] Cristescu I., Cristescu I., Peculea M. *Studies about the separation of molecular species of hydrogen's isotopes by cryogenic distillation in a plant for heavy water detritiation.* In *Hydrogen Power: Theoretical and Engineering Solutions.* Kluwer Academic Publishers. 1998. doi:10.1007/978-94-015-9054-9_73.

[35] Barron R.F. *Cryogenic systems.* Second edition. New York, Oxford University Press. 1985.

[36] Flynn T.M. *Cryogenic engineering.* Second edition. New York, Marcel Dekker. 2005.

[37] Valenti G. *Hydrogen liquefaction and liquid hydrogen storage.* In *Compendium of Hydrogen Energy – Volume 2: Hydrogen Storage, Distribution and Infrastructure.* Woodhead Publishing, Amsterdam, The Netherlands. 2015.

[38] Scott R.B., Denton W.H., Nicholls C.M. *Technology and uses of liquid hydrogen.* Pergamon Press, Oxford, UK. 1964.

[39] Busigin A., Sepa T.R., Sood S.K. 'Flosheet: Microcomputerized flowsheeting/simulation program for simulating hydrogen isotope separation processes'. *Separation Science and Technology.* 1987;**22**(2–3), 557–579. doi:10.1080/01496398708068969.

Index